NIELS BOHR
COLLECTED WORKS
VOLUME 10

NIELS BOHR IN HIS STUDY, 1942.

NIELS BOHR

COLLECTED WORKS

GENERAL EDITOR

FINN AASERUD

THE NIELS BOHR ARCHIVE, COPENHAGEN

VOLUME 10
COMPLEMENTARITY
BEYOND PHYSICS
(1928–1962)

EDITED BY

DAVID FAVRHOLDT

UNIVERSITY OF ODENSE, DENMARK

1999

ELSEVIER
AMSTERDAM · LAUSANNE · NEW YORK · OXFORD · SHANNON · SINGAPORE · TOKYO

ELSEVIER SCIENCE B.V.
Sara Burgerhartstraat 25
P.O. Box 211, 1000 AE Amsterdam
The Netherlands

Library of Congress Catalog Card Number: 70-126498
ISBN Collected Works: 0 7204 1800 3
ISBN Volume 10: 0 444 89972 3

Transferred to digital printing 2006
Printed and bound by CPI Antony Rowe, Eastbourne

GENERAL EDITOR'S PREFACE

From the time Niels Bohr formulated his complementarity viewpoint in 1927, it constituted the basis for a wide variety of his thoughts and actions. Firstly, as documented in Volumes 6 and 7 of the Niels Bohr Collected Works, the viewpoint originated from Bohr's work in physics, in which discipline he continued to refine it. Secondly, however, Bohr sought with increasing vigour to explain and promote complementarity to an ever wider audience. In the process, he applied his viewpoint in fields outside physics, such as psychology, biology and anthropology. The present volume contains Bohr's main published statements in this area, which are at the same time his main contributions to philosophical questions. Thirdly, the complementarity viewpoint was present in the background even in what broadly may be termed Bohr's political activities, a topic which will constitute the central focus of Volume 11 in this series.

Yet it must be emphasized that because Bohr regarded complementarity to be a general idea, encompassing all these areas, the sharp distinction between his applications of his complementarity viewpoint implied by the separate volumes of these Collected Works is arbitrary. Consequently, the contents of the present volume need to be considered in relation to Bohr's other writings, particularly those presented in Volumes 6, 7 and 11. Indeed, just as the General Introduction to the present volume begins with a discussion of Bohr's philosophical views as expressed in some of the articles reproduced in Volume 7, so the remarks introducing Volume 11 will necessarily take recourse to some of the articles printed in Volume 10.

Bohr's writings on philosophical questions remained scattered, and he was never able to bring to publication what some of his closest colleagues referred to as "The Book" – a comprehensive presentation of the complementarity viewpoint and its implications. Bohr's motivation to write such a book stemmed in part from his scepticism toward professional philosophers, whose insensitivity to the rapid developments in modern natural science he lamented in private conversation. At the same time, however, "The Book" would also have demonstrated that Bohr did not arrive at his philosophical standpoint from his scientific knowledge and activities without regard for philosophical tradition. As the editor of this particular volume shows in his General Introduction, Bohr's philosophical interest began at an early age. Indeed, when he formulated the complementarity viewpoint in the late 1920s, one of his old friends recognized in it ideas Bohr had put forth in his youth.

Without "The Book", we are compelled to content ourselves with searching for a synthesis in Bohr's various published statements on philosophical questions. The present collection of Bohr's philosophically oriented publications is no doubt the best source available for such a search. Considering the lack of focus and the repetition in Bohr's scattered writings, we are especially fortunate to have as our guide a philosopher who also has knowledge of the history of science. Although in the last instance we will have to rely on Bohr's own words, David Favrholdt's reading of them provides a helpful and innovative framework for interpreting and explaining the complexity of Bohr's philosophical viewpoint.

$$* * *$$

This is the first volume of the Niels Bohr Collected Works for which I alone must take full responsibility as General Editor. However, the work has been smoothed considerably by the solid groundwork laid by special editor Favrholdt – in close collaboration with former General Editor Erik Rüdinger – before I took up the task in 1989. Indeed, already at that time Favrholdt and Rüdinger had made substantial progress in selecting the material to be reproduced in the volume. Although I was not formally involved at that stage, Rüdinger taught me much about working on the Collected Works long before I had an inkling that I might succeed him. I am grateful to him for that as well as for his continued help and advice after he had taken up entirely different responsibilities. Upon being appointed to succeed Rüdinger as director of the Niels Bohr Archive, I have worked ever more closely with Favrholdt on the present volume. I am going to miss the instructive and pleasant collaboration with Favrholdt over the last several years.

The completion of the work in between various other duties has been a long-winded process involving numerous different tasks, among which the process of translation stands out in particular. While an imitation of Bohr's unique style would be bound to fail, an attempt has nevertheless been made to avoid words and phrases that Bohr would not use. To this end I have been fortunate to profit from the assistance of people with personal experience of working with Bohr. Hilde Levi, who has translated the letters originally written in German, has also contributed with her unrelenting criticism of any translation that she for some reason regarded as unsuitable. In particularly difficult cases, the help of Jørgen Kalckar and Aage Bohr has proved invaluable. Aage Bohr's help was also essential in making out Bohr's words in his sixth Compton Lecture, which has been retranscribed especially for this volume. Helle Bonaparte, our former secretary, translated some of Bohr's articles before she left the archive in 1991; her considerable language abilities meant much for the preparation of the volume at that early stage. Since then, the translations from Danish have been done in close collaboration with Felicity Pors; I am convinced that our often divergent views resulting in constructive discussions about nearly all aspects of the English language have had a positive effect on the final result.

Other tasks concern the location and choice of publications to be used as facsimiles, as well as obtaining the background information provided in the editorial footnotes. In many cases, this was part of the groundwork done by Rüdinger and Favrholdt, but a good number of questions, some of them involving several inquiries at several places, remained. Too many individuals and institutions have contributed to this process for me to name them all. Suffice it to say that Karl Grandin of the University of Uppsala has been of great assistance in finding documentation in Sweden and that Professor Giuliano Pancaldi of the University of Bologna has helped supply the facsimile for an especially difficult item.

Along the way I have had excellent collaboration with Joost Kircz and his staff at Elsevier, among whom I am pleased to direct special thanks to Betsy Lightfoot, whose efforts with regard to all aspects of the technical preparation of the present volume have been indispensable.

My closest collaboration, however, has been with the staff of the Niels Bohr Archive. Felicity Pors has participated as an equal in all phases and at all levels of the work, while Anne Lis Rasmussen has cheerfully gone through the chores of learning the complex task of computer type-setting, which is required in this new age of book production. I will always be grateful to these two, as well as Hilde Levi, for their unique combination of extreme patience and lively enthusiasm. Abraham Pais, who spends half of the year at the Niels Bohr Archive, has provided general expertise and moral support.

Finally, my thanks go to the Niels Bohr Archive's board of directors for gently, yet firmly, pushing me to complete the task.

Finn Aaserud
The Niels Bohr Archive
May 1997

FOREWORD

This volume of the Niels Bohr Collected Works, containing several lectures and articles by Bohr, is divided into four parts. The title of the volume, "Complementarity Beyond Physics", refers primarily to Part I, which is by far the largest and comprises papers discussing the fundamental questions of biology and related psychological and philosophical problems. Following the reproduction of papers brought to publication by Bohr, there is a separate Appendix to Part I including some of Bohr's most interesting and substantive unpublished contributions in this area. The papers in Part I span the last thirty years of Bohr's life and display his great interest in biological problems and his unremitting efforts to show that biology cannot be reduced to physics and chemistry. His basic viewpoint is that observation pertaining to the scientific analysis of living organisms requires a complementary perspective that excludes vitalism and traditional mechanism to an equal degree. Although it is possible in these articles to trace a certain development in Bohr's viewpoint over the thirty years, his basic viewpoint remained unchanged from first to last.

Part II contains articles of a more general cultural interest. Some of these show that Bohr regarded the complementary perspective to be of value also outside the scientific sphere.

Part III contains the articles Bohr wrote about the great Danish philosopher, Harald Høffding, of whom he always spoke with great reverence. Høffding was a close friend of Bohr's father, Christian Bohr, and in his first year as a student at the University of Copenhagen Niels Bohr was introduced to the main topics in philosophy through Høffding's lectures. These short papers are presented in a section on their own because of the continuing discussion in the history of science about Høffding's possible influence on Bohr's work in physics and his whole scientific approach.

Part IV comprises articles illuminating the history of 20th century physics. Bohr had great veneration for his predecessors and teachers, and he prepared these articles with great care.

In each of Parts I through IV the published material is arranged chronologically according to the date when an address was held or – whenever a publication was not the direct outcome of such an address – when the article was published. Thus, although printed only in 1953, Bohr's speech in 1928 at the 25th anniversary of the graduation of his class from the Danish Gymnasium (Part II, item I) is placed according to the year he presented the talk. In cases, however, where a manuscript for a talk has gone through substantial changes before it appeared in print, the publication date has been used as a basis. This is the case for the Steno Lecture (Part I, item X), which Bohr gave in 1949 but which was published only in 1957.

Any facsimile reproduced in this volume is of the first version of the article in question. The only exception is the very first article, "Light and Life", in which case the version reproduced was the one explicitly preferred by Bohr. In cases where subsequent versions differ from the original this is noted on the page immediately preceding the facsimile.

Part V contains correspondence relating to the material in Parts I through IV. As usual, an inventory of relevant unpublished manuscripts held at the Niels Bohr Archive constitutes an appendix to the whole volume.

The work was begun during Erik Rüdinger's directorship of the Niels Bohr Archive, and I am indebted to him for the first introduction to the Archive, and for general guidance about many questions as regards biography and history of science. Since Finn Aaserud took over as director in 1989, I have had a close and fruitful cooperation with him. He has with the greatest care checked and improved my work down to the smallest detail and his effort has been quite indispensable in the editing of this volume. In the work at the Archive I have had exceptionally qualified help from Hilde Levi, Felicity Pors and Helle Bonaparte, former secretary at the Archive. I am also grateful to Anne Lis Rasmussen, the present secretary, for the great assistance in the editing of the final text.

Some of the preliminary drafts to the Introductions have been read not only by Erik Rüdinger and Finn Aaserud, but also by Aage Bohr and Jørgen Kalckar. From these and from Jens Lindhard I have received much critical instruction, for which I am grateful. For a period during the work I shared an office with Abraham Pais, whose many inspiring comments and penetrating observations have been of great importance to me.

David Favrholdt
Odense University
April 1997

CONTENTS

PART I: COMPLEMENTARITY IN BIOLOGY AND RELATED FIELDS

PART II: COMPLEMENTARITY IN OTHER FIELDS

PART III: PAPERS ON HARALD HØFFDING

PART IV: HISTORICAL PAPERS

PART V: SELECTED CORRESPONDENCE

Correspondents

INVENTORY OF RELEVANT MANUSCRIPTS
IN THE NIELS BOHR ARCHIVE

INDEX

EARLIER VOLUMES OF THE
NIELS BOHR COLLECTED WORKS

After Léon Rosenfeld (1904–1974) had served as General Editor of the first three volumes of the *Niels Bohr Collected Works*, Erik Rüdinger continued the task for Volumes 5 through 9 (Volume 7 jointly with Finn Aaserud). These volumes (all published by North-Holland/Elsevier) are in the following simply referred to as "Vol. 1", "Vol. 2", etc. They are:

Vol. 1, *Early Work (1905–1911)* (ed. J. Rud Nielsen), 1972.

Vol. 2, *Work on Atomic Physics (1912–1917)* (ed. Ulrich Hoyer), 1981.

Vol. 3, *The Correspondence Principle (1918–1923)* (ed. J. Rud Nielsen), 1976.

Vol. 4, *The Periodic System (1920–1923)* (ed. J. Rud Nielsen), 1977.

Vol. 5, *The Emergence of Quantum Mechanics (mainly 1924–1926)* (ed. Klaus Stolzenburg), 1984.

Vol. 6, *Foundations of Quantum Physics I (1926–1932)* (ed. Jørgen Kalckar), 1985.

Vol. 7, *Foundations of Quantum Physics II (1933–1958)* (ed. Jørgen Kalckar), 1996.

Vol. 8, *The Penetration of Charged Particles through Matter (1912–1954)* (ed. Jens Thorsen), 1987.

Vol. 9, *Nuclear Physics (1929–1952)* (ed. Rudolf Peierls), 1986.

ABBREVIATED TITLES OF PERIODICALS

Ann. d. Phys.	Annalen der Physik (Leipzig)
Berl. Tid.	Berlingske Tidende (Copenhagen)
Chem. News	Chemical News (London)
Fys. Tidsskr.	Fysisk Tidsskrift (Copenhagen)
Handbuch der Phys.	Handbuch der Physik (Berlin)
ICSU Review	International Council of Scientific Unions Review (Amsterdam)
J. Chem. Soc.	Journal of the Chemical Society (London)
J. Mond. Pharm.	Journal Mondial de Pharmacie (The Hague)
Kgl. Dan. Vid. Selsk., Filos. Medd.	Filosofiske Meddelelser udgivet af Det Kongelige Danske Videnskabernes Selskab (Copenhagen)
Kgl. Dan. Vid. Selsk., Mat.–fys. Medd.	Matematisk–fysiske Meddelelser udgivet af Det Kongelige Danske Videnskabernes Selskab (Copenhagen)
Kgl. Dan. Vid. Selsk. Skr.	Det Kongelige Danske Videnskabernes Selskabs Skrifter. Naturvidenskabelig og mathematisk Afdeling (Copenhagen)
Month. Not. Roy. Astr. Soc.	Monthly Notices of the Royal Astronomical Society (London)

Nach. Akad. Wiss. Göttingen, Math–Phys. Kl.	Nachrichten von der Akademie der Wissenschaften zu Göttingen, Mathematisch–Physikalische Klasse
Nach. Ges. Wiss. Göttingen, Math–Phys. Kl.	Nachrichten von der Gesellschaft der Wissenschaften zu Göttingen, Mathematisch–Physikalische Klasse
Naturwiss.	Die Naturwissenschaften (Berlin)
Overs. Dan. Vid. Selsk. *Overs. Dan. Vidensk. Selsk. Forh.* *Overs. Dan. Vidensk. Selsk. Virks.*	Oversigt over Det Kongelige Danske Videnskabernes Selskabs Forhandlinger[1] (Copenhagen)
Phil. Mag.	Philosophical Magazine (London)
Phil. Sci.	Philosophy of Science (East Lansing, Michigan)
Phil. Today	Philosophy Today (Collegeville, Indiana)
Phys. Rev.	Physical Review (New York)
Phys. Z. *Phys. Zs.* *Phys. Zeitsch.*	Physikalische Zeitschrift (Leipzig)
Proc. Phys. Soc.	Proceedings of the Physical Society (London)
Proc. Roy. Soc.	Proceedings of the Royal Society of London
Sitz. Ber. Wiener Akad. d. Wiss., mat. nat. Kl.	Sitzungsberichte der Wiener Akademie der Wissenschaften, Mathematisch–Naturwissenschaftliche Klasse
Verh. deutsch. Phys. Ges.	Verhandlungen der deutschen physikalischen Gesellschaft (Braunschweig)
Z. Phys. *Z. Physik* *Zs. f. Phys.*	Zeitschrift für Physik (Braunschweig)

[1] From June 1931: Oversigt over Det Kongelige Danske Videnskabernes Selskabs Virksomhed.

OTHER ABBREVIATIONS

AHQP	Archive for History of Quantum Physics
AIP	American Institute of Physics, College Park, Maryland
Bohr MSS	Bohr Manuscripts, AHQP
BSC	Bohr Scientific Correspondence, AHQP
CERN	Centre Européen pour la Recherche Nucléaire, Geneva
DF	David Favrholdt
Mf	Microfilm
MS, MSS	Manuscript
NBA	Niels Bohr Archive, Copenhagen

ACKNOWLEDGEMENTS

For some contributions, the editor and publisher were unfortunately unable to trace the copyright holders to ask for reproduction permission. Their importance was, however, considered sufficiently high to reprint them without further delay. The effort to identify the original copyright holders will continue.

N. Bohr, "Light and Life", Nature **131** (1933) 421–423, 457–459, is reprinted by permission of Nature © Macmillan Magazines Limited.

N. Bohr, "Causality and Complementarity", Phil. Sci. **4** (1937) 289–298, is reprinted by permission of the publisher, the University of Chicago Press.

N. Bohr, "Medical Research and Natural Philosophy", Acta Medica Scandinavica (Suppl.) **142** (1952) 967–972, is reprinted by permission of the Journal of Internal Medicine.

N. Bohr, "Address at the Opening Ceremony", Acta Radiologica (Suppl.) **116** (1954) 15–18, is reprinted by permission of the publisher, Munksgaard International Publishers Ltd.

N. Bohr, "Unity of Knowledge" in "The Unity of Knowledge" (ed. L. Leary), Doubleday & Co., New York 1955, pp. 47–62, is reprinted by permission of Helen King.

N. Bohr, "Physical Science and Man's Position", Ingeniøren **64** (1955) 810–814, is reprinted by permission of the Society of Danish Engineers.

N. Bohr, "Quantum Physics and Biology", Symposia for the Society for Experimental Biology, Number XIV: "Models and Analogues in Biology",

Cambridge 1960, pp. 1–5, is reprinted by permission of the Society for Experimental Biology.

N. Bohr, "Physical Models and Living Organisms" in "Light and Life" (eds. W.D. McElroy and B. Glass), The Johns Hopkins Press, Baltimore 1961, pp. 1–3, is reprinted by permission of the publisher, The Johns Hopkins Press.

N. Bohr, "The Unity of Human Knowledge", Revue de la Fondation Européenne de la Culture, July 1961, pp. 63–66, is reprinted by permission of the European Cultural Foundation.

N. Bohr, "Light and Life Revisited", ICSU Review **5** (1963) 194–199, is reprinted by permission of the ICSU Secretariat.

N. Bohr, "Atoms and Human Knowledge", Pamphlet published by the University of Oklahoma, is reprinted by permission of Aage Bohr.

N. Bohr, "Natural Philosophy and Human Cultures", Congrès international des sciences anthropologiques et ethnologiques, compte rendu de la deuxième session, Copenhague 1938, Ejnar Munksgaard, Copenhagen 1939, pp. 86–95, is reprinted by permission of the publisher, Munksgaard International Publishers Ltd.

N. Bohr, "Dansk Kultur. Nogle indledende Betragtninger" in "Danmarks Kultur ved Aar 1940", Det Danske Forlag, Copenhagen 1941–1943, Vol. 1, pp. 9–17, is reprinted by permission of the Danish Cultural Institute.

N. Bohr, "Physical Science and the Study of Religions", Studia Orientalia Ioanni Pedersen Septuagenario A.D. VII id. Nov. Anno MCMLIII, Ejnar Munksgaard, Copenhagen 1953, pp. 385–390, is reprinted by permission of the publisher, Munksgaard International Publishers Ltd.

N. Bohr, "Atomvidenskaben og menneskehedens krise", Politiken, 20 April 1961, is reprinted by permission of Politiken and Aage Bohr.

N. Bohr, "Ved Harald Høffdings 85 Aars-Dag", Berlingske Tidende, 10 March 1928, is reprinted by permission of Aage Bohr.

ACKNOWLEDGEMENTS

The following articles by N. Bohr from Overs. Dan. Vidensk. Selsk. Virks.: "Mindeord over Harald Høffding", Juni 1931 – Maj 1932, pp. 131–136; "Harald Høffdings 100-Aars Fødselsdag", Juni 1942 – Maj 1943, pp. 57–58, are reprinted by permission of the Royal Danish Academy of Sciences and Letters.

N. Bohr, "Zeeman Effect and Theory of Atomic Constitution" in "Zeeman, Verhandelingen", Martinus Nijhoff, The Hague 1935, pp. 131–134, is reprinted by permission of Kluwer Academic Publishers.

N. Bohr, "Hans Christian Ørsted", Fys. Tidsskr. **49** (1951) 6–20, is reprinted by permission of the Danish Physical Society.

N. Bohr, "Rydberg's Discovery of the Spectral Laws", Proceedings of the Rydberg Centennial Conference on Atomic Spectroscopy, Lunds Universitets Årsskrift. N.F. Avd. 2. Bd. 50. Nr 21 (1955) 15–21, is reprinted by permission of Aage Bohr.

N. Bohr, "The Rutherford Memorial Lecture 1958: Reminiscences of the Founder of Nuclear Science and of Some Developments Based on his Work", Proc. Phys. Soc. **78** (1961) 1083–1115, is reprinted by permission of the publisher, the Institute of Physics Publishing.

N. Bohr, "The Solvay Meetings and the Development of Quantum Physics" in "La théorie quantique des champs", Douzième Conseil de physique tenu à l'Université Libre de Bruxelles du 9 au 14 octobre 1961, Interscience Publishers, New York 1962, pp. 13–36, is reprinted by permission of Aage Bohr.

GENERAL INTRODUCTION

Complementarity Beyond Physics

by

DAVID FAVRHOLDT

The articles in this volume deal mainly with Niels Bohr's ideas about topics beyond physics. Long before quantum mechanics was established, he had shown an interest in problems of description pertaining to psychology and biology, without, however, publishing his views. Only after having introduced his complementarity argument in 1927 as a clarification of the conditions of observation and description in quantum mechanics, did Bohr begin to comment publicly on other fields of knowledge with the intention of showing what could be gained from "the epistemological lesson of quantum mechanics". From 1929 onward, he discussed time and time again, in published lectures and articles, epistemological problems in psychology and biology as well as in the anthropological sciences. Having found a way of dealing with the unusual features of quantum mechanics, he apparently felt himself in a better position to shed light on the conditions for observation and description in other fields of science.

The articles in each individual part of this volume are presented chronologically in order to make apparent the development of Bohr's ideas about biology, psychology and other subjects outside physics. It will be seen that he gradually clarified his views and refined his terminology in step with the many objections to his ideas. However, as regards the basic themes in his thinking, little change can be detected from 1927 on. In this General Introduction, therefore, I have ignored chronology when presenting quotations from this period.

1. BOHR'S FIRST PUBLISHED STATEMENTS

Bohr introduced the concept of complementarity in his so-called Como Lecture, given in September 1927 at the International Congress of Physicists on the Occasion of the Centenary of the Death of Alessandro Volta. In a revised version of the lecture, published seven months later, he writes[1]:

> "The very nature of the quantum theory thus forces us to regard the space–time co-ordination and the claim of causality, the union of which characterises the classical theories, as complementary but exclusive features of the description, symbolising the idealisation of observation and definition respectively. ... Indeed, in the description of atomic phenomena, the quantum postulate presents us with the task of developing a 'complementarity' theory the consistency of which can be judged only by weighing the possibilities of description and observation."

Bohr's complementarity argument is often taken to be a philosophical interpretation of the observational situation in quantum mechanics and is spoken of as the "Copenhagen interpretation", which was supported by many others – Heisenberg, Pauli and Born, to mention a few. Bohr, however, did not consider his view an interpretation but rather a spelling out of the possibilities of description and observation within quantum mechanics. The complementarity argument deals with our conditions for description and is therefore a statement of epistemology. However, since these conditions are dictated by the existence of the quantum of action, and thus differ from those we meet in classical physics, quantum mechanics has given us an "epistemological lesson", as Bohr phrased it, which may help us throw light on fundamental problems within other fields of science as well. As we shall see, the clue is to examine constantly our conditions for observation and description, and consequently Bohr was preoccupied with reflections concerning the very nature of description and the use of concepts.

Thus, from the year 1927 onward, his old interest in fundamental psychological and biological problems was revived. The first indication of the renewed

---

[1] N. Bohr, *The Quantum Postulate and the Recent Development of Atomic Theory*, Nature (Suppl.) **121** (1928) 580–590, quotation on p. 580. The lecture has also been published, in a slightly amended version, in *Atomic Theory and the Description of Nature*, Cambridge University Press, Cambridge 1934 (reprinted 1961), pp. 52–91, quotation on pp. 54–55. The latter volume is photographically reproduced as *Atomic Theory and the Description of Nature, The Philosophical Writings of Niels Bohr, Vol. I*, Ox Bow Press, Woodbridge, Connecticut 1987. The former version of the article is reproduced in Vol. 6, pp. [148]–[158], quotation on p. [148]. Detailed bibliographical information on the various versions of Bohr's lecture is given in Vol. 6, pp. [110]–[112].

interest appears in the article quoted above. Referring to the situation within quantum mechanics it ends as follows[2]:

> "I hope, however, that the idea of complementarity is suited to characterise the situation, which bears a deep-going analogy to the general difficulty in the formation of human ideas, inherent in the distinction between subject and object."

This remark is elaborated on in an article published in "Die Naturwissenschaften" in June 1929. Here Bohr comments upon the difficulties in describing our mental activity[3]:

> "The epistemological problem under discussion may be characterized briefly as follows: For describing our mental activity, we require, on one hand, an objectively given content to be placed in opposition to a perceiving subject, while, on the other hand, as is already implied in such an assertion, no sharp separation between object and subject can be maintained, since the perceiving subject also belongs to our mental content."

He also takes up the problem of the freedom of the will and suggests that a "detailed investigation of the processes of the brain" is excluded on account of the quantum of action. We must expect, he says, that an attempt to observe the processes in the brain "will bring about an essential alteration in the awareness of volition"[4].

Shortly thereafter, Bohr advanced some more detailed views on biology in his address to the 18th Scandinavian Meeting of Natural Scientists in August 1929. Here he states that[5]:

> "With regard to the more profound biological problems, however, in which

---

[2] Bohr, *Quantum Postulate*, ref. 1, Vol. 6, p. [158].

[3] N. Bohr, *Wirkungsquantum und Naturbeschreibung*, Naturwiss. **17** (1929) 483–486. Reproduced in Vol. 6, pp. [203]–[206] and reprinted in *Atomtheorie und Naturbeschreibung*, Julius Springer Verlag, Berlin 1931, pp. 60–66. Translated into English as *The Quantum of Action and the Description of Nature* in *Atomic Theory and the Description of Nature*, ref. 1, pp. 92–101, quotation on p. 96. The English version is reproduced in Vol. 6, pp. [208]–[217], quotation on p. [212].

[4] *Ibid.*, pp. [216]–[217].

[5] N. Bohr, *The Atomic Theory and the Fundamental Principles Underlying the Description of Nature* in *Atomic Theory and the Description of Nature*, ref. 1, pp. 102–119, quotation on pp. 118–119. Reproduced in Vol. 6, pp. [236]–[253], quotation on pp. [252]–[253].

we are concerned with the freedom and power of adaptation of the organism in its reaction to external stimuli, we must expect to find that the recognition of relationships of wider scope will require that the same conditions be taken into consideration which determine the limitation of the causal mode of description in the case of atomic phenomena."

The three articles mentioned were published in November 1929 in a *Festschrift* for the University of Copenhagen[6]. In the "Introductory Survey"[7], Bohr writes that the reference to psychological problems in the book's third article[8] has a twofold purpose, i.e. to make it easier to accustom ourselves to the new situation in physics and to pave the way for a deeper understanding of the psychological problems in the light of the epistemological lesson of quantum mechanics[9]:

"As stressed in the article, it is clear to the writer that for the time being we must be content with more or less appropriate analogies. Yet it may well be that behind these analogies there lies not only a kinship with regard to the epistemological aspects, but that a more profound relationship is hidden behind the fundamental biological problems which are directly connected to both sides."

The phrase "both sides" refers to physics on the one hand and psychology on the other. The book appeared in German in 1931[10] and in English in 1934[11]. Both editions contained an additional lecture[12], and the "Introductory Survey" included a new "Addendum"[13], where it is argued that biology cannot be reduced to physics or chemistry. In a French edition of the book, published

---

[6] N. Bohr, *Atomteori og Naturbeskrivelse. Festskrift udgivet af Københavns Universitet i Anledning af Universitetets Aarsfest November 1929*, Bianco Luno, Copenhagen 1929.

[7] N. Bohr, *Indledende Oversigt* in *ibid.*, pp. 5–17, reproduced in Vol. 6, pp. [259]–[273]. Translated into English as *Introductory Survey* in *Atomic Theory and the Description of Nature*, ref. 1, pp. 1–24, which is reproduced in Vol. 6, pp. [279]–[302].

[8] Bohr, *Quantum of Action*, ref. 3.

[9] Bohr, *Introductory Survey*, ref. 7, Vol. 6, p. [298].

[10] Bohr, *Atomtheorie und Naturbeschreibung*, ref. 3.

[11] Bohr, *Atomic Theory and Description of Nature*, ref. 1.

[12] Bohr, *Atomic Theory and the Fundamental Principles*, ref. 5.

[13] Bohr, *Introductory Survey*, ref. 7, Vol. 6, pp. [299]–[302].

in 1932, a new passage is added at the end of the "Addendum". It runs as follows[14]:

> "Notwithstanding the intrinsic interest the biological and psychological questions have, even for those who like me are strangers to these fields, my primary aim in dealing with them in these articles has been to throw light on the physical and epistemological problems met with in the atomic theory. Incidentally, I hope to deal with the latter problems in a detailed exposition of the principles of atomic theory, currently under preparation, in a more thorough manner than circumstances have permitted while writing these articles."

The arguments advanced will be considered in the Introduction to Part I. The first comprehensive presentation of Bohr's views on biology is to be found in his lecture "Light and Life" which is reproduced in this volume[15].

2. EARLY ORIGINS OF BOHR'S VIEW

It has often been discussed whether Bohr developed his complementarity view on the basis of quantum mechanics alone, or whether he saw quantum mechanics in the light of an already developed viewpoint. Special interest has been attached to the question of whether, before this breakthrough in physics, he had concerned himself with the conditions of observation and description in other fields such as psychology and biology, and was thus prepared to meet the unusual observational problems which presented themselves in quantum mechanics. This question is in fact well documented.

---

[14] N. Bohr, *La théorie atomique et la description des phénomènes*, Gauthier-Villars et Cie, Editeurs, Paris 1932 (translated by Andrée Legros and Léon Rosenfeld), p. 21. "Abstraction faite de l'intérêt propre que présentent les questions biologiques et psychologiques, même pour ceux qui, comme nous, y sont étrangers, j'ai eu principalement pour but, en m'en occupant dans ces articles, de mettre en lumière les problémes physiques et épistémologiques que nous rencontrons dans la théorie atomique. J'espère d'ailleurs, dans un exposé détaillé des principes de la théorie atomique, actuellement en préparation, traiter ces derniers problèmes d'une manière plus approfondie que ne le permettaient les circonstances dans lesquelles j'ai écrit ces articles." This addition remained unnoticed until after the publication of Vol. 6 and is therefore reproduced here.

[15] N. Bohr, *Light and Life*, Nature **131** (1933) 421–23, 457–59, and in *Atomic Physics and Human Knowledge*, John Wiley & Sons, New York 1958, pp. 3–12. The latter volume is photographically reproduced as *Essays 1933–1957 on Atomic Physics and Human Knowledge, The Philosophical Writings of Niels Bohr, Vol. II*, Ox Bow Press, Woodbridge, Connecticut 1987. The former version of the article is reproduced in this volume on pp. [29]–[35].

Bohr had been preoccupied with reflections on fundamental questions of psychology and biology long before 1927. In a short autobiography he writes as follows[16]:

"My interest in the biological and psychological problems which one is thereby led to [via the epistemological problems pertaining to quantum physics – DF] stems from my early youth, when I listened to the discussions in the circles of my father and his friends, among whom I later came into contact with especially the physicist Christian Christiansen, who was my teacher at the university, and the philosopher Harald Høffding, with whom I had many instructive conversations right until his last days."

Niels Bohr's father, Christian Bohr, was a distinguished professor of physiology at the University of Copenhagen from 1890 until his death in 1911. During this period the Bohr family lived in the professorial residence at the Institute of Physiology in Bredgade, a street in Copenhagen. Thus from his childhood Niels Bohr grew up in a rich scientific milieu.

In an article from 1957, Bohr quotes his father on the position of biology and continues: "I have quoted these remarks which express the attitude in the circle in which I grew up and to whose discussions I listened in my youth ..."[17]. In the drafts for the article, we find the following note: "From my earliest youth I remember having heard discussions between Carl Lange, Chievitz and my father concerning such questions [i.e. the Vitalism–Mechanism dispute – DF]"[18]. Johan Henrik Chievitz (1850–1901) was professor of anatomy; Carl Lange (1834–1900) professor of pathological anatomy. Lange is the originator, with William James, of the well-known "James–Lange theory of emotions". Since Lange died in 1900 and Chievitz in 1901, Bohr must here be referring to his early youth.

In his years as a student Bohr had the idea that the elusiveness of the subject and the problem of the freedom of the will could be clarified by means of an analogy to the so-called Riemann surfaces. In his last oral history interview he tells us how[19]:

[16] N. Bohr, *Selvbiografi af Æresdoktoren* (*Autobiography of the Honorary Doctor*), Acta Jutlandica **28** (1956) 138.

[17] N. Bohr, *Physical Science and the Problem of Life* in *Atomic Physics and Human Knowledge*, ref. 15, pp. 94–101, quotation on p. 96. Reproduced on pp. [116]–[123], quotation on p. [118].

[18] *Fysik og biologi*, Foredrag i Biologisk Selskab, 26.3.1946. Bohr MSS, microfilm no. 17.

[19] Interview with Niels Bohr, 17 November 1962, AHQP. Transcript, p. 1. The square brackets (though not the ellipses) are in the original transcript.

"At that time I really thought to write something about philosophy, and that was about this analogy with multi-valued functions. ... If you have a square root of x, then you have two values. If you have a logarithm, you have even more. And the point is that if you try to say you have now two values, let us say of square root, then you can walk around in the plane, because, if you are in one point, you take one value, and there will be at the next point a value which is very far from it and one which is very close to it. ... If in these functions, as the logarithm or the square root, they have a singular value at the origin, then if you go round from one point and go in a closed orbit and it doesn't go round the origin, you come back to the same [value]. ... But when you go round the origin, then you come over to the other [value of the] function, and that is then a very nice way to do it, as Dirichlet [Riemann], of having a surface in several sheets and connect them in such a way that you just have the different values of the function on the different sheets."

The point is that we are able to use the word "I" with different references in a situation and yet have a clear understanding of its meaning because the different instances of "I" are arranged in different planes or surfaces. A statement such as "I don't know whether I should be ashamed of what I did, but I really couldn't help doing it" is understandable because we subconsciously arrange the different uses of "I" each on its level.

We have good reason to believe that Bohr was correct in dating this view back to his years as a student. In a letter to his brother Harald of 26 June, 1910, he writes[20]:

" ... I must confess that I don't know if I am most happy over your appointment, over the good behaviour of my electrons at the moment, or over this portfolio; probably, the only answer is that emotions, like cognition, must be arranged in planes that cannot be compared".

Niels to Harald Bohr, 26 June 10
Danish text: Vol. 1, p. [510]
Translation: Vol. 1, p. [511]

At that time, Harald Bohr was in Göttingen, where he had studied since the autumn of 1909, so it seems natural to assume that they had discussed such planes at an earlier date. Furthermore, we know that Bohr discussed psychological problems with his second cousin Edgar Rubin (1886–1951) who, after a period as associate professor of philosophy, was promoted in 1922 to professor of psychology in which position he would gain

---

[20] In Vol. 1, p. [513], the English text has "sensations" instead of "emotions" ("Følelser" in Danish).

"Family portrait", c. 1910. From the left: Harald Bohr; Poul Nørlund, historian and director (1938–1951) of the Danish National Museum; Edgar Rubin; Niels Bohr; and Niels Erik Nørlund, professor of mathematics at the University of Copenhagen from 1922. Rubin was Niels Bohr's second cousin, and the sister of the Nørlund brothers, Margrethe, married Niels Bohr in 1912.

international reputation. Their long-lasting discussion about recognition and memory is evident from a postcard Bohr sent to Rubin on 20 May 1912[21]. Later, Bohr also helped Rubin with an experiment concerning visual perception[22].

An oral history interview with Oskar Klein shortly after Bohr's death provides further testimony of Bohr's early interest in the epistemological aspects of psychology and biology. Klein comments here on Bohr's interest in the problem of the freedom of the will: "That he told already on that walk in the summer of 1918, so I think he must have thought about that a long time. Then he

[21] Postcard from Bohr to Rubin, 20 May 1912, NBA. Reproduced on pp. [575] (Danish original) and [576] (English translation).

[22] Rubin mentions this in his doctoral thesis on visually experienced figures, *Synsoplevede Figurer: Studier i Psykologisk Analyse* (Visually Experienced Figures: Studies in Psychological Analysis), Gyldendal, Copenhagen 1915, pp. 191–192.

spoke – but that was very vague, at least to me – about an analogy between that and the quantization." In the same interview Klein says: "I believe at that time already he said that his father had some ideas about biology and that one might think of quite different kinds of laws in biology than in physics, that in biology one might have finalistic laws. I believe that he mentioned that already at that time"[23].

Finally, it may be mentioned that Bohr in a letter of November 1928 to his Swedish colleague Carl W. Oseen – who became a close friend as early as 1911[24] – comments upon the final remark in his article in "Die Naturwissenschaften" cited above[25]:

> "As we already discussed years ago, the difficulty in all philosophy is the circumstance that the functioning of our consciousness presupposes a requirement as regards the objectivity of the content, while on the other hand the idea of the subject, of our own ego, forms a part of the content of our consciousness. This is exactly the kind of difficulties of which we have got such a clear example in the character of the description of nature required by the essence of the quantum postulate".

Bohr to Oseen,
5 Nov 28
Danish text: Vol. 6, p. [430]
Translation: Vol. 6, p. [189]

Bohr himself linked his reflections upon the self and the freedom of the will to an unfinished novel by the Danish author Poul Martin Møller[26]. When Bohr was a boy, this novel was already a classic, and he himself tells us that every young person received it as a Confirmation present, i.e. at the age of about fourteen[27]. He apparently read the book at that age or perhaps earlier. In any case, it made an indelible impression on him. Throughout his life he often spoke about this

---

[23] Interview with Oskar Klein, 20 February 1963, AHQP. Transcript, pp. 7 and 8–9, respectively. Klein gives a similar account in *Glimpses of Niels Bohr as Scientist and Thinker* in *Niels Bohr, His life and work as seen by his friends and colleagues* (ed. Stefan Rozental), North-Holland Publishing Company, Amsterdam 1968, pp. 74–93, on p. 76.

[24] Bohr and Oseen met each other for the first time at the Scandinavian Mathematical Congress in the summer of 1911 in Copenhagen. See Vol. 1, p. [102].

[25] Bohr, *Wirkungsquantum*, ref. 3.

[26] P.M. Møller, *En dansk Students Eventyr* (The Adventures of a Danish Student) in *P.M. Møller: Efterladte Skrifter. Vol. 3*, Reitzel, Copenhagen 1843. Poul Martin Møller (1794–1838) was one of Denmark's great poets, but also a philosopher with a background of theological education. He was professor in philosophy at Kristiania (Oslo) from 1826 to 1830 and at the University of Copenhagen from 1830 to 1838. He was probably inspired by the German philosopher J.G. Fichte concerning the problem of the self.

[27] Bohr notes this in *Samtale i Tisvilde August 1959* (Conversation in Tisvilde August 1959), unpublished tape recording, NBA.

book and introduced it to others. "Everyone of those who came into closer contact with Bohr at the Institute, as soon as he showed himself sufficiently proficient in the Danish language, was acquainted with the little book: it was a part of his initiation"[28].

We may conclude that Bohr had developed his views concerning the observational conditions in psychology long before 1927 and maybe even earlier than 1909. As for biology, the Klein interview suggests that he was inspired by his father's view – that finalistic descriptions were indispensable and that biology was therefore not reducible to physics and chemistry – long before he began writing about these matters in 1929.

As we shall see later, Bohr's ideas about the observational conditions in psychology bear a conspicuous resemblance to his concept of the observational conditions in quantum mechanics, and it has often been maintained by persons close to him that he had developed the idea of complementarity long before he faced the complementary features of quantum mechanics. In an interview from 1963, Bohr's wife Margrethe says: "Rubin was the one who understood him so well, yes. And he often said later, when Niels published his things about complementarity, 'you have spoken like that since you were 18 years old' "[29].

3. OUTLINE OF BOHR'S PHILOSOPHICAL VIEWS

Although Bohr always presented his views on complementarity beyond physics in a rather brief manner, he considered complementarity to be fundamental to the understanding of all aspects of life and reality. The following outline of his fundamental philosophical ideas is presented in a systematic fashion, which arguably is not true to the spirit of Bohr. However, in his numerous general reflections on science Bohr always returned to a few basic themes which, on further inspection, are closely related.

It may seem strange to begin with Bohr's ideas on language and description in that he neither wrote a paper nor gave a lecture on precisely this topic. However, several paragraphs in his articles and – in particular – in his manuscripts contain comments on the topic and provide a key to a deeper understanding of what he called "the epistemological lesson" of quantum mechanics.

---

[28] L. Rosenfeld, *Niels Bohr in the Thirties* in *Niels Bohr* (ed. Rozental), ref. 23, pp. 114–136, quotation on p. 121.
[29] Interview with Margrethe Bohr, 23 January 1963, AHQP. Transcript, p. 16.

LANGUAGE AND THE CONDITIONS FOR DESCRIPTION

The status of language and description constituted a central theme in Bohr's thought, as displayed, for example, in his introduction to his second collection of articles, first published in 1958 (my italics)[30]:

"The main point of the lesson given us by the development of atomic physics is, as is well known, the recognition of a feature of wholeness in atomic processes, disclosed by the discovery of the quantum of action. The following articles present the essential aspects of the situation in quantum physics and, at the same time, stress the points of similarity it exhibits to our position in other fields of knowledge beyond the scope of the mechanical conception of nature. We are not dealing here with more or less vague analogies, but with *an investigation of the conditions for the proper use of our conceptual means of expression.* Such considerations not only aim at making us familiar with the novel situation in physical science, but might on account of the comparatively simple character of atomic problems be helpful in *clarifying the conditions for objective description in wider fields.*"

Time and time again Bohr would stress the fact that all quantum mechanical experiments must necessarily be described by means of ordinary language supplemented with classical physical concepts. In his famous article "Discussion with Einstein on Epistemological Problems in Atomic Physics" he writes[31]:

"For this purpose, it is decisive to recognize that, *however far the phenomena transcend the scope of classical physical explanation, the account of all evidence must be expressed in classical terms.* The argument is simply that by the word 'experiment' we refer to a situation where we can tell others what we have done and what we have learned and that, therefore, the account of the experimental arrangement and of the results of

---

[30] N. Bohr, *Introduction* in *Atomic Physics and Human Knowledge*, ref. 15, pp. 1–2. Reproduced on pp. [111]–[112].

[31] N. Bohr, *Discussion with Einstein on Epistemological Problems in Atomic Physics* in *Albert Einstein: Philosopher–Scientist* (ed. P.A. Schilpp), Library of Living Philosophers, Vol. VII, The Library of Living Philosophers, Inc., Evanston, Illinois 1949, pp. 201–241, quotation on p. 209. Also published in *Atomic Physics and Human Knowledge*, ref. 15, pp. 32–66, quotation on p. 39. The former version of the article is reproduced in Vol. 7, pp. [341]–[381], quotation on p. [349]. The italics are Bohr's.

the observations must be expressed in unambiguous language with suitable application of the terminology of classical physics."

In the following we shall see why Bohr considered this fact important.

In the quotation just given, Bohr states that we must use unambiguous language supplemented with the terminology of classical physics. Elsewhere he declares that we must use classical physics supplemented with unambiguous language or simply makes the point that we are forced to use plain language suitably refined by the usual physical terminology[32].

Apparently, Bohr considered these statements more or less equivalent. As we shall see, he considered classical physics to be a conceptual clarification of the descriptive use of ordinary language. The common element in the descriptive use of ordinary language and that of classical physics is that in both cases we base our descriptions on the fact that a clear line of separation can be drawn between subject and object[33]. In doing so we are able to speak of objects in our surroundings without referring to our subjective experiences of them. Thus Bohr states that[34]:

"... the feature which characterizes the so-called exact sciences is, in general, the attempt to attain to uniqueness by avoiding all reference to the perceiving subject."

The rules of divalent logic and the principles of algebra provide still other conditions for unambiguous description. As shown in particular in a letter from 1938 to the Danish author H.P.E. Hansen, Bohr rejected John Stuart Mill's view that these rested on empirical generalization[35]. Bohr, of course, knew

---

[32] See, for instance, N. Bohr, *Quantum Physics and Philosophy – Causality and Complementarity* in *Philosophy in the Mid-Century, A Survey* (ed. R. Klibansky), La nuova Italia editrice, Firenze 1958, pp. 308–314. Reproduced in Vol. 7, pp. [388]–[394]. The article is reprinted in N. Bohr, *Essays 1958–1962 on Atomic Physics and Human Knowledge*, Interscience Publishers, New York 1963, pp. 1–7. The latter volume is photographically reproduced as *Essays 1958–1962 on Atomic Physics and Human Knowledge, The Philosophical Writings of Niels Bohr, Vol. III*, Ox Bow Press, Woodbridge, Connecticut 1987.

[33] See, for instance, N. Bohr, *On Atoms and Human Knowledge*, Dædalus **87** (1958) 164–175, reproduced in Vol. 7, pp. [412]–[423]. See also N. Bohr, *Physical Science and the Problem of Life*, reproduced in this volume on pp. [116]–[123]. Both articles were published in *Atomic Physics and Human Knowledge*, ref. 15, pp. 83–93, 94–101.

[34] Bohr, *Quantum of Action*, ref. 3, quotation in Vol. 6, pp. [212]–[213].

[35] Letter from Bohr to Hansen, 8 September 1938, NBA. Reproduced on pp. [501] (Danish original) and [503] (English translation).

about multi-valued logic, especially in connection with discussions concerning the logical status of the quantum-mechanical formalism. However, he seems not to have considered any multi-valued logic relevant to epistemology. In this connection Bohr wrote[36]:

"In fact, the limited commutability of the symbols by which [kinematical and dynamical] variables [required for the definition of the state of a system in classical mechanics] are represented in the quantal formalism corresponds to the mutual exclusion of the experimental arrangements required for their unambiguous definition. ... In this connection, the question has even been raised whether recourse to multivalued logics is needed for a more appropriate representation of the situation. ... it will appear, however, that all departures from common language and ordinary logic are entirely avoided by reserving the word 'phenomenon' solely for reference to unambiguously communicable information, in the account of which the word 'measurement' is used in its plain meaning of standardized comparison."

To Bohr, "unambiguity" was a fundamental concept which required no further explanation. While "unambiguity" escapes definition, we may specify the necessary conditions for unambiguous communication.

Further conditions for unambiguity are that it should be possible to identify objects in space and time and order them in causal chains. In a manuscript from 1929, Bohr writes as follows[37]:

"In order to clarify the situation dealt with, it might be relevant briefly to bring to mind the use of our forms of perception on which the traditional description of nature rests. With forms of perception we simply mean the

---

[36] Bohr, *Quantum Physics and Philosophy*, ref. 32, quotation in Vol. 7, pp. [392], [393]. See also the letter from Bohr to Pauli 16 May 1947, reproduced in Vol. 6, pp. [451]–[454].

[37] *Kausalität und Objektivität*, 1929. Bohr MSS, microfilm no. 12: "For at klargøre sig den omhandlede Situation turde det være formaalstjenligt kort at erindre om den Benyttelse af vore Anskuelsesformer, hvorpaa den sædvanlige Naturbeskrivelse hviler. Derved skal vi ved Anskuelsesformer simpelthen forstaa den Begrebsbygning, hvorpaa den tilvante Indordning af Sansefornemmelserne beror, og som ligger til Grund for vort sædvanlige Sprogbrug. Grundlaget for denne Indordning er vel Muligheden for Genkendelse og Sammenligning, og i Overenstemmelse hermed kendetegnes den sædvanlige Naturbeskrivelse ved Bestræbelsen for at udtrykke alle Erfaringer ved Stedsangivelser for materielle Legemer og disses Ændring med Tiden relativt til et paa sædvanlig Maade ved Maalestokke og Uhre defineret Koordinatsystem."

conceptual structure upon which our customary ordering of our sense-impressions depends and our customary use of language is based. The basis of this ordering is, certainly, the possibility for recognition and comparison and accordingly the usual description of nature is characterized by the attempt to express all experience by stating the locations of material bodies and changes of location with time relative to a coordinate system defined in the traditional manner by means of measuring rods and clocks."

The whole idea is simply that identification of macroscopic objects in time and space is a necessary condition for unambiguity. We cannot speak of macroscopic objects without assuming that they are located somewhere at a certain time and that it is possible to identify an object and thereby state that it is the same object with which we had to do at another point in time and space. Further reflection upon our conditions for obtaining knowledge shows

Bohr to Dirac,
24 March 28
English
Full text on p. [495]

"... that the permanency of results of measurements is inherent in the very idea of observation; whether we have to do with marks on a photographic plate or with direct sensations the possibility of some kind of remembrance is of course the necessary condition for making any use of observational results. It appears to me that the permanency of such results is the very essence of the ordinary causal space–time description."

It would appear that Bohr thought that any interpretation of quantum mechanics must be given in ordinary language supplemented with the terminology of classical physics. Within quantum mechanics, objectivity is again linked to the distinction between subject and object. The measuring apparatus will always interact with the atomic system in a way which is in principle uncontrollable, but a sharp line of separation must be drawn not only between the measuring apparatus and the atomic object but also between the experimental results and the subject, i.e. the physicist performing the experiment. The observer is, as is the case in classical physics, a "detached observer", a point which Wolfgang Pauli questioned in his correspondence with Bohr[38]. From the very outset, quantum physics is based on macroscopic, closed observations, i.e. observations that are brought to an end and are in principle irreversible. In the description of these observations as experiments, all concepts applied are to be understood in

---

[38] Letters from Pauli to Bohr, 15 February 1955, from Bohr to Pauli, 2 March 1955, from Pauli to Bohr, 11 March 1955, and from Bohr to Pauli, 25 March 1955. Reproduced on pp. [563], [567], [569] and [572].

their classical physical sense. Concepts such as frequency, wavelength, momentum and amplitude are defined in terms of classical physics, and all so-called quantum-physical experiments are described within the framework of space, time, causality and divalent logic. As a consequence of this, Bohr advocated that in quantum mechanics the word "phenomenon" should refer exclusively to the observations obtained under specified circumstances, including an account of the whole experimental arrangement.

Bohr's views of language contain yet another, very important, point. In speaking about an unambiguous description of our surroundings, we have only one language at our disposal, namely what Bohr calls "ordinary language". For descriptive purposes, there is no alternative to ordinary language[39]. Objective description belongs to language as such in whatever native tongue[40]:

"By objectivity we understand a description by means of a language common to all (quite apart from the differences in languages [i.e. tongues – DF] between nations) in which people may communicate with each other in the relevant field."

As already stated, it is a common feature of the descriptive use of ordinary language and the observational situation in classical physics that they presuppose a sharp, immovable line of separation between subject and object. But, according to Bohr, there is a still deeper connection between the two. Thus, he often reminds us of the fact that classical physics is a refinement of the descriptive use of ordinary language, i.e. that the fundamental concepts of classical physics are developed from the concepts we use in our everyday description of our surroundings. Already in the descriptive use of ordinary language we have concepts such as "velocity", "distance", "time-interval" and "acceleration". But not until Galileo and Newton did we learn exactly how these concepts relate to one another. The same is true for concepts such as "force" and "mass", the latter having its origin in the everyday concept of "weight". Another example could be "temperature" which serves as a refinement of our daily concepts of "hot" and "cold". Concepts such as "mass point" and "electromagnetical field" are, of course, not clarifications of concepts couched in ordinary language,

---

[39] See, for instance, passages such as those referred to in ref. 32.
[40] *Unity of Knowledge*, 24.9.1953, Bohr MSS, microfilm no. 21: "Ved objektivitet vil vi forstå en beskrivelse ved hjælp af et sprog, der er fælles for alle (ganske bortset fra sprogforskellighederne mellem nationer) og på hvilket mennesker kan meddele sig til hverandre på det område, hvorom talen er."

but they were introduced by means of experiments which ultimately can be accounted for in ordinary language.

It was essential for Bohr to emphasize that both relativity theory and quantum mechanics should be viewed as generalizations of classical physics – although he sometimes spoke of relativity theory as belonging to classical physics. As Bohr says in his article "Unity of Knowledge" from 1954, the main point is that all knowledge presents itself within a conceptual framework adapted to account for previous experience and that any such frame may prove too narrow to comprehend new experience. The widening of the conceptual framework opens up for "the possibility of an ever more embracing objective description"[41].

Bohr's view is that whenever we describe something unambiguously, we are operating within a conceptual framework, i.e. a set of concepts which are mutually interdependent. In the explanation of any one of these concepts, the others must necessarily be presupposed; they cannot be understood independently of each other. Physics has taught us, however, that such a conceptual framework may be too narrow to describe and explain new, unexpected experience. It always seems possible to establish unambiguity in these new fields by widening the conceptual framework through a generalization of the interplay between its fundamental concepts.

"WE ARE BOTH SPECTATORS AND ACTORS"

Bohr emphasized time and time again that as human beings and knowing subjects we are part of the world we explore. We are both spectators and actors in the great drama of existence, he often said, and much of our scientific work consists in trying to harmonize these two positions[42]. We are "in" the world and therefore we cannot see it from "without", nor even ascribe any sense to this word. Hence, we are subject to the conditions for description laid out above.

---

[41] N. Bohr, *Unity of Knowledge* in *The Unity of Knowledge* (ed. L. Leary), Doubleday & Co., New York 1955, pp. 47–62, quotation on p. 48. Reproduced on pp. [83]–[98], quotation on p. [84]. Also printed in *Atomic Physics and Human Knowledge*, ref. 15, pp. 67–82, quotation on p. 68.

[42] This "picture" appears numerous times in Bohr's articles. See, for example, N. Bohr, *Biology and Atomic Physics* in *Celebrazione del secondo centenario della nascita di Luigi Galvani, Bologna – 18–21 ottobre 1937-XV: I. Rendiconto generale* Tipografia Luigi Parma 1938, pp. 68–78. Reproduced on pp. [52]–[62]. Also published in Bohr, *Atomic Physics and Human Knowledge*, ref. 15, pp. 13–22. See also Bohr, *Unity of Knowledge*, ref. 41.

We cannot transcend them. Nor can we frame any idea of alternative conditions. We are, so to speak, suspended in language, as Bohr loved to say[43].

We may get a still deeper understanding of Bohr's view by contrasting it to the common philosophical assumption that a transcendental point of view is possible. In the discussions about the epistemological status of quantum mechanics, it has been suggested that although the indeterminacy relations prevent us from ascribing simultaneous position and momentum to, say, an electron, it may very well be that the electron *in itself* has both a definite position and momentum. Or to put it differently: If an omniscient God exists, he may know the exact position and momentum of the electron at any moment. Yet we are precluded from obtaining this knowledge.

The question is, however, whether we can make any sense at all of such a "God's Eye View". Bohr himself made a comment on this matter in the oral history interview conducted the day before he died. He here relates that in a discussion, Max Planck advanced the view that God was able to state the exact position and momentum of an electron from his divine point of observation. Planck was religious and had a firm belief in God; Bohr was not, but his objection to Planck's view had no anti-religious motive. Bohr, in the interview, says as follows[44]:

> "Planck really was religious ... he said that a God-like eye could certainly know what was the energy and the momentum [the position of the electron being known – DF]. And that was very difficult you see. ... I said to him: You have spoken about such an eye; but it is not a question of what an eye can see; it is a question of what you mean by knowing."

The idea is that we always observe our surroundings under conditions determined by the fact that we are part of the world. This means that we must apply our concepts in a definite way in order to think and speak unambiguously. Even if we tried to imagine a descriptive language different from ours in which all concepts were applied in quite a new manner, we would not be able to understand this "language". It would not be translatable into our language and therefore we would not be able to characterize it as a language.

In Bohr's terminology, the idea of a God's Eye View is sometimes called the idea of an "ultimate subject"[45]. No matter what form we give to this idea, it

---

[43] A. Petersen, *The Philosophy of Niels Bohr*, Bulletin of the Atomic Scientists **14** (1963) 8–14, on p. 10.

[44] Interview with Niels Bohr, 17 November 1962, AHQP. Transcript, p. 7.

[45] See Bohr, *Unity of Knowledge*, ref. 41, quotation on p. [95].

is, according to Bohr, unthinkable. This is the reason why Bohr time and time again emphasized that "we are both onlookers and actors in the great drama of existence"[46].

Bohr's views on language and description are in full accordance with his views concerning physics, biology, psychology, religion and cultural problems. Even though he touches in nearly every article on the use of concepts, unambiguous communication and the conditions for description, he published nothing substantial on the subject. Probably, he felt that his point of view was obvious. Aage Petersen's characterization of Bohr's attitude in this respect is as follows[47]:

"As far as I can see, the doctrine that we are, philosophically speaking, suspended in language, that we depend on our conceptual framework for unambigous communication, and that the scope of the frame may be extended by generalization in the way illustrated in mathematics, forms the general basis of Bohr's philosophy. In his writings he never gave a detailed exposition of this view. Nor did he discuss its relation to other conceptions of the philosophical status of language. He considered it completely obvious and was surprised that others felt it so difficult to understand."

To sum up, it may be said that Bohr held on to the view that the observational conditions in quantum mechanics were different *in principle* from those in classical physics – in the last resort because of Planck's constant. However, because the ultimate requirement of unambiguity must be maintained, and because unambiguous communication can only be made with ordinary language supplemented with classical physical concepts – and because quantum-mechanical effects can only be registered via amplifying mechanisms and apparatus, which must be described in terms of classical physics – we have to accept experimental arrangements that mutually exclude each other, yet are complementary. There is no possibility of looking "behind" the quantum phenomena, and no possibility of visualizing an unobserved quantum-mechanical reality. This would presume a God's Eye View to which we cannot attach any meaning, because being part of the world – as Bohr stressed with the actor–spectator metaphor – we must always be aware of our conditions for description.

---

[46] See, for instance, Bohr, *Atomic Theory and the Fundamental Principles*, ref. 5. Quotation in Vol. 6, p. [253].

[47] Petersen, *Philosophy of Niels Bohr*, ref. 43, pp. 10–11.

SUBJECT AND CONSCIOUSNESS

One of the great movements within psychology in the first half of the 20th century was behaviourism. The pioneer of this movement, John B. Watson, claimed that all mental phenomena could be described and explained solely by observing the behaviour of individuals. Before he entered the stage in 1913, psychology had to a great extent been based on introspection. According to Watson introspection was unscientific because it referred to so-called "inner" mental states which were beyond intersubjective control. Furthermore, it was redundant because descriptions of human emotions, imagination, thinking etc. actually referred to nothing but behaviour and physiological states. Watson boldly maintained that thinking was nothing but suppressed speech-movements in the larynx.

With some modifications, behaviourism became an integral part of logical empiricism and the Unity of Science movement[48]. Philosophers such as Otto Neurath, Carl Hempel and Rudolf Carnap aimed at a description of mental states in terms of a "physicalistic" language referring solely to "observables" located in space and time. As we shall see below, Bohr was in close contact with these philosophers for a brief period. Later in his life, he became acquainted with Gilbert Ryle who in his book, "The Concept of Mind"[49], maintained that there was no reason to speak of "inner" mental states and that consequently nobody had a privileged access to his own thoughts and emotions, a viewpoint that Bohr found entirely untenable.

Bohr thought behaviourism was a delusion. He would joke about it, characterizing it as an

"... ideology which by its very name indicates the limited vision of its supporters, ...".

Bohr to Pauli,
31 Dec 53
Danish text on p. [543]
Translation on p. [547]

He considered it an undeniable fact that we may observe our own thoughts, emotions and moods by means of introspection. Also, he thought it obvious that we must speak of every person as a subject capable of observing, acquiring

---

[48] The term "Unity of Science" was introduced by O. Neurath (see e.g. his *Empirische Soziologie. Der wissenschaftliche Gehalt der Geschichte der Nationalökonomie*, Vienna 1931) and became part of the logical empiricists' programme. They rejected the view that there exist different *kinds* of science corresponding to different kinds of reality or existence (such as matter, life and consciousness).

[49] G. Ryle, *The Concept of Mind*, Hutchinson's University Library, London 1949.

knowledge, thinking, performing acts of will and so forth. As he wrote late in life[50]:

> "As regards our knowledge of fellow-beings, we witness, of course, only their behaviour, but we must realize that the word consciousness is unavoidable when such behaviour is so complex that its account in common language entails reference to self-awareness."

As explained above, unambiguous description presupposes a line of separation between the subject and the object. When we describe our own mental activity via introspection, the situation is a little different because the line of separation cannot be drawn in the usual manner[51].

> "For describing our mental activity, we require, on one hand, an objectively given content to be placed in opposition to a perceiving subject, while, on the other hand, as is already implied in such an assertion, no sharp separation between object and subject can be maintained, since the perceiving subject also belongs to our mental content."

Another important insight of Bohr's is that consciousness, and consequently the perceiving subject, are inseparably connected with life. This seems to have been obvious to him, since he mentions it only *en passant*[52]:

> "Besides, the fact that consciousness, as we know it, is inseparably connected with life ought to prepare us for finding that the very problem of the distinction between the living and the dead escapes comprehension in the ordinary sense of the word."

As we shall see in the Introduction to Part I, this was of paramount importance for his conception of biology. However, he never said anything substantial about the connection between consciousness and life, although in Bohr's correspond-

---

[50] N. Bohr, *The Unity of Human Knowledge,* Revue de la Fondation Européenne de la Culture, July 1961, pp. 63–66, quotation on p. 66. The article is reproduced on pp. [157]–[160], quotation on p. [160]. Also published in *Essays 1958–1962*, ref. 32, pp. 8–16.

[51] Bohr, *Quantum of Action*, ref. 3. Quotation in Vol. 6, p. [212].

[52] Bohr, *Atomic Theory and the Fundamental Principles*, ref. 5. Quotation in Vol. 6, p. [253].

ence with Delbrück there is a clear rejection of Delbrück's idea that the subject lives on after life is extinguished[53].

PSYCHOPHYSICAL PARALLELISM

During the compulsory course in philosophy – the so-called "Filosofikum" – which Bohr attended during his first year as a student at the University of Copenhagen, he learnt about the different views concerning the mind–body problem. The "Filosofikum" was in the hands of three professors, and Bohr chose to follow Harald Høffding's lectures which were attended by approximately 150 students.

Høffding advocated a view which he named the hypothesis of identity (in Danish: "identitetshypotesen"), according to which mind and matter are but two attributes or aspects of one substance. If, for instance, I think of the number 27, a physiological process in the brain must correspond to this thought and this brain process must take place whenever I think of the number 27. Conversely, whenever the brain process in question takes place, the thought of 27 must occur. Both the thought and the brain process are aspects of the same substance.

Høffding mistakenly believed that Benedict Spinoza (1632–1677) had advocated this view, and he always spoke of Spinoza as his great precursor. However, although Spinoza certainly maintained that mind and matter were two attributes of the same substance, his conception was not at all like Høffding's. Spinoza only insisted that there is a relation between the structure of the material attribute on the one hand and the conceptual structure of human knowledge on the other.

Undoubtedly, Høffding's misinterpretation of Spinoza stems from the German philosopher and psychologist G.Th. Fechner (1801–1887). Fechner compared the relation between an idea in the mind and the corresponding brain process to the relation between the concave and the convex side of a circle. For instance, the thought about the number 27 and the corresponding brain process are simply identical. They are one and the same entity viewed from two different angles. Høffding accepted this view but read it into Spinoza's *Ethica*.

There are two other prominent viewpoints with regard to the mind–body relationship from the 17th century and onward. One is presented by René Descartes (1596–1650), who maintained that mind and matter are two fundamentally different forms of existence which interact in some way. The other is represented

---

[53] Letters from Delbrück to Bohr, 30 June 1959 (two letters), from Bohr to Delbrück, 25 July 1959, from Delbrück to Bohr, 3 August 1959, from Bohr to Delbrück, 19 November 1959. All reproduced on pp. [478], [479], [481], [482] and [484].

by G.W. Leibniz's (1656–1716) idea of a pre-established harmony: if we imagine that two clocks are set alike and started at the same time, they will strike the hours simultaneously without interacting. In the same manner mental events and brain processes are correlated without any interaction. Høffding rejected both Descartes's and Leibniz's views. The latter he named "psychophysical parallelism" in contradistinction to his own hypothesis of identity[54].

Bohr learnt about these views through the "Filosofikum". Considering Høffding's many admonitions on this subject, it is surprising that Bohr always used the label "psychophysical parallelism" about the relation between mental events and brain-processes and even spoke of Spinoza as the founder of this view. To Høffding, this must have been pure sacrilege. The deviation from Høffding's terminology stems from the fact that Bohr had other philosophical sources besides Høffding. One of them was the philosopher Anton Thomsen (1877–1915), who in the last year of his life succeeded Høffding in the chair of philosophy at the University of Copenhagen and disagreed with Høffding on the mind–body problem. Thomsen was married to Bohr's cousin Ada Adler and was a close friend of Bohr. Another source was Edgar Rubin who originally studied philosophy with psychology as his field of specialization. In his 1925 entry on "Psychology" in the major Danish encyclopedia, Rubin defined psychophysical parallelism in precisely the same way as would Bohr shortly after[55].

* * *

Bohr never spoke of parallelism as a relation between a physical and a mental "world" or "substance". His concern was the conditions for the description of mental events and for the description of the brain processes presumably corresponding to them. Already as a youngster he realized – most probably inspired by Poul Martin Møller's novel[56] – that mental events cannot be described as a series of distinct phenomena following upon each other cinematographically and observed by a passive subject. This was the view of the associationist psychologists at the end of the 19th century and the view Høffding presented in his "Outlines of Psychology"[57]. Bohr realized that thought processes are often

---

[54] H. Høffding, *Outlines of Psychology* (translated by Mary E. Lowndes), Macmillan, London 1891, pp. 66–70.

[55] *Salmonsens Konversations Leksikon, Bind XIX*, J.H. Schultz Forlagsboghandel A/S, Copenhagen 1925, p. 683.

[56] Møller, *En dansk Students Eventyr*, ref. 26.

[57] H. Høffding, *Psykologi i Omrids paa Grundlag af Erfaring*, Copenhagen 1882 (first English edition: *Outlines of Psychology*, London 1891).

discontinuous because the subject can choose between different points of view when considering a problem. When quantum mechanics had been established, Bohr realized that the brain processes allegedly corresponding to mental events most likely are not deterministic classical physical processes[58]. Hence it is Bohr's suggestion that the parallelism between the mental side and the physiological side is based on the fact that in both domains a detailed observation alters the phenomenon under investigation. Psychophysical parallelism must therefore not be understood as stringently as in traditional philosophy.

Bohr himself tells us about the insight he gained from Møller's novel[59]:

"In particular, the conditions of analysis and synthesis of so-called psychic experiences have always been an important problem in philosophy. It is evident that words like thoughts and sentiments, referring to mutually exclusive experiences, have been used in a typical complementary manner since the very origin of language. In this context, however, the subject–object separation demands special attention. Every unambiguous communication about the state and activity of our mind implies, of course, a separation between the content of our consciousness and the background loosely referred to as 'ourselves', but any attempt at exhaustive description of the richness of conscious life demands in various situations a different placing of the section between subject and object."

Bohr goes on to quote extensively from "The Adventures of a Danish Student"[60], through which he became aware of an observational situation alien to the one of classical physics. In observing our own conscious life, we draw a line of separation between the observing subject and the content of the consciousness observed. However, the line of separation is movable and shifts take place invariably.

Some mental states exclude others. As noted in the last quotation, thoughts seem to exclude feelings or emotions. Very deep concentration of thought excludes all emotional aspects, whereas great emotional excitement excludes calculated thought. As Bohr often pointed out, the use of words such as thought and feeling does not refer to a firmly interconnected causal chain, but to experiences which exclude each other because the conscious content and "the

---

[58] See Bohr, *Light and Life*, ref. 15, this volume on p. [35].
[59] Bohr, *Unity of Human Knowledge*, ref. 50. Quotation on p. [159].
[60] Møller, *En dansk Students Eventyr*, ref. 26.

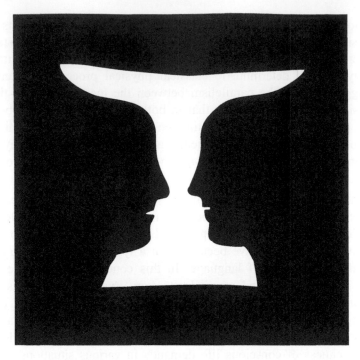

"Rubin's vase", which can also be seen as two heads in profile, was a picture Rubin constructed for his thesis "Visually Experienced Figures" (*Synsoplevede Figurer*, ref. 22, figure 3) in order to demonstrate what may be called a complementary aspect of our perception.

background loosely referred to as 'ourselves' " can be distinguished in different ways.

We may find a similar complementarity in perception. To take one of Bohr's examples, the emotional experience of a piece of music excludes conscious analysis of it, whereas analysis of the music excludes emotional experience. Yet both approaches are necessary for our understanding of what music is.

In general, we may, within certain limits, choose between a variety of approaches, thereby deliberately causing shifts in the line of separation between subject and object. This is often the case when we work with a theoretical problem. A striking example is the perception of ambiguous pictures. In Rubin's doctoral thesis, mentioned above, we are presented with the famous figure "Rubin's vase", which may be perceived as a vase or as two profiles, the one perception excluding the other.

In his popular lectures, Bohr returned throughout his life to the problem of psychophysical parallelism, primarily, one would think, in order to emphasize that in introspection the observational situation is quite different from that in classical physics and resembles more closely the observational conditions in quantum mechanics. As we shall see below, he sought a solution to the problem in our ability to choose between an object-oriented and a subject-oriented use of language.

The essential features of a quantum phenomenon are that it cannot be sub-divided without changing the entire course of events, i.e. that it is in principle impossible to analyze parts of it, and that it has meaning only as a completed, irreversible process. I believe that Bohr found an analogy to this in William James's remarks on the stream of thought[61]. A thought cannot be subdivided, and it is not recognized until it has come to an end. We cannot speak of half a thought. Moreover, it makes no sense to speak of reversing a thought. In his last articles, Bohr emphasized the irreversible character of our thoughts and knowledge. Any perception or observation must so to speak be a closed and irreversible phenomenon – corresponding to irreversible changes in the living organism at the biological level – thus presupposing an "arrow of time".

In some of Bohr's manuscripts, this consideration is linked with his own very special version of psychophysical parallelism in a wide sense. In a way, all our knowledge, also of biology, may be considered as part of our consciousness. In terms of traditional psychophysical parallelism, this would lead to a paradox: we would be forced to maintain that corresponding to our observation of the relationship between mind and body there is a biological counterpart and the experience of this must imply yet another biological counterpart, thus involving us in an infinite regress. But according to Bohr, this form of parallelism is not possible, because[62]:

"We have no possibility through physical observation of finding out what in brain processes corresponds to conscious experience. An analogy to this is

---

[61] W. James, *The Principles of Psychology*, H. Holt, New York 1980 (reprinted by Dover, New York 1950). Cf., in particular, Chapter IX, *The Stream of Thought*, Vol. 1, pp. 224–290. We know that Bohr read parts of James's book. See interview with Bohr, 17 November 1962, AHQP. Transcript, pp. 6–7.

[62] *Unity of Knowledge*, 30.8.1954. Bohr MSS, microfilm no. 21: "Vi har ingen mulighed for ad den fysiske iagttagelses vej at få oplysning om, hvad der i hjerneprocesserne svarer til de bevidste oplevelser. En analogi er her forholdet imellem de oplysninger vi kan få om cellernes struktur og de udslag denne struktur giver sig i det organiske livs udfoldelse". *Unity of Knowledge*, 22.9.1954, *ibid.*: "Det der er komplementært er ikke forestillingen om sjæl og legeme, men den del af indholdet af the mind, som drejer sig om forestillingerne om fysikken og organismerne og den situation, hvor vi bringer tanken om det observerende subjekt ind."

the relation between the information we can obtain concerning the structure of cells and the effects this structure has on the way organic life displays itself."

"What is complementary is not the idea of a mind and a body but *that* part of the contents of the mind which deals with the ideas of physics and the organisms and *that* situation where we bring in the thought about the observing subject."

In the same manuscript, Bohr underlines that[63]:

"In one way of speaking it is obvious that we are only concerned with what is in our mind (and even in our personal mind in the sense of Berkeleyian philosophy) but it is on the other hand clear that if we take such views to their extremes it is obvious we are not able to define the very word 'mind' and strictly speaking any other word either."

We may, in a certain sense, speak of a complementary relationship between situations that differ in respect of the placing of the line of separation between subject and object. We may deliberately shift between two attitudes. There are, he says[64],

"... (a) those situations where we do not 'think' of ourselves, where we in a way speak about the external world ... (b) situations in which we deal with or at least refer to ourselves. Whatever can be said about these situations presupposes the features of the wholeness of the organisms. Here we have the terminological or perhaps more correctly the dialectical clarification that we are aware of the fact that we displace the line of separation between subject and object."

According to Bohr's wider conception of parallelism, we are able to speak of our surroundings, including physical and biological facts, using language in

---

[63] *Unity of Knowledge*, 30.8.1954, *ibid.* (original quotation in English).

[64] *Steno-Forelæsning i Medicinsk Selskab*, 2.8.1957. Bohr MSS, microfilm no. 22: "Om bevidst-hedsproblemet. (1) Alt foregår i den såkaldte bevidsthed (selvfølge), men der er to situationer som vi i sproget skelner mellem (a) de situationer, hvor vi ikke 'tænker' på os selv, eller hvor vi så at sige taler om omverdenen ... (b) situationer, i hvilke der er tale om, eller i det mindste stadig henvises til, os selv. Enhver tale om sådanne situationer forudsætter organismernes helhedstræk. Her er den terminologiske eller måske rettere dialektiske afklaring, at vi er opmærksom på at vi forskyder objekt–subjekt skillelinien."

a way that does not refer to the subject. Both in physics and in biology we can speak of a "detached observer". But each of us may also "bring in the thought about the observing subject". In this case, anything we say turns into a description of the contents of our own consciousness.

* * *

Bohr's many reflections about "the epistemological lesson of quantum physics" and about philosophical topics which may seem far removed from physics, can perhaps be summarized as follows.

What do physicists do? They observe. They perform experiments in order to obtain all relevant data. They describe. They attempt to provide explanations by means of theoretical considerations. But in order to be understandable to others, all description and explanation must be unambiguous. This means that our conceptual framework depends upon the conditions for observation and description within the field of experience under investigation.

These conditions vary from one field to another. The conditions for observation and description within quantum mechanics differ from those for classical physics, and the biological, psychological and anthropological sciences each introduce quite different conditions for observation and description. Thus, there seems to be no basis for attempting to reduce psychology to biology or biology to physics – just to mention two attempts at reduction made in the 20th century.

It is characteristic for Bohr that he did not speak of unity of science in quite as strong terms as the logical empiricists, but preferred to speak of unity of knowledge. In so doing he wished to emphasize that all human knowledge originates in our everyday experience and that all use of language in science must be introduced with the help of the unambiguous, descriptive use of language already inherent in our everyday language. All scientific results must in the final analysis be communicated in unambiguous everyday language supplemented with technical terms built upon such language. Thus understood, unity of knowledge can be expected to persist in the future development of science.

PART I

COMPLEMENTARITY IN BIOLOGY AND RELATED FIELDS

INTRODUCTION

by

DAVID FAVRHOLDT

The articles in Part I concentrate for the most part on the conditions of observation and description in biology. In this introduction it hence seems appropriate, first, to discuss Bohr's efforts to shape a new viewpoint within the framework in which this discussion had traditionally taken place. Second, I will sketch the contemporaneous discussion of Bohr's views. In this way I hope to contribute to a deeper understanding of Bohr's considerations regarding the status of biology.

1. BOHR'S VIEW ON BIOLOGY

Questions about the characteristics and explanation of life were discussed already by Aristotle. The development of mechanics in the 17th century reduced these problems to the question of whether or not life could be explained in terms of mechanics. The view that living organisms were nothing but complicated mechanical machines was named "mechanism" (or "mechanicism"). In the 18th and 19th centuries, this view was refined in accordance with the development of physics and chemistry, but the name remained unaltered[1]. Mechanism of today is the view that life can be explained exhaustively in terms of physics and chemistry. In the course of the 19th and 20th centuries mechanism has been accepted by most biologists as a natural and sound basis for research.

---

[1] Many philosophers, for example René Descartes (1596–1650), Thomas Hobbes (1588–1679), Julien Lamettrie (1709–1751) and Paul Holbach (1723–1789), have defended mechanism.

The quite different viewpoint of "vitalism" has also had its followers[2]. Vitalists deny that life can be reduced to physics and chemistry, but their arguments cover a wide range of views. The most well-known view throughout the ages has been that life is due to a special vital force alien to inanimate matter. Another popular idea declares that nature as a whole is alive and that living organisms actualize latent possibilities already present in so-called inanimate matter.

Part of the controversy between mechanism and vitalism concerns the relationship between causal and teleological description. This problem was discussed by Immanuel Kant (1724–1804) in his "Kritik der Urteilskraft" (Critique of Judgment) from 1790[3]. Kant was not a vitalist in the sense just mentioned, yet he emphasized that a teleological approach is indispensable in the description of the origin, development and function of the living organism. According to Kant, the very recognition of something as a living organism entails acknowledging the purpose as a cause itself. If we describe a tree, for example, the branches and leaves require a trunk just as the trunk requires foliation. Kant stresses the remarkable capacity of an organism to preserve its wholeness, even under duress, to heal itself, and even to adjust to some new or abnormal formation when this is needed in the interest of its continued growth. The principles of mechanism and teleology, he maintains, express only two modes of thought which are incommensurable yet are both necessary if we are to grasp and investigate living nature. They are complementary in Bohr's sense of the word.

At the beginning of the 20th century the controversy was intensified, especially by the biologist Hans Driesch, who in various books[4] presented his own investigations into growth, regeneration and reproduction, which gave results that at the time were beyond any physico–chemical explanation.

Niels Bohr's father, Christian Bohr, had been taught Kant's views on teleology and causality at the "Filosofikum", the compulsory course at the University of Copenhagen introducing philosophical concepts. His teacher, the philosophy professor Rasmus Nielsen (1809–1884), was at the time very interested in precisely these ideas. Christian Bohr passed his examination as a medical

---

[2] For instance, G.E. Stahl (1660–1734), better known for his phlogiston theory, Louis Dumas (1765–1813), Lorenz Oken (1779–1851), and, in more recent times, Henri Bergson (1859–1941), J.J. von Uexküll (1864–1944) and Hans Driesch (1867–1941).

[3] I. Kant, *Kritik der Urteilskraft*, Bey Lagrande and Friederich, Berlin 1790. A useful English version is I. Kant, *Critique of Judgment* (transl. J.C. Meredith), Clarendon Press, Oxford 1952.

[4] See, for instance, H. Driesch, *Der Vitalismus als Geschichte und als Lehre,* J.A. Barth, Leipzig 1905, and *Philosophie des Organischen,* Engelmann, Leipzig 1909.

doctor in 1878 but had specialized in physiology already from 1874 under the guidance of the Danish physiologist Peter Ludvig Panum. Prior to his nomination in 1890 as professor of physiology, he had been a student of the renowned physiologist Carl Ludwig in Leipzig from 1880 to 1882, and again in 1883. Here he worked mostly on muscle physiology, but very soon he became interested in the physiology of respiration which was to become his main field of research. At the time, Carl Ludwig debated with Eduard Pflüger the problem of whether the exchange of oxygen and carbon dioxide in the lungs could be explained as a diffusion process. This inspired Christian Bohr to study gas exchange in the lungs, and on the basis of his work from 1887 and onwards, he concluded that this exchange could not be explained completely on the basis of physico–chemical laws. Christian Bohr posited a regulation process in the cells of the lungs that cannot be explained by the diffusion laws but is governed by the needs of the organism as a whole. He concluded that the sustaining of life must be taken into account in describing the functions of the lungs[5].

Christian Bohr never wrote anything substantial about the discussion between vitalists and mechanists. His only remark about the problem may be found in a short passage in a paper on the pathological dilation of lungs[6] quoted by Niels Bohr many years later[7]. Yet these questions were regularly discussed in Christian Bohr's laboratory.

Hans Driesch was the last great advocate of the doctrine of a "vital force". Other biologists rejecting mechanism tried to find an alternative. One of them was the British physiologist John Scott Haldane, who knew Christian Bohr from a visit to the latter's Copenhagen institute around 1890.

In a number of books and articles, Haldane argued that living organisms cannot be explained in physico–chemical terms. Life is an irreducible fact. A living organism, he said, is a totality and an individuality, and its continuous striving for optimal conditions regulates the life processes in a way that cannot

---

[5] C. Bohr has given a full account of his results in *Handbuch der Physiologie des Menschen, Vol. I* (ed. W. Nagel), 1909. A list of C. Bohr's publications in physiology and physics is given in R. Tigerstedt, *Christian Bohr. Ein Nachruf*, Skandinavisches Archiv für Physiologie **25** (1911) ix–xviii.

[6] C. Bohr, *Om den pathologiske Lungeudvidning (Lungeemphysem)* (On the pathological expansion of the lungs (lung emphysema)) in *Festskrift udgivet af Københavns Universitet i Anledning af Universitetets Aarsfest*, November 1910, Copenhagen 1910, pp. 5–47.

[7] N. Bohr, *Physical Science and the Problem of Life* in *Atomic Physics and Human Knowledge*, John Wiley & Sons, New York 1958, pp. 94–101, quotation on p. 96. The latter volume is photographically reproduced as *Essays 1933–1957 on Atomic Physics and Human Knowledge, The Philosophical Writings of Niels Bohr, Vol. II*, Ox Bow Press, Woodbridge, Connecticut 1987. The article is reproduced in this volume on pp. [116]–[123], quotation on p. [118].

be reduced to physics and chemistry. As regards metabolism, he maintained that "we cannot tell what exactly becomes of the atoms and molecules which pass into the body – how far, or in what sense, they are built up into the living tissue, or in what way their potential energy is immediately utilised"[8].

Haldane named his view "organicism" in order to dissociate himself from vitalism[9]. There were, however, diverging opinions with regard to the appropriate definition of "organicism". According to a book by C. Lloyd Morgan[10] published in 1926, the fundamental idea of organicism is that the parts of an organism do not have a complete existence on their own but owe their existence or being to the whole of which they form a part: if the whole is taken apart, the parts cease to be what they were.

The above remarks may serve to illustrate some tendencies of the time similar to Niels Bohr's views. We do not know whether Bohr was acquainted with the above-mentioned authors or the contemporaneous debate more generally, but there is reason to believe that he was familiar with Haldane's views[11].

From the time he introduced complementarity in the Como Lecture in 1927[12], Bohr was confident that it had application beyond physics. The most obvious analogy is found in the subject's analysis of its own states of consciousness with which Bohr was so well acquainted through reflections in his youth on the problem of the freedom of the will. Now he turned to biology, hoping to throw new light upon the old problems concerning the concept of life. He thought that "the epistemological lesson" of quantum mechanics could help us overcome the schism between vitalism and mechanism by drawing our attention to the conditions of observation in biology. Bohr rejected vital force or "entelechy" as being unscientific and formulated a number of arguments against mechanism. Bohr presented his views in an obscure manner, and it is understandable that many biologists called him a vitalist. Hence a more detailed investigation is called for.

It is reasonable to start off with his view of the "I" or the subject. Bohr considered it urgent to emphasize that all knowledge presupposes a subject that cannot be characterized exhaustively because it is itself the ultimate pre-

[8] J.S. Haldane, *Mechanism, Life and Personality*, John Murray, Albemarle Street, London 1913, quotation on p. 36.

[9] See J.S. Haldane, *Organicism and Environment as Illustrated by the Physiology of Breathing*, Oxford & New Haven 1917.

[10] C.L. Morgan, *Life, Mind and Spirit*, Williams & Norgate, London 1926.

[11] F. Aaserud, *Redirecting Science. Niels Bohr, Philanthropy, and the Rise of Nuclear Physics*, Cambridge University Press, Cambridge 1990, p. 183.

[12] See Vol. 6, pp. [118] and [129].

supposition for any analysis. Not even through introspection can the subject be "caught". Any attempt to do so demonstrates the elusive character of the subject. In his "Light and Life" lecture given in 1932, Bohr explicitly states "that any analysis of the very concept of explanation would, naturally, begin and end with a renunciation as to explaining our own conscious activity"[13].

A main trend in Western philosophy since the days of Descartes has been based on the recognition of the immaterial nature of the subject; both subject and consciousness have been spoken of as forms of existence totally different from that of matter. Descartes maintained that spatial extension is the defining property of matter and that consciousness is unextended, a mere "res cogitans". Living organisms belong to the material world. Not only plants and animals but also humans are simply machines, "inanimate" in the true sense of the word. However, human beings have an immaterial soul or spirit which is somehow connected with the body.

Even the opponents of Descartes's philosophy generally agreed that there must be two fundamentally different forms of existence, the mental and the corporeal. Leibniz and Fechner as well as Høffding assumed that the two forms of existence were attributes of one and the same substance in the sense of Spinoza.

The prevailing idea was that when we speak of inanimate matter, a, living organisms, b, and mental "stuff" such as subjects or consciousness, c, the dividing line is between $a + b$ on the one side and c on the other:

$$a + b \mid c.$$

Whether b is reducible to a is a matter of science and empirical investigation, whereas a reduction of c to b (and in turn to a) is in principle impossible. This is the view put forward in Høffding's textbook on psychology used in the elementary philosophy course that Bohr took in his first year of study. To Høffding, a and b are of the same provenance, both of them being subject to the law of causality and the law of conservation of energy.

Bohr's view was quite different. He was convinced that the subject and all mental phenomena are inseparable from the living organism and form an integral part of it. This means that he placed the dividing line between a and $b + c$:

$$a \mid b + c.$$

---

[13] N. Bohr, *Light and Life*, Nature **131** (1933) 421–23, 457–59, quotation on p. 459. The article is also published in *Atomic Physics and Human Knowledge*, ref. 7, pp. 3–12, quotation on p. 11 (which has "an analysis" instead of "any analysis"). The former version of the article is reproduced in this volume on pp. [29]–[35].

Living organisms comprise, among other things, bacteria, amoeba, plants, fish, reptiles, mammals and human beings. However, in his discussions of the riddles of life Bohr usually contrasted the inanimate and the human being, the individual. Paradoxes are sharpened when we resort to extremes. If mechanism is valid, its supporters must not only explain metabolism, adaptation and replication of micro-organisms in physico–chemical terms; they must, indeed, also be able to give a physico–chemical account of the human being, this very complex organism characterized not only by its metabolism, its growth and reproduction, but also by its ability consciously to control its own behaviour, to remember, to perceive, to think and to make decisions. Mechanism must not only explain how the different parts of the human body function in order to maintain its existence through the best possible adaptation to constantly changing conditions. It must also be able to explain the wholeness, the unity and the individuality of a person. This is the problem ultimately facing mechanism. It must be able to give a physico–chemical account even of a conscious living organism, because whatever consciousness may be, it is an inseparable feature of such an organism.

What prevents the supporters of mechanism from solving these problems? The main difficulty is that the concept of having knowledge presupposes the existence of the subject, which therefore cannot itself form part of our knowledge[14]. Whatever I perceive, I shall never be able to perceive or analyze the subject – myself. A subject perceiving itself is just as impossible to imagine as is a drawing containing itself as an element. Likewise, I am, in principle, debarred from perceiving other minds or subjects. No matter how far brain physiology may advance, it will never be possible to gain direct knowledge of another person's mind, thoughts and moods. To put it differently, words describing mental states can never be part of a physical description. And any physiological investigation of functions relating to conscious activity will interfere with the integrity of the person. For example, an investigation of a person's experience of the freedom to choose in a given situation cannot be carried out[15]:

"Indeed, from our point of view, the feeling of the freedom of the will must be considered as a trait peculiar to conscious life, the material parallel of which must be sought in organic functions, which permit neither a causal mechanical description nor a physical investigation sufficiently thorough-

---

[14] See quotation at ref. 13, above.
[15] Bohr, *Light and Life*, ref. 13. Quotation on p. [35].

going for a well-defined application of the statistical laws of atomic mechanics."

Bohr's rejection of mechanism was based on philosophical considerations and not on biological knowledge. Actually, his main argument is[16]:

"... the fact that consciousness, as we know it, is inseparably connected with life ought to prepare us for finding that the very problem of the distinction between the living and the dead escapes comprehension in the ordinary sense of the word."

Another fundamental theme in Bohr's views about biology is his ideas about the nature of description. We are able to speak unambiguously about what we intend with our actions; we are conscious of our striving for certain goals and we naturally operate with a teleological frame of description in our daily life. The concept of "consciousness" is linked with the concept of "life" and hence teleological concepts are indispensable for the description of human behaviour. Also the description of the function of organs such as lungs, kidneys and stomach necessitates teleological concepts and a teleological use of language. This language functions, so to speak, in its own right and is used independently of the exact knowledge of what goes on in the organism at the physical or chemical level. It consequently must be based on special conditions of observation[17]:

"... the general lesson of atomic theory suggests that the only way to reconcile the laws of physics with the concepts suited for a description of the phenomena of life is to examine the essential difference in the conditions of the observation of physical and biological phenomena."

---

[16] N. Bohr, *The Atomic Theory and the Fundamental Principles Underlying the Description of Nature* in *Atomic Theory and the Description of Nature*, Cambridge University Press, 1934 (reprinted 1961), pp. 102–119, quotation on p. 119. The latter volume is photographically reproduced as *Atomic Theory and the Description of Nature, The Philosophical Writings of Niels Bohr, Vol. I*, Ox Bow Press, Woodbridge, Connecticut 1987. The article is reproduced in Vol. 6, pp. [236]–[253], quotation on p. [253].

[17] N. Bohr, *Biology and Atomic Physics* in *Celebrazione del secondo centenario della nascita di Luigi Galvani, Bologna – 18–21 ottobre 1937-XV: I. Rendiconto generale*, Tipografia Luigi Parma 1938, pp. 68–78, quotation on pp. 76–77. Reproduced on pp. [52]–[62], quotation on pp. [60]–[61]. Also published in *Atomic Physics and Human Knowledge*, ref. 7, pp. 13–22, quotation on p. 20.

"Life" and "consciousness" are concepts that may be used in a consistent manner although they have no place in physical description. From a physical point of view, life is, Bohr maintained, as irrational as the quantum of action is in relation to classical physics.

Bohr would often compare this non-reductionist view about life to the situation in thermodynamics, where the definition of temperature is incompatible with the mechanical account of the molecular motion, and where "temperature" is actually defined on the basis of our lack of knowledge about the location and momentum of the molecules[18]. At the same time, however, he would contest the view that life invalidates the law of entropy. If the free energy necessary to maintain and develop organic systems is continually supplied from their surroundings through nutrition and respiration, no physical laws are violated.

Starting from these general arguments Bohr surmised that he could say something more exact about the conditions for observation and description in biology. One of his basic ideas was that although most life functions can be described in terms of classical physics, they are often interwoven with functions at an atomic level. Already in his "Introductory Survey" to a collection of articles published in 1929[19], he pointed to the fact that only a few light quanta are needed to produce a visual sensation. At the same time, however, he declared that we cannot expect that life functions as a whole can be accounted for in terms of quantum physics. As we shall see below, he dissociated himself from the view that the freedom of the will could be explained in terms of quantum processes and emphasized furthermore that quantum mechanics gives no clue to understanding how an organization of atoms is able to adapt itself to the environment, as living organisms do.

Contrary to the vitalists, Bohr maintained that although life cannot be explained in physico–chemical terms, there is no limit to the physico–chemical exploration of living organisms. All *results* of biological investigations must be stated in terms of physics and chemistry in order that unambiguous descriptions be obtained. Here is a parallel to the demand that each quantum mechanical experiment must be accounted for in ordinary language supplemented with classical physical concepts. "Life" is not to be placed on a special ontological

[18] See N. Bohr, *Chemistry and the Quantum Theory of Atomic Constitution*, J. Chem. Soc., 1932, pp. 349–384. Reproduced in Vol. 6, pp. [373]–[408].

[19] N. Bohr, *Introductory Survey* in Bohr, *Atomic Theory and the Description of Nature*, ref. 16, pp. 1–24. Reproduced in Vol. 6, pp. [279]–[302]. The last four pages constitute an "Addendum (1931)" to the original Danish version.

level as a sort of substance different from physical matter. The borderline between physics and biology is epistemological in character.

By focusing on the conditions for observation and description, according to Bohr, we encounter a complementary relation between physical analysis and characteristically biological phenomena such as self-preservation and reproduction. Bohr assumed that from the physical point of view all processes of life concern atoms that are taken up into the organism and later secreted again. Originally he seems to have believed that a physico–chemical explanation of life must presuppose that we are able to follow the fate of each individual atom in the whole metabolic process. This rigorous demand, which he later modified, led him to the idea of complementary observational situations in biology. We may call this the "Metabolism Argument": a living organism must function in a manner that demands a continuous exchange of matter and energy with its surroundings. Hence, it is meaningless to speak about which atoms form part of the living organism and which do not. In speaking of an organism as a whole, including various holistic and finalistic features, we deal with observational situations where no sharp line of separation can be drawn between the organism and its surroundings. If, conversely, we wish to subject the organism to a physico–chemical investigation, we must isolate it from its environment. But in so doing we establish a new observational situation complementary to the first one. The isolation of the organism implies that the metabolism is intercepted.

A physical investigation of all states of an organism at the atomic level is incompatible with keeping it alive. Here again the investigation causes an interruption of the metabolism, thus leading to the death of the organism. Following Joseph Needham, we may call this argument the "Thanatological Principle"[20] (from the Greek word for death: Thanatos). In "Light and Life", Bohr boldly declares that[21]

> "...we should doubtless kill an animal if we tried to carry the investigation of its organs so far that we could describe the rôle played by single atoms in vital functions. In every experiment on living organisms, there must remain an uncertainty as regards the physical conditions to which they are subjected, and the idea suggests itself that the minimal freedom we must allow the organism in this respect is just large enough to permit it, so to say, to hide its ultimate secrets from us. On this view, the existence of life must be considered as an elementary fact that cannot be explained, but must

---

[20] J. Needham, *Order and Life*, M.I.T. Press, Cambridge 1936.
[21] Bohr, *Light and Life*, ref. 13. Quotation on p. [34].

be taken as a starting point in biology, in a similar way as the quantum of action, which appears as an irrational element from the point of view of classical mechanical physics, taken together with the existence of the elementary particles, forms the foundation of atomic physics."

The Thanatological Principle was strongly criticized, and Bohr was compelled to modify it. From his very last lecture, "Light and Life Revisited", we can see that in the course of time Bohr had realized that both the Metabolism Argument and the Thanatological Principle went too far[22]. However, he did not give up his general arguments.

Among the substantial number of articles and addresses that Bohr prepared between the publications of "Light and Life" in 1932 and "Light and Life Revisited" in 1962, some are so short that they appear programmatic[23], while others merely repeat the arguments from "Light and Life", for example by presenting the Thanatological Argument in an unmodified form[24]. Yet it is possible to see a certain line of development in other articles where Bohr expands his viewpoints. Thus, in "Physical Science and Man's Position"[25], the irreversibility involved in the description of organic functions is seen as the basis for our notion of time direction, and in "The Connection Between the Sciences"[26], Bohr comments on the similarity between living organisms and automatic machines while at the same time indicating that in the account of the functioning of devices for calculation and control we can disregard the

---

[22] N. Bohr, *Light and Life Revisited*, ICSU Review **5** (1963) 194–199. Reproduced on pp. [164]–[169]. Also published in *Essays 1958–1962 on Atomic Physics and Human Knowledge*, Interscience Publishers, New York 1963, pp. 23–29. The latter volume is photographically reproduced as *Essays 1958–1962 on Atomic Physics and Human Knowledge, The Philosophical Writings of Niels Bohr, Vol. III*, Ox Bow Press, Woodbridge, Connecticut 1987.

[23] N. Bohr, *Analysis and Synthesis in Science*, International Encyclopedia of Unified Science **1** (1938) 28; *Medical Research and Natural Philosophy*, Acta Medica Scandinavica (Suppl.) **142** (1952) 967–972; *Quantum Physics and Biology*, Symposia of the Society for Experimental Biology, Number XIV: *Models and Analogues in Biology*, Cambridge 1960, pp. 1–5; *Physical Models and Living Organisms* in *Light and Life* (eds. W.D. McElroy and B. Glass), The Johns Hopkins Press, Baltimore 1961, pp. 1–3; *Address at the Second International Germanist Congress* in *Spätzeiten und Spätzeitlichkeit*, Francke Verlag, Bern 1962, pp. 9–11. Reproduced on pp. [64], [67]–[72], [127]–[131], [134]–[137], [141]–[143].

[24] N. Bohr, *Address at the Opening Ceremony*, Acta Radiologica (Suppl.) **116** (1954) 15–18. Reproduced on pp. [75]–[78].

[25] Ingeniøren **64** (1955) 810–814. Reproduced on pp. [102]–[106].

[26] Journal Mondial de Pharmacie, No. 3, Juillet–Decembre 1960, pp. 262–267. Reproduced on pp. [148]–[153]. Also published in *Essays 1958–1962*, ref. 22, pp. 17–22.

atomic constitution of matter, whereas this is not possible in the account of living organisms.

Bohr, of course, never denied that biological laws may be established without disturbing the unfolding of life. From Harvey's discovery of the circulation of the blood and onwards, the history of biology and physiology presents us with many laws based on the results of dissection as well as of investigation of living organisms. What Bohr originally maintained was that an *exhaustive* account of the essential processes in a living organism cannot be established without destroying what one wishes to investigate. Here he had in mind the features of wholeness and individuality, as well as the self-regulating mechanisms met with in the metabolism, the growth of an organism, the adjustment of an organism to its environment and the genetic control of the functions of the cells in the organism.

Undoubtedly, it was the discovery of the helical DNA structure in 1953 and the subsequent understanding of the genetic code that made Bohr modify or withdraw some of his arguments. This discovery was based partly on the isolation of DNA as the essential component of the chromosomes, partly on chemical analysis of extracts, partly on X-ray microscopical pictures, and otherwise on extensive experimental and theoretical work. The mechanism of replication was in fact discovered and explained without knowledge of what happens at the atomic level in the organisms. From an epistemological point of view the double helix was trite. It raised no paradoxes. It required no new laws. Physical and chemical explanations were found to be quite sufficient to account for one of the fundamental secrets of life.

From his correspondence with Wolfgang Pauli[27], we can see that Bohr could not believe that the double helix was the end of the story. He maintains that the fundamental unit of life must be the cell, not the chromosomes, and that the functions of the cell must be regulated by information from the organism as a whole. He suggests:

"Because of the circumstance that gametes mature only at such a late stage in the life of the individual that the requirement for adaptation to changed external conditions can already have occurred, it therefore seems natural to presume that such changes are not only reflected in the phenotype but that the constancy of the genotype is only a first approximation, and that it can

Bohr to Pauli,
31 Dec 53
Danish text on p. [543]
Translation on p. [547]

[27] Letters from Pauli to Delbrück, 16 February 1954, from Pauli to Bohr, 19 February 1954, from Pauli to Bohr, 26 March 1954, from Bohr to Pauli, 6 April 1954, from Bohr to Pauli, 7 February 1955. Reproduced on pp. [551], [553], [557], [558], [561].

undergo gradual secular changes of a character suitable for adaptation to the environment."

Bohr's correspondence with Walter Elsasser shows that to the very end Bohr maintained that even the most recent results within molecular biology did not allow biology to be reduced to physics[28]. He never gave up his view that the integrity of living organisms and the characteristics of conscious individuals present features of wholeness, the account of which implies a typically complementary mode of description.

It seems reasonable to ask what Bohr had in mind when he time and time again spoke of the unity of knowledge. The logical positivists would argue that if biology and psychology cannot be reduced to physics and chemistry, then there is no such unity. However, Bohr's ideas concerning the unity of knowledge are diametrically opposite to the views of the logical positivists. Undoubtedly, Bohr felt that if we think along ontological paths, we are soon led to speak about different forms of existence. Instead of trying to intuit the essence of reality, we should stick to epistemology and concentrate on what we can *say* about nature. What separates one scientific field from another is always the difference in the conditions for observation and description. By focusing on these conditions we realize that apparent contrasts can be resolved. The "only way to reconcile the laws of physics with the concepts suited for a description of the phenomenon of life is to examine the essential difference in the conditions of the observation of physical and biological phenomena"[29]. This is also the case when we analyze the conceptual relationship between classical physics and quantum mechanics. However, the unity is preserved, because the basis for all unambiguous description ultimately is ordinary language in the sense explained in the General Introduction[30].

2. CONTEMPORANEOUS DISCUSSIONS OF BOHR'S VIEW

Not surprisingly, Bohr's views did not receive special attention among biologists in general; only those who were interested in the fundamental problems of biology reacted to them. Bohr's views were noticed among philosophers, especially those representing the school of logical empiricism or logical positivism and originating from the "Wiener Kreis" founded in 1923. This school

---

[28] Letters from Bohr to Elsasser, 19 November 1959, from Elsasser to Bohr, 18 December 1959, from Bohr to Elsasser, 29 December 1959. Reproduced on pp. [497], [498] and [500].

[29] Bohr, *Biology and Atomic Physics*, ref. 17. Quotation on pp. [60]–[61].

[30] See above, p. XXXIII.

became strong after Moritz Schlick founded the society "Verein Ernst Mach" in 1928. From the very start the logical positivists fought against metaphysical tendencies and tried to establish a unity of science programme. Ernst Mach, who died 1916, had shown the way through his criticism of the dualism between the physical and the mental world and the dualism between object and subject. Dualism, in his opinion, implied different levels of existence and should be avoided. According to Mach, "reality" must be understood as the data we experience, and nothing else. He called them "elements" in order to indicate that they were neither physical nor mental but simply neutral entities[31]. A blue spot may be examined in a physical context, for instance by measuring the wavelength of the emitted light and, consequently, may be characterized as something physical. Or it may be analyzed in a psychological context, for instance when we estimate its degree of visually perceived saturation compared to other colours, and should then be characterized as a psychological datum. But in itself it is neither something physical nor something mental but simply a neutral element.

According to Mach, reality is nothing but such elements. A stone, for example, is nothing but a relatively permanent complex of elements. A subject or a self is nothing but a relatively stable assembly of elements. Concepts such as "atom", "electron", "force" and "electromagnetic field" do not represent anything real. They are nothing but concepts which have been introduced in order to describe the outcome of our observations and experiments in a consistent way. Discussions about the relation between mental and physical events, or about vitalism versus mechanism, are quarrels about nothing. It has no meaning to speak of different kinds or different levels of existence. These reflections brought about the idea of a unity of science[32].

The first attempt to show how Mach's programme could be realized in detail was Rudolf Carnap's extensive work, "Der Logische Aufbau der Welt", from 1928[33]. However, the logical positivists soon realized that both Mach's and Carnap's epistemological foundation was phenomenalistic in character and implied an unacceptable solipsism. From 1931 to 1936 a new basis was developed, the so-called "physicalism". The positivists maintained that all scientific concepts and statements are reducible to propositions about macroscopical objects and

---

[31] E. Mach, *Die Analyse der Empfindungen* (9th edition), Gustav Fischer, Jena 1922, pp. 35–37.

[32] See, for instance, J. Jørgensen, *The Development of Logical Empiricism*, University of Chicago Press, Chicago 1951.

[33] R. Carnap, *Der Logische Aufbau der Welt*, Weltkreis Verlag, Berlin 1928. The English version is *The Logical Structure of the World*, Routledge and Kegan Paul, London 1967.

their intersubjectively verifiable properties. This physicalism formed the basis for the Unity of Science (Einheitswissenschaft) movement, which soon was to manifest itself in conferences, books and periodicals. Among the supporters of the movement were many competent philosophers of science, such as Philipp Frank, Hans Reichenbach and Carl Gustav Hempel. According to their research programme, laws and concepts in the social sciences were to be defined in terms of psychological concepts; these in turn were to be defined on a biological basis; and all biological concepts were to be reduced to language describing observable physical objects[34].

Naturally, the logical positivists became interested in Bohr's views. Here was one of the most prominent physicists maintaining that quantum mechanics was based on observations that ultimately had to be described in terms of classical physical objects. Yet Bohr declared at the same time that biology was not reducible to physics and chemistry. One of the professors of philosophy at the University of Copenhagen, Jørgen Jørgensen, was a leading figure in the Unity of Science movement and had close contact with Edgar Rubin. Rubin, who as mentioned in the General Introduction[35] was a close friend of Bohr, discussed biological and psychological problems with Jørgensen from 1929 onwards. However, the members of the movement outside Denmark only learnt about Bohr's views through Pascual Jordan who considered himself an avowed Machist as well as an accepted member of the movement, which was hardly the case.

Jordan was well known for his important contribution to the development of quantum mechanics and shared Bohr's interest in the foundations of biology and psychology. In a letter written in January 1930 Bohr expressed his pleasure in the philosophical discussions with Jordan[36], and during a visit to Copenhagen in March 1931 he had the opportunity of discussing with Bohr the general consequences of the complementarity viewpoint, whereupon he published articles on the relevance of quantum mechanics for biology. It soon appeared, however, that Jordan's views were based on a grave misunderstanding of Bohr. Yet Jordan cited Bohr in support of his views and Bohr was very slow to realize that Jordan's interests were different from his own.

---

[34] See Jørgensen, *Development*, ref. 32.

[35] See above, *General Introduction*, p. XXIX.

[36] Letter from Bohr to Jordan, 25 January 1930. Reproduced on pp. [514] (German original) and [515] (English translation).

In May 1931, Jordan sent Bohr a manuscript[37] which was published later in an expanded version in "Die Naturwissenschaften"[38]. In this paper Jordan claims that the traditional distinction between an external world ("Aussenwelt") and an inner, subjective world ("Innenwelt") has been undermined by the circumstance that everything in the external world is dependent on the process of observation. Quantum mechanics had shown that everything we register is influenced by observation in the same manner as mental states are disturbed by self-observation. By means of this untenable argument Jordan tried to support Mach's philosophy.

Referring to Bohr, Jordan explained that psychophysical parallelism is analogous to the wave–particle dualism of quantum mechanics. As regards the problem of the freedom of the will, he put forward a spry solution: in inorganic matter, the acausality of individual atomic processes is levelled through a normal statistical distribution, whereas in living organisms they may manifest themselves in macroscopical acausal behaviour through amplification mechanisms in the organism. This is Jordan's so-called "amplification hypothesis" concerning the freedom of the will, an idea which was, in fact, suggested already in 1927 by the American biologist Ralph Stayner Lillie[39].

In his reply to Jordan, Bohr criticized in friendly terms three of the points made by his younger colleague. First of all, he did not agree with Jordan's considerations concerning the "Aussenwelt" and "Innenwelt" because classical physics must presuppose a sharp distinction between subject and object and is not nullified by quantum mechanics. Secondly, he pointed out that Jordan's considerations concerning psychophysical parallelism may cause a misunderstanding of his own view. Finally, he agreed with Jordan that acausality is a feature of life:

> "I agree of course with the point that it is above all the fundamental limitation of the applicability of the causality concept in the inorganic macrocosmos which gives us the necessary freedom, and that in this sense acausality can be regarded as a characteristic of life."

Bohr to Jordan
5 June 31
German text on p. [517]
Translation on p. [520]

Yet he declared that the reactions of the organism are intimately connected with

---

[37] Letter from Jordan to Bohr, 20 May 1931. Reproduced on pp. [516] (German original) and [517] (English translation). Jordan's manuscript, *Statistik, Kausalität und Willensfreiheit* is deposited in the NBA.

[38] P. Jordan, *Die Quantenmechanik und die Grundprobleme der Biologie und Psychologie*, Naturwiss. **20** (1932) 815–821.

[39] R.S. Lillie, *Physical Indeterminism and Vital Action*, Science **66** (1927) 139–144.

laws peculiar to biology and differ in principle from amplification devices. Bohr stated that he had not finished developing his own views.

In Jordan's next letter to Bohr, dated 22 June 1931[40], Jordan suggests that a living organism has three zones – the first comprising the fundamental "centres" of life, i.e. the acausal processes, the second the amplification devices, and the third the tool-organs. The latter zone involves bones and muscles which are describable in mechanical terms. The "amplification" is some kind of teleological integration because there must be an intermediate level where the random statistical effects of single atoms one way or another are brought into harmony. Consequently, it is impossible to observe experimentally the state of a living organism on the atomic level. It would simply mean the death of the organism.

Strangely, in a letter to Jordan of 23 June 1931[41], Bohr expresses agreement with all these speculations, which ran contrary to his views expressed in "Light and Life" only one year later. He emphasizes that his main concern is not to go into biological research as such. Only the epistemological aspects of biology have his interest.

In a letter to Bohr of 26 November 1932[42], Jordan enclosed a reprint of his article from "Die Naturwissenschaften", which apart from an added section about the nature of the organic ("Das Wesen des Organischen") is identical to the draft which Bohr had received in May 1931. In a letter of 27 December 1932, Bohr thanks Jordan for the article and writes:

Bohr to Jordan
27 Dec 32
German text on p. [530]
Translation on p. [532]

"I am very glad that your beautiful article about quantum mechanical and biological questions finally appeared, ...".

For some time, the periodical "Erkenntnis" and the conferences held by the logical positivists served as the stage for discussing Jordan's and Bohr's views. In an article from 1934[43], Jordan understandably quoted Bohr in support of his ideas. In this article he repeats the amplification hypothesis and the Thana-

---

[40] Letter from Jordan to Bohr, 22 June 1931. Reproduced on pp. [523] (German original) and [525] (English translation).

[41] Letter from Bohr to Jordan, 23 June 1931. Reproduced on pp. [527] (German original) and [528] (English translation).

[42] Letter from Jordan to Bohr, 26 November 1932. Reproduced on pp. [529] (German original) and [530] (English translation).

[43] Pascual Jordan, *Quantenphysikalische Bemerkungen zur Biologie und Psychologie*, Erkenntnis **4** (1934) 215–252.

[18]

tological Principle and declares that when the brain is exposed to penetrating X-rays, the processes of life will be destroyed owing to the Compton effect.

Jordan's views were fiercely criticized. Moritz Schlick pointed out that the amplification hypothesis did not imply freedom but accident. Freedom is not simply lack of causality and free choice is not just an amplification of a quantum leap[44]. Philipp Frank observed that Jordan's arguments do not prove biology to be an autonomous science. On the contrary they reduce biology to quantum physics[45]. In his reply to these and other authors, Jordan cited Bohr and maintained that biology could not be reduced entirely to modern physics[46]. Jordan's misleading advocacy of Bohr's views stimulated the interest of the logical positivists in having the whole matter discussed with Bohr. This may be the background for placing the 1936 Unity of Science congress in Copenhagen and inviting Bohr to give the opening address.

Apart from this discussion we find only few comments on Bohr's views. Otto Meyerhof criticized the Thanatological Principle and pointed out that when an animal dies, all its organs do not die at the same time. For instance, muscular contraction as a result of electrical stimulation may be observed after the death of the organism[47]. Another opponent was Joseph Needham who in a seminal book published in 1936 attacked the views of both Haldane and Bohr as well as Lillie's article from 1927. As to the Metabolism Argument, he writes: "Haldane wishes to inquire whether any meaning can be attached to the concept of an organism apart from its environment" and concludes: "... if no line can be drawn between organism and immediate surroundings, no better line can be drawn between immediate surroundings and far-off surroundings." He criticizes what we have called Bohr's Thanatological Principle by pointing out that "With the advance of refinement in experimental technique, the injury done to the experimental material becomes ever slighter"[48]. Many biologists found Bohr's arguments too vague and speculative. It may be characteristic for the way in which his thoughts were received that biologists tended to view him as a vitalist. Having visited Bohr in Copenhagen in 1933, the prominent biologist

---

[44] Moritz Schlick, *Ergänzende Bemerkungen über P. Jordans Versuch einer quantentheoretischen Deutung der Lebenserscheinungen*, Erkenntnis **5** (1935) 181–183.

[45] Philipp Frank, *Jordan und der radikale Positivismus*, Erkenntnis **5** (1935), p. 184.

[46] P. Jordan, *Ergänzende Bemerkungen über Biologie und Quantenmechanik*, Erkenntnis **5** (1935) 348–352.

[47] O. Meyerhof, *Betrachtungen über die naturphilosophischen Grundlagen der Physiologie*, Naturwiss. **22** (1934) 311–314.

[48] J. Needham, *Order and Life*, M.I.T. Press, Cambridge, Massachusetts 1936, pp. 9, 11, 30.

H.J. Muller wrote in a letter that he had been glad "to meet the physicist Bohr there, but then I found that his ideas in biology were hopelessly vitalistic"[49].

In a letter from 1934, Max Delbrück informed Bohr that Jordan had given a lecture to the Society for Empirical Philosophy in Berlin claiming to represent Bohr's views. Delbrück warned Bohr that Max Hartmann and other biologists were rather upset about the lecture. In a brief note directed to Hartmann, Delbrück gave a summary of Bohr's views stating that Bohr really did not think that biologists must kill organisms in order to study them and allowed that biological description might be strictly causal. Delbrück sent a copy of the note to Bohr, who agreed with its contents[50].

In 1936, the logical positivists held their second international Unity of Science congress in Copenhagen. The main theme was "The Causality Problem". The congress took place from 21 to 26 June and was opened in Bohr's home, the honorary residence at Carlsberg. Bohr gave a lecture in which he repudiated Jordan's views indirectly. He thus declared that quantum mechanics offers no solution to the problems of vitalism and the freedom of the will, and emphasized furthermore that he opposed spiritualism, i.e. mind as a special form of existence in the sense of Descartes[51]. In an extensive letter to Meyerhof dated 5 September 1936, he pursues this theme and other points from the lecture[52].

Bohr's lecture did not win the approval of his audience and obviously he realized after the congress that his views were very different from those of the logical positivists. However, not everybody was negative towards Bohr's ideas. Already in 1932 he found an enthusiastic disciple in Max Delbrück. Delbrück was 26 years old, with a Rockefeller fellowship which enabled him to work under the guidance of Bohr in Copenhagen and Wolfgang Pauli in Zurich. Delbrück had studied astronomy and physics and defended his Ph.D. thesis on the quantum mechanics of lithium at the University of Berlin in 1930.

The "Light and Life" lecture was Bohr's first presentation devoted entirely to the philosophical questions of biology. He gave it as the opening address at the Second International Light Congress which was held in August 1932 and

---

[49] Aaserud, *Redirecting Science*, ref. 11, pp. 98–99.

[50] Letters from Delbrück to Bohr, 30 November 1934 (with enclosure) and from Bohr to Delbrück, 8 December 1934. The letters and enclosure are reproduced, in the German original and English translation, on pp. [465] ff.

[51] N. Bohr, *Causality and Complementarity*, Phil. Sci. **4** (1937) 289–298. Reproduced on pp. [39]–[48].

[52] Letter from Bohr to Meyerhof, 5 September 1936. Reproduced on pp. [538] (German original) and [541] (English translation).

Werner Heisenberg, Max Delbrück and Pascual Jordan in the garden of the Carlsberg honorary mansion, 1936.

which dealt with with the role of light in biology, biophysics and therapy[53]. He hesitated to accept the invitation at first, and agreed to give a lecture only three weeks before the opening of the Congress. About this Finn Aaserud writes[54]:

> "Probably as a result of the late decision, his talk at the Light Congress is one of the very few of his public lectures of which a manuscript has not been preserved. Moreover, the lecture was subsequently published in several languages and editions containing only minor alterations of formulation. This was quite atypical for Bohr, whose writings were usually drafted several times before publication. It indicates that after his correspondence with Jordan, Bohr's thoughts on biology had matured substantially."

Indeed, the usually careful Bohr considered the differences of formulation to be

---

[53] *IIe Congrès International de la lumière, biologie, biophysique, thérapeutique – Copenhague 15–18 Août, 1932*, Engelsen & Schrøder, Copenhagen (no date), where Bohr's lecture is reproduced on pp. XXXVII–XLVI.
[54] Aaserud, *Redirecting Science*, ref. 11, p. 91.

so immaterial that when many years later he republished the article in several languages, he did not use the same version in each case[55].

Delbrück came to Copenhagen just in time to hear Bohr's lecture, which stimulated his interest in biology. He was fascinated in particular by the parallels between quantum physics and biology put forward in the lecture. The problems of genetic replication and mutation presented an enigma at that time. On the one hand, genes were remarkably stable through thousands of generations. On the other hand, mutations did occur. Despite the usefulness of the concept of the gene, in 1932 nobody had seen or isolated one. The size of genes was known approximately from experiments with *Drosophila* but their physical and chemical composition was a mystery.

In the autumn of 1932, Delbrück began work with Lise Meitner and Otto Hahn at the Kaiser Wilhelm Institute for Chemistry in Berlin, where he spent his spare time during the next five years pursuing his biological interests in seminars and discussion groups. While he worked at the KWI, Delbrück came into contact with the biologist N.W. Timoféeff-Ressovsky and the biophysicist K.G. Zimmer. They introduced him to the phenomenon that radiation induces mutations in *Drosophila*. The question was raised whether quantum mechanics can account for the stability and mutability of the gene. The starting point was H.J. Muller's experiments with the *Drosophila melanogaster*, which showed that radiation in the region from ultraviolet light to gamma-rays increases the rate of mutations proportionally to the dose of radiation.

The discussions resulted in an article written by Timoféeff-Ressovsky, Zimmer and Delbrück[56]. Each of the authors wrote his own part of this "Dreimännerarbeit", while the concluding section was written in collaboration. In his part, Delbrück states that genetics cannot be reduced to physics and chemistry and suggests that the gene plays only a catalytic part in the metabolic processes. Assuming it to be a molecule – an idea that was still no more than a guess at the time – Delbrück considers the gene as a quantum mechanical system and the mutations induced by X-rays as changes from one stable electronic state to another.

Delbrück sent a copy of the article to Bohr in 1935[57], lamenting that the article contained no arguments in favour of complementarity. As can be seen

---

[55] See below, p. [28].

[56] N.W. Timoféeff-Ressovsky, K.G. Zimmer und M. Delbrück, *Über die Natur der Genmutation und der Genstruktur*, Nach. Ges. Wiss. Göttingen, Math–Phys. Kl., Fachgruppe VI: Biologie **1** (1935) 189–245.

[57] Letter from Delbrück to Bohr, 5 April 1935. Reproduced on pp. [471] (German original) and [472] (English translation).

in a letter to Delbrück from August 1935[58], Bohr became deeply interested in the matter. He went on to organize a conference on "The Mechanism of Mutation", which was held in Copenhagen from 27 to 29 September 1936 – the first conference ever devoted to the physics and chemistry of genetics. About 35 persons participated, among them Timoféeff-Ressovsky and Delbrück.

Delbrück's aim was to investigate the genetic and cytological fields closely enough to find a paradox similar in character to the paradoxes that led to quantum physics: Planck's discovery of the quantum of action and Rutherford's discovery of the atomic nucleus. If such a paradox could be found and solved, the doors would open to a new field in science and the riddles of life would appear in a new light.

In 1935, W.M. Stanley of the Rockefeller Institute made a sensational discovery. He had crystallized the tobacco mosaic virus which has its name from its ability to form a mosaic-like rash on tobacco plants. In contact with a host organism, this virus exhibits both replication and mutation but in its crystalline form consists of protein molecules that can be precipitated as a powder. Indeed, this looked like the indivisible unit of life that Delbrück was looking for. The existence of such units of life was suggested earlier by the biologist J. Gray who had concurred in Bohr's idea from "Light and Life" that life is an elementary fact in analogy with the quantum of action[59]. Although the similarity between genes and viruses had been discussed since the 1920s, it was only now that viruses could be used in experiments as models for genes. It now became possible to study, so to speak, the pure replication mechanism without having to pay attention to a whole organism of which the mechanism was only a part, as in the case of *Drosophila*.

In 1937, Delbrück received an offer for a fellowship from the Rockefeller Foundation and chose to study genetics with the noted geneticist T.H. Morgan at the California Institute of Technology in Pasadena. Instead of joining the active *Drosophila* group at Caltech, Delbrück teamed up with the biochemist Emory Ellis who was setting up experiments to investigate bacterial viruses (bacteriophages). From that time, Delbrück gradually scaled down his efforts in radiation genetics and paved the way instead to the double helix. His studies of phage genetics proved fruitful, and later he was to cooperate successfully with the bacteriologist Salvadore Luria with whom he shared the Nobel Prize in medicine together with Alfred Hershey in 1969. However, his philosophical ex-

---

[58] Letter from Bohr to Delbrück, 10 August 1935. Reproduced on pp. [472] (German original) and [473] (English translation).
[59] J. Gray, *The Mechanical View of Life*, Nature **132** (1933) 661–664.

pectations were not fulfilled. Although the phage represented the simplest case of replication in an independent singular cell and thus could be seen as "the basic unit of life", Delbrück's investigations of the molecular mechanisms showed that they were much more complex than anything on the atomic level. As the electron microscope was improved, the photographs of the phage revealed that it was not a "fundamental" biological unit, but a complicated organism which could even replicate itself without entering the host organism. Furthermore, host bacteria could become resistant to phage attacks by undergoing mutations. This suggested that processes in the bacterium interacted with the mechanism of phage replication.

In his continued research aiming to find complementary relations, Delbrück turned to studying the sensory physiology of the fungus *Phycomyces*. Although this fungus has a sensory mechanism which functions at the quantum level, Delbrück still did not succeed in fulfilling his philosophical ambitions. These ambitions are stated very clearly in a letter from Delbrück to Bohr from December 1954[60].

In 1944, the Canadian-born bacteriologist O.T. Avery and his co-workers showed that one particular property of pneumococci depended on the presence of a specific molecule, i.e. DNA. This allowed the hypothesis that DNA was genetic material. At the beginning of the 1950s, it became clear that the answer to the replication problem resided in the structure of DNA, and through James D. Watson's, Francis Crick's and Rosalind Franklin's clarification of the double helix structure, the problem was entirely resolved. One of the great riddles of life had been answered on the basis of chemistry and physics. Delbrück's phage programme had proved its relevancy. But the result contradicted Delbrück's general conception of biology.

When Delbrück heard about the double helix, his first hope was that a paradox would appear. In a letter to Bohr from April 1953, he writes:

Delbrück to Bohr
14 April 53
English
Full text on p. [474]

"Very remarkable things are happening in biology. I think that Jim Watson has made a discovery that may rival that of Rutherford in 1911."

Bohr's reply of 2 May 1953 shows that he had already heard about this[61]. Delbrück seemingly suggests that the discovery must call for a theoretical step of an order equal to Bohr's atomic theory from 1913. But, as mentioned above, the double helix raised no paradoxes. To Bohr and Delbrück, it meant that

---

[60] Letter from Delbrück to Bohr, 1 December 1954. Reproduced on p. [475].
[61] Letter from Bohr to Delbrück, 2 May 1953. Reproduced on p. [475].

their ideas about biology were momentarily disconfirmed by the new genetics. Characteristically, Bohr was not disappointed by the discovery, which indeed increased his interest in biology.

Delbrück never lost his lively interest in Bohr's views on biology. Thus, when in June 1962 he organized the inauguration of the Institute for Genetics at the University of Cologne (Institut für Genetik der Universität zu Köln), Delbrück wanted Bohr to give the principal lecture at the event. Thus he wrote to his mentor that Bohr

"might want to take such an occasion as this to elaborate on what you said 30 years ago, in the light of the new developments that have taken place."

Delbrück to Bohr
15 April 62
English
Full text on p. [488]

The correspondence between Delbrück and Bohr[62] shows that, as always, Bohr prepared himself to the fullest possible extent. After the inauguration, Bohr became ill and ill-health prevented him from finishing his manuscript "Light and Life Revisited" for publication before his sudden death on 18 November 1962[63].

Delbrück's scientific career is an example of a frequent phenomenon in the history of science, namely that a scientist obtains results or contributes to discoveries which are in contradiction to his hopes and expectations. It also exemplifies that even very abstract and speculative ideas may have a practical effect.

Although Delbrück was enthusiastic about Bohr's views and although Bohr was enthusiastic about Delbrück's work, it can even be questioned whether Delbrück understood the depth of Bohr's epistemological considerations. Indeed, Bohr never suggested that biological investigations would reveal experimental facts similar to Rutherford's discovery in 1911, thus introducing a paradox. This was not his way of approaching the problems of biology. The point continuously stressed by him is that in order to describe living organisms, and especially conscious organisms, we must establish a conceptual framework essentially different on the one hand from classical physics and chemistry and on the other from quantum mechanics. Bohr would occasionally refer to the relationship between thermodynamical description and classical mechanics – two conceptual frameworks which in a certain sense are complementary – as an analogy to the relationship between physics and biology. The special

---

[62] Correspondence between Delbrück and Bohr from 15 April to 21 May 1962. Reproduced on pp. [488]–[494].

[63] The manuscript was completed posthumously by Niels Bohr's son, Aage, and published as *Light and Life Revisited*, ref. 22.

conditions for description which Bohr claimed were to be found in biology did not necessarily depend on the discovery of some special biological building blocks resisting all physical analysis. It is my hope that the following first full presentation of Bohr's published work (supplemented with a selection of unpublished material) on biological and related questions will serve to give an accurate picture of his views, which have been misunderstood even by some of his closest colleagues.

I. LIGHT AND LIFE

Nature **131** (1933) 421–423, 457–459

Address at the opening meeting of the International
Congress on Light Therapy in Copenhagen, 15 August 1932

See General Introduction, pp. XXVII and XLV, and Introduction to Part I, pp. [7], [11] and [20] f.

LIGHT AND LIFE (1932)

Versions published in English, Danish and German

English: Light and Life

A "IIe Congrès International de la lumière, biologie, biophysique, thérapeutique Copenhague, 15–18 août 1932", Engelsen & Schrøder, Copenhagen (no date), pp. XXXVII–XLVI

B Nature **131** (1933) 421–23, 457–59

C "Atomic Physics and Human Knowledge", John Wiley & Sons, New York 1958, pp. 3–12 (reprinted in: "Essays 1933–1957 on Atomic Physics and Human Knowledge, The Philosophical Writings of Niels Bohr, Vol. II", Ox Bow Press, Woodbridge, Connecticut 1987, pp. 3–12)

Danish: Lys og Liv

D Naturens Verden **17** (1933) 49–59

E "Atomfysik og menneskelig erkendelse", J.H. Schultz Forlag, Copenhagen 1957, pp. 11–22

F "Naturbeskrivelse og menneskelig erkendelse" (eds. J. Kalckar and E. Rüdinger), Rhodos, Copenhagen 1985, pp. 96–118

German: Licht und Leben

G Naturwiss. **21** (1933) 245–50

H "Atomphysik und menschliche Erkenntnis", Friedr. Vieweg & Sohn, Braunschweig 1958, pp. 3–12

C is a reproduction of A, apart from a few trivial corrections.

D is Bohr's translation of A.

B is Bohr's retranslation of D (see Bohr to Klein 19 January 1933, p. [533]).

D, E and F are identical.

G corresponds to B.

H is identical to G, apart from a few trivial differences of formulation.

Light and Life*

By Prof. N. Bohr, For.Mem.R.S.

AS a physicist whose studies are limited to the properties of inanimate bodies, it is not without hesitation that I have accepted the kind invitation to address this assembly of scientific men met together to forward our knowledge of the beneficial effects of light in the cure of diseases. Unable as I am to contribute to this beautiful branch of science that is so important for the welfare of mankind, I could at most comment on the purely inorganic light phenomena which have exerted a special attraction for physicists throughout the ages, not least owing to the fact that light is our principal tool of observation. I have thought, however, that on this occasion it might perhaps be of interest, in connexion with such comments, to enter on the problem of what significance the results reached in the limited domain of physics may have for our views on the position of living organisms in the realm of natural science.

Notwithstanding the subtle character of the riddles of life, this problem has presented itself at every stage of science, since any scientific explanation necessarily must consist in reducing the description of more complex phenomena to that of simpler ones. At the moment, however, the unsuspected discovery of an essential limitation of the mechanical description of natural phenomena, revealed by the recent development of the atomic theory, has lent new interest to the old problem. This limitation was, in fact, first recognised through a thorough study of the interaction between light and material bodies, which disclosed features that cannot be brought into conformity with the demands hitherto made to a physical explanation. As I shall endeavour to show, the efforts of physicists to master this situation resemble in some way the attitude which biologists more or less intuitively have taken towards the aspects of life. Still, I wish to stress at once that it is only in this formal respect that light, which is perhaps the least complex of all physical phenomena, exhibits an analogy to life, the diversity of which is far beyond the grasp of scientific analysis.

From a physical point of view, light may be defined as the transmission of energy between material bodies at a distance. As is well known, such an energy transfer finds a simple explanation in the electromagnetic theory, which may be regarded as a direct extension of classical mechanics compromising between action at a distance and contact forces. According to this theory, light is described as coupled electric and magnetic oscillations which differ from the ordinary electromagnetic waves used in radio transmission only by their greater frequency of vibration and smaller wave-length. In fact, the practically rectilinear propagation of light, on which rests our location of bodies by direct vision or by suitable optical instruments, depends entirely on the smallness of the wave-length compared with the dimensions of the bodies concerned, and of the instruments.

The idea of the wave nature of light, however, not only forms the basis for our explanation of the colour phenomena, which in spectroscopy have yielded such important information of the inner constitution of matter, but is also of essential importance for every detailed analysis of optical phenomena. As a typical example, I need only mention the interference patterns which appear when light from one source can travel to a screen along two different paths. In such a case, we find that the effects which would be produced by the separate light beams are strengthened at those points on the screen where the phases of the two wave trains coincide, that is, where the electric and magnetic oscillations in the two beams have the same directions, while the effects are weakened and may even disappear at points where these oscillations have opposite directions, and where the two wave trains are said to be out of phase with one another. These interference patterns have made possible such a thorough test of the wave nature of the propagation of light, that this conception can no longer be considered as a hypothesis in the usual sense of this word, but may rather be regarded as an indispensable element in the description of the phenomena observed.

As is well known, the problem of the nature of light has, nevertheless, been subjected to renewed discussion in recent years, as a result of the discovery of a peculiar atomistic feature in the energy transmission which is quite unintelligible from the point of view of the electromagnetic theory. It has turned out, in fact, that all effects of light may be traced down to individual processes, in each of which a so-called light quantum is exchanged, the energy of which is equal to the product of the frequency of the electromagnetic oscillations and the universal quantum of action, or Planck's constant. The striking contrast between this atomicity of the light phenomena and the continuity of the energy transfer according to the electromagnetic theory places us before a dilemma of a character hitherto unknown in physics. For, in spite of the obvious insufficiency of the wave picture, there can be no question of replacing it by any other picture of light propagation depending on ordinary mechanical ideas.

Especially, it should be emphasised that the introduction of the concept of light quanta in no way means a return to the old idea of material

* Address delivered at the opening meeting of the International Congress on Light Therapy, Copenhagen, on August 15, 1932. The present article, conforming with the Danish version (*Naturens Verden*, 17, 49), differs from that published in the Congress report only by some formal alterations.

particles with well-defined paths as the carriers of the light energy. In fact, it is characteristic of all the phenomena of light, in the description of which the wave picture plays an essential rôle, that any attempt to trace the paths of the individual light quanta would disturb the very phenomenon under investigation ; just as an interference pattern would completely disappear, if, in order to make sure that the light energy travelled only along one of the two paths between the source and the screen, we should introduce a non-transparent body into one of the paths. The spatial continuity of light propagation, on one hand, and the atomicity of the light effects, on the other hand, must, therefore, be considered as complementary aspects of one reality, in the sense that each expresses an important feature of the phenomena of light, which, although irreconcilable from a mechanical point of view, can never be in direct contradiction, since a closer analysis of one or the other feature in mechanical terms would demand mutually exclusive experimental arrangements.

At the same time, this very situation forces us to renounce a complete causal description of the phenomena of light and to be content with probability calculations, based on the fact that the electromagnetic description of energy transfer by light remains valid in a statistical sense. Such calculations form a typical application of the so-called correspondence argument, which expresses our endeavour, by means of a suitably limited use of mechanical and electromagnetic concepts, to obtain a statistical description of the atomic phenomena that appears as a rational generalisation of the classical physical theories, in spite of the fact that the quantum of action from their point of view must be considered as an irrationality.

At first sight, this situation might appear very deplorable ; but, as has often happened in the history of science, when new discoveries have revealed an essential limitation of ideas the universal applicability of which had never been disputed, we have been rewarded by getting a wider view and a greater power of correlating phenomena which before might even have appeared as contradictory. Thus, the strange limitation of classical mechanics, symbolised by the quantum of action, has given us a clue to an understanding of the peculiar stability of atoms which forms a basic assumption in the mechanical description of any natural phenomenon. The recognition that the indivisibility of atoms cannot be understood in mechanical terms has always characterised the atomic theory, to be sure ; and this fact is not essentially altered, although the development of physics has replaced the indivisible atoms by the elementary electric particles, electrons and atomic nuclei, of which the atoms of the elements as well as the molecules of the chemical compounds are now supposed to consist.

However, it is not to the question of the intrinsic stability of these elementary particles that I am here referring, but to the problem of the required stability of the structures composed of them. As a matter of fact, the very possibility of a continuous transfer of energy, which marks both the classical mechanics and the electromagnetic theory, cannot be reconciled with an explanation of the characteristic properties of the elements and the compounds. Indeed, the classical theories do not even allow us to explain the existence of rigid bodies, on which all measurements made for the purpose of ordering phenomena in space and time ultimately rest. However, in connexion with the discovery of the quantum of action, we have learned that every change in the energy of an atom or a molecule must be considered as an individual process, in which the atom goes over from one of its so-called stationary states to another. Moreover, since just one light quantum appears or disappears in a transition process by which light is emitted or absorbed by an atom, we are able by means of spectroscopic observations to measure directly the energy of each of these stationary states. The information thus derived has been most instructively corroborated also by the study of the energy exchanges which take place in atomic collisions and in chemical reactions.

In recent years, a remarkable development of the atomic theory has taken place, which has given us such adequate methods of computing the energy values for the stationary states, and also the probabilities of the transition processes, that our account, on the lines of the correspondence argument, of the properties of atoms as regards completeness and self-consistency scarcely falls short of the explanation of astronomical observations offered by Newtonian mechanics. Although the rational treatment of the problems of atomic mechanics was possible only after the introduction of new symbolic artifices, the lesson taught us by the analysis of the phenomena of light is still of decisive importance for our estimation of this development. Thus, an unambiguous use of the concept of a stationary state is complementary to a mechanical analysis of intra-atomic motions ; in a similar way the idea of light quanta is complementary to the electromagnetic theory of radiation. Indeed, any attempt to trace the detailed course of the transition process would involve an uncontrollable exchange of energy between the atom and the measuring instruments, which would completely disturb the very energy transfer we set out to investigate.

A causal description in the classical sense is possible only in such cases where the action involved is large compared with the quantum of action, and where, therefore, a subdivision of the phenomena is possible without disturbing them essentially. If this condition is not fulfilled, we cannot disregard the interaction between the measuring instruments and the object under investigation, and we must especially take into consideration that the various measurements required for a complete mechanical description may only be made with mutually exclusive experimental arrangements.

In order fully to understand this fundamental limitation of the mechanical analysis of atomic phenomena, one must realise clearly, further, that in a physical measurement it is never possible to take the interaction between object and measuring instruments directly into account. For the instruments cannot be included in the investigation while they are serving as means of observation. As the concept of general relativity expresses the essential dependence of physical phenomena on the frame of reference used for their co-ordination in space and time, so does the notion of complementarity serve to symbolise the fundamental limitation, met with in atomic physics, of our ingrained idea of phenomena as existing independently of the means by which they are observed.

(To be continued.)

Paul Ehrenfest in conversation with Max Delbrück in 1933 on the terrace at Bohr's institute in Copenhagen.

Light and Life*

By Prof. N. Bohr, For.Mem.R.S.

THIS revision of the foundations of mechanics, extending to the very question of what may be meant by a physical explanation, has not only been essential, however, for the elucidation of the situation in atomic theory, but has also created a new background for the discussion of the relation of physics to the problems of biology. This must certainly not be taken to mean that in actual atomic phenomena we meet with features which show a closer resemblance to the properties of living organisms than do ordinary physical effects. At first sight, the essentially statistical character of atomic mechanics might even seem difficult to reconcile with an explanation of the marvellously refined organisation, which every living being possesses, and which permits it to implant all the characteristics of its species into a minute germ cell.

We must not forget, however, that the regularities peculiar to atomic processes, which are foreign to causal mechanics and find their place only within the complementary mode of description, are at least as important for the account of the behaviour of living organisms as for the explanation of the specific properties of inorganic matter. Thus, in the carbon assimilation of plants, on which depends largely also the nourishment of animals, we are dealing with a phenomenon for the understanding of which the individuality of photo-chemical processes must undoubtedly be taken into consideration. Likewise, the peculiar stability of atomic structures is clearly exhibited in the characteristic properties of such highly complicated chemical compounds as chlorophyll or hæmoglobin, which play fundamental rôles in plant assimilation and animal respiration.

However, analogies from chemical experience will not, of course, any more than the ancient comparison of life with fire, give a better explanation of living organisms than will the resemblance, often mentioned, between living organisms and such purely mechanical contrivances as clockworks. An understanding of the essential characteristics of living beings must be sought, no doubt, in their peculiar organisation, in which features that may be analysed by the usual mechanics are interwoven with typically atomistic traits in a manner having no counterpart in inorganic matter.

An instructive illustration of the refinement to which this organisation is developed has been obtained through the study of the construction and function of the eye, for which the simplicity of the phenomena of light has again been most helpful. I need not go into details here, but shall just recall how ophthalmology has revealed to us the ideal properties of the human eye as an optical

instrument. Indeed, the dimensions of the interference patterns, which on account of the wave nature of light set the limit for the image formation in the eye, practically coincide with the size of such partitions of the retina which have separate nervous connexion with the brain. Moreover, since the absorption of a few light quanta, or perhaps of only a single quantum, on such a retinal partition is sufficient to produce a sight impression, the sensitiveness of the eye may even be said to have reached the limit imposed by the atomic character of the light effects. In both respects, the efficiency of the eye is the same as that of a good telescope or microscope, connected with a suitable amplifier so as to make the individual processes observable. It is true that it is possible by such instruments essentially to increase our powers of observation, but, owing to the very limits imposed by the properties of light, no instrument is imaginable which is more efficient for its purpose than the eye. Now, this ideal refinement of the eye, fully recognised only through the recent development of physics, suggests that other organs also, whether they serve for the reception of information from the surroundings or for the reaction to sense impressions, will exhibit a similar adaptation to their purpose, and that also in these cases the feature of individuality symbolised by the quantum of action, together with some amplifying mechanism, is of decisive importance. That it has not yet been possible to trace the limit in organs other than the eye, depends solely upon the simplicity of light as compared with other physical phenomena.

The recognition of the essential importance of fundamentally atomistic features in the functions of living organisms is by no means sufficient, however, for a comprehensive explanation of biological phenomena. The question at issue, therefore, is whether some fundamental traits are still missing in the analysis of natural phenomena, before we can reach an understanding of life on the basis of physical experience. Quite apart from the practically inexhaustible abundance of biological phenomena, an answer to this question can scarcely be given without an examination of what we may understand by a physical explanation, still more penetrating than that to which the discovery of the quantum of action has already forced us. On one hand, the wonderful features which are constantly revealed in physiological investigations and differ so strikingly from what is known of inorganic matter, have led many biologists to doubt that a real understanding of the nature of life is possible on a purely physical basis. On the other hand, this view, often known as vitalism, scarcely finds its proper expression in the old supposition that a peculiar vital force, quite

---

* Continued from p. 423

unknown to physics, governs all organic life. I think we all agree with Newton that the real basis of science is the conviction that Nature under the same conditions will always exhibit the same regularities. Therefore, if we were able to push the analysis of the mechanism of living organisms as far as that of atomic phenomena, we should scarcely expect to find any features differing from the properties of inorganic matter.

With this dilemma before us, we must keep in mind, however, that the conditions holding for biological and physical researches are not directly comparable, since the necessity of keeping the object of investigation alive imposes a restriction on the former, which finds no counterpart in the latter. Thus, we should doubtless kill an animal if we tried to carry the investigation of its organs so far that we could describe the rôle played by single atoms in vital functions. In every experiment on living organisms, there must remain an uncertainty as regards the physical conditions to which they are subjected, and the idea suggests itself that the minimal freedom we must allow the organism in this respect is just large enough to permit it, so to say, to hide its ultimate secrets from us. On this view, the existence of life must be considered as an elementary fact that cannot be explained, but must be taken as a starting point in biology, in a similar way as the quantum of action, which appears as an irrational element from the point of view of classical mechanical physics, taken together with the existence of the elementary particles, forms the foundation of atomic physics. The asserted impossibility of a physical or chemical explanation of the function peculiar to life would in this sense be analogous to the insufficiency of the mechanical analysis for the understanding of the stability of atoms.

In tracing this analogy further, however, we must not forget that the problems present essentially different aspects in physics and in biology. While in atomic physics we are primarily interested in the properties of matter in its simplest forms, the complexity of the material systems with which we are concerned in biology is of fundamental significance, since even the most primitive organisms contain a large number of atoms. It is true that the wide field of application of classical mechanics, including our account of the measuring instruments used in atomic physics, depends on the possibility of disregarding largely the complementarity, entailed by the quantum of action, in the description of bodies containing very many atoms. It is typical of biological researches, however, that the external conditions to which any separate atom is subjected can never be controlled in the same manner as in the fundamental experiments of atomic physics. In fact, we cannot even tell which atoms really belong to a living organism, since any vital function is accompanied by an exchange of material, whereby atoms are constantly taken up into and expelled from the organisation which constitutes the living being.

This fundamental difference between physical and biological investigations implies that no well-defined limit can be drawn for the applicability of physical ideas to the phenomena of life, which would correspond to the distinction between the field of causal mechanical description and the proper quantum phenomena in atomic mechanics. However, the limitation which this fact would seem to impose upon the analogy considered will depend essentially upon how we choose to use such words as physics and mechanics. On one hand, the question of the limitation of physics within biology would, of course, lose any meaning, if, in accordance with the original meaning of the word physics, we should understand by it any description of natural phenomena. On the other hand, such a term as atomic mechanics would be misleading, if, as in common language, we should apply the word mechanics only to denote an unambiguous causal description of the phenomena.

I shall not here enter further into these purely logical points, but will only add that the essence of the analogy considered is the typical relation of complementarity existing between the subdivision required by a physical analysis and such characteristic biological phenomena as the self-preservation and the propagation of individuals. It is due to this situation, in fact, that the concept of purpose, which is foreign to mechanical analysis, finds a certain field of application in problems where regard must be taken of the nature of life. In this respect, the rôle which teleological arguments play in biology reminds one of the endeavours, formulated in the correspondence argument, to take the quantum of action into account in a rational manner in atomic physics.

In our discussion of the applicability of mechanical concepts in describing living organisms, we have considered these just as other material objects. I need scarcely emphasise, however, that this attitude, which is characteristic of physiological research, involves no disregard whatsoever of the psychological aspects of life. The recognition of the limitation of mechanical ideas in atomic physics would much rather seem suited to conciliate the apparently contrasting points of view which mark physiology and psychology. Indeed, the necessity of considering the interaction between the measuring instruments and the object under investigation in atomic mechanics corresponds closely to the peculiar difficulties, met with in psychological analyses, which arise from the fact that the mental content is invariably altered when the attention is concentrated on any single feature of it.

It will carry us too far from our subject to enlarge upon this analogy which, when due regard is taken to the special character of biological problems, offers a new starting point for an elucidation of the so-called psycho-physical parallelism. However, in this connexion, I should like to emphasise that the considerations referred to here differ entirely from all attempts at. viewing new possibilities for a direct spiritual influence on material phenomena in the limitation set for the causal mode of description in the analysis of atomic phenomena. For example, when it has been suggested that the will might have as its field of activity the regulation of certain atomic processes within the organism, for which on the atomic theory only probability calculations may be set up, we are dealing with a view that is incompatible with the interpretation of the psycho-physical parallelism here indicated. Indeed, from our point of view, the feeling of the freedom of the will must be considered as a trait peculiar to conscious life, the material parallel of which must be sought in organic functions, which permit neither a causal mechanical description nor a physical investigation sufficiently thorough-going for a well-defined application of the statistical laws of atomic mechanics. Without entering into metaphysical speculations, I may perhaps add that any analysis of the very concept of an explanation would, naturally, begin and end with a renunciation as to explaining our own conscious activity.

In conclusion, I wish to emphasise that in none of my remarks have I intended to express any kind of scepticism as to the future development of physical and biological sciences. Such scepticism would, indeed, be far from the mind of a physicist at a time when the very recognition of the limited character of our most fundamental concepts has resulted in such far-reaching developments of our science. Neither has the necessary renunciation as regards an explanation of life itself been a hindrance to the wonderful advances which have been made in recent times in all branches of biology and have, not least, proved so beneficial in the art of medicine. Even if we cannot make a sharp distinction on a physical basis between health and disease, there is, in particular, no room for scepticism as regards the solution of the important problems which occupy this Congress, as long as one does not leave the highroad of progress, that has been followed with so great success ever since the pioneer work of Finsen, and which has as its distinguishing mark the most intimate combination of the study of the medical effects of light treatment with the investigation of its physical aspects.

The Second International Congress for the Unity of Science was held in Copenhagen 21–26 June 1936. The photograph shows the opening of the congress at the Carlsberg honorary mansion, Niels Bohr's home. Among the people present are: 1. Otto von Neurath (sociologist), 2. Nicholas Rashevsky (sociologist), 3. Carl Hempel (philosopher), 4. Piet Hein (author), 5. Franz From (psychologist), 6. Pascual Jordan (physicist), 7. Karl Popper (philosopher), 8. Harald Isenstein (sculptor), 9. Jørgen Jørgensen (philosopher), 10. Hildegard Isenstein (wife of Harald I.), 11. Max Delbrück (physicist and biologist), 12. Philipp Frank (physicist and philosopher), 13. Ragnar Spärck (zoologist), 14. Frithiof Brandt (philosopher), 15. Harald Bohr (mathematician), 16. George de Hevesy (chemist and physicist), 17. Hanna Adler (physicist and educationalist, Niels Bohr's maternal aunt), 18. Edgar Rubin (psychologist), 19. Niels Bohr.

II. CAUSALITY AND COMPLEMENTARITY

Phil. Sci. **4** (1937) 289–298

Address at the Second International Congress for the Unity of Science
in Copenhagen, 21–26 June 1936

See Introduction to Part I, p. [20].

CAUSALITY AND COMPLEMENTARITY (1936)

Versions published in English, German and Danish

English: Causality and Complementarity
A   Phil. Sci. **4** (1937) 289–298

German: Kausalität und Komplementarität
B   Erkenntnis **6** (1937) 293–303

Danish: Kausalitet og Komplementaritet
C   Naturens Verden **21** (1937) 113–122

All of these versions agree with each other.

Philosophy of Science

VOL. 4   July, 1937   NO. 3

Causality and Complementarity[1]

BY

NIELS BOHR

ON SEVERAL occasions[2] I have pointed out that the lesson taught us by recent developments in physics regarding the necessity of a constant extension of the frame of concepts appropriate for the classification of new experiences leads us to a general epistemological attitude which might help us to avoid apparent conceptual difficulties in other fields of science as well. Since, however, the opinion has been expressed from various sides that this attitude would appear to involve a mysticism incompatible with the true spirit of science, I am very glad to use the present opportunity of addressing this assembly of scientists working in quite different fields but united in their striving to find a common ground for our knowledge, to come back to this question, and above all to try to clear up the misunderstandings which have arisen.

Before entering into the problems to be discussed, I need recall only briefly how often the development of physics has taught us that a consistent application even of the most elementary concepts indispensable for the description of our daily experience, is based on assumptions initially unnoticed, the explicit consideration of which is, however,

---

[1] Address delivered before the Second International Congress for the Unity of Science, Copenhagen, June, 1936.

[2] "Atomic Theory and Description of Nature", four essays and an introductory survey; Cambridge, 1934; quoted in the text as A_I, A_{II}, A_{III}, A_{IV}, and A_E; further, "Light and Life" Nature 131, 421; 457, 1933; and "Can Quantum Mechanical Description of Physical Reality be Considered Complete?" Physical Review LXVIII, 696, 1935; quoted as B and C respectively.

[39]

essential if we wish to obtain a classification of more extended domains of experience as clear and as free from arbitrariness as possible. I also hardly need to emphasize how much this development has contributed to the general philosophical clarification of the presuppositions underlying human knowledge. Even though these acquisitions are in many respects of a lasting character, we have nevertheless received only recently an incisive admonition that the analysis of new experiences is liable to disclose again and again the unrecognized presuppositions for an unambiguous use of our most simple concepts, such as space-time description and causal connection.

It was in fact the clarification of the paradoxes connected with the finite velocity of propagation of light and the judgment of events by observers in relative motion which first disclosed the arbitrariness contained even in the concept of simultaneity, and thereby created a freer attitude toward the question of space-time coördination which finds expression in the theory of relativity. As is well known, this has made possible a unified formulation of the phenomena appearing in different frames of reference, and through this brought to light the fundamental equivalence of hitherto separate physical regularities. The recognition of the essential dependence of any physical phenomenon on the system of reference of the observer, which forms the characteristic feature of relativity theory, implies, however—as especially Einstein himself has emphasized—no abandonment whatever of the assumption underlying the ideal of causality, that the behavior of a physical object relative to a given system of coördinates is uniquely determined, quite independently of whether it is observed or not.

However, a still further revision of the problem of observation has since been made necessary by the discovery of the universal quantum of action, which has taught us that the whole mode of description of classical physics, including the theory of relativity, retains its adequacy only as long as all quantities of action entering into the description are large compared to Planck's quantum. When this is not the case, as in the region of atomic physics, there appear new uniformities which cannot be fitted into the frame of the ordinary causal description (A_r). This circumstance, at first sight paradoxical, finds its elucidation in the recognition that in this region it is no longer possible sharply to distinguish between the autonomous behavior of a physical object and its inevitable interaction with other bodies serving as measuring instruments, the direct consideration of which is excluded by the very nature of the concept of observation in itself (A_{II}).

Indeed this circumstance presents us with a situation concerning the analysis and synthesis of experience which is entirely new in physics and forces us to replace the ideal of causality by a more general viewpoint usually termed "complementarity." The apparently incompatible sorts of information about the behavior of the object under examination which we get by different experimental arrangements can clearly not be brought into connection with each other in the usual way, but may, as equally essential for an exhaustive account of all experience, be regarded as "complementary" to each other. In particular, the frustration of every attempt to analyse more closely the "individuality" of single atomic processes, symbolized by the quantum of action, by a subdivision of their course, is explained by the fact that each section in this course definable by a direct observation would demand a measuring arrangement which would be incompatible with the appearance of the uniformities considered. Notwithstanding all differences, a certain analogy between the postulate of relativity and the point of view of complementarity can be seen in this, that according to the former the laws which in consequence of the finite velocity of light appear in different forms depending on the choice of the frame of reference, are equivalent to one another, whereas, according to the latter the results obtained by different measuring arrangements apparently contradictory because of the finite size of the quantum of action, are logically compatible.

In order to give as clear an idea as possible of the new epistomological situation which we meet in atomic physics, we may briefly consider those measurements designed to obtain an account of the space-time course of some physical event. The account consists in the last analysis in the establishment of a series of unambiguous connections between the behavior of the object and the measuring rods and clocks which define the system of reference involved in the space-time description. It is thus only as long as we may completely ignore, in the description of all the important circumstances of the event, all interaction between the object and these measuring instruments, which unavoidably accompanies the establishment of any such connection, that we can speak of an autonomous space-time behavior of the object under observation, independent of the conditions of observation. In case, however—as in the region of quantum phenomena—this interaction plays an essential rôle for the appearance of the phenomena themselves, the situation is completely changed, and we are in particular forced to renounce the combination, characteristic of classical physical description, of the space-time coördination of the event with the general conservation

[41]

theorems of dynamics.   For the use of rods and clocks to fix the system of reference makes it by definition impossible to take into account the energy of momentum which might be transferred to them in the course of the phenomenon.   Conversely, those quantum laws whose formulation rests essentially on the application of the concept of energy or momentum can appear only under circumstances of investigation from which a detailed account of the space-time behavior of the object is excluded.

As is well known, a mode of description suitable to this situation has been found in the so-called quantum mechanics, in which sufficient freedom for the consistent coördination of the new regularities has been achieved by the substitution for the usual kinematical and dynamical quantities of symbols which obey laws of calculation of a novel type. There is also from the point of view an interesting formal analogy between quantum mechanics and the theory of relativity, in that it has been possible in both cases with the help of abstract concepts of arithmetic and geometry respectively, to build up strictly logical formalisms which allow a mastering of the new domains of experience.   In connection with the often discussed question whether such formalisms can be regarded as an extension of our power of visualization, it must not be forgotten that the representation of the coördination of space and time in the theory of relativity by a four dimensional manifold, as also the connecting of kinematic and dynamic quantities in quantum mechanics by non-commutative algebra, rest essentially on the old mathematical artifice of the introduction of imaginary quantities; in fact the fundamental constants, the velocity of light and the quantum of action, are introduced into the formalism as factors of the $\sqrt{-1}$, the one in the definition of the fourth coördinate, the other in the commutation laws of canonically conjugate variables.

It is of course not my intention here to go deeper into such special points; I wished only to emphasize that in these fields the logical correlations can only be won by a far-reaching renunciation of the usual demands of visualization.   It would in particular not be out of place in this connection to warn against a misunderstanding likely to arise when one tries to express the content of Heisenberg's well known indeterminacy relations—which play as important a rôle in the judgment of the consistency of the essentially statistical mode of description of quantum mechanics as the Lorentz transformation does in solving the paradoxes which appear in the theory of relativity—by such a statement as: "the position and momentum of a particle cannot simultaneously be measured with arbitrary accuracy." According to such a formulation it would appear as though we had to do with some arbitrary renunciation of the measurement of either the one or the other of the two well-defined attributes of the object, which would not preclude the possibility of a future theory taking both attributes into account on the lines of the classical physics. From the above considerations it should be clear that the whole situation in atomic physics deprives of all meaning such inherent attributes as the idealizations of classical physics would ascribe to the object. On the contrary, the proper rôle of the indeterminacy relations consists in assuring quantitatively the logical compatibility of apparently contradictory laws which appear when we use two different experimental arrangements, of which only one permits an unambiguous use of the concept of position, while only the other permits the application of the concept of momentum defined as it is, solely by the law of conservation.

We thus see that the impossibility of carrying through a causal representation of quantum phenomena is directly connected with the assumptions underlying the use of the most elementary concepts which come into consideration for the description of experience. In this connection the view has been expressed from various sides that some future more radical departure in our mode of description from the concepts adapted to our daily experience would perhaps make it possible to preserve the ideal of causality also in the field of atomic physics. Such an opinion would, however, seem to be due to a misapprehension of the situation. For the requirement of communicability of the circumstances and results of experiments implies that we can speak of well defined experiences only within the framework of ordinary concepts. In particular it should not be forgotten that the concept of causality underlies the very interpretation of each result of experiment, and that even in the coördination of experience one can never, in the nature of things, have to do with well-defined breaks in the causal chain. The renunciation of the ideal of causality in atomic physics which has been forced on us is founded logically only on our not being any longer in a position to speak of the autonomous behavior of a physical object, due to the unavoidable interaction between the object and the measuring instruments which in principle cannot be taken into account, if these instruments according to their purpose shall allow the unambiguous use of the concepts necessary for the description of experience. In the last resort an artificial word like "complementarity" which does not belong to our daily concepts serves only briefly to remind us of the epistemological situation here encountered, which at least in physics is of an entirely novel character (A_E).

The repeatedly expressed hopes of avoiding the essentially statistical character of quantum mechanical description by the assumption of some causal mechanism underlying the atomic phenomena and hitherto inaccessible to observation would indeed seem to be as vain as any project of doing justice to the increased profundity of the picture of the world achieved by the general theory of relativity by means of the ordinary conceptions of absolute space and time. Above all such hopes would seem to rest upon an underestimate of the fundamental differences between the laws with which we are concerned in atomic physics and the every day experiences which are comprehended so completely by the ideas of classical physics. Not only is the well known dilemma between the corpuscular and undulatory character of light and matter avoidable only by means of the viewpoint of complementarity, but the peculiar stability properties of atomic structures which are in obvious contrast with the properties of any mechanical model, but which are so intrinsically connected with the existence of the quantum of action, form the very condition for the existence of the objects and measuring instruments, with the behavior of which classical physics is concerned. On closer consideration, the present formulation of quantum mechanics in spite of its great fruitfulness would yet seem to be no more than a first step in the necessary generalization of the classical mode of description, justified only by the possibility of disregarding in its domain of application the atomic structure of the measuring instruments themselves in the interpretation of the results of experiment. For a correlation of still deeper lying laws of nature involving not only the mutual interaction of the so-called elementary constituents of matter but also the stability of their existence, this last assumption can no longer be maintained, as we must be prepared for a more comprehensive generalization of the complementary mode of description which will demand a still more radical renunciation of the usual claims of so-called visualization.

I hope by these remarks to have conveyed the impression that in abandoning the causal description in atomic physics we are not concerned with a hasty assertion of the impossibility of comprehending the wealth of phenomena, but with a serious effort to account for the new type of laws here encountered in conformity with the general lesson of philosophy regarding the necessity of a balance between analysis and synthesis. Just in this connection it appeared to me to be of interest to point out that also in other regions of human knowledge we meet apparent contradictions which might seem to be avoidable only from the point of view of complementarity. I am far from sharing, however, the widespread opinion that the recent development in the field of atomic physics could directly help us in deciding such questions as "mechanism or vitalism" and "free will or causal necessity" in favor of the one or the other alternative. Just the fact that the paradoxes of atomic physics could be solved not by a one sided attitude towards the old problem of "determinism or indeterminism," but only by examining the possibilities of observation and definition, should rather stimulate us to a renewed examination of the position in this respect in the biological and psychological problems at issue.

In the first place, regarding the question of the extent to which we can hope to explain the characteristic features of living organisms with the sole help of the experience acquired from the study of inanimate nature, we must above all keep in mind that even a definition of life itself contains epistemological problems. When we usually refer to a machine as dead, we mean scarcely anything else than that we are able to describe the circumstances essential for its functioning by means of the conceptions of classical physics. Still in view of the insufficiency of the classical mode of description in atomic physics, such a definition of the inanimate would hardly any longer be adequate. Yet the newly recognized possibility of inducing macroscopic effects by individual atomic process, which plays an essential part in the functioning of organisms—in any case for the sensitiveness of sense perceptions (Aiv)—has been an incentive to the taking up anew of the question of a possible "explanation" of life. But at the same time the recognition of the fact that we must descend to the domain of atomic phenomena if we wish to bridge the gulf between the living and the inanimate, should bring before our eyes in a forceful way the practical and conceptual difficulties connected with this problem.

So far as we are at all in a position to follow the behavior of atoms in organisms under similar conditions of investigation as in the fundamental experiments of atomic physics, of course we can only meet with the laws disclosed by these experiments which, in spite of their feature of individuality, foreign to classical mechanics, can give us clearly no immediate understanding of the so-called holistic or finalistic characteristics of the activities of life. The only logical possibility of avoiding any contradiction between the formulation of the laws of physics and the concepts suitable for the description of the phenomena of life ought therefore to be sought in the essentially different character of the conditions of investigation concerned. On a previous occasion (B) I have tried to express this situation by saying that every experimental arrangement suitable for following the behavior of the atoms constituting an organism in as exhaustive a way as implied by the possibilities of physical observation and definition would be incompatible with the maintaining of the life of the organism. This would in fact be quite analogous to the circumstance that all observations obtained by experimental arrangements which allow of a space-time account of the behavior of the constituents of atoms and molecules stand in a complementary relation to those obtained under conditions permitting the study of the intrinsic stability of atomic structures so essential for the physical and chemical properties of matter.

To make this view clearer, it was pointed out in the article cited, that the continuous metabolism of organisms inseparably connected with life prevents us even from distinguishing strictly which atoms belong to a living organism, and that we are thus presented with a problem the treatment of which, quite apart from its complication, is beyond the scope of the methods of atomic mechanics. These methods, which govern our entire knowledge of physics and chemistry concern, just as do those of classical mechanics, in fact only systems for which it is possible in principle to specify what are to be regarded as the elementary constituents. This situation suggests that those essential features of living organisms which are brought to light only under circumstances which exclude an exact account of their atomic constituents are laws of a nature which stands in a complementary relationship to those with which we are concerned in physics and chemistry. Thus the existence of life itself would have to be regarded in biology, both as regards the possibilities of observation and of definition, as no more subject to analysis than the existence of the quantum of action in atomic physics.

I have endeavored to make it clear that in such considerations there is no question whatever—as has been sometimes feared by philosophers and biologists—of so-called purely metaphysical speculations or of an arbitrary renunciation of the possibility by continued research, of further increasing our knowledge of the functioning of organisms. Rather, they aim at avoiding futile controversies by an analysis of the presuppositions and of the appropriateness of the conceptual structures involved. Though the viewpoint of complementarity rejects every compromise with any anti-rationalistic vitalism, it ought at the same time to be suited for revealing certain prejudices in so-called mechanism. On the one hand, any violation of physico-chemical laws in organic life—such as the often mistakenly maintained contradiction between the activities of life and the fundamental theorems of thermo-dynamics —would be excluded from this point of view; on the other hand any insistence on an analogy between the existence of life itself and such laws should be rejected as irrational. As already emphasized in the article mentioned, this situation therefore implies no limitation whatever in the application to biology of the physico-chemical methods of description and investigation; in fact, the appropriate use of such methods—just as even in atomic physics all our experiences must rest upon experimental arrangements classically described—remains our sole and inexhaustible source of information about biological phenomena.

According to its tendency to make room for the phenomena of life within the conceptions suited to the description of material systems, the viewpoint discussed stands far removed from every attempt to exploit in a spiritual sense the failure of causal description in atomic physics. On the contrary, the viewpoint toward fundamental biological questions which we have here discussed, would rather seem suited to put the old problem of psycho-physical parallelism in a new light. The considerations which I have presented on previous occasions (A_{III}, B) on questions of psychology in connection with problems of atomic physics followed indeed two essentially different aims. The one was by means of well-known examples of the difficulties of analysis and synthesis of psychic phenomena connected with introspection to remind ourselves that in this region of knowledge we had already been forced to face a situation presenting in several respects a formal similarity with that with which, to the great disquietude of many physicists and philosophers, we have met in atomic physics. The other aim was to express the hope that the epistemological attitude which had led to the clarification of the much simpler physical problems could prove itself helpful also in the discussion of psychological questions. In fact, the use which we make of words like "thought" and "feeling," or "instinct" and "reason" to describe psychic experiences of different types, shows the existence of characteristic relationships of complementarity conditioned by the peculiarity of introspection. Above all, just the impossibility in introspection of sharply distinguishing between subject and object as is essential to the ideal of causality would seem to provide the natural play for the feeling of free will.

I am afraid that the short indications to which I have been obliged to restrict myself with respect to the last and many other points of this lecture will remind you only too well that in the last resort the direct use of any word must stand in a complementary relationship to an analysis of its meaning.   I hope, however, that I have to some extent succeeded in giving you the impression that my attitude is in no way in conflict with our common endeavors to arrive at as great a unification of knowledge as possible by the combating of prejudices in every field of research.

Copenhagen.

III. BIOLOGY AND ATOMIC PHYSICS

"Celebrazione del secondo centenario della nascita di Luigi Galvani",
Bologna – 18–21 ottobre 1937-XV: I. Rendiconto generale, Tipografia Luigi
Parma 1938, pp. 68–78

Address delivered in Bologna on 19 October 1937 at the conference
celebrating the Bicentennial of Luigi Galvani's birth

See General Introduction, p. xxxviii and Introduction to Part I, pp. [9] and [14].

[49]

BIOLOGY AND ATOMIC PHYSICS (1937)

Versions published in English, Danish and German

English: Biology and Atomic Physics

A "Celebrazione del secondo centenario della nascita di Luigi Galvani", Bologna – 18–21 ottobre 1937-XV: I. Rendiconto generale, Tipografia Luigi Parma 1938, pp. 68–78

B Nuovo Cimento **15** (1938) 429–438

C "Atomic Physics and Human Knowledge", John Wiley & Sons, New York 1958, pp. 13–22 (reprinted in: "Essays 1933–1957 on Atomic Physics and Human Knowledge, The Philosophical Writings of Niels Bohr, Vol. II", Ox Bow Press, Woodbridge, Connecticut 1987, pp. 13–22)

Danish: Biologi og Atomfysik

D Naturens Verden **22** (1938) 433–443

E "Atomfysik og menneskelig erkendelse", J.H. Schultz Forlag, Copenhagen 1957, pp. 23–33

German: Biologie und Atomphysik

F "Atomphysik und menschliche Erkenntnis", Friedr. Vieweg & Sohn, Braunschweig 1958, pp. 13–22

Versions *A*, *B*, *D*, *E* and *F* agree with each other, apart from some spelling mistakes in *B* (even though *A* was indeed published first).

C contains several improvements of formulation in relation to *A*.

[50]

CELEBRAZIONE

DEL SECONDO CENTENARIO

DELLA NASCITA DI

LUIGI GALVANI

BOLOGNA - 18-21 OTTOBRE 1937-XV

I

RENDICONTO GENERALE

REDATTO DAL SEGRETARIO DEL COMITATO

PROF. G. C. DALLA NOCE

BOLOGNA
TIPOGRAFIA LUIGI PARMA - VIA 3 NOVEMBRE 7
1938 - XVI

[51]

Biology and atomic Physics. Relazione del Professore Niels Bohr

The immortal work of GALVANI which inaugurated a new epoch in the whole field of science is a most brilliant illustration of the extreme fruitfulness of an intimate combination of the exploration of the laws of inanimate nature with the study of the properties of living organisms. At this occasion, it may therefore be fitting to review the attitude which scientists through the ages have taken to the

[52]

question of the relationship between physics and biology and especially to discuss the outlook created in this respect by the extraordinary development of atomic theory in recent time (1).

From the very dawn of science, atomic theory has indeed been at the focus of interest in connection with the efforts to attain a comprehensive view of the great variety of natural phenomena. Thus already DEMOCRITUS, who with so deep intuition emphasized the necessity of atomism for any rational account of the ordinary properties of matter, attempted, as is well-known, also to utilize atomistic ideas for the explanation of the peculiarities of organic life and even of human psychology. In view of the fantastic character of such extreme materialistic conceptions, it was quite a natural reaction when a man like ARISTOTLE, with his masterly comprehension of the knowledge of his time in physics as well as in biology, chose to reject atomic theory entirely and tried to provide a sufficiently broad frame for an account of the wealth of natural phenomena on the basis of essentially teleological ideas. The exaggeration of the Aristotelian doctrine, on its side, was, however, clearly brought to light by the gradual recognition of elementary laws of nature valid as well for inanimate bodies as for living organisms.

When thinking of the establishment of the principles of mechanics, which were to become the very foundations of physical science, it is, in this connection, not without interest to realize that ARCHIMEDES' discovery of the principle of equilibrium of floating objects, which, according to a familiar tradition, was suggested to him by the sensation of uplift of his own body in a bath tub, might just as well have been based on common experience regarding the loss of weight of stones in water. Likewise it is to be regarded as quite accidental that GALILEI was led to the recognition of the fundamental laws of dynamics by observing the pendulum motion of a chandelier in the beautiful cathedral of Pisa, and not by looking at a child in a swing. Yet such purely external analogies were, of course, only of little weight for the growing appreciation of the essential unity of the principles governing natu-

---

(1) For previous discussions of the latter point, cf. *Nature*, *131*, 421 (1933), and *Philosophy of Science*, *4*, 289 (1937); or in German *Die Naturwissenschaften*, *13*, 245 (1933), and *Erkenntnis*, *XIV*, 293 (1936).

[53]

ral phenomena, as compared to the deeprooted similarities between living organisms and technical machinery that were disclosed by the studies of anatomy and physiology, pursued so intensely at the time of the Renaissance especially here in Italy.

The enthusiasm for the prospects opened by the success of the new experimental approach to natural philosophy — encouraged in equal manner by the widening of the world picture due to the vision of COPERNICUS and by the elucidation of circulation mechanisms in animal bodies, initiated by HARVEY's great achievement — found perhaps its most striking expression in the work of BORELLI, who succeeded to clarify in so fine detail the mechanical function of skeleton and muscles in animal motion. The classical character of this work is in no way impeded by attempts of BORELLI himself and his followers also to explain nervous action and glandular secretion by means of primitive mechanical models, the obvious arbitrariness and coarseness of which soon gave rise to general criticism, still remembered by the semi-ironical name of « iatro-physicists » attached to the Borellian school. Likewise the endeavours, sound in their root, to apply the growing knowledge of typically chemical transformations of matter to physiological processes, which found so enthusiastic an exponent in SYLVIUS, rapidly led, by exaggerations of superficial resemblances of digestion and fermentation with the simplest inorganic reactions and their rash application to medical purposes, to an opposition which has found its expression in the labelling of such premature endeavours as « iatro-chemistry ».

To us the reasons for the shortcomings of these pioneer efforts to utilize physics and chemistry for a comprehensive explanation of the properties of living organisms are evident. Not only had one to wait until LAVOISIER's time for the disclosure of the elementary principles of chemistry, which were to give the clue to the understanding of respiration and later to provide the basis for the extraordinary development of so-called organic chemistry, but, before GALVANI's discoveries, a whole fundamental aspect of the laws of physics lay still hidden. It is most suggestive to think that the germ which, in the hands of VOLTA, OERSTED, FARADAY, and MAXWELL, was to develop into a structure rivalling Newtonian mechanics in importance, grew out of researches with a biological aim. In fact, it is difficult to imagine that the progress from experiments with electrically charged bodies, however fruitful in FRANKLIN's hands, to the study of galvanic currents could have been achieved if the sensitive instruments necessary for the detection of such currents, afterwards so readily constructed, had not been provided by nature itself in the nervous fabric of higher animals.

It is impossible here to sketch, even in outline, the tremendous development of physics and chemistry since the days of GALVANI, or to enumerate the discoveries in all branches of biology in the last century. We need only recall the lines leading from the pioneer work, in this venerable university, of MALPIGHI and SPALLANZANI to modern embryology and bacteriology respectively, or from GALVANI himself to the recent fascinating researches on nerve impulses, about which we look forward to learn presently from a most competent side. In spite of the far-reaching understanding, thus obtained, of the physical and chemical aspect of many typical biological reactions, the marvellous fineness of structure of the organisms and their wealth of interconnected regulation mechanisms goes still so far beyond any experience about inanimate nature that we feel as removed as ever from an explanation of life itself on such lines. Indeed, when we witness the passionate scientific controversies as regards the bearing on this problem of the recent discoveries of poisoning effects and generative properties of so-called viruses, we find ourselves presented with a dilemma just as acute as that with which DEMOCRITUS and ARISTOTLE were confronted.

In this situation it is again around atomic theory that interest is concentrated, although on a very different background. Not only has this theory, since DALTON applied with such decisive success atomistic conceptions to the elucidation of the quantitative laws governing the constitution of chemical compounds, become the indispensable foundation and never-failing guide of all reasoning in chemistry; but the wonderful refinement of experimental technique in physics has even given us the means of studying phenomena which directly depend on the action of individual atoms. At the same time as this development has thus removed the last traces of the traditional prejudice that, due to the coarseness of our senses, any proof of the actual existence of atoms would for ever remain beyond the reach of human experience, it has revealed still deeper features of atomicity in the laws of nature than those expressed by the

[55]

old doctrine of the limited divisibility of matter. We have indeed
been taught that the very conceptual frame, appropriate both to
give account of our experience in everyday life and to formulate
the whole system of laws applying to the behaviour of matter in
bulk and constituting the imposing edifice of so-called classical
physics, has to be essentially widened if it is to comprehend proper
atomic phenomena. In order to appreciate the possibilities which
this new outlook in natural philosophy provides with respect to a
rational attitude towards the fundamental problems of biology, it
will, however, be necessary to recall briefly the principal lines of
the development which has led to the elucidation of the situation
in atomic theory.

The starting-point of modern atomic physics was, as is well-
known, the recognition of the atomic nature of electricity itself,
first indicated by FARADAY's famous researches on galvanic electrolysis and definitely established by the isolation of the electron
in the beautiful phenomena of electric discharges through rarefied gases, which attracted so much attention towards the end of
the last century. While J. J. THOMSON's brilliant researches soon
brought to light the essential part played by electrons in most varied physical and chemical phenomena, our knowledge of the structural units of matter was, however, not completed until RUTHER-
FORD's discovery of the atomic nucleus, crowning his pioneer work
on the spontaneous radioactive transmutations of certain heavy elements. Indeed, this discovery offered for the first time an unque-
stionable explanation of the invariability of the elements in ordinary chemical reactions, in which the minute heavy nucleus remains
unaltered, while only the distribution of the light electrons around
it is affected. Moreover, it provides an immediate understanding
not only of the origin of natural radioactivity, in which we witness
an explosion of the nucleus itself, but also of the possibility, subsequently discovered by RUTHERFORD, of inducing transmutations of
elements by bombardment with high-speed heavy particles, which
in colliding with the nuclei may cause their disintegration.

It would carry us too far from the subject of this address to
enter here further upon the wonderful new field of research opened by the study of nuclear transmutations, which will be one of
the main subjects of discussion among physicists at this meeting.
The essential point for our argument is indeed not to be found in

such new experience but in the obvious impossibility to account for common physical and chemical evidence on the basis of the well-established main features of RUTHERFORD's atomic model without departing radically from the classical ideas of mechanics and electromagnetism. In fact, notwithstanding the insight provided by Newtonian mechanics into the harmony of planetary motions expressed by the Keplerian laws, the stability properties of mechanical models like the solar system which, when disturbed have no tendency to return to their original state, have clearly no sufficient resemblance with the intrinsic stability of the electronic configurations of atoms that is responsible for the specific properties of the elements. Above all, this stability is strikingly illustrated by spectral analysis which, as is well-known, has revealed that any element possesses a characteristic spectrum of sharp lines, independent of the external conditions to such an extent that it offers a means of identifying the material composition of even the most remote stars by spectroscopic observations.

A clue to the solution of this dilemma was, however, already provided by PLANCK's discovery of the elementary quantum of action which was the outcome of a very different line of physical research. As is well-known, PLANCK was led to this fundamental discovery by his ingenious analysis of just such features of the thermal equilibrium between matter and radiation which, according to the general principles of thermodynamics, should be entirely independent of any specific properties of matter, and accordingly of any special ideas on atomic constitution. The existence of such an elementary quantum of action expresses in fact a new trait of individuality of physical processes which is quite foreign to the classical laws of mechanics and electromagnetism and limits their validity essentially to those phenomena which involve actions large compared to the value of a single quantum, as given by PLANCK's new atomistic constant. This condition, though amply fulfilled in the phenomena of ordinary physical experience, does in no way hold for the behaviour of electrons in atoms, and it is indeed only the existence of the quantum of action which prevents the fusion of the electrons and the nucleus into a neutral massive corpuscle of practically infinitesimal extension.

The recognition of this situation suggested at once the description of the binding of each electron in the field around the nucleus

[57]

as a succession of individual processes by which the atom is trans-
fered from one of its so-called stationary states to another of these
states, with emission of the released energy in the form of a single
quantum of electromagnetic radiation. This view, intimately akin
to EINSTEIN's successful interpretation of the photoelectric effect,
and borne out so convincingly by the beautiful researches of
FRANCK and HERTZ on the excitation of spectral lines by impacts of
electrons on atoms, did in fact not only provide an immediate explanation of the puzzling general laws of line spectra disentangled
by BALMER, RYDBERG, and RITZ, but, with the help of spectroscopic evidence, led gradually to a systematic classification of the
types of stationary binding of any electron in an atom, offering a
complete explanation of the remarkable relationships between the
physical and chemical properties of the elements, as expressed in
the famous periodic table of MENDELEIEV. While such an interpretation of the properties of matter appeared as a realisation, even
surpassing the dreams of the Pythagoreans, of the ancient ideal
of reducing the formulation of the laws of nature to considerations
of pure numbers, the basic assumption of the individuality of the
atomic processes involved at the same time an essential renunciation of the detailed causal connection between physical events,
which through the ages had been the unquestioned foundation of
natural philosophy.

Any question of a return to a mode of description consistent
with the principle of causality was, however, not only excluded by
unambiguous experience of the most varied kind, but it soon proved
possible to develop the original primitive attempts of taking account of the existence of the quantum of action in atomic theory
into a proper, essentially statistical atom mechanics, fully comparable in consistency and completeness with the structure of classical mechanics of which it appears as a rational generalisation.
The establishment of this new so-called quantum mechanics which,
as is well-known, we owe above all to the ingenious contributions
of the younger generation of physicists has, indeed, quite apart
from its astounding fruitfulness in all branches of atomic physics
and chemistry, essentially clarified the epistemological basis of the
analysis and synthesis of atomic phenomena. The revision of the
very problem of observation in this field, initiated by HEISENBERG,
one of the principal founders of quantum mechanics, has in fact

led to the disclosure of hitherto disregarded presuppositions for the unambiguous use of even the most elementary concepts on which the description of natural phenomena rests. The critical point is here the recognition that any attempt to analyse, in the customary way of classical physics, the « individuality » of atomic processes, as conditioned by the quantum of action, will be frustrated by the unavoidable interaction between the atomic objects concerned and the measuring instruments indispensable for that purpose.

An immediate consequence of this situation is that observations regarding the behaviour of atomic objects obtained with different experimental arrangements cannot in general be combined in the usual way of classical physics. In particular, any imaginable procedure aiming at the coordination in space and time of the electrons in an atom will unavoidably involve an essentially uncontrollable exchange of momentum and energy between the atom and the measuring agencies, entirely annihilating the remarkable regularities of atomic stability for which the quantum of action is responsible. Conversely, any investigation of such regularities, the very account of which implies the conservation laws of energy and momentum, will on principle impose a renunciation as regards the space-time coordination of the individual electrons in the atom. Far from being inconsistent, the aspects of quantum phenomena revealed by experience obtained under such mutually exclusive conditions must thus be considered as complementary to each other in quite a novel way. The view-point of « complementarity » does, indeed, in no way mean an arbitrary renunciation as regards the analysis of atomic phenomena, but is on the contrary the expression of a rational synthesis of the wealth of experience in this field, which exceeds the limits to which the application of the concept of causality is naturally confined.

Notwithstanding the encouragement given to the pursuit of such inquiries by the great example of relativity theory which, just through the disclosure of unsuspected presuppositions for the unambiguous use of all physical concepts, opened new possibilities for the comprehension of apparently irreconcilable phenomena, we must realize that the situation met with in modern atomic theory is entirely unprecedented in the history of physical science. Indeed, the whole conceptual structure of classical physics, brought to so wonderful a unification and completion by EINSTEIN's work,

[59]

rests on the assumption, well adapted to our daily experience of physical phenomena, that it is possible to discriminate between the behaviour of material objects and the question of their observation. For a parallel to the lesson of atomic theory regarding the limited applicability of such customary idealisations, we must in fact turn to quite other branches of science, such as psychology, or even to that kind of epistemological problems with which already thinkers like BUDDHA and LAO TSE have been confronted, when trying to harmonize our position as spectators and actors in the great drama of existence. Still, the recognition of an analogy in the purely logical character of the problems which present themselves in so widely separated fields of human interest does in no way imply acceptance in atomic physics of any mysticism foreign to the true spirit of science, but on the contrary it gives us an incitation to examine whether the straightforward solution of the unexpected paradoxes met with in the application of our simplest concepts to atomic phenomena might not help us to clarify conceptual difficulties in other domains of experience.

There has also been no lack of suggestions to look for a direct correlation between life or free will and those features of atomic phenomena for the comprehension of which the frame of classical physics is obviously too narrow. In fact, it is possible to point out many characteristic features of the reactions of living organisms, like the sensitivity of visual perception or the induction of gene mutation by penetrating radiation, which undoubtedly involve an amplification of the effects of individual atomic processes, similar to that on which the experimental technique of atomic physics is essentially based. Still, the recognition that the fineness of organisation and regulation mechanisms of living beings goes even so far beyond any previous expectation, does in itself in no way enable us to account for the peculiar characteristics of life. Indeed, the so-called holistic and finalistic aspects of biological phenomena can certainly not be immediately explained by the feature of individuality of atomic processes disclosed by the discovery of the quantum of action; rather would the essentially statistical character of quantum mechanics at first sight seem even to increase the difficulties of understanding the proper biological regularities. Just in this dilemma, however, the general lesson of atomic theory suggests that the only way to reconcile the laws of physics with the concepts suited for a description of the phenomena of life is to examine the essential difference in the conditions of the observation of physical and biological phenomena.

First of all we must realize that every experimental arrangement with which we could study the behaviour of the atoms constituting an organism to the extent to which this can be done for single atoms in the fundamental experiments of atomic physics will exclude the possibility of maintaining the organism alive. The incessant exchange of matter which is inseparably connected with life will even imply the impossibility of regarding an organism as a well-defined system of material particles like the systems considered in any account of the ordinary physical and chemical properties of matter. In fact, we are led to conceive the proper biological regularities as representing laws of nature complementary to those appropriate to the account of the properties of inanimate bodies, in analogy with the complementary relationship between the stability properties of the atoms themselves and such behaviour of their constituent particles as allows of a description in terms of space-time coordination. In this sense, the existence of life itself should be considered, both as regards its definition and observation, as a basic postulate of biology, not susceptible of further analysis, in the same way as the existence of the quantum of action, together with the ultimate atomicity of matter, forms the elementary basis of atomic physics.

It will be seen that such a view-point is equally removed from the extreme doctrines of mechanism and vitalism. On the one hand, it condemns as irrelevant any comparison of living organisms with machines, be these the relatively simple constructions contemplated by the old iatro-physicists, or most refined modern amplifier devices, the uncritical emphasis of which would expose us to deserve the nickname of « iatro-quantists ». On the other hand, it rejects as irrational all such attempts at introducing some kind of special biological laws inconsistent with well-established physical and chemical regularities, as have in our days been revived under the impression of the wonderful revelations of embryology regarding cell growth and division. In this connection it must be especially remembered that the possibility of avoiding any such inconsistency within the frame of complementarity is given by the very fact that no result of biological investigation can be unambiguously described otherwise than in terms of physics and chemistry, just as any account of experience even in atomic physics must ultimately rest on the use of the concepts indispensable for a conscious recording of sense impressions.

The last remark brings us back into the realm of psychology, where the difficulties presented by the problems of definition and observation in scientific investigations have been clearly recognized long before such questions became acute in natural science. Indeed, the impossibility in psychical experience to distinguish between the phenomena themselves and their conscious perception clearly demands a renunciation of a simple causal description on the model of classical physics, and the very way in which words like « thoughts » and « feelings » are used to describe such experience reminds one most suggestively of the complementarity encountered in atomic physics. I shall not here enter into any further detail but only emphasize that it is just this impossibility of distinguishing, in introspection, sharply between subject and object, which provides the necessary latitude for the manifestation of volition. To connect free will more directly with limitation of causality in atomic physics, as it is often suggested, is, however, entirely foreign to the tendency underlying the remarks here made about biological problems.

In concluding this address I hope that the temerity of a physicist venturing so far outside his restricted domain of science may be forgiven in view of the most welcome opportunity of profitable discussion offered to physicists and biologists by this gathering to honour the memory of the great pioneer to whose fundamental discoveries both branches of science owe so much.

IV. ANALYSIS AND SYNTHESIS IN SCIENCE

International Encyclopedia of Unified Science **1** (1938) 28

See Introduction to Part I, p. [12].

Reprinted for private circulation from
ENCYCLOPEDIA AND UNIFIED SCIENCE, FOUNDATIONS OF THE UNITY OF SCIENCE
Vol. I, No. 1, INTERNATIONAL ENCYCLOPEDIA OF UNIFIED SCIENCE
PRINTED IN THE U.S.A.

Analysis and Synthesis in Science

Niels Bohr

Notwithstanding the admittedly practical necessity for most scientists to concentrate their efforts in special fields of research, science is, according to its aim of enlarging human understanding, essentially a unity. Although periods of fruitful exploration of new domains of experience may often naturally be accompanied by a temporary renunciation of the comprehension of our situation, history of science teaches us again and again how the extension of our knowledge may lead to the recognition of relations between formerly unconnected groups of phenomena, the harmonious synthesis of which demands a renewed revision of the presuppositions for the unambiguous application of even our most elementary concepts. This circumstance reminds us not only of the unity of all sciences aiming at a description of the external world but, above all, of the inseparability of epistemological and psychological analysis. It is just in the emphasis on this last point, which recent development in the most different fields of science has brought to the foreground, that the program of the present great undertaking distinguishes itself from that of previous encyclopedic enterprises, in which stress was essentially laid on the completeness of the account of the actual state of knowledge rather than on the elucidation of scientific methodology. It is therefore to be hoped that the forthcoming *Encyclopedia* will have a deep influence on the whole attitude of our generation which, in spite of the ever increasing specialization in science as well as in technology, has a growing feeling of the mutual dependency of all human activities. Above all, it may help us to realize that even in science any arbitrary restriction implies the danger of prejudices and that our only way of avoiding the extremes of materialism and mysticism is the never ending endeavor to balance analysis and synthesis.

V. MEDICAL RESEARCH AND NATURAL PHILOSOPHY

Acta Medica Scandinavica (Suppl.) **142** (1952) 967–972

Address delivered at the Second International Poliomyelitis Conference
in Copenhagen on 3 September 1951

See Introduction to Part I, p. [12].

REPRINT

ACTA MEDICA
SCANDINAVICA

SUPPLEMENTUM CCLXVI (266)

PAPERS ON MEDICINE
AND
THE HISTORY OF MEDICINE

Dedicated to

ERIK WARBURG, M. D.

PROFESSOR OF INTERNAL MEDICINE
IN THE UNIVERSITY OF COPENHAGEN, DENMARK,

ON HIS SIXTIETH ANNIVERSARY
FEBRUARY 3, 1952

ACCOMPANIES VOL CXLII (142)

COPENHAGEN 1952

[66]

MEDICAL RESEARCH
AND NATURAL PHILOSOPHY
By
Niels Bohr.

Invited to contribute to this volume in honour of a great master in medical science, I have felt some difficulty since, as a physicist, I have of course no expert information to offer as regards the causes and treatment of diseases. I must therefore confine myself to some general remarks on the relationship between medical research and natural philosophy suggested by the modern development of physical science.[1])

―――――――――

From early days, medical science has been closely related to many other aspects of knowledge and philosophy, and mankind still remembers with veneration the teaching of the great *Hippocrates* as regards the duty of doctors not only to keep abreast with practical developments in their field, but also to enlighten their minds through the study of general human problems in order to be able properly to fulfil their important mission.

It is this spirit which lies behind the development of anatomy and physiology in the Renaissance, when the great pioneers, like *Vesallius* and *Harvey,* were at the same time highly honoured doctors who were given most responsible tasks in the society of those days. Fundamental progress in physics at that time, which led to the clarification of the principles of mechanics, gave a great impetus to the development of medical science. Still classical within their scope are studies of *Borelli,* inspired by his close contact with *Galileo,* of the functioning of skeleton and muscles.

Through the ages, contact with the medical sciences has also been a decisive stimulus to the development of physics and chemistry. The fundamental role which electricity plays in natural phenomena might for

―――――――――

[1]) These remarks form the substance of an address delivered at the Second International Poliomyelitis Conference in Copenhagen, September 1951.

967

[67]

long have remained hidden if nature had not presented us, in the nervous system of the higher organisms, with such refined and sensitive apparatus. Indeed, *Galvani*'s discoveries were not only basic for the development of neurophysiology, but even paved the way for the discoveries of electrochemistry and electromagnetism. We also remember how *Robert Mayer* was led to the idea of energy conservation, so central in physical science, by observations during his surgical operations in tropical climates.

In return, physical and chemical sciences have offered more and more effective tools and fruitful viewpoints for the exploration of the structure and functioning of the organisms. At the same time, however, this exploration has revealed an ever increasing variety of possibilities for regulation and adaptation, which again and again has struck us with wonder and led to a renewal of the old debate of the attitude to take to the explanation of life, or rather of the rational description of the position of living beings among the natural phenomena.

According to the weight which in such discussions has been laid on words like cause and purpose, two viewpoints, characterized as mechanicism and vitalism, have often stood sharply against each other. Mechanistic viewpoints have especially prevailed at times when the development of physics and chemistry has given new insight into organic functions, while vitalistic viewpoints have been revived when attention was focused on new discoveries regarding the surprising resources of living organisms. The problem of finding a rational and consistent way to orientate ourselves in this situation reminds in many ways of the problems with which we have been confronted in modern physics through the exploration of the world of atoms, and which have given rise to a renewed discussion regarding the principles for the description and comprehension of physical phenomena.

Although is was understood already in ancient Greece that the regularities which, inspite of all variety, manifest themselves in the properties of material substances, suggest a limited divisibility of matter, it was up to our century believed that the coarseness of our senses would for ever prevent the observation of individual atoms. Still, the marvellous development of the art of experimentation, and especially of the technique of amplification, has allowed us to observe effects produced by single atoms and even to study the constitution of the atoms themselves.

The information which we have thus received is in many ways of a very simple character and, as everyone knows, our picture of the atom consists of a cluster of electrons held together by the attraction from a central charged nucleus. While, in many physical and chemical phenomena, the electrons may be displaced or even removed from the nucleus, the latter will, under ordinary physical conditions, including those existing in the organisms, remain uninfluenced. Moreover, the fact that the properties of an atom are to a high degree determined only by the electric charge of the nucleus — and not by its internal constitution — has made it possible by

968

application of so-called isotopic indicators to label the atoms and thereby study metabolic processes in far greater detail than was hitherto possible.

Notwithstanding the simplicity in such respects of our ideas concerning atomic constitution, it is equally clear that it is not possible, on the basis of the principles which have proved so fruitful in the description and comprehension of large scale physical phenomena, to account in detail for the properties of atoms. In particular, the stability of atomic systems presents us with an entirely new feature of physical regularities bound up with the universal quantum of action, discovered by *Planck* in the first years of this century.

This discovery has initiated a whole new epoch in natural philosophy in revealing that the classical principles of mechanics and electrodynamics are idealizations which hold only for phenomena where we have to do with actions which are large compared with the individual quantum. In atomic phenomena, where this condition is no longer fulfilled, we meet with regularities of a quite new kind, which reject pictorial interpretation and even defy a detailed causal description.

The clarification of the apparent paradoxes with which we are here confronted has demanded a revision of the principles for the comprehension of physical experience. A decisive point has been the recognition that, in atomic physics, we can no longer uphold the idea of a behaviour of objects, independent of the circumstances under which the phenomena are observed. It is here not a question of a practical limitation of the accuracy of measurements, but of an aspect of the laws of nature, associated with the quantum of action, which sets a lower limit to the interaction between the objects and the measuring instruments.

This very circumstance presents us with an entirely new situation, since any attempt to control the interaction between objects and measuring instruments will imply that the bodies so far used for fixing the experimental conditions will now themselves become objects under investigation. Additional measuring instruments with new uncontrollable interaction with the objects would therefore be demanded, and all which could be achieved will be the replacement of the original system by a new, more complicated one, for the description of which quite similar conditions will hold.

In this situation, it is not surprising that evidence obtained under different experimental arrangements cannot be combined in the accustomed manner. In particular, specific quantum processes cannot be represented as a continuous causal chain of events, since any subdivision into well-defined steps would demand a change in the experimental arrangement, which would be incompatible with the appearance of the phenomenon we want to study. We are here presented with a peculiar feature of wholeness in atomic processes, quite foreign to classical physics, and the manifestation of which is inherently associated with the latitude involved in the definition of the behaviour of atomic objects.

Apparently contrasting phenomena, appearing under different experimental conditions, are termed complementary in order to emphasize that only together they exhaust the knowledge obtainable about the objects. By the ingenious formalism of quantum mechanics, which abandons pictorial representation and aims directly at a statistical account of quantum processes, it has actually been possible along the lines of complementary description to bring logical order within a rich field of experience regarding fundamental physical and chemical properties of matter.

This whole development has forcefully reminded us that even principles so well established by earlier experience as the causal description in classical physics may fail to comprehend new fields of experimental evidence. Far from implying an arbitrary renunciation on detailed scientific explanation, the concept of complementarity points to a widening of our conceptual framework demanded when dealing with experience where the interaction between the objects under investigation and the tools of observation cannot be eliminated.

When, on this background, we return to the basic biological problems, we must realize that the stability of the atoms composing the tissues, and not least the fine molecular structures in the cell with which the hereditary properties of the organism are associated, bring us right into the domain of quantum physics. In this connection, the statistical laws governing the occurrence of individual quantum processes have, as is well known, found interesting and promising application in the interpretation of experience regarding artificial mutations. Also the fineness of our senses, like visual perception, has been found to go down to the atomic level, and we must assume that amplification processes analogous to those applied in the registration of atomic phenomena play a decisive role in the mechanism of nervous messages.

These remarks, however, do in no way imply that, even if quantum theory is indispensable for dealing with biological phenomena, it should in itself suffice for an explanation of life. On the contrary, the point which I want to emphasize is the wider implication of the lesson atomic physics has taught us about our position as observers of natural phenomena and, in particular, about the rational use of words like cause and purpose.

Although science will of course strive for ever more detailed knowledge of the physical mechanism underlying the functions of organisms, a description of life corresponding to the ideal of mechanism will only constitute one line of approach. In fact, we must recognize that experimental conditions demanded for an exhaustive description conforming with this ideal would involve a control of the organism to an extent which would preclude the display of life. In actual biological research, a vitalistic approach is equally indispensable, since the primary object must often be the studies of the reaction of the organism as a whole for the purpose of upholding life, a point of which we are not least reminded in medical research.

970

We are here neither speaking of any crude attempt of tracing an analogy to life in simple machinery, nor of the old idea of a mystic life force, but of two scientific approaches which only together exhaust the possibilities of increasing our knowledge. In this sense, mechanistic and vitalistic viewpoints may be considered as complementary, and in the harmonious balance between their applications we find the basis for the practical and rational use of the word life. Such an attitude implies no limit to the obtainable knowledge of the detailed structure of the organism nor to our hopes of exploiting our potentialities and combating diseases.

It is true that, up to this stage, we have as always in physical science considered ourselves merely as observers, and this attitude applies equally to classical and to quantum physics. It may be stressed, however, that the concept of complementarity is also suited to characterize essential aspects of the consciousness associated with our life. I do not only think of the unity of the personality which presents a similarity to the wholeness of the organism, but also of many features of introspection which strikingly remind of the conditions for comprehension of experience in atomic physics.

Corresponding to the relationship between object and measuring instruments in quantum phenomena, we have, in the various psychical experiences, to do with a different placing of the line of section between the object on which attention is focused and the observing and judging subject. Any attempt to avoid this section by including subject as well as object in the consciousness would clearly imply the introduction of a new subject to keep account of the original one. If in this situation the normal balance is lost, symptoms well known in psychology and psychiatry and referred to as confusion of the "ego"s and dissolution of personality may occur.

A clear example of psychological situations of mutually exclusive character we meet in the problem of the free will, which through the ages has given rise to eager debate. The decisive point is here the recognition that a state of consciousness, in the description of which words like "I will" find application, is complementary to a state where we are concerned with an analysis of motives for our actions. It is on this background that we may speak in a rational manner about volition as an indispensable element in any attempt at an exhaustive description of consciousness.

Just as no one would attempt to avoid the use of the word life, notwithstanding all difficulty in distinguishing sharply between inanimate and living, so will hardly anyone be able to deny that in speaking of freedom of will we are expressing ourselves as adequately as verbal communication allows. In this connection, we must recognize that, in all use of language, the practical application of any word is complementary to attempts at its strict definition.

Rather than with sophistical argumentation we are here concerned with sober analysis of the proper use of words in the description of our position as actors as well as spectators in the drama of existence. Indeed, how shall

971

we decide whether a man can do something because he will or whether he wants to do it because he is able to. In particular, how can we, in thinking of the mastership of *Erik Warburg* in the art of medicine, separate his ability, based on scientific insight, to assist ailing fellow-men in restoring their strength and regaining their vitality, from the power of his versatile and vigorous personality to stimulate their will to live and with his help to recover their health.

VI. ADDRESS AT THE OPENING CEREMONY

Acta Radiologica (Suppl.) **116** (1954) 15–18

Address delivered at the Opening Ceremony of the Seventh International
Congress of Radiology in Copenhagen on 19 July 1953

See Introduction to Part I, p. [12].

Seventh International Congress of Radiology

COPENHAGEN, DENMARK

19th — 24th July 1953

Patron:

HIS MAJESTY KING FREDERIK IX

Honorary President:

HIS EXCELLENCY PROFESSOR NIELS BOHR

Danish Executive Committée

PRESIDENT:

Prof. P. FLEMMING MØLLER

VICE-PRESIDENTS:	ADMINISTRATIVE ADVISER:
Dr. SV. A. CHROM (Diagnosis)	Mr. POUL HARRIS
Dr. JENS NIELSEN (Therapy)	LEGAL ADVISER:
Prof. CARL KREBS (Biology)	Mr. KAI STORM
Prof. H. M. HANSEN (Physics)	TREASURER:
Dr. E. DE FINE LICHT (Technology)	Mr. AAGE VON BENZON

SECRETARY-GENERAL:

Prof. FLEMMING NØRGAARD

ADDRESS AT THE OPENING CEREMONY

by

the Honorary President His Excellency Professor
Niels Bohr

It is an honour and pleasure to me to follow the invitation of the Organization Committee to say some words at this opening meeting of The Seventh International Congress of Radiology. Still, feeling most acutely my lack of special knowledge and experience in radiative diagnosis and therapy, it is with no small hesitation that I venture to address this assembly of prominent radiologists, and I shall confine myself to some general remarks pointing to the multifarious fields of knowledge relevant for the progress and success of your great task in the service of mankind.

I need hardly recall how already in FINSEN's pioneer work on the curative effect of rays from the sun and the electric arc on skin diseases refined physical technique proved of decisive importance. Above all, however, it was the discovery by RÖNTGEN of the penetrating radiation accompanying electric discharges in evacuated containers, which gave us the means of so-to-say looking through our bodies and of combatting diseases of the internal organs. A further great advance was the discovery of natural radio-active substances and, in particular, should Madame CURIE's isolation of radium so largely improve the means for the radiative treatment of malignant tumours.

In our days, physical technology has provided radiologists with ever more effective tools. On the one hand, the construction of high tension generators and special discharge tubes allow the convenient and

[75]

finely regulated application of very penetrating radiation. On the other hand, the progress of nuclear physics, which has created the possibility of releasing atomic energy on a large scale, supplied us with an abundance of new radioactive substances suited to treatments of very different kind.

At the same time, the study of the physico-chemical effects of radiation on the organisms is rapidly proceeding. In this connection, reference is hardly necessary to the ingenious method of labelling the atoms by the use of isotopic indicators, which was originally conceived and developed by HEVESY, and which offers so wide perspectives as regards detailed insight into the metabolism of organic matter under the most various circumstances.

During the Congress, all these problems will certainly be much debated and elucidated from expert side. At this occasion I shall, therefore, only remind of the new lesson regarding our position as observers of natural phenomena, which the development of atomic physics has impressed upon us, and point to the bearing of this lesson on our attitude to fundamental problems in biological and medical sciences.

This development originates from PLANCK's epoch-making discovery of the elementary quantum of action which disclosed an unsuspected limitation of the ideas of natural philosophy relied upon since the birth of science. In fact, we have learned that the accustomed pictorial representation of physical events applies only to phenomena in the analysis of which all actions involved are large compared with the individual quantum. In elementary atomic processes, however, we meet with a novel feature of wholeness which even prevents unambiguous separation between the behaviour of the objects under investigation and their interaction with the measuring instruments necessary for the fixation of the circumstances under which the phenomena appear.

In this situation, it is not surprising that atomic phenomena, observed under different experimental conditions, cannot be combined in a single picture and, at first sight, may even appear as contrasting with each other. Still, such phenomena represent equally important parts of the accessible evidence regarding the objects and are therefore adequately referred to as complementary. Far from implying any arbitrary renunciation on detailed analysis of physical phenomena, this viewpoint of complementarity represents a wider frame for the description, allowing us to embrace regularities beyond the scope of ordinary physical explanation.

Through the development of ingenious methods, adapted to the complementary mode of description, it has actually proved possible to attain a rational generalization of the so-called classical physical theories. These methods do not only remove the paradoxes involved in the attempts of picturing the course of radiative processes, but permit us also to account for the remarkable stability of atomic structures, exhibited not least by the complicated molecules of which the tissues in living organisms are built up.

With these remarks it is in no way meant to imply that, in atomic physics, we possess a clue to the explanation of life, but only that we have gained an insight into the circumstances under which organic life displays itself. Of course, the problems with which we are here confronted are immeasurably more complex than the elementary atomic processes where we are dealing with phenomena occurring under well defined and reproducible conditions. Indeed, in physical phenomena, we perceive in many ways greater simplicity the more the analysis proceeds, while the structure of the living organisms and their reactions appear ever more complicated with the advance of biological science.

The main point, however, is that the lesson which atomic research has given us naturally influences our attitude towards the study of the entities which the living organisms represent. Although no ultimate limit is in sight for the knowledge we may gradually reach about the physical and chemical processes responsible for the organic functions, we must all the time be aware that the conditions for obtaining such knowledge may be incompatible with the sustainment of the life of the organism. At any rate in biological, and not least in medical research, the actual approach to the problems must take the entity of the organism into account, and progress depends on upholding the proper balance between so-called mechanistic and vitalistic argumentations which, in a certain sense, are complementary to each other.

I hope it will be understood that the tendency of such utterances is equally remote from mysticism and sophistication and that they aim only at a sober logical description of the situation. Especially, I think that we all agree that, in the art of medicine, we are not attempting at repairing the organism like technicians who replace some outworn piece of machinery, but at establishing proper conditions for enabling the organism to restore itself and regain health.

In concluding, I wish to join our president in expressing the hope that this Congress will prove a landmark in the advance of radiology. We may not least base this hope on the inspiration the Congress will offer its members by the opportunity of exchanging knowledge and viewpoints derived from their individual experience in the many fields of research which comprise such rich promises for furthering the great goal of radiology.

VII. UNITY OF KNOWLEDGE

"The Unity of Knowledge" (ed. L. Leary), Doubleday & Co.,
New York 1955, pp. 47–62

Address delivered at a conference celebrating the Bicentennial of
Columbia University, New York, on 28 October 1954

See General Introduction, pp. XXXVIII f.

[79]

UNITY OF KNOWLEDGE (1954)

Versions published in English, Danish and German

English: Unity of Knowledge
A   *Science and the Unity of Knowledge*, in "The Unity of Knowledge" (ed. L. Leary), Doubleday & Co., New York 1955, pp. 47–62
B   "Atomic Physics and Human Knowledge", John Wiley & Sons, New York 1958, pp. 67–82 (reprinted in: "Essays 1933–1957 on Atomic Physics and Human Knowledge, The Philosophical Writings of Niels Bohr, Vol. II", Ox Bow Press, Woodbridge, Connecticut 1987, pp. 67–82)

Danish: Kundskabens Enhed
C   "Atomfysik og menneskelig erkendelse", J.H. Schultz Forlag, Copenhagen 1957, pp. 83–99
D   "Naturbeskrivelse og menneskelig erkendelse" (eds. J. Kalckar and E. Rüdinger), Rhodos, Copenhagen 1985, pp. 19–39

German: Einheit des Wissens
E   "Atomphysik und menschliche Erkenntnis", Friedr. Vieweg & Sohn, Braunschweig 1958, pp. 68–83

   B is an improved version of *A* (see next page).
   B, *C*, *D* and *E* agree with each other.

[80]

CHANGES FROM VERSION *A* TO VERSION *B*

The title is changed to *Unity of Knowledge* and the numbering of the four main sections is omitted. Otherwise, the following non-trivial changes have been made (page and line numbers refer to version *A*):

Page	Line	Description of change
53	33	The word "logical" is omitted in front of "impossibility".
54	11	The rest of the sentence after "experience" is omitted.
55	33	"reminding us of the recording devices used in experiments of atomic physics" is omitted.
57	31	The rest of the sentence after "formalism" is omitted.
59	2	After "definable" is inserted "separately".
59	28	"the renunciation" is replaced with "the ever more extensive renunciation".
60	37	The rest of the sentence following "epistemological lesson" is omitted.
61	34	The sentence "When we compare ..." is changed to: "In comparing different cultures resting on traditions fostered by historical events, we meet with the difficulty of appreciating the culture of one nation on the background of traditions of another".
62	9	The first sentence in section IV is replaced with the following two sentences: "In concluding this address, I feel that I ought to apologize for speaking on such general topics with so much reference to the special field of knowledge represented by physical science. I have tried, however, to indicate a general attitude suggested by the serious lesson we have in our day received in this field and which to me appears of importance for the problem of unity of knowledge".

Principal Contributors

Felice Battaglia

Niels Bohr

Detlev W. Bronk

Robert L. Calhoun

Etienne Gilson

Julian Huxley

Frank H. Knight

Alfred L. Kroeber

Harry Levin

Sir Richard Livingstone

Archibald MacLeish

Charles H. Malik

Gardner Murphy

Willard Van Orman Quine

John Herman Randall, Jr.

John von Neumann

Robert Penn Warren

Hermann Weyl

5

Science and the Unity of Knowledge

NIELS BOHR

Director of the Institute for Theoretical Physics
University of Copenhagen

Before trying to answer the question of to what extent we may speak of unity of knowledge, we must consider the meaning of the word knowledge itself. It is not my intention to enter directly into an academic philosophical discourse for which I hardly possess the required scholarship. Every scientist, however, is constantly confronted with the problem of objective description, of communication in unambiguous terms. Our basic tool is plain language, which serves the needs of practical life and social intercourse. We shall not be concerned here with origins of such language, but with its scope in scientific communication, and especially with the problem of how objectivity of description may be retained during the growth of experience beyond events of daily life.

It is important, first, to realize that all knowledge is originally represented within a conceptual framework adapted to account for previous experience, and that any such frame may prove too narrow to comprehend new experiences. Scientific research in many domains of knowledge has indeed, time after time, proved the necessity of abandoning or remolding viewpoints which, due to their fruitfulness and apparently unrestricted applicability, were regarded as indispensable for rational explanation. Although such developments have been initiated by special studies, they entail a general lesson of importance for the problem of unity of knowledge. In fact, the widening of the conceptual framework has not only served to restore order within the respective branches of knowledge, but also has disclosed analogies in our position as regards analysis and synthesis of experience in apparently

47

separate domains of knowledge, suggesting the possibility of an ever more embracing objective description.

When speaking of a conceptual framework, we merely refer to an unambiguous logical representation of relations between experience. This attitude is also apparent in the historical development in which formal logic is no longer sharply distinguished from studies of semantics or even philological syntax. A special role is played by mathematics which has contributed so decisively to the development of logical thinking, and by its well-defined abstractions offers invaluable help in expressing harmonious relationships. Still, in our discussion we shall not consider pure mathematics as a separate branch of knowledge, but rather as a refinement of general language, supplementing it with appropriate tools for representing relations for the communication of which ordinary verbal expression is either not sufficiently sharp or becomes too cumbersome. In this connection it may be stressed that, in avoiding reference to the conscious subject with which daily language is so largely infiltrated, the use of mathematical symbols serves to secure the unambiguity of definition required for objective description.

In the development of the so-called exact sciences, characterized by the establishment of numerical relationships between measurements, it has indeed been of decisive importance to be able to take recourse to abstract mathematical methods often developed without reference to such applications, but originating from detached pursuit of generalizing logical constructions. This situation is especially illustrated in physics, which was originally understood as all knowledge concerning that nature of which we ourselves are part, but which gradually came to mean the study of the elementary laws governing the properties of inanimate matter. The necessity, even within this comparatively simple theme, to pay constant attention to the problem of objective description has deeply influenced the attitude of philosophical schools through the ages. In our day, the exploration of new fields of experience has disclosed unsuspected presuppositions for the unambiguous application of some of our most elementary concepts and thereby has given us an epistemological lesson with bearings on problems far beyond the domain of physical science. It may be convenient therefore to start with a brief account of this development.

I

It would carry us too far to recall in detail how, with the elimination of mythical cosmological ideas and arguments referring to purposes of our own actions, a consistent scheme of mechanics was built up on the basis of Galileo's pioneering work, and how it reached completion through Newton's mastership. Above all, the principles of Newtonian mechanics meant a far-reaching clarification of the problem of cause and effect in permitting, through observation of the state of a physical system at a given instant, defined by measurable quantities, prediction of its state at any subsequent time. It is well known how a deterministic or causal account of this kind led to the mechanical conception of nature and came to stand as an ideal of scientific explanation in all domains of knowledge, irrespective of the way experience is obtained. It is therefore important to recall that the study of wider fields of physical experience has revealed the necessity of a closer consideration of the observational problem.

Within its large field of application, classical mechanics presents an objective description in the sense that it is based on a well-defined use of pictures and ideas referring to events in daily life. Still, however rational the idealizations used in Newtonian mechanics might appear, they actually went far beyond the range of experience to which our elementary concepts are adapted. Thus, the adequate use of the very notions of absolute space and time is inherently connected with the practically instantaneous propagation of light, which allows us to locate the bodies around us independently of their velocities and to arrange events in a unique time sequence. The attempt at developing a consistent account of electromagnetic and optical phenomena revealed, however, that observers moving relative to each other with great velocities will coordinate events differently. Such observers may, in fact, not only take a different view of shapes and positions of rigid bodies, but events at separate points of space, which to one observer appear as simultaneous, may to another be judged as occurring at different times.

Far from giving rise to confusion and complication, the exploration of the extent to which the account of physical phenomena depends on the standpoint of the observer proved an invaluable guide in tracing general physical laws common to all observers. Retaining the idea of determinism, but relying only on relations between unambiguous measurements referring ultimately to coincidences of events, Albert Einstein succeeded in remolding and generalizing the whole edifice of classical physics, and in lending to our world picture a unity surpassing all previous expectations. In the general theory of relativity, the description is based on a curved four-dimensional space-time metric which automatically accounts for gravitational effects and the singular role of the speed of light signals representing an upper limit for any consistent use of the physical concept of velocity. The introduction of such unfamiliar but well-defined mathematical abstractions in no way implies any ambiguity. It offers, rather, an instructive illustration of how a widening of the conceptual framework affords the appropriate means of eliminating subjective elements and enlarging the scope of objective description.

New unsuspected aspects of the observational problem should be disclosed by the exploration of the atomic constitution of matter. The ideas of a limited divisibility of substances, introduced to explain the persistence of their characteristic properties in spite of the variety of natural phenomena, go back to antiquity. Still, almost to our day, such views were regarded as essentially hypothetical in the sense that they seemed inaccessible to direct confirmation by observation, because of the coarseness of our sense organs and of the tools we use, which themselves are composed of innumerable atoms. Nevertheless, with the great progress in chemistry and physics in the last centuries, atomic ideas proved increasingly fruitful, and it was found possible to obtain a general understanding of the principles of thermodynamics through a direct application of classical mechanics to the interaction of atoms and molecules in incessant motion.

In this century, the study of newly discovered properties of matter, like natural radioactivity, has convincingly confirmed the fundaments of atomic theory. In particular, through the development of amplification devices, it has been possible to study phenomena essentially dependent on single atoms, and even to obtain extensive knowledge of the structure of atomic systems. The first step was the recognition of the electron as a common constituent of all substances, and an essential completion of our ideas of atomic constitution was obtained by Ernest Rutherford's discovery of the atomic nucleus containing almost the whole mass of the atom within an extremely small volume. As a result of this, the invariability of the elements in ordinary physical and chemical processes is directly explained by the circumstance that in such processes, although the electron binding may be largely influenced, the nucleus remains unaltered. With his demonstration of the transmutability of atomic nuclei by more powerful agencies, Rutherford, moreover, opened a new field of research, often referred to as modern alchemy, which eventually should lead to the possibility of releasing immense amounts of energy stored in atomic nuclei.

Although many fundamental properties of matter were explained by this simple picture of the atom, it was from the beginning evident that classical ideas of mechanics and electromagnetism did not suffice to account for the essential stability of atomic structures exhibited by the specific properties of the elements. A clue to the elucidation of this problem was afforded, however, by the discovery of the universal quantum of action to which Max Planck was led in the first year of our century by his penetrating analysis of the laws of thermal radiation. This discovery revealed a feature of wholeness in atomic processes quite foreign to the mechanical conception of nature, and made it evident that the classical physical theories are idealizations valid only in the description of phenomena, in the analysis of which all actions are sufficiently large to permit the neglect of the quantum. While this condition is amply fulfilled in phenomena on the ordinary scale, we meet in atomic phenomena regularities of quite a new kind, defying deterministic pictorial description.

A rational generalization of classical physics, allowing for the existence of the quantum but retaining the unambiguous interpretation of experimental evidence defining inertial mass and electric charge of the electron and the nucleus, presented a very difficult task. By concerted efforts of a whole generation of theoretical physicists, a consistent and, within a wide scope, exhaustive description of atomic phenomena, however, was gradually developed. This description makes use of a mathematical formalism in which the variables in the classical physical theories are replaced by symbols subject to a non-commutable algorism involving Planck's constant. Due to the very character of such mathematical abstractions, the formalism does not allow for pictorial interpretation on accustomed lines, but aims directly at establishing relations between observations obtained under well-defined conditions. Corresponding to the circumstance that different individual quantum processes may take place in a given experimental arrangement, these relations are of an inherently statistical character.

By means of the quantum mechanical formalism, a detailed account of an immense amount of experimental evidence regarding the physical and chemical properties of matter has been achieved. Moreover, by adapting the formalism to the exigencies of relativistic invariance, it has been possible to a large extent to order the rapidly growing new experience concerning the properties of elementary particles and the constitution of atomic nuclei. Notwithstanding, the astounding power of quantum mechanics, the radical departure from accustomed physical explanation, and especially the renunciation of the very idea of determinism, have given rise to doubts in the minds of many physicists and philosophers as to whether we are here dealing with a temporary procedure of expediency or whether we are confronted with an irrevocable step as regards objective description. The clarification of this problem has actually demanded a radical revision of the foundation for the description and comprehension of physical experience.

Therefore we must above all recognize that even when the phenomena transcend the scope of classical physical theories, the account of the experimental arrangement and the recording of the observations must be given in plain language, suitably supplemented by technical physical terminology. This is a clear logical demand since the very word experiment refers to a situation where we can tell others what we have done and what we have learned. The fundamental difference between the analysis of the phenomena in classical and in quantum physics is that in the former the interaction between the objects and the measuring instruments may be neglected or compensated for, while in the latter this interaction forms an integral part of the phenomena. The essential wholeness of a proper quantum phenomenon finds logical expression in the circumstance that any attempt at its well-defined subdivision would require a change in the experimental arrangement incompatible with the appearance of the phenomenon itself.

The impossibility of a separate control of the interaction between the atomic objects and the instruments indispensable for the definition of the experimental conditions prevents in particular the unrestricted combination of space-time coordination and dynamical conservation laws on which the deterministic description in classical physics rests. In fact, any unambiguous use of the concepts of space and time refers to an experimental arrangement involving a transfer of momentum and energy, uncontrollable in principle, to the instruments—like fixed scales and synchronized clocks—required for the definition of the reference frame. Conversely, the account of phenomena governed by conservation of momentum and energy involves, in principle, a renunciation of detailed space-time coordination. These circumstances find quantitative expression in Heisenberg's indeterminacy relations which specify the reciprocal latitude for the fixation of kinematical and dynamical variables in the definition of the state of a physical system. In accordance with the character of the quantum mechanical formalism such relations cannot be interpreted, however, in terms of attributes of objects referring to classical pictures. We are here dealing with the mutually exclusive conditions for the unambiguous use of the very concepts of space and time, on the one hand, and of dynamical conservation laws, on the other.

In this context, one sometimes speaks of the "disturbance of phenomena by observation" or the "creation of physical attributes to atomic objects by measurements." Such phrases, however, are apt to cause confusion, since words like phenomena and observation, and attributes and measurements, are here used in a way incompatible with common language and practical definition. In objective description, it is indeed more appropriate to use the word phenomenon only to refer to observations obtained under specified circumstances, including an account of the whole experimental arrangement. In this terminology, the observational problem in quantum physics is deprived of any special intricacy and we are directly reminded that every well-defined atomic phenomenon is closed in itself, since its observation implies a permanent mark on a photographic plate left by the impact of an electron or similar recordings obtained by suitable amplification devices of essentially irreversible functioning. Moreover, it is important to realize that the quantum mechanical formalism allows of well-defined applications only when referring to phenomena closed in such sense, and also that in this respect it represents a rational generalization of classical physics in which every stage of the course of events is described by measurable quantities.

The freedom of experimentation presupposed in classical physics of course is retained and corresponds to the free choice of experimental arrangements for which the mathematical structure of the quantum mechanical formalism offers the appropriate latitude. The circumstance that, in general, one and the same experimental arrangement may yield different recordings, is sometimes picturesquely described as a "choice of nature" between such possibilities. Needless to say, one is here not in any way alluding to a personification of nature, but rather pointing to the logical impossibility of ascertaining directives on accustomed lines for the course of a closed indivisible phenomenon. As regards such directives, logical approach cannot go beyond the deduction of the relative probabilities for the appearance of the individual phenomena under given experimental conditions, and in this respect quantum mechanics fulfills all requirements as a consistent generalization of deterministic mechanical description which it embraces as an asymptotic limit in the case of physical phenomena on a scale sufficiently large to neglect the quantum of action.

A most conspicuous characteristic of atomic physics is the novel relationship between phenomena observed under experimental conditions demanding different elementary concepts for their description. Indeed, however contrasting such phenomena might appear when attempting to picture a course of atomic processes on classical lines, they have to be considered as complementary in the sense that they represent equally essential aspects of well-defined knowledge about atomic systems and together exhaust this knowledge. Far from indicating a departure from our position as detached observers, the notion of complementarity represents the logical expression for our situation as regards objective description in this field of experience, which has demanded a renewed revision of the foundation for the unambiguous use of our elementary concepts. The recognition that the interaction between the measuring tools and the physical systems under investigation constitute an integral part of quantum phenomena has not only revealed an unsuspected limitation of the mechanical conception of nature characterized by attribution of separate properties to physical systems, but has forced us in the ordering of experience to pay proper attention to the conditions of observation.

Returning to the much debated question of what has to be demanded of a physical explanation, it must be borne in mind that classical mechanics already implied the renunciation of cause of uniform motion and especially that relativity theory has taught us how arguments of invariance and equivalence must be counted as categories of rational explanation. Similarly, in the complementary description of quantum physics, we have to do with a further self-consistent generalization which permits us to include regularities decisive for the account of fundamental properties of matter but transcending the scope of deterministic description. The history of physical science thus demonstrates how the exploration of ever wider fields of experience, in revealing unsuspected limitations of accustomed ideas, indicates new ways of restoring logical order. The epistemological lesson contained in the development of atomic physics reminds us of similar situations in description and comprehension of experience far beyond the borders of physical science, and allows us to trace common features promoting the search for unity of knowledge.

II

The first problem with which we are confronted when leaving the proper domain of physics is the question of the place of living organisms in the description of natural phenomena. Originally, no sharp distinction between animate and inanimate matter was made, and it is well known that Aristotle, in stressing the wholeness of individual organisms, opposed the views of the atomists, and even in the discussion of the foundations of mechanics retained ideas like purpose and potency. However, with the great discoveries in anatomy and physiology at the time of the Renaissance, and especially with the advent of classical mechanics in deterministic description in which any reference to purpose is eliminated, a completely mechanistic conception of nature suggested itself, and a large number of organic functions could in fact be accounted for by the same physical and chemical properties of matter which found explanation in simple atomic ideas. It is true that the structure and functioning of organisms involve an ordering of atomic processes which has sometimes been felt difficult to reconcile with the laws of thermodynamics, implying a steady approach towards disorder among the atoms constituting an isolated physical system. If, however, sufficient account is taken of the circumstance that the free energy necessary to maintain and develop organic systems is continually supplied from their surroundings by nutrition and respiration, it becomes clear that there is here no question of any violation of general physical laws.

In the last few decades, great advances have been achieved in our knowledge of the structure and functioning of organisms, and it has become evident that quantum regularities in many respects here play a fundamental role. Not only are such regularities basic for the remarkable stability of the complex molecular structures which form the essential constituents of the cells responsible for the hereditary properties of the species, but research into mutations which can be produced by exposure of the organisms to penetrating radiation offer a striking application of the statistical laws of quantum physics. Also, the sensitivity of perceptive organs, so important for the integrity of the organisms, has been found to approach the level of individual quantum processes, and amplification mechanisms, reminding us of the recording devices used in experiments of atomic physics, play an important part especially in the transmission of nervous messages. This development has again, although in a novel manner, brought the mechanistic approach to biological problems to the foreground, but at the same time the question has become acute whether a comparison between organisms and highly complicated physical systems, like modern industrial constructions or electronic calculation machines,

offers the proper basis for an objective description of such self-regulating entities which living organisms present.

Returning to the general epistemological lesson which atomic
physics has given us, we must in the first place realize that the closed
processes studied in quantum physics are no direct analogue to biological functions for the sustaining of which a continual exchange of
matter and energy between the organism and the environments is required. Moreover, any experimental arrangement which would allow
a control of such functions to the extent demanded for their well-
defined description in physical terms, would be prohibitive to the free
display of life. This very circumstance suggests an attitude to the problem of organic life, providing appropriate balance between mechanistic and finalistic approach. In fact, just as the quantum of action in
the account of atomic phenomena appears as an element for which an
explanation is neither possible nor required, the notion of life is elementary in biological science where, in the existence and evolution of
living organisms, we are concerned with manifestations of possibilities
in that nature to which we belong rather than with outcomes of experiments which we can ourselves perform. Actually, we must recognize
that the requirements of objective description, at least in tendency are
fulfilled by the characteristic complementary way in which arguments
based on the full resources of physical and chemical science, and concepts directly referring to the integrity of the organism transcending
the scope of these sciences are practically used in biological research.
The main point is that only in the renunciation of an explanation of
life in the ordinary sense do we gain the possibility of taking its characteristics into account.

Of course, in biology just as in physics we retain our position as
detached observers and the question is only of the different conditions
for the logical comprehension of experience. This also applies to the
study of such innate and conditioned behavior of animals and man
to the account of which psychological concepts most readily lend
themselves. Even in an allegedly behavioristic approach, it is hardly
possible to avoid such concepts, and the very idea of consciousness
presents itself when we are dealing with behavior of so high a degree
of complexity that its description virtually involves introspection on
the side of the individual organism. We have here to do with mutually
exclusive applications of the words instinct and reason, illustrated by
the extent to which instinctive behavior is suppressed in human societies. Quite apart from the academic philosophical discussion of the

consciousness of other persons, presumed in all human communication, we surely meet in trying to account for the state of our mind and its ever greater difficulties in attaining observational detachment. Still, it is possible to uphold the requirements of objective description widely, even in human psychology. In this connection it is interesting to note that, while in the early stages of physical science one could rely directly on such features of daily life as permitted simple causal account, an essentially complementary description of the content of our mind has been used since the origin of languages. In fact, the rich terminology adapted to such communication does not point to an unbroken course of events, but rather to separate mutually exclusive experiences characterized by different placings of the separation between the content on which attention is focused and the background indicated by the word ourselves.

An especially striking example is offered in the relationship between situations where we are pondering on motives for our actions, and where we experience a feeling of volition. In normal life, such shifting of the separation is more or less intuitively recognized, but symptoms characterized as "confusion of the egos," which may lead to dissolution of the personality, are well known in psychiatry. The use of apparently contrasting attributes referring to equally important aspects of the human mind presents indeed a remarkable analogy to the situation in atomic physics, where complementary phenomena for their definition demand different elementary concepts. Above all, a comparison between conscious experiences and physical observations suggests itself by the circumstance that the very word conscious refers to experiences capable of being retained in the memory, like the permanent recordings of atomic phenomena. In such an analogy, the vagueness inherent in the idea of the subconsciousness corresponds to the impossibility of pictorial interpretation of the quantum mechanical formalism which, in principle, serves merely to order well-defined observations. Incidentally, medical use of psychoanalytical treatment in curing neurosis may be said to restore balance in the content of the memory of the patient by bringing him new conscious experience rather than by helping him to fathom the abysses of his subconsciousness.

From a biological point of view, we can hardly interpret the characteristics of psychical phenomena except by concluding that every conscious experience corresponds to a residual impression in the organism, amounting to an irreversible recording of the outcome of processes in the nervous system which are not open to introspection and hardly adapted to exhaustive definition by mechanistic approach. Certainly these recordings in which the interplay of numerous brain cells is involved are essentially different from such structures in the single cells which are connected with genetic reproduction. We need emphasize not only the usefulness of residual impressions in their influence on our reactions to subsequent stimuli, but also the importance of the fact that later generations are not encumbered by the actual experiences of individuals but merely rely on the reproduction of such properties of the organism as have proved serviceable for the collection and utilization of knowledge. In any attempt at pursuing such inquiry we must be prepared to meet with increasing difficulties at every step. It is suggestive that the simple concepts of physical science to an ever higher degree lose their immediate applicability the more we approach the features of living organisms related to the characteristics of our mind.

To illustrate such argument, we may briefly refer to the old problem of free will. From what has already been said, it is evident that the word volition is indispensable in an exhaustive description of psychical phenomena, but the problem is how far we can speak of a freedom to act according to our possibilities. As long as unrestricted deterministic views are held, the idea of such freedom is excluded. However, the general lesson of atomic physics—in particular of the limited scope of mechanistic description of biological phenomena—suggests that the potency of organisms to adjust themselves to environment includes the power of selecting the most appropriate way to make this adjustment. In view of the impossibility of judging such questions on a purely physical basis, it is most important to recognize that psychical experience may offer more pertinent information about the problem. The decisive point is that if we attempt to predict what another person will decide to do in a given situation, we must not only strive to know his whole background, including the story of his life in all respects which may have contributed to form his character, but we must realize that what we are ultimately aiming at is to put ourselves in his place. Of course, it is impossible to say whether a person wants to do something because he believes he can, or whether he can because he will, but it is hardly disputable that we have the feeling of being able to make the best out of the circumstances. From the point of view of objective description nothing can be added here or taken away. In this sense we may both practically and logically speak of freedom of will in a way which leaves the proper latitude for the use of words like responsibility and hope, which themselves are as little definable as other words indispensable to human communication.

The essence of such considerations is to point to the epistemological implications of the lesson regarding our observational position which the development of physical science has impressed upon us. In return for the renunciation of accustomed demands on explanation, it offers logical means of comprehending wider fields of experience, necessitating proper attention to the placing of the object-subject separation. Since, in philosophical literature, reference is sometimes made to different levels of objectivity or subjectivity or even reality, it may be stressed that the notion of an ultimate subject, as well as conceptions like realism and idealism, find no place in objective description as we have defined it, but this circumstance does not imply any limitation of the scope of such inquiry as that with which we are concerned.

III

Having touched upon some of the problems in science which relate to the unity of knowledge, I shall turn to the further question of whether there is a poetical or spiritual or cultural truth distinct from scientific truth. With all the reluctance of a scientist to enter such fields, I shall venture an attitude similar to that which I have already put forward. When we consider the relation between our means of expression and the field of experience with which we are concerned, we are indeed directly confronted with the relationship of science and art. The enrichment which art can give us originates in its power to remind us of harmonies beyond the grasp of systematic analysis. Literary, pictorial, and musical art may be said to form a sequence of modes of expression where the renunciation of definition characteristic of scientific communication permits us to allow fantasy a freer display. In poetry this purpose is achieved by a juxtaposition of words related to shifting observational situations, and which as a result emotionally unite manifold aspects of human knowledge.

Notwithstanding the inspiration required in all works of art, it may not be irreverent to remind ourselves that even at the climax of his work, the artist relies on the common human foundation on which we stand. We must realize especially that a word like improvisation, so readily on our tongue when speaking of artistic achievement, points to a feature essential in all communication. Not only are we more or less unaware in ordinary conversation of the verbal expressions we choose in communicating what is on our mind, but even in written papers, where we have the opportunity of reconsidering every word, the question of whether we shall allow it to stand or change it demands for its answer a final decision essentially equivalent to an improvisation. In the balance between seriousness and humor characteristic of all great works of art, we are reminded of complementary aspects conspicuous in children's play and no less appreciated in mature life. Indeed, if we endeavored always to speak quite seriously, we would run the risk of appearing very soon ridiculously tedious to our listeners and ourselves. But if we try to joke all the time, we rapidly bring ourselves and our listeners too—if the witticisms have point—to the desperate mood which Shakespeare with such genius has pictured for us in the role of the jesters in his immortal dramas.

In a comparison between science and art, we must not forget that in the former we have to do with systematic efforts at augmenting experience and developing appropriate concepts for its comprehension, resembling the carrying and fitting of stones to a building, while in the latter we are presented with more intuitive individual endeavors of evoking sentiments which recall the wholeness of our situation. We are here at a point where the question of the unity of knowledge contains ambiguity similar to that of the word truth itself—and this holds also for spiritual or cultural values. However, in regard to the communication of such values, we are reminded of epistemological problems related to the proper balance between our desire for an all-embracing way of looking at life in its multifarious aspects and our power of expressing ourselves in a logically consistent manner.

Essentially different starting points are taken here by science, which aims at the development of general methods for the ordering of common human experience, and religions, which originate in endeavors to further harmony of outlook and behavior within communities. Of course, in any religion, all knowledge shared by the members of the community was included in the general framework within which the values and the ideals emphasized in cult and faith constituted a primary content. The inherent relation between content and frame therefore hardly called for attention before the subsequent progress of science entailed a cosmological or epistemological lesson pointing beyond the common horizon at the time of the foundation of the religion. The course of history presents many illustrations of this. Outstanding among them was the schism between science and religion which accompanied the development of the mechanical conception of nature at the time of the European Renaissance. On the one hand, many phenomena hitherto regarded as manifestations of divine providence, appeared then as consequences of general immutable laws of nature. On the other hand, the physical methods and viewpoints of science were far remote from the emphasis on human values and ideals essential to religion. Common to the schools of so-called empirical and critical philosophy, an attitude therefore prevailed of a more or less vague distinction between objective knowledge and subjective belief.

In emphasizing the necessity of paying proper attention to the placing of the object-subject separation in unambiguous communication, the modern development of science has created a new basis for the use of such words as knowledge and belief. Above all, the recognition of inherent limitations in the notion of causality has offered a frame in which the idea of universal predestination is replaced by the concept of natural evolution. In the organization of human societies, the description of the position of the individual within the community to which he belongs presents typically complementary aspects related to the shifting separation between the appreciation of values and the background on which they are judged. Surely every stable human society demands fair play specified in judicial rules, but at the same time, life without attachment to kinsfolk and friends would obviously be deprived of some of its most precious values. Though the closest possible combination of justice and charity presents a common goal in all cultures, it may be recognized that any occasion which calls for the strict application of law has no room for display of charity and that, conversely, benevolence and compassion may conflict with all ideas of justice. This point, which in many religions is illustrated mythically in the fight between deities personifying such ideals, in old Oriental philosophy is stressed in the admonition never to forget as we search for harmony in human life, that on the scene of existence we are ourselves actors as well as spectators.

When we compare different cultures resting on traditions fostered by historical events, we have obviously to do with mutually exclusive relationships impeding unprejudiced appreciation of the values in cultural traditions of one nation against the background of traditions of another. In this respect, the relation between national cultures has sometimes been described as complementary, although this word cannot here be taken in the strict sense in which it is used in atomic physics or in psychological analysis, where we are dealing with invariable characteristics of our situation. In fact, not only has contact between nations often resulted in the fusion of cultures, retaining valuable elements of national traditions, but anthropological research is steadily becoming a most important source for the illumination of common features of cultural developments. Indeed, the problem of unity of knowledge can hardly be separated from the striving for universal understanding as a means of elevating human culture.

IV

In presuming to speak on topics so general, and with so much reference to the special field of knowledge represented by the physical sciences, I have tried to indicate a general attitude suggested by important recent developments in my field which seem significant in consideration of the problem of the unity of knowledge. This attitude may be summarized as an effort at harmonious comprehension of ever more aspects of our situation, recognizing that no experience is definable without a logical frame and that any apparent disharmony can only be removed by an appropriate widening of the conceptual framework.

VIII. PHYSICAL SCIENCE AND MAN'S POSITION

Ingeniøren **64** (1955) 810–814

Address delivered at the opening session of the United Nations Conference
on Peaceful Uses of Atomic Energy in Geneva in August 1955

See Introduction to Part I, p. [12].

PHYSICAL SCIENCE AND MAN'S POSITION (1955)

Versions published in English

A   Ingeniøren **64** (1955) 810–814.
B   Proceedings of the International Conference on the Peaceful Uses of Atomic Energy, Geneva 1955. Vol. 16, "Record of the Conference", New York 1956, pp. 57–61.
C   Phil. Today **1** (1957) 65–69.

B is identical to *A*, apart from minor differences of formulation.
In *C*, paragraph numbers 3–17 and the last two paragraphs are omitted.

Professor Bohr on the platform in Geneva.

Physical Science and Man's Position

by Niels Bohr

53:172

It is a great privilege to be given the opportunity of addressing this assembly convened by the United Nations Organization in order to promote international co-operation on the use, to the benefit of humanity, of the vast new energy sources made accessible by the exploration of the world of atoms. I am also grateful for the invitation, as an introduction to these evening lectures at which some broader aspects of our great subject will be discussed, to speak about the general lesson, regarding our position as observers of that nature of which we ourselves are part, gained by the study of this new field of experience.

It shall not be my task to enlarge upon the great practical consequences of the development forming the main theme of our conference, but we are, of course, all deeply aware of the responsibility associated with any advance of our knowledge resulting in an increased mastery of the forces of nature. Indeed, in the present situation, our whole civilization is confronted with a most serious challenge demanding an adjustment of the relationship between nations to ensure that unprecedented menaces can be eliminated and all people in common can strive for the fulfilment of the promises offered by the progress of science for promotion of human welfare all over our globe.

For the appreciation of the common interests and for the furthering of a spirit of mutual confidence,

it may be considered a good omen that we are dealing with consequences of endeavours which know no national borders. Truly, the fruits of scientific research, which through the ages have so largely enriched our life, are a common human inheritance. Moreover, the exploration of new fields of knowledge has steadily illuminated man's position, and in atomic science, so removed from ordinary experience, we have received a lesson which points far beyond the domain of physics. In stressing the necessity for a widening of our conceptual framework for the harmonious comprehension of apparently contrasting phenomena, this lesson may even contribute to a broadening of our attitude to the relationship between human socities with different cultural traditions.

As is well known, the first vague ideas of the atomic constitution of matter go back to antiquity, but it was the great progress of physics and chemistry, which followed the Renaissance, that gave these ideas a firmer basis. Even close up to our days, however, the atomic theory was generally considered as a hypothesis for which no direct proof could ever be given. Indeed, it was believed that our sense organs and tools, themselves composed of innumerable atoms, were far too coarse to allow the detection of individual atomic particles. Still, the marvellous development of experimental technique has not only made it possible to observe effects of single atoms but has

even given us a far-reaching insight into the structure of the atoms themselves.

The development of modern atomic science is the outcome of a most intense international co-operation, in which the progress has been so rapid and the intercourse so intimate that it is often impossible to disentangle the contributions of individuals to the common enterprise. In this address I shall abstain from mentioning names of living scientists, but I feel that we are united in paying tribute to the memory of Ernest Rutherford who with such vigour explored the new field of research opened by the great discoveries by Roentgen, Thomson, Becquerel, and the Curies. We think not merely of Rutherford's fundamental discoveries of the atomic nucleus and its transmutability, but above all of the inspiration with which through so many years he guided the development of this new branch of physical science. At this occasion we also deeply miss Enrico Fermi whose name for all times will be associated with the advent of the »atomic age«.

Notwithstanding the measure to which it has been possible by familiar physical approach to further and utilize our knowledge about atoms, we have at the same time been confronted with unsuspected limitations of the ideas of classical physics demanding a revision of the foundation for the unambiguous application of some of our most elementary concepts.

The first decisive step in this direction was, as is wellknown, the establishment of the relativity theory, by which Albert Einstein, whose recent death all the world deplores, widened the horizon of mankind and gave our world picture a unity surpassing all previous expectation. In return for the abandoning of customary ideas of absolute space and time the relativity principle offered means of tracing general physical laws independent of the standpoint of the observer, and in this connection we think not least of Einstein's recognition of the equivalence of mass and energy which proved an unerring guide in nuclear research.

Still, to cope with the experience concerning atomic particles, further departures from the mechanical conceptions which since Newton's days had been the foundation for the account of physical phenomena, have been necessary and have even led to the recognition of the limited applicability of deterministic description. Here I do not think merely of the recourse to statistical considerations in accounting for the thermodynamical properties of physical systems containing large numbers of atoms, but above all of the discovery of the universal quantum of action to which Max Planck, in the first year of this century, was led by his penetrating analysis of the laws of thermal radiation. In revealing an essential feature of wholeness in elementary processes pointing far beyond the old doctrine of the limited divisibility of matter, this epoch-making discovery showed that the theories of classical physics, according to which physical phenomena are described as a continuous chain of events, are idealizations applicable only to phenomena where the actions involved are sufficiently large to permit the neglect of the individual quantum.

While this condition is amply fulfilled for phenomena on the ordinary scale and even in the interpretation of the experiments permitting measurement of the masses and charges of atomic particles, we meet in proper quantum phenomena with regularities of quite a novel kind, responsible for fundamental properties of matter. The extent to which these regularities transgress the possibility of an analysis in classical physical terms is illustrated by the necessity of using such contrasting pictures as waves and corpuscles in attemps on classical lines to visualize different aspects of the behaviour of atomic objects.

Especially after the discovery of the atomic nucleus the accumulated evidence regarding the properties of the chemical elements became available for the further exploration of atomic processes. From the outset it was evident that the quantum of action afforded a clue to the understanding of the peculiar stability of the electron binding in the atoms, resisting explanation within the framework of classical mechanics. Still, the establishment of a consistent interpretation of atomic phenomena presented a very difficult task which was only gradually accomplished by the concerted efforts of a whole generation of theoretical physicists.

In the mathematical formalism of quantum mechanics, which contains the classical physical theories as a limiting case, the kinematical and dynamical variables are replaced by symbolic operators subject to a noncommutative algorithm involving Planck's constant. The formalism thus defies pictorial representation and aims directly at the prediction of observations appearing under well-defined conditions. Corresponding to the circumstance that in a given experimental arrangement a number of different quantum processes may in general occur, these predictions are of essentially statistical character. In contrast to previous application of statistics in the account of mechanical systems with many degrees of freedom, the use of prohability considerations in quantum physics presents a direct departure from deterministic description inherently connected with the indivisibility of the elementary processes.

By means of the quantum mechanical formalism it has, as is well known, been possible to account in detail for an immense amount of experimental evidence concerning the physical and chemical properties of matter depending on the binding of the electrons to the atomic nuclei. I particular, the peculiar periodic variation of these properties, the discovery of which we owe to Mendelejef's penetrating intuition, has found a complete elucidation. Also as regards the constitution and properties of the nuclei themselves, great progress in the interpretation of the rapidly increasing experimental evidence has steadily been obtained. In this connection we may especially recall how the remarkable early discovered law governing spontaneous radioactive decay is most harmoniously incorporated in the statistical quantum mechanical description.

In spite of the power of quantum theoretical methods, the renunciation of accustomed demands on physical explanation has given rise to doubts in many minds as to whether we are dealing with an exhaustive description of atomic phenomena. Thus, it has been argued that the statistical treatment should be regarded as a temporary approach, eventually to be replaced by a more detailed deterministic theory. The vivid discussion of this basic issue has greatly stimulated the analysis of our position as observers of nature and especially stressed the caution necessary in the application to a new domain of knowledge of concepts adapted to our orientation under ordinary conditions.

Of course, even when the phenomena transcend the scope of classical physical theories, the account of the experimental arrangement and the recording of the observations must be given in plain language, suitably supplemented by technical terminology. This is a pure logical demand, since the very word »experiment« refers to a situation where we can tell others what we have done and what we have learned. In the proper quantum phenomena, however, it is not possible to make the sharp separation, characteristic of the mechanical conception of nature, between the behaviour of the objects under investigation and their interaction with the measuring instruments. Not only does the recording of atomic phenomena involve some amplification device of essentially irreversible functioning, like the production of a permanent mark on a photographic plate left by the impact of an electron. Especially, however, any attempt at controlling the interaction between the atomic objects and the instruments serving to specify the experimental arrangement would imply a change in the observational conditions incompatible with the appearance of the very phenomenon in question.

The essential wholeness of the quantum phenomena makes it impossible to speak in an unambiguous way of attributes of the objects independent of the conditions under which they are observed. Thus, the evidence obtained under different experimental conditions may exhibit a type of relationship quite foreign to classical physics. Still, however contrasting the different phenomena may appear from the classical point of view, they must be regarded as complementary in the sense that only together they exhaust all obtainable knowledge regarding the atomic objects.

Within its scope the quantum mechanical theory gives the appropriate mathematical formulation of the notion of complementarity. Thus, the non-commutability of the operators symbolizing the mechanical quantities of classical physics which imply the indeterminacy relations of conjugate variables corresponds to the mutually exclusive experimental conditions which permit unambiguous use of the corresponding classical concepts. In particular any quantum phenomenon which includes the registration of the position of an atomic particle at a given moment will involve an exchange of momentum and energy, uncontrollable in principle, between the particle and the instruments like fixed scales and synchronized clocks serving to define the reference frame. Conversely, the account of phenomena governed by conservation of momentum and energy involves necessarily a limitation of space-time description.

The freedom of experimentation, presupposed in classical physics, is of course retained and corresponds to the variability of the experimental conditions provided for in the mathematical structure of the quantum mechanical formalism. While, however, in classical physics, the assumed unlimited subdivisibility of the phenomena entails the possibility by suitable experiments in unrestricted measure to interfere with and even to reverse the course of events, the wholeness of each quantum phenomenon implies a restriction of such interference. In particular, all features, of reversibility are reduced to the statistical balancing implied in thermodynamical argumentation.

The renunciation in quantum physics of customary demands on physical explanation reminds of the abandoning in relativity theory of the concepts of absolute space and time, the application of which is restricted by the upper limit of the rate of propagation of alle physical signals, represented by the velocity of light. Similarly, the unsuspected lower limit for the unambiguous use of the mechanical concept of action excludes the unrestricted combination of space-time coordination and momentum and energy balance on which the deterministic description of classical physics rests. I both cases we have to do with irrevocable steps as regards the description of physical experience, based on the recognition of essential features of our situation as observers, which have demanded wider frames for the analysis and synthesis of natural phenomena.

The importance of the epistemological lesson which the exploration of the world of atoms has given us must be seen on the background of the impact of the mechanical conception of nature on general thinking through the centuries. Above

all, the recognition of an inherent limitation in the scope of the deterministic description within a field of experience concerned with fundamental properties of matter, stimulates the search in other domains of knowledge for similar situations in which the mutually exclusive application of concepts, each indispensable in a full account of experience, calls for a complementary mode of description.

When leaving the proper domain of physics we meet at once with the old and much debated question of the place of the living organisms in the description of natural phenomena. Originailly, no sharp distinction between animate and inanimate matter was made, and it is well known that Aristotle, in stressing the wholeness of the individual organisms, opposed the views of the atomists, and even in the discussion of the foundations of mechanics retained ideas like purpose and potency. However, with the great discoveries in anatomy and physiology at the time of the Renaissance, and especially the advent of classical mechanics in the deterministic description of which any reference to purpose is eliminated, a completely mechanistic conception of nature suggested itself.

It is true that the structure and functioning of the organisms involve an ordering of atomic processes which has sometimes been felt difficult to reconcile with the laws of thermodynamics, which imply a steady approach towards disorder among the atoms constituting an isolated physical system. If, however, sufficient account is taken of the circumstace that the free energy necessary to maintain and develop organic systems is continually supplied from their surroundings by nutrition and respiration, it becomes clear that there is in such respect no question of any violation of general physical laws. Still, as remarked already by Boltzman, it is important to recognize that the essential element of irreversibility involved in the description of the organic functions is the very basis for our notion of time direction.

In the last decades, great advances have been achieved in our knowledge of the structure and functioning of the organisms, and it has in particular become evident that quantum regularities in many respects here play a fundamental role. Such regularities are indeed basic for the remarkable stability of the highly complex molecular structures which form the essential constituents of the cells responsible for the hereditary properties of the species. Moreover, the induced mu-

tations resulting from the exposure of the organisms to penetrating radiation offer a striking application of the statistical laws of quantum physics. Also the sensitivity of perceptive organs, so important for the integrity of the organisms, has been found to approach the level of individual quantum processes, and clearly amplification mechanisms reminding of the recording devices used in experiments of atomic physics play an important part not least in the transmission of nervous messages.

The whole development has again, although in a novel manner, brought mechanistic approach to biological problems to the foreground, but at the same time the question has become acute whether a comparison between the organism and highly complex and refined physical systems, like modern industrial constructions or electronic calculation machines, offers the proper basis for an adequate description of such self-regulating entities as living organisms present.

Returning to the general epistemological lesson which atomic physics has given us, we must in the first place realize that the isolated phenomena studied in quantum physics are no direct analogue to biological processes involving a continual exchange of matter and energy between the organism and its environments. Moreover, any experimental arrangement, which would allow a control of biological functions to the extent demanded for their exhaustive description in physical terms, would be prohibitive to the free display of life. Thus, notwithstanding the ever improving technique of studying metabolism especially by means of the ingenious atomic tracer method applying radioative isotopes now so abundantly available, we must realize that as regards organic life the possibilities of interference with or reversion of the course of events are still more limited than in the study of individual atomic processes. Incidentally, we may also recall that medical treatment, however efficient it may prove, essentially aims at supporting the organism to recover its health or resume its normal functions.

The emphasis on this point suggests an attitude to the problem of organic life providing an appropriate balance between mechanistic and finalistic approach. In fact, just as the quantum of action in the account of atomic phenomena appears as an element which cannot be defined in classical mechanical terms, the notion of life is elementary in biological science in the sense that it applies to situations where the conditions for an exhaustive physical analysis are not fulfilled. Actually, we must recognize that the practical approach in biological research is characterized by the complementary way in which arguments, based on the full resources of physical and chemical science and concepts directly referring to the integrity of the organism transcending the scope of these sciences are employed.

Similar situations regarding the comprehension of experience are met with in the study of such innate and conditioned behaviour of animals and man, which call for the application of psychological concepts. Even in an allegedly behaviouristic approach it is indeed hardly possible to avoid such concepts, and the very idea of consciousness impresses itself when we are dealing with behaviour of a so high degree of complexity that its description virtually involves introspection on the side of the individual organism. In this connection it is interesting to note that, while in the early stages of physical science one could directly refer to such features of daily life events which permitted simple causal account, an essentially complementary description of the states of our mind has been used since the origin of language. In fact, the rich terminology adapted to this purpose does not point to an unbroken course of events, but rather to separate mutually exclusive experiences reminding of the complementary phenomena in atomic physics. Just as these phenomena for their definition demand different experimental arrangements, the various psychological experiences are characterized by different placings of the separation between the content on which attention is focused and the background indicated by the word »ourselves«.

From a purely biological point of view, we can hardly interpret the characteristics of psychical phenomena, except by concluding that every conscious experience, capable of retainment in the memory, corresponds to a residual impression in the organism, amounting to an irreversible recording of the outcome of processes in the nervous system. Certainly, such recordings in which the interplay of numerous brain cells are involved are essentially different from the permanent structures in the single cells connected with genetic reproduction. On finalistic approach, however, we may not only stress the usefulness of permanent recording in their influence on our reactions to subsequent stimuli, but equally the importance that later generations

are not encumbered by the actual experiences of individuals, but only rely on the reproduction of such properties of the organism which have proved serviceable for the collection and utilization of knowledge. In any attempt at pursuing the enquiry we must, of course, be prepared to meet increasing difficulties in every step, and it is suggestive that the simple concepts of physical science to an ever higher degree lose their immediate applicability, the more we approach the aspects of organic life related to the characteristics of our mind.

To illustrate such argumentation, we may refer to the old problem of the freedom of will. In an unrestricted deterministic approach this concept, of course, finds no place, but it is evident that the word volition is indispensable in an exhaustive description of psychical phenomena. Not only do we have the feeling of so to say being able to make the best out of the circumstances but also if we attempt to predict what another person will decide to do in a given situation, we must strive to know his whole background to such an extent that we shall actually be placing ourselves in his position. A logically consistent basis for speaking of the »freedom of our will« is provided by the recognition that the psychological situations in which we have a feeling of volition and those in which we ponder over the motives for our actions offer a typical example of complementary relationships. Thus, the proper latitude is also left for the use of words like aspirations and responsibility which seperately are as little definable as other words indispensable in the account of the variety and potentiality of our situation.

When I have entered on such general biological and psychological problems familiar to all, the intention has only been to remind of common features of scientific enquiry and point to an attitude characterized by the striving for harmonizing apparently contrasting experiences by their incorporation in a wider conceptual framework. Such an approach may perhaps also contribute to the promotion of the mutual understanding between human societies with different cultural traditions, and before concluding this adress I shall allow myself to add some remarks concerning this question.

In this connection it may be pertinent to refer to the similarities and differences between animal and human societies. In animal life we meet with communities of very diverse kinds corresponding to the

needs of the different species. Especially among insects we are sometimes confronted with a division of functions among the individuals carried to such extremes that the whole society in various respects resembles a single organism, while in many species of birds and mammals who live in flocks more or less divided into families we have rather to do with an innate behaviour which reminds us of many habits in human communities serving the support of the individuals as well as the protection of the society.

The essential difference between such animal societies and the human communities is however that in our cultural traditions we have to do not with biologically inherited behaviour but with modes of reaction if the adult individuals carried from generation to generation by more or less consciously directed education. In this connection it is decisive to realize the extent to which instinctive behaviour is suppressed in human life. In the terminology suggested by modern science we may even say that words like »instinct« and »reason« have mutually exclusive complementary applications.

Human cultures developed in isolation from each other often exhibit deep-rooted differences not only as regards the adjustment to external conditions like climate and natural resources but also as regards the traditions which they have fostered and which often stand in the way of mutual understanding. Sometimes one has compared the various cultures with the different ways in which physical phenomena are described according to the standpoint of the observer. Still, the great scientific advance marked by relativity theory implies the possibility for any observer to predict, in terms of common concepts, how any other observer will account for physical experience. Just the difficulty of appreciating the traditions of other nations on the basis of one's own national tradition suggests that the relationship between cultures may rather be regarded as complementary. In all such comparisons it is, however, not taken into account that every culture is continually developing. Especially contacts between different cultural communities may influence the attitude of each to an extent which may even lead to a common culture with a more embracing outlook.

As an unifying element of human cultures, the development of science plays an ever more important role. Not only is any advance of knowledge, wherever gained, of benefit for all humanity, but the co-opera-

[105]

tion in scientific research offers perhaps more than anything else opportunities for the furthering of close contacts and common understanding. These opportunities have a special significance at the present crucial stage of history. Indeed, the establishment of a co-operation in confidence between all peoples which is now so urgently needed depends essentially on the free access to all information and unhampered discussions of all problems of human interest.

We are all united in the hope that this Conference where representatives for so many nations are assembled for the exchange of knowledge will come to stand as a landmark for scientific and technological co-operation. We trust that the opportunity of intercourse and acquaintance offered us at this great occasion will essentially promote the common striving for the elevation of culture in all its aspects.

IX. PREFACE and INTRODUCTION

"Atomic Physics and Human Knowledge", John Wiley & Sons,
New York 1958, pp. v–vi, 1–2

PREFACE and INTRODUCTION (1957)

Versions published in Danish, English and German

Danish: Forord and *Indledning*
A   "Atomfysik og menneskelig erkendelse", J.H. Schultz Forlag, Copenhagen 1957, pp. 5, 9–10

English: Preface and *Introduction*
B   "Atomic Physics and Human Knowledge", John Wiley & Sons, New York 1958, pp. v–vi, 1–2 (reprinted in: "Essays 1933–1957 on Atomic Physics and Human Knowledge, The Philosophical Writings of Niels Bohr, Vol. II", Ox Bow Press, Woodbridge, Connecticut 1987, pp. v–vi, 1–2)

German: Vorwort and *Einleitung*
C   "Atomphysik und menschliche Erkenntnis", Friedr. Vieweg & Sohn, Braunschweig 1958, pp. V, 1–2

All of these versions agree with each other.

Preface

This collection of articles, written on various occasions within the last 25 years, forms a sequel to earlier essays edited by the Cambridge University Press, 1934, in a volume titled *Atomic Theory and the Description of Nature*. The theme of the papers is the epistemological lesson which the modern development of atomic physics has given us and its relevance for analysis and synthesis in many fields of human knowledge. The articles in the previous edition were written at a time when the establishment of the mathematical methods of quantum mechanics had created a firm foundation for the consistent treatment of atomic phenomena, and the conditions for an unambiguous account of experience within this framework were characterized by the notion of complementarity. In the papers collected here, this approach is further developed in logical formulation and given broader application. Of course, much repetition has been unavoidable, but it is hoped that this may serve to illustrate the gradual clarification of the argumentation, especially as regards more concise terminology.

In the development of the views concerned, discussions with former and present collaborators at the Institute for Theoretical Physics in the University of Copenhagen have been most valuable to me. For

v

assistance in the elaboration of the articles in this volume, I am especially indebted to Oskar Klein and Léon Rosenfeld, now in the universities of Stockholm and Manchester, as well as to Stefan Rozental and Aage Petersen at the Copenhagen Institute.   Also I should like to extend my thanks to Mrs. S. Hellmann for her most effective help in the preparation of the articles and the present edition.

NIELS BOHR

Copenhagen
August 1957

Introduction

The importance of physical science for the development of general philosophical thinking rests not only on its contributions to our steadily increasing knowledge of that nature of which we ourselves are part, but also on the opportunities which time and again it has offered for examination and refinement of our conceptual tools. In our century, the study of the atomic constitution of matter has revealed an unsuspected limitation of the scope of classical physical ideas and has thrown new light on the demands on scientific explanation incorporated in traditional philosophy. The revision of the foundation for the unambiguous application of our elementary concepts, necessary for comprehension of atomic phenomena, therefore has a bearing far beyond the special domain of physical science.

The main point of the lesson given us by the development of atomic physics is, as is well known, the recognition of a feature of wholeness in atomic processes, disclosed by the discovery of the quantum of action. The following articles present the essential aspects of the situation in quantum physics and, at the same time, stress the points of similarity it exhibits to our position in other fields of knowledge beyond the scope of the mechanical conception of

I

nature. We are not dealing here with more or less vague analogies, but with an investigation of the conditions for the proper use of our conceptual means of expression. Such considerations not only aim at making us familiar with the novel situation in physical science, but might on account of the comparatively simple character of atomic problems be helpful in clarifying the conditions for objective description in wider fields.

Although the seven essays here collected are thus closely interconnected, they fall into three separate groups originating from the years 1932–1938, 1949, and 1954–1957, respectively. The first three papers, directly related to the articles in the previous edition, discuss biological and anthropological problems referring to the features of wholeness presented by living organisms and human cultures. Of course, it is in no way attempted to give an exhaustive treatment of these topics, but only to indicate how the problems present themselves against the background of the general lesson of atomic physics.

The fourth article deals with the discussion among physicists of the epistemological problems raised by quantum physics. Owing to the character of the topic, some reference to the mathematical tools has been unavoidable, but the understanding of the arguments demands no special knowledge. The debate led to a clarification of the new aspects of the observational problem, implied by the circumstance that the interaction between atomic objects and measuring instruments forms an integral part of quantum phenomena. Therefore, evidence gained by different experimental arrangements cannot be comprehended on accustomed lines, and the necessity of taking into account the conditions under which experience is obtained calls directly for the complementary mode of description.

The last group of articles is closely related to the first, but it is hoped that the improved terminology used to present the situation in quantum physics has made the general argument more easily accessible. In its application to problems of broader scope, emphasis is laid especially on the presuppositions for unambiguous use of the concepts employed in the account of experience. The gist of the argument is that for objective description and harmonious comprehension it is necessary in almost every field of knowledge to pay attention to the circumstances under which evidence is obtained.

X. PHYSICAL SCIENCE AND THE PROBLEM OF LIFE

"Atomic Physics and Human Knowledge", John Wiley & Sons,
New York 1958, pp. 94–101

Article completed in 1957, based on a Steno Lecture in the
Danish Medical Society, Copenhagen, February 1949

See General Introduction, p. xxviii and Introduction to Part I, p. [5].

Brothers Niels and Harald Bohr and their father Christian Bohr. The photograph was probably taken c. 1906.

PHYSICAL SCIENCE AND THE PROBLEM OF LIFE (1957)

Versions published in Danish, English and German

Danish: Fysikken og Livets Problem

A   "Atomfysik og menneskelig erkendelse", J.H. Schultz Forlag, Copenhagen 1957, pp. 115–124

English: Physical Science and the Problem of Life

B   "Atomic Physics and Human Knowledge" John Wiley & Sons, New York 1958, pp. 94–101 (reprinted in: "Essays 1933–1957 on Atomic Physics and Human Knowledge, The Philosophical Writings of Niels Bohr, Vol. II", Ox Bow Press, Woodbridge, Connecticut 1987, pp. 94–101)

German: Die Physik und das Problem des Lebens

C   "Atomphysik und menschliche Erkenntnis" Friedr. Vieweg & Sohn, Braunschweig 1958, pp. 96–104

All of these versions agree with each other.

Physical Science
and
the Problem of Life

1957

It has been a pleasure to accept the invitation of the Medical Society of Copenhagen to give one of the Steno lectures by which the Society commemorates the famous Danish scientist whose achievements are admired in ever greater measure, not only in this country but in the whole scientific world. As my theme I have chosen a problem which has occupied the human mind through the ages and with which Niels Stensen himself was deeply concerned, namely how far physical experience can help us to explain organic life in its rich and varied display. As I shall try to show, the development of physics in recent decades and in particular the lesson regarding our position as observers of that nature of which we are part, received through the exploration of the world of atoms so long closed to us, have created a new background for our attitude to this question.

Even in the philosophical schools of ancient Greece, we find divergent opinions regarding the conceptual means suited to account for the striking differences between living organisms and other material bodies. As is well known, the atomists considered a limited divisibility of all matter necessary to explain not only simple physical phenomena but also the functioning of organisms and the related

94

psychical experiences. Aristotle, on the other hand, refuted atomic ideas and, in view of the wholeness exhibited by every living organism, maintained the necessity of introducing into the description of nature such concepts as perfection and purposefulness.

For almost 2000 years the situation remained essentially unchanged, and not until the Renaissance did there occur the great discoveries in physics as well as in biology, which were to give new incentives. Progress in physics consisted above all in the liberation from the Aristotelian idea of driving forces as the cause of all motion. Galileo's recognition that uniform motion is a manifestation of inertia and his emphasis on force as a cause of change of motion were to become the foundation of the development of mechanics, which Newton to the admiration of succeeding generations endowed with a firm and completed form. In this so-called classical mechanics all reference to purpose is eliminated, since the course of events is described as automatic consequences of given initial conditions.

The progress of mechanics could not avoid making the strongest impression on all contemporary science. In particular, the anatomical studies of Vesalius and Harvey's discovery of the circulation of the blood suggested the comparison between living organisms and machines working according to the laws of mechanics. On the philosophical side, it was especially Descartes who stressed the similarity between animals and automata, but ascribed to human beings a soul interacting with the body in a certain gland in the brain. However, the insufficiency of contemporary knowledge of such problems was emphasized by Steno in his famous Paris lecture on the anatomy of the brain, which bears witness of the great observational power and open-mindedness characteristic of all Steno's scientific work.

Subsequent developments in biology, especially after the invention of the microscope, revealed an unsuspected fineness in organic structure and regulatory processes. At the same time that mechanistic ideas thus found ever wider applications, so-called vitalistic or finalistic points of view, inspired by the wonderful power of regeneration and adaptation in organisms, were repeatedly expressed. Rather than returning to primitive ideas of a life force acting in the organisms, such views emphasized the insufficiency of physical approach in accounting for the characteristics of life. As a sober presentation of the situation as it stood in the beginning of this century, I should like to refer to the following statement by my father, the physiologist Christian Bohr, in the introduction to his paper "On Pathological Lung Expansion" which appeared in the anniversary publication of the Copenhagen University in 1910.

[117]

As far as physiology can be characterized as a special branch of natural sciences, its specific task is to investigate the phenomena peculiar to the organism as a given empirical object in order to obtain an understanding of the various parts in the self-regulation and how they are balanced against each other and brought into harmony with variations in external influences and inner processes. It is thus in the very nature of this task to refer the word purpose to the maintenance of the organism and consider as purposive the regulation mechanisms which serve this maintenance. Just in this sense we shall in the following use the notion "purposiveness" about organic functions. In order that the application of this concept in each single case should not be empty or even misleading it must, however, be demanded that it be always preceded by an investigation of the organic phenomenon under consideration, sufficiently thorough to illuminate step by step the special way in which it contributes to the maintenance of the organism. Although this demand, which requires no more than the scientific demonstration that the notion of purposiveness in the given case is used in accordance with its definition, might appear self-evident, it may nevertheless not be unnecessary to stress it. Indeed, physiological investigations have brought to light regulations of utmost fineness in a multitude so great that it is a temptation to designate every observed manifestation of life as purposive without attempting an experimental investigation of its detailed functioning. By means of analogies which so easily present themselves among the variety of organic functions, it is merely the next step to interpret this functioning from a subjective judgement about the special character of purposiveness in the given case. It is evident, however, how often, with our so narrowly limited knowledge about the organism, such a personal judgement may be erroneous, as is illustrated by many examples. In such cases, it is the lacking experimental illumination of the details of the process which is the cause of the erroneous results of the procedure. The *a priori* assumption of the purposiveness of the organic process is, however, in itself quite natural as a heuristic principle and can, due to the extreme complication and difficult comprehension of the conditions in the organism, prove not only useful, but even indispensable for the formulation of the special problem for the investigation and the search of ways for its solution. But one thing is what may be conveniently used by the preliminary investigation, another what justifiably can be considered an actually achieved result. As regards the problem of the purposiveness of a given function for the maintenance of the whole organism, such a result can, as stressed above, be secured only by a demonstration in detail of the ways in which the purpose is reached.

I have quoted these remarks which express the attitude in the circle in which I grew up and to whose discussions I listened in my youth, because they offer a suitable starting point for the investigation of the place of living organisms in the description of nature. As I shall try to show, modern development of atomic physics, at the same time as it has augmented our knowledge about atoms and their constitution of more elementary parts, has revealed the limitation in principle of the so-called mechanical conception of nature and thereby created a new background for the problem, most pertinent to our subject, as to what we can understand by and demand of a scientific explanation.

In order to present the situation in physics as clearly as possible, I shall start by reminding you of the extreme attitude which, under the impact of the great success of classical mechanics, was expressed in Laplace's well-known conception of a world machine. All interactions between the constituents of this machine were governed by the laws of mechanics, and therefore an intelligence knowing the relative positions and velocities of these parts at a given moment could predict all the subsequent events in the world, including the behaviour of animals and man. In this whole conception which has, as is well known, played an important role in philosophical discussion, due attention is not paid to the presuppositions for the applicability of the concepts indispensable for communication of experience.

In this respect the later development of physics has given us an urgent lesson. Already the far-reaching interpretation of heat phenomena as incessant motion of molecules in gases, liquids, and solids has called attention to the importance of the conditions of observation in the account of experience. Of course, there could be no question of a detailed description of the motions of the innumerable particles among each other, but only of deducing statistical regularities of heat motion by means of general mechanical principles. The peculiar contrast between the reversibility of simple mechanical processes and the irreversibility typical of many thermodynamical phenomena was thus clarified by the fact that applications of such concepts as temperature and entropy refer to experimental conditions incompatible with complete control of the motions of single molecules.

In the maintenance and growth of living organisms one has often seen a contradiction to that tendency, implied by the thermodynamical laws, towards temperature and energy equilibrium in an isolated physical system. However, we must remember that organisms are continually supplied with free energy by nutrition and respiration, and the most thorough physiological investigation has never revealed any departure from thermodynamical principles. Yet, recognition of such similarities between living organisms and ordinary power engines is of course in no way sufficient to answer the question about the position of organisms in the description of nature, a question obviously demanding deeper analysis of the observational problem.

This very problem has indeed been brought to the foreground in an unexpected way by the discovery of the universal quantum of action which expresses a feature of wholeness in atomic processes that prevents the distinction between observation of phenomena and independent behaviour of the objects, characteristic of the mechanical conception of nature. In physical systems on the ordinary scale, representation of events as a chain of states described by measurable quantities rests on the circumstance that all actions are here large enough to permit neglect of the interaction between the objects and the bodies used as measuring tools. Under conditions where the quantum of action plays a decisive part and where such an interaction is therefore an integral part of the phenomena, there cannot to the same extent be ascribed a mechanically well-defined course.

The breakdown of ordinary physical pictures which here confronts us is strikingly expressed in the difficulties in talking about properties of atomic objects independent of the conditions of observation. Indeed, an electron may be called a charged material particle, since measurements of its inertial mass always give the same result, and any transmission of electricity between atomic systems always amounts to a number of so-called unit charges. Yet, the interference effects appearing when electrons pass through crystals are incompatible with the mechanical idea of particle motion. We meet analogous features in the well-known dilemma about the nature of light, since optical phenomena require the notion of wave propagation, while the laws of transmission of momentum and energy in atomic photo-effects refer to the mechanical conception of particles.

This situation, novel in physical science, has demanded a renewed analysis of the presuppositions for the application of concepts used for orientation in our surroundings. Of course, in atomic physics we retain the freedom by experimenting to put questions to nature, but we must recognize that the experimental conditions which can be varied in numerous ways are defined only by bodies so heavy that in the description of their functions we can disregard the quantum. Information concerning atomic objects consists solely in the marks they make on these measuring instruments, as, for instance, a spot produced by the impact of an electron on a photographic plate placed in the experimental arrangement. The circumstance that such marks are due to irreversible amplification effects endows the phenomena with a peculiarly closed character pointing directly to the irreversibility in principle of the very notion of observation.

The special situation in quantum physics is above all, however, that the information gained about atomic objects cannot be comprehended along the lines of approach typical of the mechanical conception of nature. Already the fact that under one and the same experimental arrangement there may in general appear observations pertaining to different individual quantum processes entails a limitation in principle of the deterministic mode of description. The demand of unrestricted divisibility on which classical physical description rests is also clearly incompatible with that feature of wholeness in typical quantum phenomena which involves that any definable subdivision requires a change of the experimental arrangement giving rise to new individual effects.

In order to characterize the relation between phenomena observed under different experimental conditions, one has introduced the term complementarity to emphasize that such phenomena together exhaust all definable information about the atomic objects. Far from containing any arbitrary renunciation of customary physical explanation, the notion of complementarity refers directly to our position as observers in a domain of experience where unambiguous application of the concepts used in the description of phenomena depends essentially on the conditions of observation. By a mathematical generalization of the conceptual framework of classical physics it has been possible to develop a formalism which leaves room for the logical incorporation of the quantum of action. This so-called quantum mechanics aims directly at the formulation of statistical regularities pertaining to evidence gained under well-defined observational conditions. The completeness in principle of this description is due to the retention of classical mechanical ideas to an extent including any definable variation of the experimental conditions.

The complementary character of the quantum-mechanical description is clearly expressed in the account of the composition and reactions of atomic systems. Thus, the regularities regarding the energy states of atoms and molecules, responsible for the characteristic spectra of the elements and the valences of chemical combinations, appear only under circumstances where a control of the positions of the electrons within the atom and the molecule is excluded. In this connection, it is interesting to note that fruitful application of the structural formulae in chemistry rests solely on the fact that the atomic nuclei are so much heavier than the electrons. However, with respect to the stability and transmutations of the atomic nuclei themselves, quantum-mechanical features are again decisive. Only in a complementary description transcending the scope of the mechanical conception of nature is it possible to find room for the fundamental regularities responsible for the properties of the substances of which our tools and our bodies are composed.

Progress in the field of atomic physics has, as is well known, found wide application in the biological sciences. In particular, I may mention the understanding we have gained of the peculiar stability of chemical structures in the cells responsible for the hereditary properties of the species, and of the statistical laws for the occurrence of mutations produced by exposing organisms to special agencies. Furthermore, amplification effects similar to those permitting observation of individual atomic particles play a decisive role in many functions of the organism. In this way is stressed the irreversible character of typical biological phenomena, and the time direction inherent in the description of the functioning of organisms is strikingly marked by their utilization of past experience for reactions to future stimuli.

In this promising development we have to do with a very important and, according to its character, hardly limited extension of the application of purely physical and chemical ideas to biological problems, and since quantum mechanics appears as a rational generalization of classical physics, the whole approach may be termed mechanistic. The question, however, is in what sense such progress has removed the foundation for the application of so-called finalistic arguments in biology. Here we must realize that the description and comprehension of the closed quantum phenomena exhibit no feature indicating that an organization of atoms is able to adapt itself to the surroundings in the way we witness in the maintenance and evolution of living organisms. Furthermore, it must be stressed that an account, exhaustive in the sense of quantum physics, of all the continually exchanged atoms in the organism not only is infeasible but would obviously require observational conditions incompatible with the display of life.

However, the lesson with respect to the role which the tools of observation play in defining the elementary physical concepts gives a clue to logical application of notions like purposiveness foreign to physics, but lending themselves so readily to the description of organic phenomena. Indeed, on this background it is evident that the attitudes termed mechanistic and finalistic do not present contradictory views on biological problems, but rather stress the mutually exclusive character of observational conditions equally indispensable in our search for an ever richer description of life. Here, there is of course no question of an explanation akin to the classical physical account of the functioning of simple mechanical constructions or of complicated electron calculation machines, but we are concerned with wider pursuit of that analysis of the presuppositions and scope of our conceptual means of communication which has become so characteristic of the newer development of physics.

Apart from all differences with respect to observational conditions, communication of biological experiences contains no more reference to the subjective observer than does the description of physical evidence. Thus, so far it has not been necessary to enter more closely into the conditions of observation characteristic for the account of psychological phenomena, for which we cannot rely on the conceptual frame developed for our orientation in inanimate nature. However, the fact that conscious experience can be remembered and therefore must be supposed to be connected with permanent changes in the constitution of the organism points to a comparison between psychical experiences and physical observations. With respect to relationships between conscious experiences we also encounter features reminiscent of the conditions for the comprehension of atomic phenomena. The rich vocabulary used in the communications of the states of our mind refers indeed to a typical complementary mode of description corresponding to a continual change of the content on which attention is focused.

Compared to the extension of the mechanical mode of description demanded by the account of the individuality of atomic phenomena, the integrity of the organism and the unity of the personality confront us of course with a further generalization of the frame for the rational use of our means of communication. In this respect, it must be emphasized that the distinction between subject and object, necessary for unambiguous description, is retained in the way that in every communication containing a reference to ourselves we, so-to-speak, introduce a new subject which does not appear as part of the content of the communication. It need hardly be stressed that it is just this freedom of choosing the subject-object distinction which provides room for the multifariousness of conscious phenomena and the richness of human life.

The attitude to general problems of knowledge to which the development of physics in this century has led us differs essentially from the approach to such problems at Steno's time. This does not mean, however, that we have left the road to the enrichment of knowledge followed by him with such great results, but we have realized that the striving for beauty and harmony which marked Steno's work demands a steady revision of the presuppositions and scope of our means of communication.

XI. QUANTUM PHYSICS AND BIOLOGY

Symposia of the Society for Experimental Biology, Number XIV:
"Models and Analogues in Biology", Cambridge 1960, pp. 1–5

See Introduction to Part I, p. [12].

PREFACE

A symposium of this title could well have been held in Ancient Greece, but it would then have had the assistance of a philosopher: this it was denied through the indisposition of Professor K. R. Popper, who had hoped to contribute. It could have been held in 1900, but again, the particular subject of anatomical homologue and analogue which might have dominated it then, is not now included. The contributors certainly represent a very much wider cross-section of every field of science than has been the case with any previous volume in this series; if it appeared from the programme of the Conference that the titles of contributions bore little relation to one another, the Conference itself certainly concluded with an appreciation of the diversity of approaches which can be made towards a fundamental feature of science, which is particularly critical to the present state of biology. It is consequently hoped that this volume may serve two purposes: to attract the attention of scientists of all disciplines to the problems which are central in biological investigation and communication and to indicate to biologists the forms of approach made to analogous problems in other sciences; to provide a variety of pathways, one of which may lead the student towards these problems of thought, language and biology.

The Conference was held in Queen's Building, University of Bristol from 6 to 12 September, and owed much to many members of that University for their hospitality; Dr R. B. Clark undertook the arduous duties of Local Secretary. Professor T. Weis-Fogh flew from Copenhagen at extremely short notice to deliver Professor Bohr's Paper.

I am much indebted to contributors and colleagues for their suggestions during the planning of the Symposium, and must particularly acknowledge the help of Professor C. F. A. Pantin and Dr K. E. Machin. Finally, it is a pleasure to record the help and co-operation of the Cambridge University Press in the preparation of the volume.

J. W. L. BEAMENT

Editor of the fourteenth Symposium of the
Society for Experimental Biology

QUANTUM PHYSICS AND BIOLOGY†

By NIELS BOHR

Institute for Theoretical Physics, University of Copenhagen

I

The significance of physical science for philosophy does not merely lie in the steady increase of our experience of inanimate matter, but above all in the opportunity of testing the foundation and scope of some of our most elementary concepts. Notwithstanding refinements of terminology due to accumulation of experimental evidence and developments of theoretical conceptions, all account of physical experience is, of course, ultimately based on common language, adapted to orientation in our surroundings and tracing of relationships between cause and effect. Indeed, Galileo's programme to base the description of physical phenomena on measurable quantities has afforded a solid foundation for the ordering of an ever larger field of experience.

In Newtonian mechanics, where the state of a system of material bodies is defined by their instantaneous positions and velocities, it proved possible, by the well-known simple principles, to derive, uniquely from the knowledge of the state of the system at a given time and of the forces acting upon the bodies, the state of the system at any other time. A description of this kind, which evidently represents an ideal form of causal relationships, expressed by the notion of *determinism*, was found to have still wider scope. Thus, in the account of electromagnetic phenomena, in which we have to consider a propagation of forces with finite velocities, a deterministic description could be upheld by including in the definition of the state not only the positions and velocities of the charged bodies, but also the direction and intensity of the electric and magnetic forces at every point of space at a given time.

A new epoch in physical science was inaugurated by Planck's discovery of the *elementary quantum of action*, which revealed a feature of *wholeness* inherent in atomic processes going far beyond the ancient idea of the limited divisibility of matter. Indeed, it became clear that the pictorial description of classical physical theories represents an idealization valid only for phenomena in the analysis of which all actions involved are sufficiently large to permit the neglect of the quantum. While this condition is amply fulfilled in phenomena on the ordinary scale, we meet, in experimental evidence concerning atomic particles, with regularities of a novel type, incompatible with deterministic analysis. These quantal laws are determining for the peculiar

---

† The author has abbreviated his recent article from *Survey of Philosophy in the Mid-Century*, Firenze, 1958, to form the first portion of his paper, while the remainder he specially prepared for this Symposium.

stability and reactions of atomic systems, and thus ultimately responsible for the properties of matter on which our means of observation depend.

In spite of the power of quantum mechanics as a means of ordering an immense amount of evidence regarding atomic phenomena, its departure from accustomed demands of causal explanation has naturally given rise to the question whether we are here concerned with an exhaustive description of experience. The answer to this question evidently calls for a closer examination of the conditions for the unambiguous use of the concepts of classical physics in the analysis of atomic phenomena. The decisive point is to recognize that the description of the experimental arrangement and the recording of observations must be given in plain language, suitably refined by usual physical terminology. This is a simple logical demand, since by the word experiment we can only mean a procedure about which we are able to communicate to others what we have done and what we have learnt.

In actual experimental arrangements, the fulfilment of such requirements is secured by the use as measuring instruments of rigid bodies sufficiently heavy to allow a completely classical account of their relative positions and velocities. In this connexion, it is also essential to remember that all unambiguous information concerning atomic objects is derived from the permanent marks—such as a spot on a photographic plate, caused by the impact of an electron—left on the bodies which define the experimental conditions. Far from involving any special intricacy, the irreversible amplification effects on which the recording of the presence of atomic objects rests rather remind us of the essential irreversibility inherent in the very concept of observation. The description of atomic phenomena has in these respects a perfectly objective character, in the sense that no explicit reference is made to any individual observer and that therefore, with proper regard to relativistic exigencies, no ambiguity is involved in the communication of information.

As regards all such points, the observation problem of quantum physics in no way differs from the classical physical approach. The essentially new feature in the analysis of quantum phenomena is, however, the introduction of a *fundamental distinction between the measuring apparatus and the objects under investigation*. This is a direct consequence of the necessity of accounting for the functions of the measuring instruments in purely classical terms, excluding in principle any regard to the quantum of action. On their side, the quantal features of the phenomenon are revealed in the information about the atomic objects derived from the observations. While, within the scope of classical physics, the interaction between object and apparatus can be neglected or, if necessary, compensated for, this interaction, in quantum physics, thus forms an inseparable part of the phenomenon. Accordingly, the unambiguous account of proper quantum phenomena must, in principle, include a description of all relevant features of the experimental arrangement.

The very fact that repetition of the same experiment, defined on the lines described, in general yields different recordings pertaining to the object, immediately implies that a comprehensive account of experience in this field must be expressed by statistical laws. It need hardly be stressed that we are not here concerned with an analogy to the familiar recourse to statistics in the description of physical systems of too complicated a structure to make practicable the complete definition of their state necessary for a deterministic account. In the case of quantum phenomena, the unlimited divisibility of events implied in a deterministic account is, in principle, excluded by the requirement to specify the experimental conditions. Indeed, the feature of wholeness typical of proper quantum phenomena finds its logical expression in the circumstance that any attempt at a well-defined subdivision would demand a change in the experimental arrangement incompatible with the definition of the phenomenon under investigation.

Within the scope of classical physics, all characteristic properties of a given object can in principle be ascertained by a single experimental arrangement, although in practice various arrangements are often convenient for the study of different aspects of the phenomena. In fact, data obtained in such a way simply supplement each other and can be combined into a consistent picture of the behaviour of the object under investigation. In quantum physics, however, evidence about atomic objects obtained by different experimental arrangements exhibits a novel kind of complementary relationship. Indeed, it must be recognized that such evidence, which appears contradictory when combination into a single picture is attempted, exhausts all conceivable knowledge about the object. Far from restricting our efforts to put questions to nature in the form of experiments, the notion of *complementarity* simply characterizes the answers we can receive by such inquiry in the case when the interaction between the measuring instruments and the objects forms an integral part of the phenomena.

The question has been raised whether recourse to multivalued logics is needed for a more appropriate representation of the situation. From the preceding argumentation it will appear, however, that all departures from common language and ordinary logic are entirely avoided by reserving the word 'phenomenon' solely for reference to unambiguously communicable information, in the account of which the word 'measurement' is used in its plain meaning of standardized comparison. Such caution in the choice of terminology is especially important in the exploration of a new field of experience, where information cannot be comprehended in the familiar frame which in classical physics found such unrestricted applicability.

In general philosophical perspective, it is significant that, as regards analysis and synthesis in other fields of knowledge, we are confronted with situations reminding of that in quantum physics. Thus, the integrity of living organisms and the characteristics of conscious individuals and human cultures present features of wholeness, the account of which implies a typical complementary mode of description (Bohr, 1958). Due to the diversified use of the rich vocabulary available for communication of experience in those wider fields,

1–2

and above all to the varying interpretations, in philosophical literature, of the concept of causality, the aim of such comparisons has sometimes been misunderstood. However, the gradual development of an appropriate terminology for the description of the simpler situation in physical science indicates that we are not dealing with more or less vague analogies, but with clear examples of logical relations which in different contexts are met with in wider fields.

II

The discussion of the position of living organisms in a general description of physical phenomena has, in the development of science, passed through a number of stages. In Antiquity, the obvious difficulties inherent in a comparison between organisms and primitive machinery deeply influenced the attitude towards mechanical problems and even led to the attribution of vital characteristics to all matter. With the abandonment of these views, at the time of the Renaissance, through the clarification of the principles of classical mechanics, the problem entered into another stage, stimulated by the great anatomical and physiological discoveries at that period.

Recent advances in technology, and especially the development of automatic control of industrial plants and calculation devices, have given rise to a renewed discussion of the extent to which it is possible to construct mechanical and electrical models with properties resembling the behaviour of living organisms. Indeed, it may be feasible to design models reacting in any prescribed manner, including their own reproduction, provided that they have access to the necessary materials and energy sources. Still, quite apart from the suggestive value of such comparisons, we must realize that, in the study of models of given structure and functions, we are very far from the situation in which we find ourselves in the investigation of living organisms, where our task is gradually to unravel their constitution and capacities.

In any model on the ordinary scale, we can essentially disregard the atomic constitution of matter and confine ourselves to the account of the mechanical and electrical properties of the materials used for the construction of the machine and to the application of the simple laws governing the interaction between its parts. From biological research, however, it is evident that fundamental characteristics of living organisms, and in particular genetic reproduction, depend primarily on processes on the atomic scale, where we are faced with essential limitations of the applicability of the concepts of classical physics.

As is well known, quantum physics offers a frame sufficiently wide for the account of properties of atoms entirely beyond the grasp of classical approach. A main result of this development is the recognition of a peculiar stability of atomic and molecular structures, which implies a degree of order incompatible with the unlimited use of mechanical pictures. The deterministic account of classical physics which implies that any disturbance of a system composed of an immense number of parts invariably leads to chaotic disorder, is in quantum physics replaced by a description according to which the result of any interaction between atomic systems is the outcome of a competition between various individual processes by which the states of the new systems, like those of the original systems, in a simple way are defined by the atomic particles they contain. With suitable adjustments, this description directly corresponds to the chemical kinetics which has found extensive application in molecular biology.

Quite novel prospects of a gradual elucidation of biological regularities on the basis of well-established principles of atomic physics have been opened in later years by the discoveries of the remarkably stable specific structures carrying genetic information, and the increasing insight into the processes by which this information is transferred. Indeed, the view suggests itself that, in metabolism, the formation and regeneration of the persistent constituents of the organism are to be regarded as processes of essentially irreversible character which at any step secure the greatest possible stability compatible with the prevailing conditions as regards material and energy exchange.

Although, thus, we have no reason to expect any inherent limitation of the application of elementary physical and chemical concepts to the analysis of biological phenomena, the peculiar properties of living organisms, which have resulted from the whole history of organic evolution, reveal potentialities of immensely complicated material systems, which have no parallel in the comparatively simple problems with which we are concerned in ordinary physics and chemistry. It is on this background that notions referring to the behaviour of organisms as entities, and apparently contrasting with the account of the properties of inanimate matter, have found fruitful application in biology.

Even though we are here concerned with typical complementary relationships as regards the use of appropriate terminology, it must be stressed that the argument differs in essential aspects from that concerning exhaustive objective description in quantum physics. Indeed, the distinction demanded by this description between the measuring apparatus and the object under investigation, which implies mutual exclusion of the strict application of space-time co-ordination and energy-momentum conservation laws in the account of individual atomic processes, is already, as indicated above, taken into account in the use of chemical kinetics and thermodynamics. The complementary approach in biology is rather required by the practically inexhaustible potentialities of living organisms entailed by the immense complexity of their structures and functions.

REFERENCE

BOHR, N. (1958). *Atomic Physics and Human Knowledge*. John Wiley, N.Y.

EDITOR'S NOTE

As stated in the footnote on its first page, section I of the previous paper is an abbreviated version of *Quantum Physics and Philosophy – Causality and Complementarity*, "Philosophy in the Mid-Century, A Survey" (ed. R. Klibansky), La nuova Italia editrice, Firenze 1958, pp. 308–314, reproduced in Vol. 7, pp. [388]–[394]. Section II, however, was prepared especially for the symposium. The following paper, *Physical Models and Living Organisms*, is an expansion of that section, as noted in the first footnote.

XII. PHYSICAL MODELS AND LIVING ORGANISMS

"Light and Life" (eds. W.D. McElroy and B. Glass),
The Johns Hopkins Press, Baltimore 1961, pp. 1–3

See Introduction to Part I, p. [12].

PREFACE

A Symposium on "Light and Life" was held at The Johns Hopkins University under the sponsorship of the McCollum-Pratt Institute on March 28-31, 1960. This volume contains the papers and informal discussions presented during the Symposium.

In the planning of the present Symposium we attempted, as in previous symposia, to bring together scientists from a number of different disciplines. Unfortunately time and printing limitations prevented the consideration of a number of interesting photobiological problems.

I would like to acknowledge the active participation of the members of the McCollum-Pratt Institute and of the Department of Biology in the planning of the Symposium. Many of the speakers on the program were helpful in suggesting areas or specific topics that should be discussed. It is also a pleasure to acknowledge the valuable contributions of the following moderators: Dr. Albert Szent-Györgyi, Dr. James Franck, Dr. William Arnold, Dr. C. B. van Niel and Dr. H. K. Hartline.

The support of a limited number of foreign investigators as participants in the Symposium was made possible through the generous aid of the National Science Foundation. Unfortunately, insufficient funds prevented the attendance of a larger number of participants from great distances.

We hope, however, that the published book will be of value to all investigators interested in photobiological processes.

W. D. McELROY
Director, McCollum-Pratt Institute

[134]

INTRODUCTION

PHYSICAL MODELS AND LIVING ORGANISMS[1]

NIELS BOHR

Copenhagen, Denmark

The discussion of the position of living organisms in a general description of physical phenomena has, in the development of science, passed through a number of stages. In Antiquity, the obvious difficulties inherent in a comparison between organisms and primitive machinery deeply influenced the attitude towards mechanical problems and even led to the attribution of vital characteristics to all matter. With the abandonment of these views, at the time of the Renaissance, through the clarification of the principles of classical mechanics, the problem entered into another stage, stimulated by the great anatomical and physiological discoveries at that period.

Recent advances in technology, and especially the development of automatic control of industrial plants and calculation devices, have given rise to a renewed discussion of the extent to which it is possible to construct mechanical and electrical models with properties resembling the behavior of living organisms. Indeed, it may be feasible to design models reacting in any prescribed manner, including their own reproduction, provided that they have access to the necessary materials and energy sources. Still, quite apart from the suggestive value of such comparisons, we must realize that, in the study of models of given structure and functions, we are very far from the situation in which we find ourselves in the investigation of living organisms, where our task is gradually to unravel their constitution and potentialities.

As regards this problem it is essential to realize from the very beginning that in organic life we are dealing with further resources of nature than in the construction of machines. Indeed, for this

[1] Except for a few small additions the substance of these remarks was presented in a contribution to the symposium on Models in Biology, Bristol, 1959.

1

purpose we can essentially disregard the atomic constitution of matter and confine ourselves to the account of the mechanical and electrical properties of the materials used and to the application of the simple laws governing the interaction between the parts of the machine. From biological research, however, it is evident that fundamental characteristics of living organisms, and in particular genetic reproduction, depend primarily on processes on the atomic scale, where we are faced with new problems.

On the ground of classical physics, the very question of maintaining a high degree of order of such immensely complicated systems presents serious difficulties. In fact, the incessant encounters between the atoms with a more or less liquid phase like the cytoplasm would lead to rapidly increasing disorder. Doubts have even been expressed about the compatibility of the existence and stability of living organisms with the laws of thermodynamics, but thorough investigation of the exchanges of energy and entropy accompanying the metabolism and movements of the organisms has never disclosed any departure from these laws.

A whole new background has, however, been created by the development of quantum physics, which, at the same time as it has revealed an essential limitation of the deterministic description of classical mechanics, has offered a proper basis for the account of the stability of atomic and molecular structures. As is well known, no picture on classical lines can be given of the electronic constitution of atoms or of the behavior of the electrons responsible for the bindings between atoms in chemical combinations. Owing to the large masses of the atomic nuclei compared with that of the electron, it is possible, however, effectively to retain a pictorial representation of the relative positions of the atoms in accordance with the structure formulae of chemistry, which have proved so adequate even for the highly complicated molecules with which we are concerned in organic metabolism.

In spite of the multifarious enzymatic processes involved in this metabolism, the problem of the stability of the organisms presents a fundamental simplicity, since, in the range of temperatures within which life can be upheld, the thermal fluctuations in the states of vibration and rotation of the molecules are in general far from sufficient to break the chemical bonds. Such fluctuations rather effect the rapid disappearance of correlations between all secondary characters of the states of the reacting systems and permit us to account for the primary features of their constitution merely by a specification of the atoms of which they are composed and the configuration in which these atoms are bound together.

The discoveries in recent years of the specific molecular structures carrying genetic information, and the increasing insight into the processes by which this information is transferred, have opened quite new prospects for the gradual elucidation of biological regularities on the basis of well-established principles of chemical kinetics. In particular, the almost unlimited possibilities of probing our metabolic transformations lend support to the view that the formation and regeneration of the structural constituents of the organisms are to be regarded as processes of not immediately reversible character, which at any step secure the greatest possible stability under the conditions maintained by nutrition and respiration.

Thus, there appears to be no reason to expect any inherent limitation of the application of elementary physical and chemical concepts to the analysis of biological phenomena. Yet, the characteristic properties of living organisms, which have resulted from the whole history of organic evolution, reveal potentialities of immensely complicated material systems, which have no parallel in the comparatively simple phenomena studied under reproducible experimental conditions. It is on this background that notions referring to the behavior of organisms as entities, and apparently contrasting with the account of the properties of inanimate matter, have found fruitful application in biology.

Even though we are here concerned with typical complementary relationships as regards the use of appropriate terminology, it must be stressed that the argumentation differs in essential aspects from that concerning exhaustive objective description in quantum physics. Indeed, the distinction demanded by this description between measuring apparatus and object under investigation, which implies mutual exclusion of the strict application of space-time coordination and energy-momentum conservation laws in the account of individual quantum processes, is already taken into account in the use of chemical kinetics and thermodynamics. Thus, the dual approach in biology does not seem to be conditioned by an interference with the properties of the specific molecular structures, inherently involved in their identification, but is rather required by the practically inexhaustible potentialities of living organisms entailed by the immense complexity of their constitution and functions.

XIII. ADDRESS AT THE SECOND INTERNATIONAL GERMANIST CONGRESS

"Spätzeiten und Spätzeitlichkeit", Francke Verlag, Bern 1962, pp. 9–11

Address at the Second International Germanist Congress,
Copenhagen, 22 August 1960

See Introduction to Part I, p. [12].

SPÄTZEITEN
UND SPÄTZEITLICHKEIT

Vorträge, gehalten auf dem II. Internationalen
Germanistenkongreß 1960 in Kopenhagen

Herausgegeben im Auftrage der Internationalen Vereinigung
für Germanische Sprach- und Literaturwissenschaft
von Werner Kohlschmidt

FRANCKE VERLAG BERN
UND MÜNCHEN

Address delivered by Professor Niels Bohr
at the Second International Germanist Congress

Copenhagen, August 22, 1960

It is a great pleasure for me on behalf of the Royal Danish Academy of Science and Letters to extend a most hearty welcome to the many distinguished philologists who have gathered here in Copenhagen for this international Germanist Congress.

Language is indeed our main tool for the expression and communication of knowledge and views and hopes, and researches into the development of languages and their mutual influence on each other present one of the richest sources for the illumination of many aspects of human culture. Linguistic studies therefore present an important theme for such international scientific co-operation, which through the ages has inspired the growth of knowledge and understanding.

As a physicist I have, of course, no special scholarship in your subject, but in spite of the different directions of our studies we must not forget that they are both part of man's age-old endeavours to clarify his position in that nature of which he is himself a part. While originally these endeavours could be pursued in their entirety by a single individual, the growth of knowledge and ever more elaborate methods of enquiry have called for a division of labour among research workers. It ist true that increasing specialization entails the danger that we loose sight of the unity of human knowledge, but the studies of language would appear especially suited to remind us of the common foundation on which we stand.

Of course, it is not possible to distinguish between language as such and its use in social intercourse and common orientation in our surroundings. Indeed, poetry and philosophy have in the course of time enriched and refined language by creating means of expression for the appreciation of ethical and aesthetic values and for the formulation of epistemological enquiry. Similarly, the development of the sciences has steadily augmented our vocabulary with words pointing to new experience which, through its technological applications, has gradually changed the frame of our daily life.

A special role is played by mathematics which has contributed so decisively to the development of logical thinking, and which by its well-defined abstractions offers invaluable help in expressing harmonious relationships. Rather than a separate branch of knowledge, pure mathematics may be considered as a refinement of general language, supplementing it with appropriate tools to represent relations for which ordinary verbal expression is unprecise or cumbersome.

From the beginning, the use of mathematics has been essential for the development of the physical sciences. While Euclidian geometry sufficed for Archimedes' elucidation of many problems of static equilibrium, the detailed account of the motion of ordinary bodies demanded the development of the infinitesimal calculus on which the whole edifice of Newtonian mechanics rests. In our days, still more abstract tools of mathematics have proved indispensable for ordering new fields of experience, made accessible by the development of the art of experimentation, and at the same time have thrown new light on the scope of human language.

In this connection I think in the first place of the general theory of relativity by which Einstein, through his recognition of the inherent limitations of accustomed ideas of absolute space and time, has given our world picture a unity and harmony surpassing all previous dreams. As often emphasized, the disclosure of the extent to which our approach to even the comparatively simple problems of co-ordinating physical experience depends on the standpoint of the observer entails a forceful reminder, in dealing with more complex aspects of human life, to be aware of the specific background from which they are viewed.

A further development, which has disclosed an unsuspected limitation of the unambiguous applications of the very concepts of cause and effect, and which has given rise to much discussion among physicists and philosophers about the range of logic and language, was, as well known, initiated by Planck's discovery, in the first year of this century, of the universal quantum of action, which revealed a feature of wholeness in individual atomic processes defying causal description in space and time.

This discovery proved a clue to the peculiar stability of atomic systems, on which all properties of matter ultimately depend, and taught us that the wide applicability of so-called classical physics to the description of the behaviour of matter in bulk rests entirely on the circumstance that the action involved in usual physical phenomena is so large, that the quantum can be completely neglected. Modern development of experimental technique has, however, given us direct information concerning single atoms and proper quantum effects.

The comprehension of the new rich experience in atomic physics has presented great difficulties, and proved only possible by means of a suitable mathematical symbolism. The characteristic feature of this so-called quantum mechanics is the novel relationship, exhibited by phenomena observed under different experimental conditions, expressed by the notion of complementarity. Indeed, however contrasting such phenomena may at first sight appear, it must be realized that they together exhaust all knowledge about the atomic object which can be unambiguously expressed.

This novel situation in physical science has its origin in the circumstance that the interaction between the objects under investigation and the tools of observation, which in usual experience can be neglected or compensated for, in the domain of quantum physics represents an inseparable part of the phenomena. Analogies to such a state of affairs may be recognized in many other fileds of human knowledge, where we meet with similar conditions for analysis and synthesis of experience.

In particular, the vocabulary employed for the description of our state of mind is composed of words which have been used in a complementary manner since the very origin of language. Thus, words like thoughts and sentiments, or contemplation and volition, refer to psychical experiences which – due to the different direction of our attention – are mutually exclusive, but which together point to the richness of conscious life. Also as regards social intercourse, many aspects of the relation between the individual and the community entail typically complementary features. Further, the well known difficulties of an unprejudiced comparison between different national cultures may be traced to the impossibility of appreciating foreign traditions on the background of own social customs.

In this assembly I shall not venture to enlarge on these topics with which scholars in humanistic research are so well acquainted, but I wanted to stress the important role which enquiries into our means of communication by language plays in all branches of knowledge, and to point to the help which a comparison between the lessons in different fields of research may offer for the recognition of the unity of human knowledge. Not least, in face of the serious challenge with which the whole civilization is presented due to our increasing mastery of the forces of nature, and which can be met only by goodwill and understanding between all people, the appreciation of our common position is more necessary than ever before.

Striving together for growth and harmony of human knowledge offers unique opportunities for the promotion of mutual confidence. With the hope that your Congress may contribute to this great goal, I express my warmest wishes that it will be an inspiring experience to all participants.

XIV. THE CONNECTION BETWEEN THE SCIENCES

Journal Mondial de Pharmacie,
No. 3, Juillet–Decembre 1960, pp. 262–267

Address delivered at the International Congress of Pharmaceutical Sciences
in Copenhagen on 29 August 1960

See Introduction to Part I, p. [12].

[145]

Bohr studying the programme for the Pharmaceutical Congress together with Hugh Linstead, president of the International Pharmaceutical Union.

THE CONNECTION BETWEEN THE SCIENCES (1960)

Versions published in English, German and Danish

English: The Connection Between the Sciences
A [Untitled,] Address delivered at the International Congress of Pharmaceutical Sciences of the Fédération Internationale Pharmaceutique, Copenhagen, August 29, 1960. Journal Mondial de Pharmacie, No. 3, Juillet–Decembre 1960, pp. 262–267
B "Essays 1958–1962 on Atomic Physics and Human Knowledge", Interscience Publishers, New York 1963, pp. 17–22 (reprinted in: "Essays 1958–1962 on Atomic Physics and Human Knowledge, The Philosophical Writings of Niels Bohr, Vol. III", Ox Bow Press, Woodbridge, Connecticut 1987, pp. 17–22)

German: Die Verbindung zwischen den Wissenschaften
C *Über die Einheit unseres Wissens*, Universitas **16** (1961) 835–840
D "Atomphysik und menschliche Erkenntnis II", Friedr. Vieweg & Sohn, Braunschweig 1966, pp. 17–22

Danish: Forbindelsen mellem videnskaberne
E "Atomfysik og menneskelig erkendelse II", J.H. Schultz Forlag, Copenhagen 1964, pp. 29–35

B is identical to A, apart from minor improvements of formulation.

The introductory sentence, as well as the sentence beginning with "On this background" (A, p. 265, line 8) is omitted in C. Otherwise, C and D are seemingly independent translations of A and B, respectively.

E corresponds to B.

Address delivered by Professor Niels Bohr
at the International Congress of Pharmaceutical Sciences
of the Fédération Internationale Pharmaceutique

Copenhagen, August 29, 1960.

It is with great pleasure, although not without hesitation, that
I have accepted the kind invitation to speak at the opening of this
International Congress of Pharmaceutical Sciences. As a physicist
I have, of course, no such insight in the field of pharmacy as is
possessed in richest measure by the many distinguished scientists
from different countries who are here assembled. At this occasion,
however, it may be appropriate to comment on the intimate connection between our knowledge in all branches of science. This connection was indeed emphasized with vigor and enthusiasm by
Hans Christian Orsted, who established the first regulated pharmaceutic examination in Denmark, and it was to him a constant source
of inspiration in his fundamental scientific researches and his
many-sided and fruitful activity in the Danish community.

* * *

Experience of the help which substances occuring in nature
can afford for the cure of human disease goes back to the infancy
of civilization, when the conception of rational scientific enquiry
was still unknown. Yet, it is interesting to recall how much the
search for medical herbs in woods and meadows has stimulated the
development of systematic botany. Furthermore, the preparation of
medicaments and the study of their effects were to prove of essential
importance for the progress of chemistry.

For a long time the study of the properties and transformations of substances stood conspicuously apart from the endeavours,
characteristic for the approach in physics, to account for the
behaviour of the bodies in our surroundings in terms of space and
time and cause and effect. Indeed this was the foundation for the
whole edifice of Newtonian mechanics and even electromagnetic
theory based on Orsted's and Faraday's discoveries, which through
their technological applications have so largely changed the frame
of our daily life.

The development in the former century of the ancient ideas of
the atomic constitution of matter stimulated the search for at closer
connection between chemistry and physics. On the one hand the

clarification of the concept of chemical elements led to the under-standing of the laws governing the proportions in which these elements enter in chemical combinations. On the other hand the study of the remarkably simple properties of gases led to the develop-ment of the mechanical theory of heat, offering an explanation of the general laws of thermodynamics which have found such fruitful application, not least in physical chemistry.

Studies of thermal radiative equilibrium, based on electro-magnetic theory, were, however, to disclose a feature of wholeness in atomic processes, irreconcilable with the ideas of classical physics. Indeed, Planck's discovery of the universal quantum of action taught us that the wide applicability of the accustomed description of the behaviour of matter in bulk rests entirely on the circumstance that the action involved in phenomena on the ordinary scale is so large that the quantum can be completely neglected. In individual atomic processes, however, we meet with regularities of a novel kind, responsible for the peculiar stability of atomic systems on which all properties of matter ultimately depend.

The ordering of this new rich field of experience has demanded a radical revision of the foundation for the unambiguous use of our most elementary physical concepts. To account for what we actually do and learn in physical experimentation, it is of course necessary to describe the experimental arrangement and the recording of observations in common language. In the study of atomic phenomena, however, we are presented with a situation where the repetition of an experiment with the same arrangement may lead to different recordings, and experiments with different arrangements may give results which at first sight seem contradictory to each other.

The elucidation of these apparent paradoxes has been brought about by the recognition that the interaction between the objects under investigation and our tools of observation, which in ordinary experience can be neglected or taken into account separately, forms, in the domain of quantum physics, an inseparable part of the phenomena. Indeed, under such conditions experience cannot be combined in the accustomed manner, but the phenomena must be considered as complementary to each other in that sens that they together exhaust all information about the atomic objects which can be unambiguously expressed.

The proper mathematical tools for a comprehensive description on complementary lines have been created by the so-called quantum formalism by which we have to such an extent been able to account for the physical and chemical properties of matter. The humorous dispute between physicists and chemists as to whether chemistry has been swallowed by physics or physics become chemistry illus-trates the character and scope of this progress.

[149]

It will carry us much too far from our theme to mention in detail the great development of atomic science in our days, and I shall only briefly remind that in the binding of the electrons to the nucleus of the atom and in the role they play for the joining of the atoms in molecules of chemical compounds we have to do with typical quantum effects, resisting accustomed pictorial representation. Due to the large mass of the nuclei compared with that of the electrons, it is, however, possible to account with high approximation for the atomic configuration in the molecules, corresponding to the well known structural formulae which have proved so indispensable in the ordering of chemical evidence.

The whole approach is not only in complete accordance with the usual chemical kinetics, but even stresses the simple assumptions on which it is based. Thus, in any process resulting in chemical combinations, the properties of the new molecules do not primarily depend on the composition of the molecules by whose interaction they were formed, but only on the relative placing of the atoms of which they consist. Any secondary characteristics of the state of such molecules, corresponding to oscillations left from their formation, will indeed not essentially affect their chemical properties and will even, due to the general thermal agitations in the medium, rapidly loose all connection with their previous history.

The general understanding of the specific properties of matter, to which the quantum of action provided a clue, has initiated a period of rapid growth of the natural sciences, reminiscent in many respects of the scientific revolution in the sixteenth and seventeenth century. Among the most impressive of these developments is the modern rise of biochemistry which has been equally beneficial to physiology and pharmacology. In particular the widegoing obliteration of the distinction between organic and inorganic chemistry has raised anew the old problem of the extent to which the physical sciences can account for the display of life.

The gradual recognition, through the development of anatomy and physiology, of the immense complexity of the structure of living organisms and the multifarious refined regulative mechanisms governing their function, has often led to doubts whether the maintenance of order in the organism is compatible with the general laws of thermodynamics. Still, from the standpoint of modern chemical kinetics, no such departure is to be expected, and thorough investigations into the exchange of energy and entropy accompanying the metabolism and movements of the organisms have in fact never disclosed any restriction of thermodynamical principles.

Great progress has in the last years been achieved as regards our knowledge of the complicated molecular structures in living cells and especially of the specific molecular chains which carry the genetic information from generation to generation. Further, our insight in the enzymatic processes by which this information serves to direct the formation of other specific molecular structures, like the proteins, is steadily increasing. In fact, for all we know, we have here to do with a steady increase in the stability of the constitution of the cells with an expenditure of free energy corresponding to the increase of entropy in usual irreversible chemical processes.

On this background the view suggests itself that, in the whole life of the organism, we have to do with processes of not immediately reversible character corresponding to an ever increasing stability under the prevailing conditions maintained by nutrition and respiration. In spite of all differences of scale and function we are here faced with a far-reaching similarity between living organisms and automatic machines. Indeed, on the basis of recent advances in technology it is possible to design machines reacting in any prescribed manner, including their own repair and reproduction, provided that they have access to the necessary materials and energy sources.

Still, as regards the much debated question concerning the comparison between organisms and machines, it is essential to keep in mind that organic life is a manifestation of nature's resources far beyond those used for the construction of machines. In fact, in the account of the functioning of devices for calculation and control we can essentialy disregard the atomic constitution of matter and confine ourselves to the account of the mechanical and electrical properties of the materials used and to the application of the simple physical laws governing the interaction between the parts of the machine. However, the whole history of organic evolution presents us with the results of the trying out in nature of the immense possibilities of atomic interactions.

Because of their immense complexity it is not surprising that the organisms reveal properties and potentialities, which are in striking contrast with those exhibited by so-called inanimate matter under simple reproducible experimental conditions. It is on this background that such notions as purpose-fulness and self-preservation, referring to the behaviour of organisms as entities, have found fruitful application in biological research.

In the discussions of the foundations of biology the question of the rôle of notions beyond the language of physics has formed a main topic. From the one side the view has been expressed that such concepts, despite their evident fertility, would eventually prove superfluous. From the other side it has been argued that we have here to do with irreducible elements in any account of the display of life.

The lesson as regards our position as observers of nature which quantum physics has taught us, has given a new background to

[151]

such discussions. Indeed, this lesson suggests that the situation as regards objective description of biological phenomena reflects different approaches in ordinary physiology and modern biochemistry. The basis for the complementary mode of description in biology is not connected with the problems of controlling the interaction between object and measuring tool, already taken into account in chemical kinetics, but with the practically inexhaustible complexity of the organism.

This situation can hardly be regarded as being of temporary character, but would rather seam to be inherently connected with the way in which our whole conceptual framework has developed from serving the more primitive necessities of daily life to coping with the growth of knowledge gained by systematic scientific research. Thus, as long as the word « life » is retained for practical or epistemological reasons, the dual approach in biology will surely persist.

In our discussion we have so far considered living organisms as objects under investigation, in a way similar to that in which we strive to comprehend experience of any other part of nature. When we approach the problems of psychology we enter into a new domain of knowledge where the question of analysis and synthesis has attracted vivid interest through the ages. The language which we use in social intercourse to communicate our state of mind is indeed very different from that usually employed in the physical sciences. Thus words like contemplation and volition, referring to situations which are mutually exclusive, but equally characteristic of conscious life, have been used in a typical complementary manner since the very origin of language.

The close relationship between psychical experiences and physical and chemical processes in our body is evidenced not least by the application of medicaments in mental disease. The irreversible character of the physiological processes concerned is also clearly reflected by the degree to which all that has ever come to consciousness can be remembered. It is of course tempting to pursue such considerations, but at each further step new difficulties turn up, inherently connected with the limited scope of the concepts available for such inquiry.

* * *

In this address I have tried to show how researches into the world of atoms have offered new opportunities of tracing that harmony in nature of which Orsted spoke, but which we perhaps would rather refer to as the unity of human knowledge. It is indeed only the appreciation of such harmony or unity which can help us to keep a balanced attitude to our position and avoid that confusion which the tumultuous progress of science and technology in almost every field of human interest may so easily produce. The program of this congress bears witness to the fact that pharmaceutic and pharmacological science represent an integral part of that inquiry into the wonders of nature by which we strive to promote human understanding and welfare. With the hope that your meeting will contribute to this great goal, I want to express my warmest wishes that it will be an inspiring experience to you all.

XV. THE UNITY OF HUMAN KNOWLEDGE

Revue de la Fondation Européenne de la Culture, July 1961, pp. 63–66

Address delivered at the International Congress
"Caractère et culture de l'Europe"
in Copenhagen on 21 October 1960

See General Introduction, pp. XLII and XLV.

THE UNITY OF HUMAN KNOWLEDGE (1960)

Versions published in Danish, English and German

Danish: Den menneskelige erkendelses enhed
A   Berlingske Tidende, 22 October 1960
B   "Atomfysik og menneskelig erkendelse II", J.H. Schultz Forlag, Copenhagen 1964, pp. 19–28

English: The Unity of Human Knowledge
C   Revue de la Fondation Européenne de la Culture, July 1961, pp. 63–66
D   "Essays 1958–1962 on Atomic Physics and Human Knowledge", Interscience Publishers, New York 1963, pp. 8–16 (reprinted in: "Essays 1958–1962 on Atomic Physics and Human Knowledge, The Philosophical Writings of Niels Bohr, Vol. III," Ox Bow Press, Woodbridge, Connecticut 1987, pp. 8–16)

German: Die Einheit menschlicher Erkenntnis
E   "Europa", Monatsschrift für Politik, Wirtschaft und Kultur, August 1961, pp. 45–48
F   "Atomphysik und menschliche Erkenntnis II", Friedr. Vieweg & Sohn, Braunschweig 1966, pp. 8–16

All of these versions agree with each other, apart from minor improvements of formulation from *C* to *D* and from *E* to *F*.

THE UNITY OF HUMAN KNOWLEDGE

PROFESSOR NIELS BOHR

The unity of human knowledge is perceptible only to a lofty mind. Professor Bohr, working for the clarification of the principles of quantum physics and their epistomological implications, has stressed and analysed logical analogies between basic problems in this field and in the biological, psychological and social sciences. Director of the Copenhagen Institute for Theoretical Physics, he received the Nobel Prize (1922) for his theory of the hydrogen atom and the first Atoms for Peace Award (1957).

The question alluded to in the title is as old as civilization itself but has acquired renewed attention in our days with the increasing specialization of studies and social activities. From various sides concern has been expressed with the widespread confusion arising from the apparently divergent approaches taken by humanists and scientists to human problems and there has even been talk about a cultural rift in modern society. We must, however, not forget that we are living in times of rapid developments in many fields of knowledge, reminiscent in such respect of the age of European Renaissance.

However great the difficulties of liberation from the medieval world view were felt at that time, the fruits of the so-called Scientific Revolution are now a part of the common cultural background. In our century the immense progress of the sciences has not only greatly advanced technology and medicine, but has at the same time given us an unsuspected lesson about our position as observers of that nature of which we are part ourselves. Far from implying a schism between humanism and physical science, this development entails a message of importance for our attitude to common human problems, which has given the old question of the unity of knowledge new perspective.

The pursuit of scientific enquiry with the aim of augmenting and ordering our experience of the world around us has through the ages proved fertile, not least for the progress of technology which has changed the frame of our daily life. While early developments of astronomy, geodesy and metallurgy in Egypt, Mesopotamia, India and China were primarily directed to serve requirements of the community, it is in ancient Greece that we first meet with systematic endeavours to clarify the basic principles for the description and ordering of knowledge.

We admire the Greek mathematicians who in many respects laid the firm foundation on which later generations have built. It is important to realize that the definition of mathematical symbols and operations is based on simple logical use of common language. Mathematics is therefore not to be regarded as a special branch of knowledge based on the accumulation of experience but rather as a refinement of general language, supplementing it with appropriate tools to represent relations for which ordinary verbal expression is imprecise or too cumbersome.

In view of the apparent remoteness of mathematical abstractions, it may be noted that even elementary mathematical training allows school disciples to see through the famous paradox of the race between Achilles and the tortoise. How could the fleet-footed hero ever catch up with and pass the slow reptile if it were given even the smallest handicap? Indeed, at his arrival at the starting point of the turtle Achilles would find that it had moved to some further point along the race track, and this situation would be repeated in an infinite sequence.

The logical analysis of situations of this type was to play an important rôle for the development of mathematical concepts and methods.

From the beginning, the use of mathematics has been essential for the progress of the physical sciences. While Euclidean geometry sufficed for Archimedes' elucidation of fundamental problems of static equilibrium, the detailed description of the motion of material bodies demanded the development of the infinitesimal calculus on which the imposing edifice of Newtonian mechanics rests. Above all the explanation of the orbital motion of the planets in our solar system, based on simple mechanical principles and the law of universal gravitation, deeply influenced the general philosophical attitude in the following centuries and strengthened the view that space and time as well as cause and effect had to be taken as *a priori* categories for the comprehension of all knowledge.

The extension of physical experience in our days has, however, necessitated a radical revision of the foundation for the unambiguous use of our most elementary concepts, and has changed our attitude to the aim of physical science. From our present standpoint physics is to be regarded not so much as the study of something *a priori* given, but rather as the development of methods for ordering and surveying human experience. In this respect our task must be to account for such experience in a manner independent of individual subjective judgement and therefore objective in that sense, that it can be un-

63

ambiguously communicated in the common human language.

As regards the very concepts of space and time reflected in the primitive use of words as *here* and *there* and *before* and *after*, it is to be remembered how essential the immense speed of light propagation, compared with the velocities of the bodies in our neighbourhood, is for our ordinary orientation. However, the surprise that it proved impossible even by the most refined measurements to ascertain, in laboratory experiments, any effect of the orbital motion of the earth around the sun, revealed that the shape of rigid bodies and their mutual distances would be differently perceived by observers swiftly moving relative to each other, and that even events, which by one observer would be judged as simultaneous, could be reckoned by another as occurring at different moments. The recognition of the extent to which the account of physical experience depends on the standpoint of the observer proved fertile in tracing fundamental laws valid for all observers.

Indeed, the general theory of relativity, by which Einstein, in renouncing all ideas of absolute space and time, gave our world picture a unity and harmony surpassing any previous dreams, offered an instructive lesson as regards the consistency and scope of plain language. Although the convenient formulation of the theory involves mathematical abstractions as four-dimensional non-Euclidean geometry, its physical interpretation rests fundamentally on every observer's possibility of maintaining a sharp separation between space and time and of surveying how any other observer, in his frame, will describe and coordinate experience by means of a common language.

New fundamental aspects of the observational problem, entailing a revision of the very foundation for the analysis of phenomena in terms of cause and effect, were to be uncovered by the development initiated by Planck's discovery of the universal quantum of action in the first year of this century. This discovery proved that the wide applicability of so-called classical physics rests entirely on the circumstance that the action involved in any phenomena on the ordinary scale is so large that the quantum can be completely neglected. In atomic processes, however, we meet with regularities of a novel kind, defying causal pictorial description but nevertheless responsible for the peculiar stability of atomic systems on which all properties of matter ultimately depend.

In this new field of experience, we have met with many great surprises and even been faced with the problem of what kind of answers we can receive by putting questions to Nature in the form of experiments. Indeed, in the account of ordinary experience it is taken for granted that the objects under investigation are not interfered with by observation. It is true that when we look at the moon in a telescope we receive light from the sun reflected from the moon-surface, but the recoil from this reflection is far too small to have any effect on the position and velocity of a body as heavy as the moon. If, however, we have to do with atomic systems, whose constitution and reactions to external influence are fundamentally determined by

the quantum of action, we are in a quite different position. Faced with the question of how under such circumstances we can achieve an objective description, it is decisive to realize that however far the phenomena transcend the range of ordinary experience, the description of the experimental arrangement and the recording of observations must be based on common language. In actual experimentation this demand is satisfied with the specification of the experimental conditions through the use of heavy bodies like diaphragms and photographic plates, the manipulation of which is accounted for in terms of classical physics. Just this circumstance, however, excludes any separate account of the interaction between the measuring instruments and the atomic objects under investigation.

In particular this situation prevents the unlimited combination of space-time coordination and the conservation laws of momentum and energy on which the causal pictorial description of classical physics rests. Thus, an experimental arrangement aiming at ascertaining where an atomic particle, whose position at a given time has been controlled, will be located at a later moment, implies a transfer, uncontrollable in principle, of momentum and energy to the fixed scales and regulated clocks necessary for the definition of the reference frame. Conversely, the use of any arrangement suited to study momentum and energy balance—decisive for the account of essential properties of atomic systems—implies a renunciation of detailed space-time coordination of their constituent particles.

Under these circumstances it is not surprising that with one and the same experimental arrangement we may obtain different recordings corresponding to various individual quantum processes for the occurrence of which only statistical account can be given. Likewise we must be prepared that evidence, obtained by different, mutually exclusive experimental arrangements, may exhibit unprecedented contrast and even at first sight appear contradictory.

It is in this situation that the notion of complementarity is called for to provide a frame wide enough to embrace the account of fundamental regularities of nature which cannot be comprehended within a single picture. Indeed, evidence obtained under well-defined experimental conditions—and expressed by adequate use of elementary physical concepts—exhausts in its entirety all information about the atomic objects which can be communicated in common language.

A detailed account on complementary lines of a new wide domain of experience has been possible by the gradual establishment of a mathematical formalism, known as quantum mechanics, in which the elementary physical quantities are replaced by symbolic operators subject to an algorism, involving the quantum of action and reflecting the non-commutativity of the corresponding measuring operations. Just by treating the quantum of action as an element evading customary explanation—similar to the rôle of the velocity of light in relativity theory as a maximum speed of signals—this formalism can be regarded as a rational generalization of the conceptual

64

H.R.H. the Prince of the Netherlands receiving Professor Niels Bohr at the Copenhagen Congress of the Fondation Européenne de la Culture

framework of classical physics. For our theme, however, the decisive point is that the physical content of quantum mechanics is exhausted by its power to formulate statistical laws governing observations obtained under conditions specified in plain language.

The fact that in atomic physics, where we are concerned with regularities of unsurpassed exactness, objective description can be achieved only by including in the account of the phenomena explicit reference to the experimental conditions, emphasizes in a novel manner the inseparability of knowledge and our possibilities of enquiry. We are here concerned with a general epistemological lesson, illuminating our position in many other fields of human interest.

In particular the conditions of analysis and synthesis of so-called psychic experiences have always been an important problem in philosophy. It is evident that words like thoughts and sentiments, referring to mutually exclusive experiences, have been used in a typical complementary manner since the very origin of language. In this context, however, the subject-object separation demands special attention. Every unambiguous communication about the state and activity of our mind implies of course a separation between the content of our consciousness and the background loosely referred to as "ourselves", but any attempt at exhaustive description

of the richness of conscious life demands in various situations a different placing of the section between subject and object.

In order to illustrate this important point, I shall quote a Danish poet and philosopher, Poul Martin Møller, who lived about a hundred years ago and left behind an unfinished novel called "The Adventures of a Danish Student", in which the author gives a remarkably vivid and suggestive account of the interplay between the various aspects of our position, illuminated by discussions within a circle of students with different characters and divergent attitudes to life.

I shall refer to a conversation between two cousins, one of whom is very soberly efficient in practical affairs, of the type which is known among students as a philistine, whereas the other, called the licentiate, is addicted to remote philosophical meditations detrimental to his social activities. When the philistine reproaches the licentiate for not having made up his mind concerning a practical job, offered him by the kindness of his friends, the poor licentiate apologizes most sincerely, but explains the difficulties in which his reflections have brought him.

Thus he says: "My endless enquiries make it impossible for me to achieve anything. Furthermore, I get to think about my own thoughts of the situation in which I find

65

[159]

myself. I even think that I think of it, and divide myself into an infinite retrogressive sequence of 'I's who consider each other. I do not know at which 'I' to stop as the actual, and the moment I stop at one, there is indeed again an 'I' which stops at it. I become confused and feel a dizziness as if I were looking down into a bottomless abyss, and my ponderings finally result in a terrible headache.''

In his reply the cousin says: "I cannot in any way help you in sorting your many 'I's. It is quite outside my sphere of action, and I should either be or become as mad as you, if I let myself in for your superhuman reveries. My line is to stick to palpable things and walk along the broad highway of common sense; therefore my 'I's never get tangled up."

Quite apart from the fine humour with which the story is told, it is certainly not easy to give a more pertinent account of essential aspects of the situation with which we all are faced.

The complementary way in which words like *contemplation* and *volition* are used has especially to be taken into account when turning to the problem of the freedom of the will, discussed by philosophers through the ages. Even if we cannot say whether we want to do something because we gather that we can, or we can only do it because we will, the feeling of being able to "make the best out of circumstances" is a common human experience. Indeed, the notion of volition plays an indispensable part in human communication, similarly to words like *hope* and *responsibility*, in themselves equally undefinable outside the context in which they are used.

The flexibility of the subject-object separation in the account of conscious life corresponds to a richness of experience so multifarious that it involves a variety of approaches. As regards our knowledge of fellow-beings, we witness of course only their behaviour, but we must realize that the word consciousness is unavoidable when such behaviour is so complex that its account in common language entails reference to self-awareness. It is evident, however, that all search for an ultimate subject is at variance with the aim of objective description which demands the contraposition of subject and object.

Such considerations involve no lack of appreciation of the inspiration which the great creations of art offer us by pointing to features of harmonious wholeness in our position. Indeed, in renouncing logical analysis to an increasing degree, and in turn allowing the play on all strings of emotion, poetry, painting and music contain possibilities of bridging between extreme modes as those characterized as *pragmatic* and *mystic*. Conversely, already ancient Indian thinkers understood the logical difficulties in giving exhaustive expression to such wholeness. In particular they found escape from apparent disharmonies in life by stressing the futility of demanding an answer to the question of the meaning of existence; realizing that any use of the word "meaning" implies comparison, and with what can we compare the whole existence?

The aim of our argumentation is to emphasize that all experience, whether in science, philosophy or art, which may be helpful to mankind, must be capable of being communicated by human means of expression, and it is on this basis that we shall approach the question of unity of knowledge. Confronted with the great diversity of cultural developments, we may therefore search for those features in all civilizations which have their roots in the common human situation. Especially we recognize that the position of the individual within the community exhibits in itself multifarious, often mutually exclusive aspects.

When approaching the age-old problem of the foundation of so-called ethical values we shall in the first place ask about the scope of such concepts as *justice* and *charity*, the closest possible combination of which is attempted in all human societies. Yet it is evident that a situation permitting unambiguous use of accepted judicial rules leaves no room for the free display of charity. But, as stressed especially by the famous Greek tragedians, it is equally clear that compassion can bring everyone in conflict with any concisely formulated idea of justice. We are here confronted with complementary relationships inherent in the human position, and unforgettably expressed in old Chinese philosophy, reminding us that in the great drama of existence we are ourselves both actors and spectators.

In comparing different national cultures we meet with the special difficulty of appreciating the culture of one nation in terms of traditions of another. In fact, the element of complacency inherent in every culture corresponds closely to the instinct of self-preservation characteristic of any species among the living organisms. In such context it is, however, important to realize that the mutually exclusive characteristics of cultures, resting on traditions fostered by historical events, cannot be immediately compared to those met with in physics, psychology and ethics, where we are dealing with intrinsic features of the common human situation.

In fact, contact between nations has often resulted in the fusion of cultures, retaining valuable elements of the original national traditions. The question of how to ameliorate the so-called cultural rift in modern societies is after all a more restricted educational problem, the attitude to which would seem to call not only for information but also for some humour. A most serious task is, however, to promote mutual understanding between nations with very different cultural background.

The rapid progress of science and technology in our days, which entails unique promises for the promotion of human welfare, and at the same time imminent menaces to universal security, presents our whole civilization with a veritable challenge. Certainly, every increase in knowledge and potentialities has always implied a greater responsibility, but at the present moment, when the fate of all peoples is inseparably connected, a collaboration in mutual confidence, based on appreciation of every aspect of the common human position, is more necessary than ever before in the history of mankind.

66

[160]

XVI. LIGHT AND LIFE REVISITED

ICSU Review **5** (1963) 194–199

Address delivered at the inauguration of the Institute for Genetics,
Cologne, on 21 June 1962

See Introduction to Part I, pp. [12] and [25].

[161]

Max Delbrück (centre) and Niels Bohr (standing on the right) at the inauguration of the Institute for Genetics at the University of Cologne, June 1962.

LIGHT AND LIFE REVISITED (1962)

Versions published in English, German and Danish

English: Light and Life Revisited

A   ICSU Review **5** (1963) 194–199

B   "Essays 1958–1962 on Atomic Physics and Human Knowledge", Inter-science Publishers, New York 1963, pp. 23–29 (reprinted in: "Essays 1958–1962 on Atomic Physics and Human Knowledge, The Philosophical Writings of Niels Bohr, Vol. III", Ox Bow Press, Woodbridge, Connecticut 1987, pp. 23–29)

German: Licht und Leben – noch einmal

C   Naturwiss. **50** (1963) 725–727

D   "Atomphysik und menschliche Erkenntnis II", Friedr. Vieweg & Sohn, Braunschweig 1966, pp. 23–29

Danish: Lys og liv påny

E   "Atomfysik og menneskelig erkendelse II", J.H. Schultz Forlag, Copenhagen 1964, pp. 36–42

F   "Naturbeskrivelse og menneskelig erkendelse" (eds. J. Kalckar and E. Rüdinger), Rhodos, Copenhagen 1985, pp. 110–118

All of these versions agree with each other, apart from minor improvements of formulation from A to B. A and C have a foreword by Aage Bohr. A postscript by Ronald Fraser appears only in A.

LIGHT AND LIFE REVISITED*

NIELS BOHR

(Late Director of the Institute for Theoretical Physics, The University, Copenhagen (Denmark)

Foreword

The present article is a draft of a manuscript prepared by Niels Bohr as the basis of a lecture he delivered in Cologne in June 1962.

The relationship between physics and biology had through the years deeply interested my father, and he first discussed his ideas on the subject in the article '*Light and Life*', published in 1933[1]. He felt that some of his remarks from that time had not always been properly interpreted, and he was anxious to give an account of his views as they had developed since then, in particular under the stimulation of the great new discoveries in the field of molecular biology, which he had followed with such enthusiasm.

My father was hoping to prepare a more detailed article based on his lecture in Cologne, but shortly afterwards he became ill and, although he was well on his way to recovery and had resumed work on the article, it was not completed when he died, suddenly, on November 18, 1962.

It is only with considerable hesitation that the manuscript is being published. Those familiar with my father's way of working will know what great efforts he devoted to the preparation of all his publications. The text would always be rewritten many times while the matter was being gradually elucidated, and until a proper balance was achieved in the presentation of its various aspects. Although a great deal of work had been done on the present manuscript, it was still far from completion. With regard to a few passages containing comments on specific biological problems, the author had planned major revisions. These passages have therefore been omitted from the text; their substance is indicated in notes inserted in small print. Furthermore, a few minor alterations of a formal character have been introduced.

Aage Bohr

It is a great pleasure for me to follow the invitation of my old friend Max Delbrück to speak at the inauguration of the new Institute of Genetics at the University of Cologne. Of course, as a physicist I have no first-hand knowledge of the extensive and rapidly developing field of research to which this institute is devoted, but I welcome Delbrück's suggestion to comment upon some general considerations about the relationship between biology and atomic physics, which I presented in an address entitled 'Light and Life', delivered at an International Congress on Radiation Therapy in Copenhagen thirty years ago. Delbrück, who at that time was working with us in Copenhagen as a physicist, took great interest in such considerations which, as he has been kind enough to say, stimulated his interest in biology and presented him with a challenge in his successful researches in genetics.

The place of living organisms within general physical experience has through the

---

* Unfinished manuscript.

ages attracted the attention of scientists and philosophers. Thus, the integrity of the organisms was felt by Aristotle to present a fundamental difficulty for the assumption of a limited divisibility of matter, in which the school of atomists sought a basis for the understanding of the order reigning in nature in spite of the variety of physical phenomena. Conversely, Lucretius, summing up the arguments for atomic theory, interpreted the growth of a plant from its seed as evidence for the permanence of some elementary structure during the development, a consideration strikingly reminiscent of the approach in modern genetics.

Still, after the development of classical mechanics in the Renaissance and its subsequent fruitful application to the atomistic interpretation of the laws of thermodynamics, the upholding of order in the complicated structure and functions of the organisms was often thought to present unsurmountable difficulties. A new background for the attitude towards such problems was, however, created by the discovery of the quantum of action in the first year of our century, which revealed a feature of individuality in atomic processes going far beyond the ancient doctrine of the limited divisibility of matter. Indeed, this discovery provided a clue to the remarkable stability of atomic and molecular systems on which the properties of the substances composing our tools as well as our bodies ultimately depend.

The considerations in my earlier address were inspired by the recent completion of a logically consistent formalism of quantum mechanics. This development has essentially clarified the conditions for an objective account in atomic physics, involving the elimination of all subjective judgement. The crucial point is that, even though we have to do with phenomena outside the grasp of a deterministic pictorial description, we must employ common language, suitably refined by the terminology of classical physics, to communicate what we have done and what we have learned by putting questions to nature in the form of experiments. In actual physical experimentation this requirement is fulfilled by using as measuring instruments rigid bodies like diaphragms, lenses, and photographic plates sufficiently large and heavy to allow an account of their shape and relative positions and displacements without regard to any quantum features inherently involved in their atomic constitution.

In classical physics we assume that phenomena can be subdivided without limit, and that especially the interaction between the measuring instruments and the object under investigation can be disregarded or at any rate compensated for. However, the feature of individuality in atomic processes, represented by the universal quantum of action, implies that in quantum physics this interaction is an integral part of the phenomena, for which no separate account can be given if the instruments shall serve their purpose of defining the experimental arrangement and the recording of the observations. The circumstance that such recordings, like the spot produced on a photographic plate by the impact of an electron, involve essentially irreversible processes presents no special difficulty for the interpretation of the experiments, but rather stresses the irreversibil'ty which is implied in principle in the very concept of observation.

The fact that in one and the same well-defined experimental arrangement we generally obtain recordings of different individual processes thus makes indispensable the recourse to a statistical account of quantum phenomena. Moreover, the impossibility of combining phenomena observed under different experimental arrangements into a single classical picture implies that such apparently contradictory phenomena must be regarded as complementary in the sense that, taken together, they exhaust all well-defined knowledge about the atomic objects. Indeed, any logical contradiction in these respects is excluded by the mathematical consistency of the formalism of quantum mechanics, which serves to express the statistical laws holding for observations made under any given set of experimental conditions.

For our theme it is of decisive importance that the fundamental feature of complementarity in quantum physics, adapted as it is to the clarification of the well-known paradoxes concerning the dual character of electromagnetic radiation and material particles, is equally conspicuous in the account of the properties of atomic and molecular systems. Thus, any attempt at space–time location of the electrons in atoms and molecules would demand an experimental arrangement prohibiting the appearance of spectral regularities and chemical bonds. Still, the fact that the atomic nuclei are very much heavier than the electrons allows the fixation of the relative positions of the atoms within molecular structures to an extent sufficient to give concrete significance to the structural formulae which have proved so fruitful in chemical research. Indeed, renouncing pictorial description of the electronic constitution of the atomic systems and only making use of empirical knowledge of threshold and binding energies in molecular processes, we can within a wide field of experience treat the reactions of such systems by ordinary chemical kinetics, based on the well-established laws of thermodynamics.

These remarks apply not least to biophysics and biochemistry, in which in our century we have witnessed such extraordinary progress. Of course, the practically uniform temperature within the organisms reduces the thermodynamical requirements to constancy or steady decrease of free energy. Thus, the assumption suggests itself that the formation of all permanently or temporarily present macromolecular structures represents essentially irreversible processes which increase the stability of the organism under the prevailing conditions kept up by nutrition and respiration. Also, the photo-synthesis in plants is, of course, as recently discussed by Britten and Gamow, accompanied by an overall increase in entropy.

Notwithstanding such general considerations, it appeared for a long time that the regulatory functions in living organisms, disclosed especially by studies of cell physiology and embryology, exhibited a fineness so unfamiliar to ordinary physical and chemical experience as to point to the existence of fundamental biological laws without counterpart in the properties of inanimate matter studied under simple reproducible experimental conditions. Stressing the difficulties of keeping the organisms alive under conditions which aim at a full atomic account, I therefore suggested that the very existence of life might be taken as a basic fact in biology in the same sense as the quantum of action has to be regarded in atomic physics as a fundamental element irreducible to classical physical concepts.

In reconsidering this conjecture from our present standpoint it must be kept in mind that the task of biology cannot be that of accounting for the fate of each of the innumerable atoms permanently or temporarily included in a living organism. In the study of regulatory biological mechanisms the situation is rather that no sharp distinction can be made between the detailed construction of these mechanisms and the functions they fulfil in upholding the life of the whole organism. Indeed, many terms used in practical physiology reflect a procedure of research in which, starting from the recognition of the functional role of the parts of the organism, one aims at a physical and chemical account of their finer structures and of the processes in which they are involved. Surely, as long as for practical or epistemological reasons one speaks of life, such teleological terms will be used in complementing the terminology of molecular biology. This circumstance, however, does not in itself imply any limitation in the application to biology of the well-established principles of atomic physics*.

To approach this fundamental question it is essential to distinguish between separate atomic processes taking place within small spatial extensions and completed within short time intervals, and the constitution and functions of larger structures formed by the agglomeration of molecules keeping together for periods comparable to or exceeding the cycle of cell division. Even such structural elements of the organism often display properties and a behaviour which imply an organization of a more specific kind than that exhibited by the parts of any machine we are able to construct. Indeed, the functions of the building blocks of modern mechanical and electromagnetic calculation devices are determined simply by their shape and such ordinary material properties as mechanical rigidity, electric conductivity and magnetic susceptibility. As far as the construction of machines is concerned, such materials are formed once and for all by more or less regular crystalline accumulations of atoms, while in the living organisms we have to do with a remarkable rhythm of another kind where molecular polymerization which, if carried on indefinitely, would make the organism as dead as a crystal, is time and again interrupted.

A paragraph commenting on the isotopic tracer investigations by Hevesy, which showed that a major part of the calcium atoms incorporated in the skeleton of a mouse at the foetal stage remains there for the whole life of the animal, is omitted here. The author discussed the problem of how the organism is able to economize with its calcium to such a remarkable extent during the growth of the skeleton.

The application of physical methods and viewpoints has led to great progress in many other fields of biology. Impressive examples are the recent discoveries of the fine structure of muscles and of the transport of the materials used for the activity of the

* In his lecture in Cologne (which was given in German) Bohr inserted the following phrase: In the last resort, it is a matter of how one makes headway in biology. I think that the feeling of wonder which the physicists had thirty years ago has taken a new turn. Life will always be a wonder, but what changes is the balance between the feeling of wonder and the courage to try to understand. (Translated from the transcript of the tape recording.)

nerves. At the same time as these discoveries add to our knowledge of the complexity of the organisms, they point to possibilities of physical mechanisms which hitherto have escaped notice. In genetics, the early studies by Timofjeev-Ressofskij, Zimmer and Delbrück of the mutations produced by penetrating radiation permitted the first approximate evaluation of the spatial extensions within the chromosomes critical for the stability of the genes. A turning point in this whole field came, however, about ten years ago with Crick and Watson's ingenious proposal for an interpretation of the structure of the DNA molecules. I vividly remember how Delbrück, in telling me about the discovery, said that it might lead to a revolution in microbiology comparable with the development of atomic physics, initiated by Rutherford's nuclear model of the atom.

In this connection I may also recall how Christian Anfinsen in his lecture at a symposium in Copenhagen a few years ago started by saying that he and his colleagues had hitherto considered themselves learned geneticists and biochemists, but that now they felt like amateurs trying to make head and tail of more or less separated biochemical evidence. The situation he pictured was, indeed, strikingly similar to that which confronted physicists by the discovery of the atomic nucleus, which to so unsuspected a degree completed our knowledge about the structure of the atom, challenging us to find out how it could be used for ordering the accumulated information about the physical and chemical properties of matter. As is well known, this goal was largely achieved within a few decades oy the cooperation of a whole generation of physicists, and which, in its intensity and scope resembles that now taking place in genetics and molecular biology.

A section commenting on the problem of the rhythm in the process of growth of a cell is omitted here. The author discussed in particular the control of the DNA duplication and the role which the structure of the chromosomes may play in this process as well as in the stability of the genetic material. He further considered the possibility that the duplication process is intimately associated with the transfer of information from DNA.

Before I conclude, I should like briefly to call attention to the source of biological knowledge which the so-called psychical experience connected with life may offer. I need hardly stress that the word 'consciousness' presents itself in the description of a behaviour so complicated that its communication implies reference to the individual organism's awareness of itself. Moreover, words, like 'thought' and 'sentiments' refer to mutually exclusive experiences and have, therefore, since the origin of human language, been used in a typically complementary manner. Of course, in objective physical description, no reference is made to the observing subject, while in speaking of conscious experience we say 'I think' or 'I feel'. The analogy to the demand of taking all essential features of the experimental arrangement into account in quantum physics is, however, reflected by the different verbs we attach to the pronoun.

The fact that every thing which has come into our consciousness is remembered points to its leaving permanent marks in the organism. Of course we are only here concerned with novel experiences of importance for action or contemplation. Thus.

we are normally unconscious of our respiration and the beating of the heart, and hardly aware of the working of our muscles and bones during the motion of our limbs. However, by the reception of sense impressions on which we act at the moment or later, some irreversible modification occurs in the nervous system, resulting in a new adjustment. Without entering on any more or less naïve picture of the localization and integration of the activity of the brain, it is tempting to compare such adjustment to irreversible processes by which stability in the novel situation is restored. Of course, only the possibility of such processes but not their actual traces are hereditary, leaving coming generations unencumbered by the history of thinking, however valuable it may be for their education.

Postscript

We have no call to shy from the use of a cliché in writing about the posthumous publication of this, one of the last works of Niels Bohr, when we say that we have been proud to present it in the pages of *ICSU Review of World Science* ahead of its appearance in a collection of essays from Bohr's later years, to be published in the fall of this year by Interscience Publishers Inc. Bohr himself never hesitated to rope in a cliché if he thought that its use might contribute to his life-long devotion to the clearest possible exposition of his own and other people's discoveries.

Niels Bohr, with his unparalleled natural facility and acquired skill in dialectics, had long pondered on the resolution of the dichotomy so plainly evident in the dual approach of experimenters to the discipline of molecular biology: on the one hand those cell biologists who insist on the autonomy of the living cell *per se*, and in so doing all too often losing sight and the touch of the molecular processes within it; on the other, those biological chemists who would reduce life's many-coloured dome to a handful of chemical equations, forgetting that the full achievement of their goal involves the death of a living unit. And here we have, as it were, Bohr's last will and testament in the matter: so that we may 'feel, who have laid our groping hands away, And see, no longer blinded by our eyes[2].   *R.F.*

REFERENCES

[1] N. BOHR, *Nature*, 131 (1933) 421.
[2] This quotation is from a sonnet by Rupert Brooke dated 1913, and is taken from *The Poetical Works of Rupert Brooke*, edited by GEOFFREY KEYNES, Faber and Faber, London, 1946.

APPENDIX. SELECTED UNPUBLISHED WRITINGS

CAUSALITY AND COMPLEMENTARITY – SUMMARY OF THE GIFFORD LECTURES

Talk prepared for the British Broadcasting Corporation, 1950

Unpublished Manuscript

THE KARL TAYLOR COMPTON LECTURES – THE PHILOSOPHICAL LESSON OF ATOMIC PHYSICS, SIXTH LECTURE

Last of six lectures given at the
Massachusetts Institute of Technology, 5–26 November 1957

Unpublished Transcript

ATOMS AND HUMAN KNOWLEDGE

Public lecture delivered on 13 December 1957,
in Holmberg Hall, Norman, Oklahoma

Pamphlet published by
the University of Oklahoma Public Lectures Committee and
the Frontiers of Science Foundation of Oklahoma, Inc.

ADDRESS AT THE UNIVERSITY OF BRUSSELS

Address prepared for the occasion of the nomination of *Doctor Honoris Causa* at the University of Brussels on 11 October 1961

Unpublished Manuscript

See Foreword, p. IX.

[171]

EDITOR'S REMARKS

This Appendix contains four items which cannot be regarded as proper publications. They have nevertheless been deemed worthy of inclusion for various reasons.

The first item, "Summary of the Gifford Lectures", is a manuscript brought to completion by Bohr and may as such be considered to be a finished work. The background is as follows. From 21 October to 11 November 1949, Bohr gave a series of ten lectures – the distinguished Gifford Lectures – at the University of Edinburgh, Scotland[1]. The item reproduced below is a summary of the lectures intended as a talk for the British Broadcasting Corporation (BBC). However, the talk was never transmitted. The typewritten manuscript contains some handwritten corrections, which, whenever readable, have been taken into account below. The manuscript is furthermore paginated, with a date provided on each page. This information is given in the margin in the following reproduction. In transcribing the manuscript, the editors have followed the original text, including punctuation, except for obvious spelling mistakes.

The second item is the sixth (and last) of the Karl Taylor Compton Lectures given by Bohr in November 1957 at the Massachusetts Institute of Technology. It was Bohr's intention to develop these lectures into a book, and for this purpose his assistant, Aage Petersen, edited the transcript of the lectures[2]. However, Petersen omitted significant phrases and passages in his transcript, and the version printed below is a new transcription prepared specially for this volume. Even this version, however, contains some blank passages, marked by ellipses, due to the inaudibility of Bohr's voice. In some cases, the editors have added a word in square brackets to enhance the meaning. The item is included because of the philosophical considerations it contains about instinct and common sense, which, although not contained in his published writings, were frequently expressed by Bohr in informal communication with friends and colleagues. The considerations in the lecture are thus similar to ideas discussed

---

[1] Bohr's notes for the Gifford Lectures themselves are part of the Bohr MSS. In addition, the NBA holds an incomplete audiotape prepared from the original wire recording, as well as an unmicrofilmed transcript of the lectures. The circumstances of the lectures are described in some detail in S. Rozental, *Schicksalsjahre mit Niels Bohr, Erinnerungen an den Begründer der modernen Atomtheorie*, Deutsche Verlags-Anstalt, Stuttgart 1991, pp. 100–107. More extensive bibliographic information on Rozental's book is provided in the Introduction to Part II, p. [220], ref. 3.

[2] The tape recordings of the Compton Lectures (with the exception of the fifth lecture, which has been lost) are deposited in the NBA. The original transcript of the lectures is part of the Bohr MSS.

by Bohr in letters to some of his close colleagues, such as Oskar Klein and Wolfgang Pauli[3].

The lecture "Atoms and Human Knowledge", given at the University of Oklahoma on 13 December 1957, is included because Bohr originally consented to its publication. Yet, although the published article (reproduced in facsimile below) is based on a transcription of the tape-recorded lecture, it contains several mistakes. Indeed, Bohr did not receive the completed manuscript for approval, and never included the published article in his list of publications[4]. For these reasons, the lecture is reproduced in this appendix rather than among Bohr's publications proper.

The last item is a manuscript for a guest lecture Bohr was going to give at the University of Brussels on the occasion of his nomination there in October 1961 as honorary doctor. However, Bohr was prevented from attending because of illness. The manuscript is typewritten and paginated, and some of the pages are provided with dates. This information is given in the margin of the reproduction below. As in the case of the Gifford Lectures, the manuscript was brought to completion by Bohr, for which reason it is printed here.

---

[3] See, in particular, the letters from Bohr to Klein, 6 March 1940, and from Bohr to Pauli, 31 December 1953. Reproduced on pp. [535], [543] (Danish originals) and [537], [547] (English translations).

[4] See the relevant correspondence with Jens Rud Nielsen, NBA.

CAUSALITY AND COMPLEMENTARITY.

MS, p. 1, 16/1/50.

The course of Gifford Lectures which I had the honour to deliver at the University of Edinburgh last autumn dealt with the lesson derived from the recent progress of atomic physics concerning the logical aspects of man's position in existence.

The development of physical science, which step by step has augmented our knowledge of that nature of which we ourselves are part and afforded ever increasing opportunities of employing such knowledge in the service of civilization, has time after time reminded us that, however useful a logical frame has proved for the comprehension of earlier experience, we must always be prepared for the necessity of taking recourse to a wider frame when confronted with new experience.

Already in ancient Greece, where the attempts of basing science on well-defined logical principles met with such remarkable success in the field of mathematics, great difficulties were encountered in finding a rational basis for the comprehension of physical phenomena. In fact, many accustomed ideas, like cause and effect, developed to deal with daily-life experience lacked obviously the precision necessary for such purpose.

It is true that the Greek philosophers who laid the first foundation of atomic theory showed great intuition in concluding that a rational account of the regularities of nature demanded the assumption of an ultimate limit for the divisibility of matter. Still, not only was the art of experimentation yet far from the stage of development required for a verification of such views but, above all, the principles of mechanics were not sufficiently clarified to serve as a basis for their further exploration.

MS, p. 2, 16/1/50.

In view of the limited knowledge at that time it is quite understandable that Aristotle did not share the views of the atomists and found it necessary to endue matter with properties like "natural" positions of heavy and light substances and "natural" motions of the planetary bodies round the Earth. In particular, the phenomena of organic life and the associated psychical experience were felt by Aristotle to demand the introduction of the ideas of purpose and potentiality.

A new epoch was initiated in the Renaissance, when a rational approach to the fundamental problems of mechanics was initiated by Galilei who renounced all explanation of translation itself and only in the departure from uniform rectilinear motion of bodies looked for the action of forces. On this basis Newton built the great edifice of classical mechanics by means of which a comprehensive account of all physical evidence known at that time could be given and, in conjunction with his conception of universal gravitation, the Keplerian laws for the motion of planets found so beautiful an explanation.

[174]

Due to the simultaneous great advance in mathematics represented by the development of infinitesimal calculus it became, in fact, possible on well-defined principles to describe the continuous alteration of a mechanical system from instant to instant and, in particular, by means of the knowledge of its state at a given time to predict the state at any subsequent time. The concept of causality formulated in this way came for a long time to stand as the basis and the very ideal of physical explanation.

MS, p. 3, 16/1/50.

The establishment of classical mechanics, together with the flourishing of anatomical and physiological science, could not fail to give new incitement to attempting a comparison between living beings and machines. Still, the extreme view of mechanicism, especially advocated by Descartes, was by him combined with views of pure spiritualism and with specific ideas of the interaction between soul and body.

A foundation for a more harmonious attitude was, however, soon offered by Spinoza's idea of a psycho-physical parallelism according to which soul–body interaction is replaced by the view that in physical and in psychical experience we have to do with two aspects of existence each of which in itself exhibits continuity and causality. The germ to progress as regards the survey of our position, which this attitude contains, has, as we shall see, not been impeded by subsequent developments, even though to-day we have been led to a different approach to the analysis and synthesis of physical as well as psychical experience.

Meanwhile the scope and fertility of the viewpoints of classical physics were extended and enhanced in the following centuries especially by the successful co-ordination of the great discoveries in the domain of electromagnetic phenomena. Here again a description conforming with the principle of causality could be achieved by extending the specification of the state of a physical system to include not merely the location and velocities of the material bodies, but also the electric and magnetic forces at every point of space. A most important consequence of this development was the prediction and subsequent experimental realisation of the propagation of electromagnetic waves in free space. This remarkable advance, which initiated the wonderful development of radio-communication, led in particular to a wide-going explanation of optical phenomena, according to which light is regarded as electromagnetic waves of very high frequency.

MS, p. 4, 16/1/50.

Nevertheless, as a consequence of new experimental evidence, we have in our century been forced to subject the foundation for the applicability of our most elementary concepts, space–time coordination and causality, to a radical revision. One line of progress in such respect started from refined measurements which did not exhibit the expected dependence of optical phenomena on the motion of the earth through space. These experiments revealed in fact that observers moving relatively to each other will not only judge the dimensions and shape of rigid bodies differently, but that even events at two separate places which by one observer will be regarded as simultaneous to another will appear as happening one before the other.

The recognition of the extent to which the description of physical phenomena depends on the standpoint of the observer has, as is well known, in Einstein's hand proved a powerful guidance in tracing physical laws of most general validity. For this purpose, the abstract mathematical methods of four-dimensional non-euclidian geometry were found remarkably well-adapted for incorporation in a unified description of physical evidence of the most various kinds, including even gravitation effects. Notwithstanding its novelty in abandoning the deep-rooted ideas of absolute space and time, accustomed from our orientation in daily-life events, the theory of relativity may indeed be regarded as providing a most harmonious completion of the whole system of classical physics, ["characterized by a consistent use of the principle of causality" is crossed out in the manuscript].

MS, p. 5, 16/1/50.

A new epoch in physical science, which should disclose the limited scope of the notion of causality itself, was, however, initiated in the first year of this century by Planck's discovery of the universal quantum of action. This discovery in fact revealed a feature of indivisibility of atomic phenomena, which is quite foreign to the whole framework of classical physics and goes far beyond the old doctrine of the limited divisibility of matter.

Atomic theory was revived in the course of the last century, and came to the foregound especially as a result of Dalton's explanation of the remarkably simple laws governing the proportions of elements in chemical compounds. Indeed, since that time, pictures of the atomic composition of molecules have been an indispensable and most helpful tool in chemical research. To begin with, however, there was no question of a departure from the ideas of classical physics, and the practical difficulties of dealing with mechanical properties of systems containing immense numbers of atoms or molecules were succesfully overcome by means of statistical considerations based on ordinary mechanics.

In this way, it was above all found possible to obtain a wide-going interpretation of the laws of thermodynamics which together with energy conservation govern the action of all machines used for converting heat into work. Still, it was just the exploration, on this basis, of the phenomena of temperature radiation which first should disclose the insufficiency of the methods of classical physics, and lead to the discovery of the quantum of action. Shortly afterwards, it was also realized that the results of measurements of specific heats at low temperatures exhibit a similar departure from the results predicted by statistical application of classical mechanics. MS, p. 6, 16/1/50.

The situation was brought into especially sharp relief by a number of discoveries towards the end of the nineteenth and the beginning of the twentieth century, which for the first time made it possible to study the properties of single atoms, and to obtain a detailed knowledge of their constitution. The first decisive step in this development was the recognition of the electron as a constituent of all atoms, and a remarkable completion of our ideas of atomic structure was achieved by Rutherford's discovery of the nucleus, in which practically the whole mass of the atom is concentrated within a very small part of its extension. These discoveries brought us indeed a great step forward towards the goal, formulated already by Pythagoras, of accounting for physical phenomena by considerations of whole numbers. At the same time, however, as the picture of the nucleus atom meant a great advance in the coordination of a large amount of physical and chemical evidence, it presented us with insurmountable difficulties when trying to account on the basis of ordinary mechanics and electrodynamics for the peculiar stability of the atom, exhibited by such evidence.

The clue to coping with this situation was afforded by the very feature of indivisibility of atomic processes embodied in the quantum of action. In fact, the stability of atoms just demands the impossibility of changing the state of an atomic system except through a complete transition from one of its so-called stationary states to another such state. Since, however, no description of the state of a physical system by means of mechanical concepts can determine the choice between different individual processes of transition to other states, we have obviously to do with a situation which lies beyond the scope of the classical idea of causality. In particular, it must be realized that the notion of an a priori probability of occurrence of complete transition processes, indispensable in accounting for atomic phenomena, has no analogue in the recourse to statistical methods for the practical dealing with complicated systems assumed to obey the laws of classical mechanics.

The elucidation of the situation as regards analysis and synthesis of experience in atomic physics has necessitated a renewed revision of the foundation of the description and explanation of physical phenomena. A main point is here that, however far phenomena transcend the scope of classical physical explanation, the description of the experimental arrangement and the record of the observations must always be expressed in common language, supplemented only with the terminology of classical physics. This demand is purely logical, since the word "experiment" can only be rationally used under circumstances in which we are able to tell others what we have done and what we have learnt. MS, p. 7, 18/1-50.

[177]

It is equally clear that when we have to do with phenomena involving typical quantum effects, there can be no question of tracing their course between the initial state determined by the experimental arrangement and the observations ascertaining the state reached by the system as a result of the transition process. In such phenomena it is even impossible to control the interaction of the objects under investigation with the measuring instruments which serve to fix the conditions under which the phenomena appear. In fact, any experimental arrangement suited for the location of atomic objects in space and time will involve an uncontrollable exchange of momentum and energy between the objects and the scales and clocks defining the reference frame. Conversely, no arrangement suitable for the control of momentum and energy balance will admit precise description of the phenomena as a chain of events in space and time.

Under such circumstances, the assigning of conventional attributes to atomic objects involves an essential ambiguity strikingly illustrated by the well-known dilemma concerning the corpuscular and wave properties of material particles as well as of electromagnetic radiation. We are here faced with the impossibility to comprise within a single picture observations regarding atomic objects obtained by means of different mutually exclusive experimental arrangements. Such empirical evidence exhibits a novel type of relationship without analogue in classical physics and may conveniently be termed "complementary" in order to stress that in the contrasting phenomena we are concerned with equally essential aspects of all possible knowledge about the objects.

The scope of such knowledge is, indeed, set by the circumstance that any attempt at subdividing quantum phenomena demands an experimental arrangement which introduces new sources of interaction, uncontrollable in principle, between objects and measuring instruments. In this situation, even the old question of an ultimate determinacy of natural phenomena has lost its conceptual basis, and it is against this background that the viewpoint of complementarity presents itself as a rational generalization of the very ideal of causality.

Far from implying any arbitrary renunciation of rational demands of physical explanation, the argument stresses, on the contrary, the necessity of looking for a wider frame for the co-ordination of experimental evidence, which cannot be comprised within the framework of classical physics. By a uniquely fruitful collaboration between contemporary physicists with most ingenious contributions from the younger generation, it has also been possible step by step to develop proper mathematical methods adapted to this novel situation. These methods, known as quantum mechanics and quantum electrodynamics, make use of a formalism which does not allow interpretation by means of accustomed

MS, p. 8, 20.1.1950.

physical pictures and aim only at deriving probabilities for the occurrence of individual events observable under well-defined experimental conditions.

The result of this remarkable progress was not only the complete co-ordination of a wealth of spectroscopical evidence pertaining to the binding of the electrons to the nucleus in the atom, but even a detailed explanation of the laws governing the spontaneous disintegration of atomic nuclei exhibited by the phenomena of radioactivity. The interpretation of chemical evidence was also most decisively furthered by the quantum-mechanical treatment of exchange of electrons between atoms, which is responsible for the valency bonds in chemical combinations. Here we meet with a typical feature of quantum statistics related to the limited possibility in the interpretation of the phenomena to trace the part played by a single among a number of identical entities, like electrons.

In this way, the complementary mode of description has overbridged the gap which through the ages has separated the attitude of physicists trying to explain all natural phenomena by motion of particles, and the endeavours of chemists to reveal the secrets of nature by studies of such transmutations of substances as defy a visualization in terms of displacements. The radical departure from the pictorial description in classical theories, characteristic of the modern development of atomic physics, can indeed not be regarded as a transitory stage, but represents an irrevocable step forced upon us to avoid logical inconsistencies in the account of most elementary experience.

MS, p. 9, 21-1-50.

* * *

The theme of the first part of the Gifford Lectures was the epistemological lesson derived from the exploration of the world of atoms, where, notwithstanding its novelty, the comparatively simple character of the evidence allowed a clear formulation of principles. On this background, the second part of the lectures was devoted to an emphasis of the guidance which this lesson may afford to clarify our attitude to problems of less accessible character in other fields of human knowledge and interests.

One such problem already touched upon in the preceding is the attitude to be taken to the old question of the physical explanation of organic life. Here, the situation is, of course, essentially changed by the progress of our knowledge of atomic phenomena, which has offered a new basis for the account of the immense variety of properties of matter involved in the structure and functioning of organisms. Indeed, the old comparison between organisms and machines rested entirely upon the laws of classical physics supposed to govern the behaviour of even the smallest parts of the machines constructed for practical purposes. The development of microscopic anatomy and especially

MS, p. 10, 21-1-50.

of biochemistry has taught us, however, that many functions characteristic of the organisms rest upon the behaviour of complicated structures of minute size which, like the location of the genes responsible for hereditary properties, reach down to extensions of single molecules.

In this connection, it is important to realize the difference between crystals and living organisms. On the one hand, the regular shapes and growth of crystals exhibit analogies with organisms, on the other hand, crystals differ by their properties most markedly from the typical features of organic life. The structure of living organisms and of crystals represents, indeed, two poles of atomic organization. While in the latter we have to do with an order characterized by a simple repetition of arrangement throughout the whole structure, the former is characterized by a co-operation of all parts to achieve the sustainment of life and the adjustment of the organism to the external conditions. Nothing in biological experience gives us reason to doubt that the interaction between the atoms in the organism follows the same laws as those exhibited by simple physical and chemical evidence. Rather recent progress in the study of spontaneous and artificial mutations seems to show convincingly that the stability of the structures to which the genes are connected possess a stability just of the same kind as expressed in the quantum-mechanical treatment of atomic or molecular constitution. Moreover, in the fineness of sense organs and in the mechanism of many other organic functions it has been possible to trace wide-going similarities with the amplification devices used for the detection of individual particles in physical research.

MS, p. 11, 21-1-50.

In our days, there can surely no more be question of accounting for life within the frame of classical physics, and the problem with which we now are confronted is whether the essentials of life can be accounted for on the basis of the development of atomic physics, initiated by the discovery of the quantum of action. For our attitude to this problem it is, however, necessary to examine most closely the similarities and differences in the basic conditions of researches in the domain of physics and biology, respectively. Notwithstanding the recognition of the limitation inherent in the notion of causality, physical research may still be characterized by the observation of events occurring under reproducible conditions. In the study of the phenomena of life we are not only excluded from speaking about single objects and measuring instruments of our own construction, the situation is not only in this respect essentially different, but any attempt of confirming the investigation with the aim of approaching the conditions of research in atomic physics will obviously exclude the proper display of life. This situation does certainly not set a limit to progress in biology, accompanied by ever better possibilities of medical treatment, or even in ever greater detail to trace the steps through which organic life has developed

[180]

from such simple atomic processes which for long have formed the object of chemical research. The essential point is, however, that the actual procedure in biological research is characterized by an intuitive combination of experiences about the functioning of organisms and their physical structure, none of which is exhaustive to the degree characteristic of the demands in the so-called exact sciences. We have, indeed, here to do with a balance between approach to analysis and synthesis, respectively, which reminds about that relation between description and comprehension of atomic phenomena embodied in the notion of complementarity. Instead of asking for an explanation of organic life, we must in fact be content with a logical attitude which leaves room for the existence of life. The tendency of such a sentence will perhaps appear more clearly when we consider the physical experiences related to organic life.

MS, p. 12, 21-1-50.

THE KARL TAYLOR COMPTON LECTURES
THE PHILOSOPHICAL LESSON OF ATOMIC PHYSICS

Sixth Lecture   November 26, 1957

Atomic Physics and the Position of Man

The title of these lectures has been "The Philosophical Lesson of Atomic Physics", and this last lecture has been entitled "The Position of Man". Now one might ask [whether] that seems a subject for humanists [with] great scholarship and not for a physicist, but, when I venture to speak about – to bring some remarks about – this problem, it is because ... even in physics – a very simple, perhaps the simplest field of knowledge, very removed from the playfield of human aspirations and passions – ... we have also a human problem. The lesson of the development of physics has been that we are not simply recording experience arranged in given general categories for human thinking, as one might like to say in the epoch of critical philosophy, but we have learned that our task, most essentially, is to develop human concepts, [to] find a way of speaking suited to bring order in new experience and, so to say, be able to put questions to nature in a manner where, by experiments, we can get some help [to obtain] the answer.

Now, as I tried to say some words about in the first lecture, such developments we already witnessed in the so-called classical physics, where it was great trouble – an achievement of generations – to eliminate references to aspects of life irrelevant for this field, ... which allowed [us] to build up a wonderful consistent edifice, which, until the very latest years, has been the foundation for all technology. In classical physics, one could still – thinking through the ages – ... gradually use, in a comparatively simple manner, the simple concepts developed for the orientation in the surroundings and ... connect ... events by the idea of cause and effect. Really, the whole chain of events was considered to be unrestrictively divisible and all the chains were linked together by physical laws called the dynamical conservation laws, of which the conservation of energy is the most essential.

The exploration of the world of atoms, which until recently was closed to us, has revealed not only a restricted divisibility of matter which was the foundation, of course, for all atomic ideas, but even the restricted divisibility of physical processes. It is of course something which is entirely outside the whole idea of the mechanical conception of nature, and even the word "process" is an old-fashioned word which is implying somehow that things develop in steps. But we have seen that this necessitated a radical revision of all possibilities of communicating experience, and we had to, in a quite unaccustomed way, ... distinguish between measuring instruments and the atomic objects under investigation. ... Our whole basis for communication was that the experimental arrangement had to be described in the ordinary, unambiguous manner of classical physics. And as you remember, I just tried to say that this is a simple logical demand, because by an experiment we cannot understand anything else than something about which we can tell others what we have done and what we have learned.

But this brings with it that, in the field of quantum physics, ... the interaction between the measuring instruments and the objects ... is an integral part of the phenomena that cannot be separated from the description of the phenomena, which have a feature of wholeness in themselves. Just also [remember] that it brought with it that experience gained by different experimental arrangements ... had to be described by different physical concepts. I shall of course not repeat everything, only remind [you] that such phenomena, ... governed by energy–momentum conservation, for instance, ... demand in their description a renunciation of a detailed description in space and time, because this would refer to the experimental arrangement where the necessary use of measuring rods [and] clocks would involve an in principle uncontrollable exchange of energy–momentum between the object and these bodies, preventing an accurate use of conservation laws. Now, we just called such phenomena complementary to each other in the sense that each of them contains definite information relating to the objects and that they, together and only together, exhaust the definable information we can obtain about the objects.

I might also, before we go on, just shortly remind [you] how complementary in that way our description of the world of atoms is today – that ... all such properties of atoms and molecules [such] as spectra and chemical bonds, where conservation laws are so essential, actually only take place under conditions where it is impossible to locate – [or give] a kinematical description of – the electrons in the atoms and molecules. And the whole great practical use of chemical structural formulas depends on the fact that the nuclei in the atoms are so very much heavier than electrons that we, to a great approximation, need not in such phenomena take the quantum mechanics into account. But, of course, when we are considering such properties of nuclei [as] their transmutations, which are playing – are going to play – so large [a] part in technology, we have just to play the same game again. And the most conspicuous properties of nuclei appear only under conditions where we can give no meaning to the actual location of the constituents of the nuclei, the so-called nucleons – the protons and the neutrons.

And if we go on, it is the same thing all the time. If we go on to properties of matter at large, then the question of the rigidity of bodies cannot be accounted for by classical mechanics, and such properties will not occur under conditions where we had a complete description in the classical sense of the various particles of which the bodies are built up. And that is, of course, very interesting, because ... such fundamental properties of matter ... are the basis for our use of tools and even of the measuring instruments, [whose] existence, of course, is also basic in classical physics. But that is just the kind of description to which we have been led and to which we have been forced and which allows [us] to account for a richness of details and properties of matter which are quite beyond the reach of classical physics.

Now, last time I made a few short remarks about the lesson we have got from atomic physics [with regard to] biological and psychological problems – in what light they appear. Very shortly said, ... just by the great progress in the knowledge of the behavior of atomic systems we have new and most powerful tools in understanding many processes in living organisms, where we have, in all probability, unlimited applications of what one might call mechanistic ideas, but, [where,] at the same time – when one thinks of the actual research in biology – we deal with the word "life", which finds no counterpart in physics, and [which] we express, in a way, by referring to properties of the organism as a whole. ...What I tried to point out was that this necessitates, in practical research, applications of a mechanistic approach as well as what one might call a finalistic approach. But in these we see no kind of contradiction but are just referring to different observational conditions. Now, this is a point which just may appear, as I last time said, so sophisticated, but I hope that some of you will find in that a harmonious reference to the place of living organism[s] among the physical bodies.

Now, as regards psychology, there were some quite different problems we pointed to. We were just pointing to [the circumstance] that in the way in which we speak of the states of our minds – speaking quite loosely – we use many words which have been used with such wonderful power by great poets through the ages, and [which] do not refer in any way directly to physical pictures, ... [or] to any kind of chain of events like that we are contemplating in classical physics, but [refer instead] to situations – separated and mutually exclusive situations – when we use words as "thought", "sentiment", "volition" or "pondering". ...One sees [that] the possibility of a kind of communication by means of such words ... – useful and consistent communication – lies in [the circumstance] that their mutual exclusion just depends [on the fact] that we in the various situations make a different separation between what we pay attention to, what we talk about, what we might loosely [call] the conscious content of our mind and the background from which we judge it and which we quite loosely call ourselves. Of course, in this field, where we use a word like "ourselves", it is obviously more difficult to have an objective description. But it is just [a way of] pointing to [the circumstance] that we still have the possibility of bringing unambiguous communications to others with this whole kind of description, ... where the richness in the terminology as well as the richness of conscious life just depend on the shifting of the separation line between what we call, so to say, object and subject. It is [a way of pointing to] – and that's the only word I want to add to it – [that] we will never actually have to do with a last subject. Whenever we refer to ourselves, then ... we just take something [as] the object and we just put a new subject behind [it]. Of course, the whole kind of communication ... is only [un]ambiguous by some separation line between subject and object.

And I shall just say that at the end I just referred to how we practically and consistently in communication speak of "volition" or "free will". And we do it in a quite similar manner as we use other words [such] as "responsibility" and "hope", [which are] just as indispensable for describing what we may call the richness of conscious experience. There we are nearing, of course, just what we might call social problems – problems referring to the relations between individuals in a society.

Now, in order just to near this problem, I might like to say a few words about the position [of] animal [and] man and especially as regards the ... way we separate in various fields between such words as "instinct" and "reason". It's not the meaning to go into animal psychology in any details but only just to remind [you] of the situation. ...One has, not least, ... the wonderful study of insects brought out – told to the world – with such wisdom and beauty by the famous Fabre[1]. To take an example, he speaks about an insect, I think those you might call a bug and have a curious kind of life circle: it starts by being deposited from its mother, from the female bug, on the bark of a tree. Then small larvae develop and it goes into the wood. And it's nourished by the pulp. It's a very curious, and, from the outside, a very simple thing. It mainly consists of some very powerful jaws and a stomach. It has just to begin with almost the same activity as a mining engineer [laughter]. And it lives in the dark. It's secure enough in its retreat as a mining engineer. The hole gets, as [the bug] grows, wider and wider. It never gets very wide, but starts with being very

---

[1] [J.H. Fabre, *Les Merveilles des Instincts chez les Insectes*, Delagrave, Paris 1913. Bohr read the Danish translation, *Instinktets Mysterier hos Insekter og Edderkopper*, Gyldendalske Boghandel/Nordisk Forlag, Copenhagen/Christiania 1915.]

small, and [the bug] can go for years in the wood. But the point is that this little organism, although it has these very few activities – it has no eyes, there is nothing to see – ... has, so to say, in it, the memory of the whole species. When it nears the time for a transmutation, when it nears the time for the, I think, pupal stage we might call it, then the new insects which will eventually come out [will] have many other faculties, but it has not those of such a little engineer. It will not be able to work itself out of the tree. Therefore, it goes out to the very edge of the tree and leaves only a paper-thin sheet which the insect can break through. [Undeciphered passage.] ... It is fitted out in a much more smooth way than the fairly rougher mine it has so far gone in.

Now, that is just a very wonderful – but of course [one of] very many, a typical [one] – example of what instinct[s] such an organism can have. If we come more near to other words, I should also like ... in a simple manner to get [to] the words we're going to use ... about the most wonderful behavior of what we call sometime higher animals. And we need not go so very high up. [We may talk] about the fishes, and not least the salmon, which has been studied for many reasons, also for some fun and sporting interest. Now, such a salmon lives in lakes in the mountains in Western America or Norway. It starts its life in a lake, sometimes small, sometimes large. When it's got a certain development, a certain maturity, it starts going out of the lake, through the small streams, into the small rivers, to the big rivers, and out into the ocean, where it then gets such nourishment that it gets very fat and powerful, and then returns, goes up to the mountain lakes. Now, of course, we all know what marvellous force is behind such [an] instinct, how such a salmon can jump the waterfalls. But that is not the most wonderful part of it. The wonderful part of it is that it has been discovered by marking of the fishes that they practically always find back to the same lake where they started. And when one thinks of the innumerable divisions and confluxes of streams in Norway, that is a very curious thing. And that would certainly not be easy for a human to do. If one had some kind of frogman, or some other [person], working under water, he would need some underwater lights and, at any rate, some charts with him [laughter]. But how does one speak about the possibilities of such a salmon? One would like to say that it has certainly a sixth sense, but it is easy to count from five to six, but not so easy to say what that sense should be [laughter]. And the more reasonable – it may look strange – explanation is this: that the salmon can do it because it doesn't know itself how it does it [laughter].

The point is just they're different from our positions. We have developed a language. We can tell even a fairly small child to go out in Cambridge or Boston to a certain shop and buy something. The point, of course, is that we could have told him to go to another shop and buy something else. That is what, only in a rough way, ... language is used for. But for these far more simple and less varied tasks, then one need not [have] such differentiation, and the faculty of memory by all living beings can ... do such feats as it's almost unbelievable to us.

But now, if we just come and speak about humans, our position is so very different. First of all, we are only speaking in the simplest outside manner about our possibility of using language, and – what do we call it? – mastering language. And of course, [the] most wonderful language [that] has [been] studied [is] how children learn – by asking "how" and "why" – ... to master a language.

Now, we may in some way say that, of course, there are also instincts in human life, but they are so very much suppressed due to this: that in a society one develops customs and traditions which in almost all – really what we may call human – problems take the part of instincts and rule them out just because the situations exclude each other. One may even say that a small child, of course, belongs to the human race, belongs to the species of homo sapiens, and ... has in it its potentiality to acquire a human civilization and become a fellow man in such a society. But in each society we have a wonderful structure of traditions and customs and one can obtain a wonderful degree of harmony under very different customs and traditions, as the ethnological explorer knows so well. But now the question is how one speaks of relationships between such cultures. Then one might, just if one liked, by reference to the great surprise in physics, say that the difference is similar in various ways to that with which different observers describe physical phenomena. As we have spoken about, ... observers moving relative to each other will record their experiences in a very different manner. But what is essential in such relativistic description is this: that however large such [a] difference of orientations [is], it is possible for an observer to predict accurately how another observer will describe experience. That, of course, is just the power and the use of relativity theory which ... allowed Einstein to discover physical laws [more] general than one had perceived before and common to all observers.

But in cultures it is different, because any culture contains an element of complacency. [This] is not meant as any criticism, but something which quite corresponds to the instinct of self-preservation of any living organism. And that makes it not so simple to appreciate the prejudices of one culture on the basis – on the background – of the prejudices of others. And the situation is actually this: that if one studies a culture – I do not speak here about present difficulties in the world, I speak about explorer work of cultures which, let us say, in the South Sea Islands, had lived for a long time in comparative isolation ... – an explorer ... gets the suspicion that they are also humans. [When one] sees what

[187]

kind of harmony they can have, then one has changed, then one is no longer an American or Dane, because we have lost some of the prejudices which just – not as a criticism – ... make us so.

Therefore, [considering] just this difficulty, one has sometimes said that such cultures are most appropriately described as complementary to each other. Of course, in some way, I suppose it is the nearest kind of resemblance and analogy with physics we can have. But there is this very great difference, that in the problems of atomic physics or in psychology, we have to do with situations which, in principle, exclude each other. As regards cultures, they can be changed by intercourse, and we know from history so many examples of conflux, sometimes called, of cultures in which valuable elements of the different cultures are retained.

You see, now I have used quite a new word. I have hitherto at all not used the word of "value". We have only been interested in the conditions for objective unambiguous descriptions in a different field of knowledge. But now, when we speak of ... values, then we are reminded, of course – even if we go back to old times in philosophy – of the beautiful way in which Socrates analyzed the use of the word "justice". But there have been many different attitudes of philosophical schools. One may have been to say that here we have to do with elementary virtues connected with the nature of man. Other schools, not for [reasons of] inhumanity but for [reasons of] caution of position, have thought it was best to leave all evaluation out of description. But it isn't so simple, it is not so simple, to leave appreciation [and] depreciation out. And I think if we just joke, then I would say that any philosophical school, if it exists, which would try to prevent the use of "good" and "bad", would be very tempted to say that to do that was "good" philosophy [laughter]. So that is only, in a joking manner, just to say that it's a thing which is not so simply left out.

Now, if we just speak a moment more seriously, then we may just analyze briefly – not to bring something new in, but only [to] remind ourselves – how and for what purpose such words as "justice" and "charity" are used. Now, to speak quite briefly, in any society which has stability, the individuals will demand some kind of fair play. And that is developed and crystallized – embodied – in what we call judicial rules of the law of the society. But on the other hand, it is equally clear that we would deprive ourselves of a large part, which is essential, of the richness of life if we would abstain from using words as "sympathy", "love", and so on. Now the point is, when we look upon it, that these words are used in some odd manner, because if we have a situation which allows of an unambiguous use of the word "justice" – well-defined judicial rules – then there is no room for application, in such a situation, of the word "charity". But, on the other hand, life tells us, [as] brought out so wonderfully by the great poets – in the old days by Sophocles and Aeschylus, in this part of the world and more modern times by Shakespeare – that compassion and good will can bring us in conflict with any idea of justice. That is just what all great poetical work reminds us about. Now, it is clear that in all societies one wants, to the utmost part, to combine justice and charity. But it has likeness ... to how such useful concepts as space and time and conservation laws were combined in classical physics, but do not allow a combination to that extent if we shall account for the finer regularities of nature which are quite fundamental for the properties of matter. And it is similar [to the circumstance] that we use these words in a complementary manner in a way [that we] actually do not forget what they stand for, [but] just see how much we in human life can make out of them.

In order not to make any misunderstanding, I shall just briefly remind [you of] the point in quantum physics. It was just parallel, but there the sharpness of the logical parallel is just this: that in the quantum physics the question was the situation brought about by the interaction between the measuring instruments and the object, which forced us to a description which had a wider frame. Here, of course, it is the analysis of the relationship between the individual and the community which is the point in using such words. But [this analysis] also just shows that to get the factual richness of the human possibilities, these words are not to replace each other, as one might sometimes think by rough analogies, but are to complement each other in the full description.

I might try, quite briefly, just to summarize the lesson I tried, and perhaps in an unclear manner, to bring before you. The general lesson is this: that in any communication of experience, we have to rely on some conceptual frame which is common for those between [whom] the communication takes part, but that we must be prepared – as we see in science and in all regions of life that new experience can demand in order to remove inconsistencies or disharmonies – to take recourse to a wider frame. That's just a general lesson of mathematics, but I hope I just have given you an impression how one is led, not to say something new, ... to represent things in such a manner by the very forceful lesson we have received in physics in our days.

I want just to end these lectures by saying that I have felt it a great honor to introduce these courses of Karl Taylor Compton Lectures. I hope that I in some way have been able to give you an impression of the conviction of the significance of science, scientific research, as a means of human enlightenment and mutual understanding. It was just that conviction which was the basis for the great achievement and high aspirations of Karl Compton. And I want to add that it has been a very great pleasure and encouragement to me to find the same attitude among the faculty and the students of the Massachusetts Institute of Technology, and just at this time, where science is called upon to play such a role and entails such promises, I want to express my warmest wishes for the success of your endeavors in the service of mankind.

Atoms

and

Human

Knowledge

A Public Lecture By

Professor Niels Bohr
Director, Institute of Theoretical Physics,
University of Copenhagen, Copenhagen, Denmark

Delivered on December 13, 1957,
in Holmberg Hall, Norman, Oklahoma,
under the Auspices of the
University of Oklahoma Public Lectures Committee and the
Frontiers of Science Foundation of Oklahoma, Inc.

NIELS BOHR

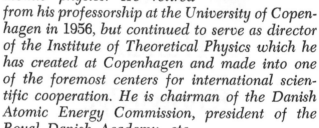

Since 1913, when he published his epoch-making theory of the hydrogen atom, Niels Bohr has been the leading pioneer in atomic physics. He retired from his professorship at the University of Copenhagen in 1956, but continued to serve as director of the Institute of Theoretical Physics which he has created at Copenhagen and made into one of the foremost centers for international scientific cooperation. He is chairman of the Danish Atomic Energy Commission, president of the Royal Danish Academy, etc.

For his great contributions to science Bohr has received virtually all the honors open to a scientist. He received the Nobel Prize in 1922 and the Atoms for Peace Award in 1957. He has been awarded the Guldberg, Hugh, Oersted, Barnard, Mateucci, Franklin, Faraday, Planck, Copley, and Bohr medals. Some 25 universities have honored him with doctor's degrees, and more than 60 scientific societies have elected him to honorary or regular membership. He has been decorated by several governments, and the Danish King has made him a Knight of the Elephant, an honor normally reserved for royalty.

* * * * * * *

This biographical sketch and the transcriptions of Professor Bohr's address were prepared by Dr. J. Rud Nielsen, Research Professor of Physics, The University of Oklahoma.

Ladies and gentlemen! I am deeply moved by your warm welcome and by the kind words of my old friend and your distinguished physicist, Mr. Jens Rud Nielsen. It is a great honor for me to speak to you at this great University of Oklahoma, where such vigorous and enthusiastic endeavors in education and scientific research are taking place. It is a very great pleasure for my wife and me to come back here, after the visit of twenty years ago, to see old friends and to see how the activities of this University are developing. This morning I had the great pleasure of seeing in your physics department many-sided researches in atomic physics which play, or will play, a very important role in technology.

Tonight I shall not try to tell you anything new about atomic physics, but with the theme for this lecture, "Atoms and Human Knowledge",* I want to describe how we in this new field of experience, the explorational world of atoms—where, so to speak we wander in an unknown land on paths hitherto untrod by man—have got a very forceful reminder of our position as observers of that nature of which we ourselves are part.

In order to give you an impression of what this new lesson teaches us, I want to remind you of the development of the great edifice which, until most recently, has been the basis for all technology and which we usually refer to as classical physics. To develop this has been, in itself, a great and truly human endeavor. It is not so that in physics we merely record measurements and are able to order them or put them together directly by means of the notions with which we are equipped from orientation in everyday life. Rather, it has been a continual

* This text was transcribed from magnetic tape recording. It has not been corrected by Professor Bohr.

[193]

endeavor through the years to develop human concepts or views suited to the ordering and comprehension of our increasing experience. As everyone knows, this has been a long task. In ancient Greece, where, to the admiration of later generations, a spirited effort was made to erect science on well-established, clearly formulated, logical principles, and where wonderful contributions were made to mathematics which were to become the foundations for later developments, it was not found simple or easy to separate one's self from such experiences as the exertion required for the motion of our bodies or even from the motives for the actions which these motions serve. It was as long as two thousand years later, at the time of the Renaissance, that it became possible for Galileo to liberate himself and renounce any explanation of motion itself, taking uniform motion, instead, as something elementary and asking only for the cause of changes in motion in terms of forces. Out of this grew, as you know, the science of classical mechanics, completed by the genius of Newton. The description thus achieved was a so-called deterministic description. This means that it proved possible, from a knowledge of the state of a physical system as defined by the positions and motions of its parts, together with a knowledge of the forces between these parts, to calculate or predict the state of the system at any later time. This great achievement, the foremost expression of which was Newton's explanation of Kepler's laws governing the motions of the planets around the sun and of the satellites around the planets, made an overwhelming impression at the time. This kind of description, therefore, came to stand as the ideal for scientific explanation.

But what we have learned in the exploration of the new field of atomic physics is that most of the phenomena we meet there cannot be pictured in this way: they defy any casual, deterministic description. The starting point was the discovery of an element of wholeness, so to speak, in the physical processes, a feature going far beyond the old doctrine of the restricted divisibility of matter. This element is called the universal quantum of action. It was discovered by Max Planck in the first year of this century and came to inaugurate a whole new epoch in physics and natural philosophy. We came to understand that the ordinary laws of physics, i.e, classical mechanics and electrodynamics, are idealizations that can only be applied in the analysis of phenomena in which the action involved at every stage is so large compared to the quantum that the latter can be completely disregarded.

This condition is amply fulfilled by experience on the ordinary scale, but for phenomena that depend on individual atoms we need entirely new physical laws. Now, what could one do in this situation? I can say at once that it proved possible, by the most active co-operation of a whole generation of experimental and theoretical physicists, to achieve a generalization of classical physics called quantum mechanics, or quantum physics, which helps us in expressing these laws. But this new kind of description is in principle a statistical one, and this fact has given rise to a great deal of discussion. The question is whether we are really dealing with something that represents an irrevocable step in the description of nature or merely with an expediency that we can give up later to achieve a deterministic description. This problem has been essentially (to my mind, fully) clarified by a radical revision of the very foundations for the use of our most elementary physical concepts.

We have to ask ourselves: "How do we communicate physical experience at all?" It is clear that, even if we are quite beyond the scope of classical physics, we can only speak about experience obtained under experimental conditions described in the ordinary way. By an experiment, we must understand a situation in which we can tell others what we have done and what we have learned. That actually means that what every physicist does—and he couldn't do anything else—is to find a condition or experimental arrangement under which a phenomenon occurs and then study it with the aid of such heavy measureing instruments that he can completely disregard the quantum of action.

What observations can we make? Observations are only the marks left on the bodies used as measuring instruments after their interaction with the atomic objects, such as the spot on a photographic plate developed after the impact of an electron. Now, so far we have described everything in the ordinary manner; but how are we to analyze the phenomena? Here we find that in the atomic field, where the quantum of action is essential for the phenomena, the interaction between the measuring instrument and the object plays an essential part. The crucial point is that this part cannot be separated from the phenomena. It is in no way possible to control it separately, since in describing and accounting for the measuring instruments we must neglect the quantum of action. This means that we have lost the basis for a deterministic description. The deterministic description in Newtonian mechanics rests entirely on a combination of the co-ordination, or location, of objects in space and time and the application of such laws as those of the conservation of energy and momentum which allow us to join together, so to speak, the single links of the chain of events. But now, in the field of quantum phenomena, when we have any unambiguous information about atomic objects expressed as location in space or in time, we understand that an experiment has been carried out in which there occurred an exchange of energy and momentum between the measuring instruments, such as fixed scales or synchronized clocks, and the atomic objects, an exchange which, in principle, cannot be controlled. On the other hand, in the description of many other properties of atoms, we make extensive use of the conservation of energy and momentum. But for such phenomena we must renounce a detailed description in space and time. This is the essence of the lesson, that two kinds of concepts which are equally important in the description of experience, and which classical theory thought could be combined unrestrictedly, cannot be so combined. If, in atomic physics, phenomena are observed under different experimental conditions and are described by different physical concepts, they cannot be combined into a simple picture. If the attempt is made, we get apparent contradictions. Such phenomena we call complementary to each other, in the sense that each of them offers unambiguous information about the atomic objects under observation. Together they exhaust the knowledge about the objects that can be defined in human words or concepts.

The lesson, to state it more philosophically, is this: we have learned that, in this simple field of atomic physics, we have to pay attention to the conditions under which experience is obtained, and to the conditions under which the words we use can have precise meaning.

But the fact that, in many other fields of human interest, we meet situations requiring similar caution or attention is a general lesson from human experience. It is thus not a new thing, but just because we have learned this lesson in such a simple field as atomic physics—so far removed from the aspects of life where human aspirations and passions play their part—it may be helpful to apply it to other fields.

I would like to say a few words first about the old question of the position of living organisms among the other physical objects. At the time of the great triumphs of classical mechanics, organisms were often compared to machines, and to certain good purpose, although it could not be a complete description. Today we know that, to account for the properties of living organisms, we must have essential recourse to what we have learned in atomic physics. It is quite clear that very complicated molecular structures of great stability are responsible for the hereditary properties of the species. However, it is quite impossible to understand their stability on the basis of classical physics. Through quantum mechanics, on the other hand, this is possible. Next, in order to give just one more simple example, the empirical laws for the production of mutations by the influence of penetrating radiations are exactly the same as the laws describing the reactions between radiation and atoms and molecules which we study in atomic physics, but these laws have aspects quite different from what would be expected from classical physics.

Now, in this field of biology, great progress is taking place as a consequence of the advances in atomic physics. We may call this a mechanistic approach, because quantum mechanics is, after all, a rational generalization of classical mechanics. This approach holds very great promise and probably will have no limitations; but the question is: "What bearing does it have on the old problem of explaining life?" Here we must realize, first of all, that "life" is not a word that finds any application in physics. We find no reason to describe the phenomena we deal with in classical physics in terms of life, and exactly the same is true of atomic physics. Life is, in a sense, an irreducible element in biology, just as the quantum of action is now an element in physical science; and, at any rate for the quantum of action, the essential fact is that no explanation can possibly be given for it on the basis of classical physical ideas.

Biologists use the word "life" to remind us of such properties of living organisms as their self-regulation, their adaptability to environment, and so on. Now, the lesson we have learned in atomic physics strongly suggests that such different approaches as the mechanistic and the so-called finalistic are in no way contradictory to each other. They are, rather, expressions for complementary situations of observation where we either try to study details or think of the organism as a whole.

These remarks should not be taken to imply that physicists can directly be of help in developing biology. However, they point to an attitude which, may contribute to a better understanding among various groups of biologists by urging them to think of how words are actually used.

I would like to go on to something very different from physics, to the way we describe the state of our mind; what we call psychic conscious experience. Here we have developed a very rich vocabulary, by means of which we are able to communicate, even to pour out, information to one another. This is important when a man tries to say that he is dissatisfied or that he is contented. Now, all such words as thought, sentiment, volition, conscience, hope, and so on, do not, of course, refer to physical features.

Pairs of such words point to mutually exclusive situations, and the complementary use of words in this field has been common since the beginning of civilization. For example, a situation which calls for a description of our feeling of volition and a situation demanding that we ponder on the motives for our actions have quite different conscious contents. We pay attention to different objects in these two situations and make different separations of these objects from the background from which we, so to speak, judge them and which we loosely call our "self". In general, just as different phenomena in atomic physics require different experimental arrangements for their definition and observation, so the various psychic experiences are characterized by different placements of the separation between the content on which attention is focused and the background indicated by the word "self."

To point out how the lesson we have learned in physics may be helpful, I should like to recount a humorous tale from Danish literature "The Adventures of a Danish Student". This student has a very open mind and also lays himself open to all kinds of adventure. He has two cousins. One is very practical—what you might call dry—and the other is very philosophical. Now, far along in the tale, this cousin makes all kinds of trouble for himself and demands much forbearance for the plight he is in. His practical cousin comes to visit him and talks to him about how his position has completely deteriorated, and explains to him that it is absolutely necessary to do something; he must, as it is spoken of here in America, "look for a job." And everything is accordingly arranged for him. The next day he is to go to a neighboring mansion and get a position as teacher of children, to which he agrees. Then, after some weeks, the practical cousin comes back and finds that things have gone from bad to worse and that absolutely nothing can stop it. The erring cousin says: "I am very sorry, but it is impossible for me, for I have become confused about my different selves. You can so easily speak about me, but I must think of the self that controls me, and as soon as I do that I am equally aware that there is another self that controls and thinks of the self who controls what you call me, and if I start on this line of thought, my trouble becomes worse and worse. If I try to get out of it in any way, I get a terrible headache and have to give everything up".

Now, of course, the whole thing is meant by the writer to describe something very general, the different aspects of any human being. The author has made quite clear and has expressed very beautifully a situation in which all of us may find ourselves. Every healthy child has the possibility of living a normal life; but in case of certain diseases, as the psychiatrist well knows, it runs the risk of a splitting of its personality.

I would now like to say a few words about the old discussion of the freedom of the will. This, of course, is not related in any way to physics. Moreover, though attempts have been made through the ages, it is quite impossible to relate this problem to determinism; for a rigorous deterministic approach leaves no room for the concept of free will. On the other hand, it is clear that, like many other commonly used words, the word freedom is quite necessary to describe the richness of conscious life. Now, what do we use it for? In some situations we like to say that we have the feeling that it is possible for us, so to speak, to make the best of things. Speaking very loosely, it is simply a problem of cause; it is not possible to say whether we have the feeling that we are going to do something because we have the feeling that we can, or whether we can only because we will.

The problem is this, to see that we use the words "free will" to describe our situation in just as clear a manner as we use such words as "responsibility", "hope", and the like, all of which cannot be applied or defined unambiguously, except on the basis of the situations in which they are used.

I might get still further away from physics for a moment and say a few words about what we call ethical values. These are words that are very far from classical physics and equally far from quantum physics. Various philosophies, or philosphical schools, sometimes say, not out of cynicism but for definite reasons and in all honesty, that we must avoid evaluations or value judgments. But this is simpler said than done, for if there should be (and I don't think there are) philosophical schools that say that we must avoid the use of the words "good" and "bad", I think they would be tempted to say that it is good philosophy to do so.

But let us for a moment go right into the matter and ask how such words as "justice" and "charity" are actually used. We know, first of all, that they are necessary in speaking of human conditions in all stable societies that demand provisions for fair play. They get incorporated in judicial rules and become the law of the land, as, in this country, Oliver Wendell Holmes has so wonderfully explained and made clear in his book on the common law. On the other hand, human life would be deprived of much beauty and richness if we could not speak of sympathy, friendship, love, and so on.

It is clear that in all cultures the attempt is made to combine these two things, justice and charity, to the utmost extent. On the other hand, it is also clear that in a situation where there is a clear basis for the unambiguous use of the word "justice" after these rules, in such a situation the word "charity" would find no place whatever. But it is equally true, of course, as has been emphasized by great writers and philosophers, that compassion and good will can bring any of us into conflict with our ideals of justice. Now, the point that I wish to emphasize is that here we have two words that can be combined to a very large extent, as space-time description and conservation laws can be combined in classical physics, while for the finer and very varied regularities of atomic physics they must be used in a complementary manner. To my mind, the words charity and justice must be used in a complementary manner in many situations in which we express the richness of life and emphasize human values.

I want to emphasize that what we have learned in physics arose from a situation where we could not neglect the interaction between the measuring instrument and the object. In psychology, we meet a quite similar situation involving the decision as to where to draw the separation line between subject and object. A similar state of affairs is inherent in social problems, where we have to do with the relationship of the individual and the community to which he belongs, a relationship that is very rich and many-sided, and where again, different words correspond to different situations.

Before I close I should like to say a few words about the comparison of human cultures. I am not thinking directly of the difficulties we are having in the world today which we surely must hope to overcome; but I am thinking of the kind of experiences ethnographers have on expeditions to peoples who have lived for a long time in comparative isolation, say, on one of the beautiful islands in the Pacific Ocean. What they find, as everybody knows, are traditions and customs very different from what we are used to, so different that it appears as a great surprise to us that under such conditions there can exist a certain harmony within the population, as these explorers find to be the case. Now, when we attempt to compare cultures, we may be tempted to think that the differences between them are analogous to the different ways in which different observers co-ordinate and describe physical phenomena, ways which have been made so clear in the theory of relativity. But there is a great and fundamental difference. In fact, with the aid of the theory of relativity—through which Albert Einstein succeeded in giving physics great unity and discovered new physical laws common for all observers—however differently various observers may describe experience, any of them can foresee how the other observers will co-ordinate experience. On the other hand, if we compare human cultures the situation is different, because every culture contains an element of complacency. This is not something to deplore or be critical about, for it is quite analogous to the instinct of self-preservation found in any living organism. However, this complacency makes it very difficult, if not impossible, to appreciate the traditions of one culture on the basis of the traditions of another culture. One may be inclined to think that there is such a mutual exclusion among cultures that they are complementary to one another, as has sometimes been claimed; but there is a very great difference between the logical and necessary exclusion that holds for the use of concepts in atomic physics and in psychology and the relationship of cultures to one another, because cultures are only mutually exclusive as long as they are isolated. By intercourse, cultures can develop and change; cultures can flow together and merge, as we know so well from history, and progress can be attained thereby.

I want to say just a few words about how important it is today to promote intercourse among nations. We must understand that the situation into which the progress of science has brought us, with very great new promises for the promotion of human welfare, and at the same time the grave dangers resulting from the increase of our mastery of the forces of nature, is similar to the situations brought about by previous increases in knowledge and abilities, in that it places upon us a greater responsibility. At present, we find ourselves in a situation that constitutes a most forceful challenge to civilization. Nevertheless, this situation holds out a hope that is quite unique in some ways because, quite apart from all present difficulties, we must clearly realize now that there cannot be another great war without human suicide. When we think about how, through the long history of civilization, every conflict so far has been settled by armed force, we have in some ways greater promises today for a peaceful future. The problem is how to go about securing peace; and that is, of course, very difficult to say. But the first point I wish to mention is that the present situation is a radically new one to mankind and certainly demands a novel approach, just as in science, when we met new problems it was necessary time after time to modify our viewpoints and approaches. None of us underrates the difficulties, but the question we must ask ourselves is, what resources do we have? And I should like to say that the resources, and therefore the responsibilities, should be greatest in countries like your great country, where through happy historical developments there is so large a freedom for the individual, and where it is so much easier to speak openly with everyone about all possible procedures than in countries where unhappily such freedom does not exist in temporary (we hope) epochs of dictatorship.

Another point I would like to make is that it ought to be the best omen for the future that we are dealing with the results of a development that arose from purely scientific endeavors, having as their sole aim the augmentation of our knowledge and understanding of that nature of which we are part. The very fact that this development was brought about by close international co-operation strongly emphasizes the importance of the closest possible international co-operation for the achievement of a peaceful world. At any rate, such co-operation is essential in science, if peace is to be secured.

I know that much of what I have said in this lecture may sound superficial, but by pointing to something that is quite common to all human beings, I hope modestly that what I have said may lead to better understanding. I believe that, however difficult the problems are at the moment, the goal where every nation can attain prominence only by the help it can offer others and by what it can contribute to common human culture may be nearer today than it has ever been in the history of mankind.

It has been a great honor and pleasure for me to speak at the University of Oklahoma, where the question of the education of the younger generation is considered so seriously and where education in our day can mean not only education for citizenship in a great country but education for the service in the great cause of all mankind.

Address by Professor Niels Bohr
delivered[1] on October 11, 1961
at the University of Brussels

MS, p. 1

It is a great pleasure to me on behalf of the graduants to express our deep appreciation of the honour which the University of Brussels has bestowed upon us. We are happy in this way to be connected with your illustrious university to which all participants in the Solvay conferences have through the years felt so strongly attached.

Being kindly invited on this occasion to say some words about the development of physical science, to which the discussions at the Solvay meetings have afforded so valuable a stimulus, I have thought it might be appropriate to choose a topic which formed a central issue in these discussions. Thus, I shall briefly recall the deepgoing lesson which the development of atomic physics has given us about our position as observers of that nature of which we ourselves are part, and to indicate the message this lesson contains as regards our orientation in wider fields of human knowledge and interest.

As is well known, the origin of atomistic ideas goes back to thinkers in ancient Greece who, confronted with the multifarious variety of natural phenomena, sought a foundation for an understanding of the specific properties of the substances, of which our tools as well as our bodies are built up, in the assumption of a limited divisibility of matter. Indeed, the apparently miraculous change of state of a substance like water by freezing or evaporation, and the reverse process of melting or condensation by which the water is again brought back to its original liquid state, can be understood only when we assume that the water molecules remain unaltered during these transformations, in which merely their state of motion and ordering in space are changed. Also it is of interest to recall how Lucretius two thousand years ago, by contemplating the development of a plant from a seed placed in the soil, emphasized the necessity of atomistic conceptions in terms which remind of our present description of the persistence during the organic metabolism of molecular structures responsible for the characteristics of the species.

MS, p. 2, 30.9.61.

In spite of the pertinence of such ideas, it was for a long time taken for granted that the coarseness of our senses would forever prevent the direct observation of individual atoms or molecules. Still, the wonderful progress in the art of experimentation, especially as regards the construction of amplification devices, has in our time made it possible to study effects originating from

---

[1] [The address was actually *not* delivered. See Editor's Remarks, p. [173].]

single atomic particles. In this situation, however, we have got new reason to admire the wisdom of ancient thinkers in expressing caution as regards the application to new fields of experience of concepts which have proved sufficient for the orientation in more limited domains of human knowledge. Indeed, the modern development of atomic science has revealed unsuspected limitations of principles which were thought to be fundamental for rational physical explanation.

The wide validity of these principles was in fact most clearly demonstrated by the imposing progress in physical science through the ages. Thus had the analysis of problems of static equilibrium, initiated in Antiquity by Archimedes and pursued in the Renaissance by Stevin, proved basic for the understanding of the functioning of many kinds of machines and for ever more refined constructions of windmills, sluices and bridges. Above all, the pioneer work in dynamics by Galileo, which was brought to such completion by Newton, made it possible to explain the regularities exhibited by the motion of the planets around the sun by the same simple principles governing the fall of bodies on the earth.

The whole edifice of this so-called classical mechanics rested on the assumption that the changes in the state of a physical system, defined by the positions and velocities of its parts, can be represented as a continuous chain of events coupled uniquely to each other by the conservation of momentum and energy. The triumph of such pictorial deterministic description, which so closely corresponded to familiar conceptions of the relationship between cause and effect, exerted decisive influence on general philosophical thinking in the following centuries and has sometimes even been considered as the only adequate basis for scientific analysis.

The conception of the atomic constitution of matter was given a firmer foundation by the great advances in chemistry, which through the work of Lavoisier and Dalton led to an unambiguous definition of the elements and the elucidation of the laws governing their chemical combinations. Stimulated by the exploration of the simple physical properties of gases great progress was also achieved by applying to the atomistic view the principles of classical mechanics. In particular, an interpretation of heat phenomena and of the laws of thermodynamics was obtained by developing kinetic pictures of the internal state of bodies regarded as mechanical systems of molecules. Even though the immense complication of such systems demanded the use of statistical methods, and a close relationship between the concepts of entropy and probability was revealed, the recourse to such considerations was not regarded as an infringement of the deterministic conception underlying the whole approach.

In the meantime the scope of deterministic description was further extended

MS, p. 3

MS, p. 4

by the analysis of the new phenomena of electromagnetism through the work of Ørsted and Ampère, and above all by Faraday's discovery of electromagnetic induction, basic for Maxwell's work, in which the electromagnetic theory received a temporary completion rivalling Newtonian mechanics in perfection. In this theory the definition of a state of a system at a given moment involved not only the positions and velocities of the electrified and magnetized bodies, but also the specification of the intensity and direction of the electric and magnetic forces at any point of space. As is well known, Maxwell was in this way able to predict the possibility of propagation of electromagnetic waves in free space, which did not only offer an explanation of optical phenomena on the lines introduced by Huygens, and especially developed by Fresnel, but which also was to form the basis for modern radio communication.

I need hardly remind you how the absence of the expected influence on optical phenomena of the motion of the earth around the sun challenged Einstein to take up for revision the foundation for the unambiguous use of our most elementary concepts of space and time. This analysis, which disclosed the essential influence of the standpoint of the observer on the way he will coordinate experience, opened possibilities of tracing fundamental regularities common to all observers and thereby establishing relationships between phenomena which so far had remained unconnected. In spite of the use for such purpose of mathematical abstractions like four dimensional non-Euclidean geometry, the physical content of the theory of relativity depends essentially on the possibility for any observer to separate sharply between space and time and connect experience in terms of cause and effect. In all such respects Einstein's work may be considered as a harmonious completion of classical physics.

MS, p. 5, 2.10.61.

Still, a new era in natural philosophy had been inaugurated already in the first year of this century by Planck's discovery of the universal quantum of action which revealed a feature of wholeness in atomic processes, far transcending the old doctrine of the limited divisibility of matter. Thus it became evident that the causal pictorial description of classical physics represents an idealization, valid only for the analysis of phenomena for which the action involved is sufficiently large to permit the neglect of the individual quantum. While this condition is amply fulfilled in phenomena on the ordinary scale, we meet in the study of atomic processes novel regularities which defy deterministic description, but are nevertheless responsible for the peculiar stability of atomic systems on which all properties of matter ultimately depend.

Such problems came more and more to the foreground by the increasing knowledge of the constituent particles of the atom. The first step was the discovery of the electron demonstrating the atomistic character of electricity itself, but the intrinsic stability of which was clearly inexplicable on classical

MS, p. 6, 2.10.61.

electrodynamics. While such stability of so-called elementary particles has hardly as yet found satisfactory explanation, the constitution of composite atomic systems offered a far more direct opportunity for approach. Of decisive importance was here Rutherford's discovery of the atomic nucleus, by which we learned that almost the entire mass of the atom is concentrated within a region of extension far smaller than the size of the whole atom as defined by the system of electrons bound to the nucleus by its opposite electric charge.

This simple model of the atom, resembling a microscopic solar system, allowed an immediate explanation of many important properties of matter. Thus, owing to the smallness of the atomic nucleus and its great mass compared with that of an electron, the constitution of the surrounding electron system, which determines the ordinary physical and chemical properties of the elements, will to a high approximation depend only on the number of unit charges on the nucleus and not on its mass and internal structure. The relevance of such views was indeed clearly brought out by the discovery of isotopes of nearly all elements differing only in atomic weights and radioactive properties. Notwithstanding the adequacy of the Rutherford model in such respects, it was from the beginning obvious that its stability was clearly irreconcilable with the principles of classical physics.

Indeed, no system of point charges will on classical mechanics admit of a stable static equilibrium, and any motion of the electrons around the nucleus would, according to electromagnetic theory, give rise to a dissipation of energy through radiation accompanied by a rapid contracting of the atom, until the electrons and the nucleus would combine into a system far smaller than atomic dimensions as deduced from physical and chemical evidence. Altogether, it must be realized how deeply the behaviour of an atomic system, in spite of any apparent resemblance, differs from that of a solar system, which certainly exhibits simple regularities as long as it is left undisturbed, but which possesses no intrinsic stability of the kind characteristic of the atoms of the elements and the chemical molecules.

MS, p. 7

In fact, if we imagine a comet of great mass one day penetrating into our solar system and colliding with the earth, we had to be prepared that – quite apart from what else might happen – the orbit of the earth and thus the length of the year from that day would have been permanently changed. In contrast, the atoms or molecules in a gas retain all characteristics under their mutual collisions and through the spectral analysis, which discloses the existence of the ordinary chemical elements in the most distant stars, we have obtained striking evidence of the conservation of the specific properties of the atoms even under influences so violent that the electrons are temporarily removed from the nucleus. Similarly holds for the molecules of chemical compounds which, whatever chemical processes are involved in their formation, exhibit the same characteristic properties.

The recognition of the fundamental role played by the quantum of action in all such questions of atomic stability was prepared already a few years after Planck's discovery through Einstein's ingenious analysis of the paradoxes involved in the exchange of momentum and energy by radiative processes, by which he introduced the concept of the light quantum, or photon. Of course there could be no question of reverting to the old corpuscular idea of light propagation, since the very definition of the momentum and energy of the photon involves the wavelength and frequency of the radiation, which directly refer to the characteristics of a wave picture.

Especially it may be stressed that no escape from this dilemma is to be found, as sometimes suggested, by carrying the analysis further and taking the atomic constitution of optical instruments like mirrors, prisms, and lenses into account. In fact, in complete agreement with the classical representation of the propagation of radiation, the functioning of such instruments depends primarily on the smallness of the atomic spacing compared to the spatial extension of the radiation field, necessary for the definition of its spectral composition.   MS, p. 8

The impossibility of comprising in any single picture the mechanism for the appearance of the individual radiative processes and the equally well established laws governing the building up of visual images as well as photographic records by the accumulation of such individual effects has indeed deprived us of any basis for exhaustive deterministic account of radiative phenomena. We are here confronted with a novel situation in physical science characterized by the necessity of even in the account of the most elementary physical processes imaginable, to take recourse to an essentially statistical description of experience.

Thus a new background was created for the attempts of coordinating the physical and chemical properties of matter with the new information about the structure of atoms, where one met with similar problems, arising from the fact that although no consistent pictorial description on classical lines was possible, all information, for instance about the magnitude of the inertial mass and electric charge of the electrons or the nucleus, was derived from experiments analyzed in terms of classical concepts. Still, a preliminary orientation about the binding process of the electrons in the atom was offered by the comparison between the individual emission and absorption processes responsible for the appearance of atomic spectra and the creation and annihilation of an Einstein photon. Although obviously incomplete such considerations offered, nevertheless, a starting point for the gradual establishment of a consistent description   MS, p. 9

[211]

of atomic phenomena based on a harmonious incorporation of the quantum in a rational generalization of classical physics.

As is well known, this great and difficult task, which demanded the application of new abstract mathematical methods, was achieved through the cooperation of a whole generation of physicists from many different countries. On this occasion I need not enlarge upon the pioneer work and constant leadership of Werner Heisenberg or the important contributions to the statistical interpretation of the formalism by Max Born, whose absence, due to illness, we all deplore. Every physicist also knows how Louis de Broglie's original comparison of the motion of a particle with wave propagation inspired the great work of Erwin Schrödinger, whose death last year was such a grievous loss to science. The differential equation which bears his name proved indeed an invaluable basis for the subsequent development in which Dirac's relativistic quantum theory of the electron was to elucidate in a novel manner symmetry arguments characteristic of the classical approach.

In spite of the power of quantum mechanics and electrodynamics in ordering an immense variety of experience regarding the properties of atomic systems, the extent to which such account renounces accustomed demands of physical explanation has raised doubts, especially among philosophers, whether we are concerned with a complete description of natural phenomena. To clarify this point, it was indeed necessary to examine what kind of answers we can receive by so to say putting questions to nature in the form of experiments. In order that such answers may contribute to objective knowledge, independent of subjective judgement, it is an obvious demand that the experimental arrangement as well as the recording of observations be expressed in the common language, developed for our orientation in the surroundings.

MS, p. 10

In actual experimentation this demand is met by the specification of the experimental conditions by means of bodies like diaphragms and photographic plates so large and heavy that the statistical element in their description can be neglected. The observations consist in the recording of permanent marks on these instruments, and the fact that the amplification devices used in the production of such marks involves essentially irreversible processes presents no new observational problem, but merely stresses the element of irreversibility inherent in the definition of the very concept of observation.

This situation is, of course, common to all physical enquiry, but is often overlooked when dealing with phenomena within the scope of classical physics, where the interaction between the measuring instruments and the object under investigation can be neglected or at any rate compensated for. In quantum physics, however, such interaction forms an integral part of the phenomena evading separate account if the instruments shall serve their purpose of defining the experimental arrangement.

It is this circumstance which makes the recourse to a statistical account of the observations imperative in the objective description of quantum phenomena since in general different individual effects can appear within one and the same experimental arrangement. Equally, we must be prepared that observations obtained under different experimental conditions may seem to contradict each other if one attempts to combine them in a single picture. Still, however contrasting such phenomena may at first sight appear, they are to be regarded as complementary to each other in the sense that, taken together, they exhaust all information about the atomic object which can be unambiguously derived and communicated.

The physical content of the formalism of quantum mechanics and electro- MS, p. 11
dynamics is indeed exhausted by the establishment of the statistical laws governing observations to be obtained under well defined experimental conditions and the completeness of such description rests upon the character and scope of the formalism which permits its application to experimental conditions chosen in any conceivable manner. In this context, the unambiguous use of pictures is restricted to such apparently trivial statements as, for instance, that when the presence of an electron or photon has been ascertained on one side of an impenetrable diaphragm with only one hole and subsequently is observed on the other side of the diaphragm, we are permitted to conclude that it has passed through the hole. In case the diaphragm is provided with a shutter which is open only within a certain time interval, we likewise say that the passage of the object has taken place during this interval.

The circumstance that any such use of the measuring instruments is ultimately connected with an incontrollable transfer of momentum and energy to the fixed scales and synchronized clocks serving to define the reference frame, implies limitations in principle in the possibilities for that unrestricted combination of pictorial representation and the application of the conservation laws for momentum and energy on which the deterministic description of classical physics rests. In fact, it must be realized that the wide scope of applicability of these conservation laws in quantum physics depends entirely on the extent to which definite values for momentum and energy can be ascribed to the states of atomic systems brought in isolation with corresponding loss of connection with the space–time frame.

In particular, the intricacies of atomic stability find their elucidation in a MS, p. 12
description along complementary lines. Thus, any experimental arrangement suited to ascertain the position and displacement of the electrons in an atom or molecule is incompatible with the conditions permitting the appearance and

[213]

study of spectral regularities and chemical bonds manifesting properties of the atomic systems which defy such pictorial account. A special point, of decisive importance for the description of molecular structures in chemistry, is the fact that the large masses of the atomic nuclei compared to that of the electrons leaves room for the fixation of the relative positions of the atoms in undisturbed molecules to a degree allowing us to retain the use of structural formulae which have offered so invaluable a guidance in chemical analysis and synthesis.

Moreover, the analysis of the quantum features, responsible for molecular stability, provided a firm basis for the use of well established thermodynamical principles in chemical kinetics. Whereas at ordinary temperature the thermal agitation of the liquid media in which the molecules are dispersed will only seldom be able to break the specific chemical bonds, the heat motion will cause frequent changes in the rotational and vibrational states of the molecules. Such effects, which do not radically influence the characteristic properties of the composite molecules, will rather serve to a rapid wiping out of any peculiarities in the state of the system arising from the way it was formed.

The remarkable stability of chemical compounds is especially conspicuous in biochemistry where in later years great progress has been made as regards our knowledge of the complex constituents of the cells, especially the so-called DNA chains, carrying genetic information. Moreover, it is being gradually understood that the very processes connected with the formation of the permanent cell structures take place with expenditure of free energy, corresponding to an ever increasing stability under the prevailing conditions in the organism upheld by nutrition and respiration or photo-assimilation. Not only does the particular firmness and shielding of the atomic pattern in the DNA chains mean that the duplication of these chains, connected with the rhythm of cell division, corresponds to a more stable configuration of the components of the system, but the same applies to all the enzymatic processes, which take part in such duplication and the further transfer of information by the building up of the specific proteins.

Modern development of biochemistry has also disclosed that many of the enzymes or even hormones are relatively simple in their atomic constitution, and to a rapidly increasing extent they are being synthesized by ordinary chemical methods. Of course it is practically excluded to perform such chemical synthesis of the structures carrying specific information determining the difference between the species ranging from microbes to vertebrates. In fact, such differences are the result of the whole history of organic evolution through which systems like living beings, of extreme stability in spite of their immense complexity, have been developed step by step.

Even if all present evidence indicates that in no aspects of organic life we are

MS, p. 13

[214]

confronted with deviations from the principles of atomic physics, we cannot be surprised that, due to their extraordinary complexity of structure and function, the organisms display properties in such striking contrast to the properties of matter exhibited under simple reproducible experimental conditions. Indeed, notions like self-preservation and regeneration, referring to the organisms as entities, have through the ages presented themselves and remain indispensable MS, p. 14 for an adequate account of biological experience. However, the connection between such notions and physical and chemical concepts cannot be immediately compared with the complementary relationships in quantum theory concerning the restricted combination of space–time account and the application of the conservation laws of momentum and energy.

In fact, when we take our starting point in empirical chemical kinetics, we are not beforehand confronted with any limitation in the scope of pictorial description arising from the necessity of taking due regard of the conditions under which evidence is obtained. It is in this connection to be remembered that even in the description of the most complicated organic phenomena, like the shift of generation through fertilized seeds or eggs, all details are accounted for in pictures, for the unambiguous use of which no such reference is involved. The use in biology of concepts foreign to purely physical enquiry is thus imposed directly by the extreme complexity of the constitution and behaviour of the living organisms.

This situation is especially evident when we consider the conscious experiences connected with life. Indeed, the very word "consciousness" presents itself when the behaviour of an organism exhibits a degree of complexity which makes reference to self-awareness unavoidable in any attempt at exhaustive description. In the account of psychical experience there is of course no question of direct comparison with physical phenomena, but it is interesting to note that the vocabulary used for communicating the state of our mind has been applied MS, p. 15 in a typically complementary manner since the very origin of language. Thus, words like thoughts and sentiments evidently refer to psychical experiences which exclude each other.

In such context it must be remembered that the objective character of the description in atomic physics depends on the detailed specification of the experimental conditions under which evidence is gained, and that therefore no reference to the observing subject is required. In the communication of psychical experience, where we express ourselves by saying "I think" or "I will", the mutual exclusion of the situations referred to is explicitly indicated by the attachment of different verbs to the same pronoun. We are here confronted with the old question of the ego, well known from serious or humorous discussions about our many I's which often seem to be in conflict with each other.

The possibility of upholding the integrity of the personality and to retain the conception of free will rests indeed on the different placing of the partition between subject and object in situations which we characterize with words like contemplation or volition. With similar complementary relationships we are confronted in the intercourse between individuals in a community, where in particular the limited compatibility of ethical ideals as justice and charity is conspicuous. Evidence of the multifarious ways, which offer themselves in the search for harmony within the life of a human community, is not least presented by a comparison between different national cultures which, in spite of deep-rooted differences in attitudes and traditions, reveal common features of the human situation.

Although in some way such cultures may be regarded as complementary to each other, we have of course not to do with immutable relationships as in atomic physics and psychology. Indeed history offers many examples of how contact between nations has led to cultural developments combining valuable elements of separate origin. We are here reminded of the contributions from different schools of learning to that growth of science, for the promotion of which such meetings as those convened by the Solvay Institute have in our time been so fruitful.

MS, p. 16

[216]

PART II

COMPLEMENTARITY IN
OTHER FIELDS

INTRODUCTION

by

DAVID FAVRHOLDT

Niels Bohr wrote numerous articles for special occasions and held many lectures on subjects outside science. The articles reproduced in Part II have been chosen because they express the general approach to daily life and human conditions that was typical for Bohr, and reflect at the same time some of the ideas that characterized his purely scientific work.

With regard to the first article[1], it should be noted that in Bohr's time Danish education was organized so that in order to qualify for the university, one had to attend ordinary school for nine years followed by a three-year high school (Gymnasium) education. The Gymnasium was intended as preparation for studies at the university or other schools of higher education, and was completed with a so-called student examination. There is still the tradition that those who pass the examination meet 25 years later to celebrate the so-called student jubilee.

In the second article[2], Bohr points out that the observational situation in the anthropological sciences has similarities to that in quantum mechanics because a complete assimilation in a foreign culture and a concise, analytic description

---

[1] N. Bohr, *Tale ved Studenterjubilæet 1903–1928* (Speech given [on 21 September 1928] at the 25th anniversary reunion of the Student Graduation Class). Printed by C. Johansen's printing press on the occasion of the 50th anniversary reunion, 15 October 1953. Reproduced on pp. [227]–[232] (Danish original) and pp. [233]–[236] (English translation).

[2] N. Bohr, *Natural Philosophy and Human Cultures* in *Congrès international des sciences anthropologiques et ethnologiques, compte rendu de la deuxième session, Copenhague 1938*, Ejnar Munksgaard, Copenhagen 1939, pp. 86–95. Reproduced on pp. [240]–[249]. Also published in *Nature* **143** (1939) 268–272, and in *Atomic Physics and Human Knowledge*, John Wiley & Sons, New York 1958, pp. 23–31. The latter volume is photographically reproduced as *Essays 1933–1957 on Atomic Physics and Human Knowledge*, The Philosophical Writings of Niels Bohr, Vol. II, Ox Bow Press, Woodbridge, Connecticut 1987.

of that culture would seem to exclude one another. Bohr regarded each individual to be formed by his culture in all important respects. He regarded as absurd any theory holding that moral or intellectual qualities should be racially determined, and he emphasized this point in the lecture, undoubtedly because of the contemporary flourishing in Germany of Aryan race theories, which was the reason for the lecture being mentioned in the New York Times. During the lecture, which was held at Kronborg Castle in Elsinore, some Nazis in the audience left the room in protest against these statements[3].

The third article[4] should be seen in the light of the political conditions at the time it was written. During World War II, German soldiers occupied Denmark on 9 April 1940. The Danish army was unable to offer notable resistance, and the Germans described the occupation as a friendly protection of Denmark against a possible invasion by Britain. Denmark was allowed to keep the royal family, its parliament, armed forces and police, and no measures were taken with respect to the Jewish population. During the first years the government tried to establish a policy of cooperation which aimed at maintaining peace in the country while at the same time offering Germany the least possible support. During this period, a Danish resistance movement grew up along with a strong patriotism, and the policy was abandoned after the public revolt on 29 August 1943. Bohr's article was written as an introduction to a work describing the current state of Danish culture. This work, which was published in eight volumes between 1941 and 1943, was but one indication of the new patriotism.

Bohr was an obvious choice to write the introduction[5]. Not only was he one of the great cultural personalities in Denmark, but it was well known in intellectual circles that he had rejected Nazism early on and was an active member of the Danish Committee for the Support of Refugee Intellectual Workers, which was founded in 1933 to aid individuals forced to flee Nazi persecution in Germany. The chairman of the board of the committee was

---

[3] S. Rozental, *NB, Erindringer om Niels Bohr*, Gyldendal, Copenhagen 1985, p. 136, republished as *Niels Bohr, Erindringer om et samarbejde*, Christian Ejlers Forlag, Copenhagen 1994, p. 165. Translated into German as *Schicksalsjahre mit Niels Bohr, Erinnerungen an den Begründer der modernen Atomtheorie*, Deutsche Verlags-Anstalt, Stuttgart 1991, p. 158. The relevant New York Times article is reproduced in facsimile on p. [238].

[4] N. Bohr, *Dansk Kultur. Nogle indledende Betragtninger* (Danish Culture. Some Introductory Reflections) in *Danmarks Kultur ved Aar 1940*, Det Danske Forlag, Copenhagen 1941–1943, Vol. 1, pp. 9–17. Reproduced on pp. [253]–[261] (Danish original) and [262]–[272] (English translation).

[5] See M. Pihl, *Niels Bohr and the Danish Community* in *Niels Bohr. His life and work as seen by his friends and colleagues* (ed. S. Rozental), North-Holland Publ. Co., Amsterdam 1967 (reprinted 1985), pp. 290–300.

Aage Friis, a history professor at the University of Copenhagen specializing in Danish–German relations. The other two members of the board were Bohr and Thorvald Madsen, director of the Serum Institute, an institution established by the Danish State in 1901. Barrister Albert V. Jørgensen, Bohr's classmate from high school, was the treasurer of the committee, which among its 49 members also counted Bohr's brother Harald[6]. Regrettably, most papers, correspondence etc. relating to the committee's work were burnt immediately upon the German occupation of Denmark[7].

In his introduction, it was natural that Bohr would try to capture several aspects of Danish culture, a difficult task which led to a large number of rewritings. Stefan Rozental relates that the writing of the short article required about two months of full-time work, whereupon as many as seven proofs were necessary before Bohr was satisfied with the result[8].

The fourth article[9] in this part, "Physical Science and the Study of Religions" was written in honour of Johannes Pedersen (1883–1977), Danish historian of religion and orientalist. Already before Pedersen was appointed professor of Semitic and oriental philology at the University of Copenhagen in 1922, he was internationally renowned for his original studies in early Christianity, the Jewish religion and Islam. He became a member of the Royal Danish Academy of Sciences and Letters in 1924. As President of the Academy, it was natural that Bohr contributed to the *Festschrift* written on the occasion of Pedersen's 70th birthday in 1953.

The last article[10] is the lecture Bohr gave in response to receiving the Sonning Prize, a Danish award instituted in 1949. Candidates for the prize are intellectual leaders of international repute, and the choice is made by the Senate of the University of Copenhagen.

---

[6] "Opfordring" (Appeal) dated 31 October 1933, Archive of the Carlsberg Breweries, Copenhagen.

[7] S. Rozental, *The Forties and the Fifties* in *Niels Bohr* (ed. S. Rozental), ref. 5, pp. 149–190, on p. 155.

[8] *Ibid.*, pp. 163–164.

[9] *Physical Science and the Study of Religions*, Studia Orientalia Ioanni Pedersen Septuagenario A.D. VII id. Nov. Anno MCMLIII, Ejnar Munksgaard, Copenhagen 1953, pp. 385–390. Reproduced on pp. [275]–[280].

[10] N. Bohr, *Atomvidenskaben og menneskehedens krise* (Atomic Science and the Crisis of Humanity), *Politiken*, 20 April 1961. Reproduced on pp. [283]–[288] (Danish original) and [289]–[293] (English translation).

I. SPEECH GIVEN AT THE 25TH ANNIVERSARY REUNION OF THE STUDENT GRADUATION CLASS

TALE VED STUDENTERJUBILÆET 1903–1928
Speech given on 21 September 1928. Printed on the Occasion
of the 50th Reunion, 15 October 1953

PRIVATE PRINT

TEXT AND TRANSLATION

See Foreword, p. x and Introduction to Part II, p. [219].

Bohr's class of students, 1903. Upper picture, standing, from the left: Albert V. Jørgensen, Aage Berlème, Niels Bohr and Ole Chievitz; front row, from the left: Axel Fremming, Svend Hertz and Fridtjof Bang. The lower photograph was taken 25 years later, in 1928, with the same positions. Bohr retained close contact with several of his classmates throughout his life. As but one example, in 1918 Berlème took the initiative to obtain financial help from private sources for the establishment of Bohr's Institute for Theoretical Physics, which was inaugurated in 1921.

SPEECH GIVEN AT THE 25TH ANNIVERSARY REUNION OF THE STUDENT GRADUATION CLASS (1928)

Versions published in Danish

Tale ved Studenterjubilæet 1903–1928

A   Private print, 1953
B   "Naturbeskrivelse og menneskelig erkendelse" (eds. J. Kalckar and E. Rüdinger), Rhodos, Copenhagen 1985, pp. 259–263

The two versions agree with each other.

NIELS BOHR

TALE VED STUDENTERJUBILÆET
1903—1928

21. SEPTEMBER 1928

Jeg maa tilstaa, at det kun er med stor Betænkelighed, jeg tager Ordet i Aften. Lige saa stærkt som jeg deler vor fælles Glæde over, at saa mange af os Studenterkammerater er samlede for at fejre vort 25 Aars Jubilæum, lige saa ringe føler jeg min Evne til at slaa paa de muntre Strenge, hvis Toner fremfor nogen ikke maa savnes ved en rigtig Studenterfest. Kun i Visheden om at andre, der i rigere Maal besidder denne Evne, i Aftenens Løb vil lade os nyde godt deraf, drister jeg mig til at prøve i jævne Ord at skildre nogle af de Stemninger, der har fyldt mit Sind i disse Dage, naar jeg tænkte paa, at vi atter skulde mødes og forny vore fælles Ungdomsminder.

Vor Samhørighed har jo dybe Rødder. Hvor langt end Skæbnen siden maa have skilt vore Veje, kan de af os, der har fulgtes Skolen igennem, aldrig glemme, hvor meget vi har delt med hverandre, og naar vi er samlede, føler vi ikke alene, hvor umiddelbart vi er knyttet til hinanden, men vi vil ogsaa sende vore Lærere, som vi dengang maaske saa det som vor Opgave at bekrige, en taknemmelig Tanke. I Aften vil vi dog først og fremmest mindes vor fælles Jubel over at slippe for, hvad der stod for os som Skolens Tvang, og gaa ind til det nye frie Liv med dets, som det syntes os, ubegrænsede Muligheder. Hvor bevæges vi ikke alle ved Tanken om Rustiden, da vort Mod stod i saa omvendt Forhold til Klarheden

3

af vore Maal, og vi i Betagelse levede i de nye Verdener, der under vore Universitetslæreres Vejledning aabnedes for os. Især til Samværet med Studenterkammeraterne, med hvem vi med vore uprøvede Kræfter mødtes i fri Kappestrid, og som vi gjorde delagtige i vor Forventning og Begejstring, vil dog for altid nogle af vore smukkeste Minder være knyttede.

Egentlig havde jeg tænkt at dvæle lidt mere ved Erindringerne fra Studieaarene, der, selv om Forholdene jo efterhaanden maatte blive mere dagligdags og Forskellen fra Skoletiden vel sommetider føltes mindre stærkt, skulde skænke os alle saa mange rige Oplevelser. Med Rector magnificus hjertelige Velkomstord paa Universitetet og de følte og rammende Udtryk, hvormed vor Talsmand i sit Svar tolkede vor Taknemmelighed over for vor gamle Højskole, er Studielivet imidlertid fremmanet for os med saa friske Farver, at jeg næppe kan tilføje noget nyt Træk. Heller ikke skal jeg komme nærmere ind paa den vidt forgrenede Forbindelse, der selv efter Studietiden og det daglige Samværs Ophør bestod mellem os, der hver paa sit Felt som Borgere i samme Land skulde forsøge at frugtbargøre vor Uddannelse. Jeg vil blot gerne minde om, hvorledes Verdenskrigen, med dens Rystelse af Grundlaget for vort ydre og indre Liv, for alle af vor Generation ikke alene skulde blive Paamindelsen om de stærke Baand, der knytter os til vort eget Folks Skæbne i Fortid og Nutid, men tillige en Manen om alle Menneskers dybe Samhørighed.

Den umiddelbare gensidige Forstaaelse mellem os Studenterkammerater bunder dog ikke alene i Minderne om de mange fælles Oplevelser. Hvor stærkt vort Syn paa stort og smaat er farvet af, at det var paa samme Baggrund vi modtog Indvielsen til saa meget

4

nyt, kan vi jo ikke undgaa at mærke, naar vi efter siden at have vandret ad forskellige Veje atter mødes. Vist nok er de Kundskaber, som kræves til Studentereksamen og læres paa Universitetet, egnede til at gøre os fortrolige med Skæbner og Stræben af andre Tiders og Steders Mennesker, at gøre os til hvad man har kaldt aandelige Verdensborgere; men Opfattelsen af alt aandeligt Indhold er betinget af Forudsætningerne, og intet er maaske mere tids- og stedsbestemt end netop Modtagelsen af Kundskaber. Først naar vi stifter Bekendtskab med Mennesker med samme Interesser som vi, men som har faaet deres akademiske Uddannelse under andre Vilkaar, føler vi rigtig, hvor dybt vort hele aandelige Liv er præget af vor særlige Indstilling,

Det er ingen let Opgave nærmere at udrede vort aandelige Fællesskab, her er vi jo ikke alene Iagttagere men selv Deltagere. Alligevel tror jeg, at vi tør sige, at Forstaaelsen af, at enhver Sag kan ses ud fra mange Synspunkter, og at dybere Erkendelse kun kan opnaas ved at lade Modsætningerne komme til deres Ret, er et Grundtræk hos dansk Studenteraand. Saa almindeligt udtrykt er denne Forstaaelse jo kun en almenmenneskelig Erfaring, der umiddelbart opfattet er Kunstens Livskilde og bevidst udformet Videnskabens Fanemærke. Jeg mener derfor heller ikke andet end, at en intim Forbindelse mellem vort danske Folkelune og vor akademiske Tradition har skabt en Jordbund, der særlig begunstiger Udviklingen af et Livssyn, hvis Kendetegn er Angsten for de Farer, som Ensidighed rummer, og som derfor hellere frivilligt giver Afkald paa den Styrke, som en saadan tilbyder os. Alle genkender vi denne Stemning i vor Digtekunst og filosofiske Litteratur. Renest og stærkest kommer den maaske til Orde hos den Mand, som man

5

har kaldt den danskeste af alle danske Digtere og Filosoffer, Poul Martin Møller, der i sit stadig lige friske Værk paa fineste og dybsindigste Maade, netop ved at give Udtryk for Kampen mellem Modsætningerne i vort eget Sind, har givet os et Billede af dansk Studenteraand, hvori vi endnu kan spejle os.

Dobbelt stærkt virker dog dette gamle Billede paa os, fordi vi i dets Grundstemning erkender et nært Slægtskab med de aandelige Rørelser, der for Tiden bevæger Menneskers Sind Verden over. Særegen for vor Tid og ikke mindst for dens travle og vidtspændende videnskabelige Virksomhed er den vaagne Bevidsthed om, at naar vore Forudsætninger ikke paaagtes og stadig uddybes, bliver de til Fordomme, der kan skjule den større Sammenhæng for vore Øjne. Dette gælder Aandsarbejdet paa alle Omraader, lige fra hvad man vel ofte vil kalde det abstrakteste af det abstrakte, vore Tanker om vor egen Tankevirksomhed, til hvad vi kunde være tilbøjelige til at kalde det konkreteste af det konkrete, vore Iagttagelser over den livløse Natur. Siden vi blev Studenter, har jo netop paa det sidste Omraade, som alle ved, Erkendelsen af forskellige Udgangspunkters Ligeberettigelse afsløret upaaagtede Fordomme ved den tilvante Indordning af Naturfænomenerne i Rum og Tid og samtidig aabenbaret uanede Sammenhæng mellem gammelkendte Ting. Denne Udvikling er ingenlunde afsluttet med Relativitetsteoriens Triumf, der utvivlsomt er en af de store Mærkepæle i Menneskeaandens Historie, men Naturvidenskabens seneste Berigelse med helt nye Erfaringsomraader har afdækket yderligere Brist i det Grundlag, hvorpaa vi troede trygt at kunne stole. Atter her synes vi at skulle opleve, at tilsyneladende uovervindelige Modsæt-

6

ninger forsvinder ved den fortsatte Udforskning af Forudsætningerne for vore Anskuelsesformer.

Det er ikke Erkendelsen af vor menneskelige Begrænsning, men Bestræbelserne paa at udforske denne Begrænsnings Natur, der præger vor Tid. Det vil kun give os en fattig Forestilling om vore Muligheder, om vi vilde sammenligne vor Begrænsning med en uoverstigelig Mur, der paa bestemte Veje er sat for vore Tanker. Overalt aabner sig nye Udblik, og nye Sammenhæng lader sig erkende, men enhver Vej, som vi slaar ind paa, deler sig atter og atter og krummer sig, saa at vi hurtigt taber Retningen og før eller senere kommer tilbage til vort Udgangspunkt. Alligevel kan vi altid vende hjem med Udbytte, og for Rigdommen af det Indhold, som vi kan samle og ordne, ser vi ingen Grænser. Ud fra dybere og dybere udforskede Forudsætninger erkendes større og større Sammenhæng. Saaledes opfattet lever vi under et stadig rigere Indtryk af en evig og uendelig Harmoni; vel at forstaa, Harmonien selv lader sig kun ane, men aldrig gribe; ved ethvert Forsøg derpaa svinder den efter sit Væsens Natur ud af vore Hænder. Intet er fast, hver Tanke, ja selv hvert Ord er kun egnet til at understrege en Sammenhæng, der i sig selv aldrig kan fuldt beskrives, men altid uddybes. Saadan er jo nu engang Vilkaarene for menneskelig Tænken, og alle maa vi føle vort Slægtskab med den stakkels Licentiat, som den krøllede Frits mødte i Møllerkroen. Som der, er det her atter kun Erkendelsen af Modsætningernes Vekselspil, der kan afrunde vort Billede af Livets Kaar. Netop i vor Svimlen over for det uendelige genoplever vi de uklare Anelser, der danner Baggrunden for Ungdommens umiddelbare Begejstring.

Vi vil dog ikke følge Licentiaten saa vidt, at vi helt glemmer

7

Tid og Sted. Som jeg begyndte, vil jeg blot slutte med at sige, hvor stærkt jeg føler, at hvad jeg har prøvet at give Udtryk for højst kan blive en Afklang af en enkelt Tone af det Stemningsspil, som skal fylde vort Sind i Aften, medens Minderne drager forbi os; en Tone, som nu engang min daglige Dont har lagt det nær for mig at forsøge at slaa an. Jeg haaber imidlertid, at vi alle maa rives ud af vore vante Folder, og at i rigt Maal ogsaa de muntre Toner, der endnu efter 100 Aar klinger os i Møde fra en dansk Students Æventyr, maa lyde ved vor Fest, og at vi alle maa gribes af den fulde Genklang af den Aand, der beherskede vort Sind for 25 Aar siden. I det Haab vil jeg gerne udbringe et „Leve den danske Studenteraand fra 1903".

———

TRANSLATION

I must confess that I take the floor this evening only with considerable hesitation. The joy I share with you at seeing so many of our fellow-students gathered here to celebrate our 25th anniversary is as great as the inability I feel to strike the cheerful chords, the notes of which – more than anything else – should not be missed at a proper reunion celebration. Only in the conviction that others, who to a greater extent than I possess this talent, will let us benefit from it in the course of the evening, I venture to try to describe some of the thoughts and feelings that have crowded in on me during these last few days when I began to think of the forthcoming reunion and the renewing of shared memories from our youth.

Our feeling of fellowship has deep roots. No matter how far apart destiny has taken us, those of us who have spent the years in school together can never forget just how much we have shared with one another. And so, when we meet again, we not only feel the immediacy of being bound together, but would also like to send a grateful thought to our teachers, whom we then perhaps felt it was our task to war against. This evening, however, we wish to remember most of all our shared delight at being released from what appeared to us the constraint of school attendance and at the prospects of entering a new, free life with its, as we saw it, unlimited possibilities. Who among us is not moved by the memory of our first days as undergraduates, when our courage was inversely proportional to the clarity of our goals, and when we lived enthralled in the new worlds which opened up for us under the guidance of our university teachers. Some of our fondest memories will always be of the comradeship with our fellow students, with whom we matched our untried wits in free rivalry and with whom we shared our expectations and enthusiasm.

I had actually intended to dwell a little longer on memories from student years, which, even though the situation gradually became more commonplace and the contrast to our schooldays at times seemed less pronounced, gave us all so many rich experiences. But the warm words of welcome at the university from the *Rector magnificus* and the deeply-felt, well-chosen words of reply with which our spokesman expressed our gratitude to our old *Alma mater*, have brought our years of study into such sharp relief that I can scarcely add anything new. Nor will I enter upon the wide-ranging network of contacts which remained even after our student days and after our daily meetings had come to an end, when we, as fellow countrymen each in his own field, were to turn our education to good account. I should only like to remind you of the effect the World War had on our generation, with its shaking of the very foundations of our public and private lives; it not only reminds us of the strong links to the destiny, both the past and the present, of our people, but also urges us to remember the profound fellowship of all human beings.

The immediate mutual understanding between us fellow students does not only arise from memories of many shared experiences. Now as we meet again, having travelled along different paths, we cannot help but notice how strongly our outlook on things great and small has been coloured by the fact that we had a common background on which to base our initiation into so many new things. True enough, the knowledge that is required to qualify as a student and that is taught at the university is suited to make us familiar with the destinies and strivings of people from times and places other than our own, to turn us into what has been called "world citizens of the intellect"; but the comprehension of all intellectual content is dependent on the existing preconditions, and nothing is perhaps so determined by time and place as precisely the acquisition of knowledge. It is only when we learn to know people with interests similar to our own, yet who have received their academic training under different circumstances, that we begin fully to realize how profoundly our intellectual life is marked by our particular outlook.

It is no easy task to shed light on our intellectual fellowship, for here we are not just observers, but are participants ourselves. Even so, I think we dare claim that the understanding that any matter can be seen from many different angles and also that more profound insight can only be gained by giving opposing views their proper weight, is a basic feature of Danish student mentality. Expressed in such general terms this understanding is of course no more than a universal human experience that, immediately perceived, is the wellspring of art and, consciously elaborated, is the hallmark of science. It therefore seems to me that an intimate link between our Danish national temper and our academic tradition has created fertile soil most suitable for nurturing a philosophy of life whose main characteristic is fear of the dangers of one-sidedness and which therefore would rather voluntarily renounce the strength that such one-sidedness offers. All of us recognize this attitude in our imaginative and philosophical literature. It finds perhaps its purest and strongest expression in the writings of the man who has been called the most Danish of all Danish writers and philosophers, Poul Martin Møller, who in his work, fresh as the day it was written, has, precisely by expressing the struggle between opposites within our own mind, given us, in such a fine and profound manner, a picture of the Danish student spirit in which we can recognize ourselves even today.

This old picture has a doubly strong effect on us now, because we recognize in its basic mood a close kinship with the intellectual movements which at present are stirring people's minds the world over. Typical for our age, and not least for its busy and wide-ranging scientific activity, is the vigilant consciousness of the fact that if we do not pay due attention to and constantly deepen our preconceptions, they will turn into prejudices that may conceal the greater whole from our eyes. This is true of all areas of intellectual activity, from what is often called the most abstract of the abstract – our thoughts about our own thought processes – to what we might be inclined to call the most concrete of the concrete – our observations of inanimate nature. As we all know, since the time when we graduated as students, it has been precisely in this latter field that the recognition of the equal validity of different starting points has revealed unnoticed prejudices concerning the customary ordering of natural phenomena in time and space and has at the same time disclosed unsuspected connections between familiar things. This development has by no means come to an end with the triumph of the theory of relativity, which is undoubtedly one of the great milestones in the history of human thought, but the recent enrichment of science with entirely new fields of experience has uncovered yet more flaws in the foundation in which we confidently put our trust. Once more we seem to be witnessing how apparently insuperable contradictions disappear with continued exploration into the preconditions for our forms of perception.

It is not the recognition of our human limitations, but the striving to investigate the nature of these limitations, which marks our time. We would only get a poor idea of our possibilities if we were to compare our limitations with an insurmountable wall blocking our thoughts along certain paths. New vistas are constantly appearing and new relationships are perceived, but each route taken divides again and again and winds, so that we soon lose our way and sooner or later return to our starting point. Nevertheless, we always return home rewarded, and we see no end to the wealth of content that we can collect and systematize. From an ever closer scrutiny of preconceptions, an ever greater interconnection will be perceived. We are thus subject to a constantly growing impression of an eternal, infinite harmony; the harmony itself can, of course, only be dimly perceived, never grasped; at any attempt to do so, it slips according to its very nature through our fingers. Nothing is constant, each thought, indeed each word uttered, serves only to underline an interconnection that in itself can never be fully described, but always deepened. Such are, after all, the conditions for human thinking, and all of us must surely feel a kinship with the poor licentiate whom curly Frits met at the Miller's Inn. Here, as there, it is once more only the recognition of the interplay of contradictions that can round off our picture of the conditions of life. And it is precisely in our dizziness when confronted with the infinite that we re-experience the vague inklings that form the background of the spontaneous enthusiasm of youth.

However, we shall not follow the licentiate so far as to completely forget both time and place. As I started, I shall finish simply by saying how strongly

I feel that what I have tried to express can be no more than the faint echo of a single note in the concert of moods that will fill our hearts this evening, as memories draw by; a note which my daily work makes it natural for me to try to strike. My hope is, though, that all of us will be torn out of our usual ways and that the cheerful notes which, even after a hundred years, greet us from the tale of a Danish student, may also be heard at our celebration, and that all of us may be moved by the full echo of the spirit that filled our minds twenty-five years ago. In that hope, I should like to propose the toast: "Long live the Danish student spirit of 1903!"

II. NATURAL PHILOSOPHY AND HUMAN CULTURES

Congrès international des sciences anthropologiques et ethnologiques, compte rendu de la deuxième session, Copenhague 1938, Ejnar Munksgaard, Copenhagen 1939, pp. 86–95

Address at the International Congress of Anthropological and Ethnological Sciences, delivered at a meeting in Kronborg Castle, Elsinore, 4 August 1938

See Introduction to Part II, p. [219].

PLEADS FOR TOLERANCE

Noted Danish Scientist Warns Against 'Racist' Theories

Special Cable to THE NEW YORK TIMES.

ELSINORE, Denmark, Aug. 4.—
Pleading for tolerance as a "basic
element in human life," Dr. Niels
H. D. Bohr, noted Danish physicist
and winner of the Nobel Prize,
warned the International Anthropological and Ethnographical Sciences Convention here today against
nationalist theories of race culture.

"It is most difficult to draw safe
conclusions, even in the physical
sciences," said Dr. Bohr, adding
that this is all the more reason to
beware of prejudiced conclusions regarding matters tinged by "personal bias and inherited points of
view."

Dr. Bohr declared that conceptions of morality and culture vary
from one country to another and
that he could define them only as
"the flower of human life."

Today's sessions of the convention
were transferred here from Copenhagen so that they could be held in
"Hamlet's" castle of Kronborg.
This afternoon Eskimos from Greenland demonstrated seal hunting and
acrobatics in their tiny kayaks for
the benefit of the 500 scientists attending.

Niels Bohr's statements at the International Congress of Anthropological and Ethnological Sciences attracted attention, as can be seen from this cutting from the New York Times of 5 August 1938.

[238]

NATURAL PHILOSOPHY AND HUMAN CULTURES

Versions published in English, Danish and German

English: Natural Philosophy and Human Cultures
A "Congrès international des sciences anthropologiques et ethnologiques, compte rendu de la deuxième session", Copenhague 1938, Ejnar Munksgaard, Copenhagen 1939, pp. 86–95
B Nature **143** (1939) 268–272
C "Atomic Physics and Human Knowledge", John Wiley & Sons, New York 1958, pp. 23–31 (reprinted in: "Essays 1933–1957 on Atomic Physics and Human Knowledge, The Philosophical Writings of Niels Bohr, Vol. II", Ox Bow Press, Woodbridge, Connecticut 1987, pp. 23–31)

Danish: Fysikkens Erkendelseslære og Menneskekulturerne
D Tilskueren **56** (1939) 1–10
E "Atomfysik og menneskelig erkendelse", J.H. Schultz Forlag, Copenhagen 1957, pp. 35–44
F "Naturbeskrivelse og menneskelig erkendelse" (eds. J. Kalckar and E. Rüdinger), Rhodos, Copenhagen 1985, pp. 119–130

German: Erkenntnistheoretische Fragen in der Physik und die menschlichen Kulturen
G "Atomphysik und menschliche Erkenntnis", Friedr. Vieweg & Sohn, Braunschweig 1958, pp. 23–31

All of these versions agree with each other.

NATURAL PHILOSOPHY AND HUMAN CULTURES

By Prof. *Niels Bohr* (Copenhagen).

Conférence faite le jeudi 4 août.

IT is only with great hesitation that I have accepted a kind invitation to address this assembly of distinguished representatives of the anthropological and ethnographical sciences of which I, as a physicist, have of course no first-hand knowledge. Still, on this special occasion when even

86

the historical surroundings speak to everyone of us about aspects of life other than those discussed at the regular congress proceedings, it might perhaps be of interest to try with a few words to draw your attention to the epistemological aspect of the latest development of natural philosophy and its bearing on general human problems. Notwithstanding the great separation between our different branches of knowledge, the new lesson which has been impressed upon physicists regarding the caution with which all usual conventions must be applied as soon as we are not concerned with everyday experience, may, indeed, be suited to remind us in a novel way of the dangers, well known to humanists, of judging from our own standpoint cultures developed within other societies.

Of course it is impossible to distinguish sharply between natural philosophy and human culture. The physical sciences are, in fact, an integral part of our civilization, not only because our ever increasing mastery of the forces of nature has so completely changed the material conditions of life, but also because the study of these sciences has contributed so much to clarify the background of our own existence. What has it not meant in this respect that we no more consider ourselves as privileged in living at the centre of the universe, surrounded by less fortunate societies inhabiting the edges of the abyss, but that through the development of astronomy and geography we have realized that we are all sharing a small spherical planet of the solar system which again is only a small part of still larger systems. How forceful an admonition about the relativity of all human judgements have we not also in our days received through the renewed revision of the presuppositions underlying the unambiguous use of even our most elementary concepts such as space and time, which, in disclosing the essential dependence of every physical phenomenon on the standpoint of the observer, has contributed so largely to the unity and beauty of our whole world-picture.

While the importance of these great achievements for our general outlook is commonly realized, it is hardly yet so as regards the unsuspected epistemological lesson which the opening of quite new realms of physical research has given us in the latest years. Our penetration into the world of atoms, hitherto closed to the eyes of man, is indeed an adventure which may be compared with the great journeys of discovery of the circumnavigators and the bold explorations of astronomers into the depths of celestial space. As is well known, the marvellous devel-

87

opment of the art of physical experimentation has not only removed the last traces of the old belief that the coarseness of our senses would for ever prevent us from obtaining direct information about individual atoms, but has even shown us that the atoms themselves consist of still smaller corpuscles which can be isolated and the properties of which can be investigated separately. At the same time we have, however, in this fascinating field of experience been taught that the laws of nature hitherto known, which constitute the grand edifice of classical physics, are only valid when dealing with bodies consisting of practically infinite numbers of atoms. The new knowledge concerning the behaviour of single atoms and atomic corpuscles has, in fact, revealed an unexpected limit for the subdivision of all physical actions extending far beyond the old doctrine of the limited divisibility of matter and giving every atomic process a peculiar individual character. This discovery has, in fact, yielded a quite new basis for the understanding of the intrinsic stability of atomic structures, which, in the last resort, conditions the regularities of all ordinary experience.

How radical a change in our attitude towards the description of nature this development of atomic physics has brought about is perhaps most clearly illustrated by the fact that even the principle of causality, so far considered as the unquestioned foundation for all interpretation of natural phenomena, has proved too narrow a frame to embrace the peculiar regularities governing individual atomic processes. Certainly everyone will understand that physicists have needed very cogent reasons to renounce the ideal of causality itself; but in the study of atomic phenomena we have repeatedly been taught that questions which were believed to have received long ago their final answers had most unexpected surprises in store for us. You will surely all have heard about the riddles regarding the most elementary properties of light and matter which have puzzled physicists so much in recent years. The apparent contradictions which we have met in this respect are, in fact, as acute as those which gave rise to the development of the theory of relativity in the beginning of this century and have, just as the latter, only found their explanation by a closer examination of the limitation imposed by the new experiences themselves on the unambiguous use of the concepts entering into the description of the phenomena. While in relativity theory the decisive point was the recognition of the essentially different ways in which observers moving relatively to each other will

88

describe the behaviour of given objects, the elucidation of the paradoxes of atomic physics has disclosed the fact that the unavoidable interaction between the objects and the measuring instruments sets an absolute limit to the possibility of speaking of a behaviour of atomic objects which is independent of the means of observation.

We are here faced with an epistemological problem quite new in natural philosophy, where all description of experiences has so far been based upon the assumption, already inherent in ordinary conventions of language, that it is possible to distinguish sharply between the behaviour of objects and the means of observation. This assumption is not only fully justified by all everyday experience but even constitutes the whole basis of classical physics, which, just through the theory of relativity, has received such a wonderful completion. As soon as we are dealing, however, with phenomena like individual atomic processes which, due to their very nature, are essentially determined by the interaction between the objects in question and the measuring instruments necessary for the definition of the experimental arrangements, we are, therefore, forced to examine more closely the question of what kind of knowledge can be obtained concerning the objects. In this respect we must, on the one hand, realize that the aim of every physical experiment — to gain knowledge under reproducible and communicable conditions — leaves us no choice but to use everyday concepts, eventually refined by the terminology of classical physics, not only in all accounts of the construction and manipulation of the measuring instruments but also in the description of the actual experimental results. On the other hand, it is equally important to understand that just this circumstance implies that no result of an experiment concerning a phenomenon which, in principle, lies outside the range of classical physics, can be interpreted as giving information about independent properties of the objects, but is inherently connected with a definite situation in the description of which the measuring instruments interacting with the objects also enter essentially. This last fact gives the straightforward explanation of the apparent contradictions which appear when results about atomic objects obtained by different experimental arrangements are tentatively combined into a self-contained picture of the object.

Information regarding the behaviour of an atomic object obtained under definite experimental conditions, may, however, according to a terminology often used in atomic physics, be adequately characterized as

89

complementary to any information about the same object obtained by some other experimental arrangement excluding the fulfilment of the first conditions. Although such kinds of information cannot be combined into a single picture by means of ordinary concepts, they represent indeed equally essential aspects of any knowledge of the object in question which can be obtained in this domain. Just the recognition of such a complementary character of the mechanical analogies by which one has tried to visualize the individual radiative effects has, in fact, led to an entirely satisfactory solution of the riddles of the properties of light alluded to above. In the same way it is only by taking into consideration the complementary relationship between the different experiences concerning the behaviour of atomic corpuscles that it has been possible to obtain a clue to the understanding of the striking contrast between the properties of ordinary mechanical models and the peculiar laws of stability governing atomic structures which form the basis for every closer explanation of the specific physical and chemical properties of matter.

Of course I have no intention, on this occasion, of entering more closely into such details, but I hope that I have been able to give you a sufficiently clear impression of the fact that we are here in no way concerned with an arbitrary renunciation as regards the detailed analysis of the almost overwhelming richness of our rapidly increasing experience in the realm of atoms. On the contrary, we have to do with a rational development of our means of classifying and comprehending new experience which, due to its very character, finds no place within the frame of causal description that is only suited to account for the behaviour of objects as long as this behaviour is independent of the means of observation. Far from containing any mysticism contrary to the spirit of science, the view-point of "complementarity" forms indeed a consistent generalization of the ideal of causality.

However unexpected this development may appear in the domain of physics, I am sure that many of you will have recognized the close analogy between the situation as regards the analysis of atomic phenomena, which I have described, and characteristic features of the problem of observation in human psychology. Indeed, we may say that the trend of modern psychology can be characterized as a reaction against the attempt at analyzing psychical experience into elements which can be associated in the same way as are the results of measurements in classical

90

physics. In introspection it is clearly impossible to distinguish sharply between the phenomena themselves and their conscious perception, and although we may often speak of lending our attention to some particular aspect of a psychical experience, it will appear on closer examination that we really have to do, in such cases, with mutually exclusive situations. We all know the old saying that, if we try to analyze our own emotions, we hardly possess them any longer, and in that sense we recognize
between psychical experiences, for the description of which words such
as "thoughts" and "feelings" are adequately used, a complementary relationship similar to that between the experiences regarding the behaviour of atoms obtained under different experimental arrangements
and described by means of different analogies taken from our usual
ideas. By such a comparison it is, of course, in no way intended to
suggest any closer relation between atomic physics and psychology,
but merely to stress an epistemological argument common to both fields,
and thus to encourage us to see how far the solution of the relatively
simple physical problems may be helpful in clarifying the more intricate
psychological questions with which human life confronts us, and which
anthropologists and ethnologists so often meet in their investigations.

Coming now closer to our subject of the bearing of such view-points
on the comparison of different human cultures, we shall first stress
the typical complementary relationship between the modes of behaviour
of living beings characterized by the words "instinct" and "reason". It
is true that any such words are used in very different senses; thus
instinct may mean motive power or inherited behaviour, and reason
may denote deeper sense as well as conscious argumentation. What
we are concerned with is, however, only the practical way in which
these words are used to discriminate between the different situations in which animals and men find themselves. Of course, nobody
will deny our belonging to the animal world, and it would even be
very difficult to find an exhaustive definition characterizing man among
the other animals. Indeed, the latent possibilities in any living organism
are not easily estimated, and I think that there is none of us who has
not sometimes been deeply impressed by the extent to which circus animals can be drilled. Not even with respect to the conveyance of information from one individual to another would it be possible to draw a
sharp separation between animals and man; but of course our power
of speech places us in this respect in an essentially different situation,

91

not only as regards the exchange of practical experience, but above all as regards the possibility of transmitting through education to children the traditions concerning behaviour and reasoning which form the basis of any human culture.

As regards reason compared with instinct, it is, above all, essential to realize that no proper human thinking is imaginable without the use of concepts framed in some language which every generation has to learn anew. This use of concepts is, in fact, not only to a large extent suppressing instinctive life, but stands even largely in an exclusive relationship of complementarity to the display of inherited instincts. The astonishing superiority of lower animals compared with man in utilizing the possibilities of nature for the maintenance and propagation of life has certainly often its true explanation in the fact that, for such animals, we cannot speak af any conscious thinking in our sense of the word. At the same time the amazing capacity of so-called primitive people to orientate themselves in forests or deserts, which, though apparently lost in more civilized societies, may on occasion be revived in any of us, might justify the conclusion that such feats are only possible when no recourse is taken to conceptual thinking, which on its side is adapted to far more varied purposes of primary importance for the development of civilization. Just because it is not yet awake to the use of concepts, a new-born child can hardly be reckoned as a human being; but belonging to the species of man, it has, of course, though more helpless a creature than most young animals, the organic possibilities of receiving through education a culture which enables it to take its place in some human society.

Such considerations confront us at once with the question whether the widespread belief that every child is born with a predisposition for the adoption of a specific human culture is really well-founded, or whether one has not rather to assume that any culture can be implanted and thrive on quite different physical backgrounds. Here we are of course touching a subject of still unsettled controversies between geneticists, who pursue most interesting studies on the inheritance of physical characters. In connection with such discussions, however, we must above all bear in mind that the distinction between the concepts genotype and phaenotype, so fruitful for the clarification of heredity in plants and animals, essentially presupposes the subordinate influ-ence of the external conditions of life on the characteristic proper-

92

ties of the species. In the case of the specific cultural characters of human societies the problem is, however, reversed in the sense that the basis for the classification is here the traditional habits shaped by the histories of the societies and their natural environments. These habits, as well as their inherent presuppositions, must therefore be analyzed in detail before any possible influence of inherited biological differences on the development and maintenance of the cultures concerned can be estimated. Indeed, in characterizing different nations and even different families within a nation, we may to a large extent consider anthropological traits and spiritual traditions as independent of each other, and it would even be tempting to reserve by definition the adjective "human" for just those characters which are not directly bound to bodily inheritance.

At first sight, it might perhaps appear that such an attitude would mean unduly stressing merely dialectic points. But the lesson which we have received from the whole growth of the physical sciences is that the germ of fruitful development often lies just in the proper choice of definitions. When we think for instance of the clarification brought about in various branches of science by the argumentation of relativity theory we see indeed what advance may lie in such formal refinements. As I have already hinted at earlier in this address, relativistic view-points are certainly also helpful in promoting a more objective attitude as to relationships between human cultures, the traditional differences of which in many ways resemble the different equivalent manners in which physical experience can be described. Still, this analogy between physical and humanistic problems is of limited scope and its exaggeration has even led to misunderstandings of the essence of the theory of relativity itself. The unity of the relativistic world picture, in fact, just implies the possibility for any one observer to predict within his own conceptual frame how any other observer will coordinate experience within the frame natural to him. The main obstacle to an unprejudiced attitude towards the relation between various human cultures is, however, the deep-rooted differences of the traditional backgrounds on which the cultural harmony in different human societies is based and which exclude any simple comparison between such cultures.

It is above all in this connection that the view-point of complementarity offers itself as a means of coping with the situation. In fact,

93

when studying human cultures different from our own, we have to deal with a particular problem of observation which on closer consideration shows many features in common with atomic or psychological problems, where the interaction between objects and measuring tools, or the inseparability of objective content and observing subject, prevents an immediate application of the conventions suited to accounting for experiences of daily life. Especially in the study of cultures of primitive peoples, ethnologists are, indeed, not only aware of the risk of corrupting such cultures by the necessary contact, but are even confronted with the problem of the reaction of such studies on their own human attitude. What I here allude to is the experience, well known to explorers, of the shaking of their hitherto unrealized prejudices through the experience of the unsuspected inner harmony human life can present even under conventions and traditions most radically different from their own. As a specially drastic example I may perhaps here remind you of the extent to which in certain societies the roles of men and women are reversed, not only regarding domestic and social duties but also regarding behaviour and mentality. Even if many of us, in such a situation, might perhaps at first shrink from admitting the possibility that it is entirely a caprice of fate that the people concerned have their specific culture and not ours, and we not theirs instead of our own, it is clear that even the slightest suspicion in this respect implies a betrayal of the national complacency inherent in any human culture resting in itself.

Using the word much as it is used, in atomic physics, to characterize the relationship between experiences obtained by different experimental arrangements and visualizable only by mutually exclusive ideas, we may truly say that different human cultures are complementary to each other. Indeed, each such culture represents a harmonious balance of traditional conventions by means of which latent potentialities of human life can unfold themselves in a way which reveals to us new aspects of its unlimited richness and variety. Of course, there cannot, in this domain, be any question of such absolutely exclusive relationships as those between complementary experiences about the behaviour of well-defined atomic objects, since hardly any culture exists which could be said to be fully self-contained. On the contrary, we all know from numerous examples how a more or less intimate contact between different human societies can lead to a

94

gradual fusion of traditions, giving birth to a quite new culture. The importance in this respect of the mixing of populations through emigration or conquest for the advancement of human civilization needs hardly be recalled. It is, indeed, perhaps the greatest prospect of humanistic studies to contribute through an increasing knowledge of the history of cultural development to that gradual removal of prejudices which is the common aim of all science.

As I stressed in the beginning of this address, it is, of course, far beyond my capacities to contribute in any direct way to the solution of the problems discussed among the experts at this congress. My only purpose has been to give an impression of a general epistemological attitude which we have been forced to adopt in a field as far from human passions as the analysis of simple physical experiments. I do not know, however, whether I have found the right words to convey to you this impression, and before I conclude, I may perhaps be allowed to relate an experience which once most vividly reminded me of my deficiencies in this respect. In order to explain to an audience that I did not use the word prejudice to imply any condemnation of other cultures, but merely to characterize our necessarily prejudiced conceptual frame, I referred jokingly to the traditional prejudices which the Danes cherish with regard to their Swedish brothers on the other side of the beautiful Sound outside these windows, with whom we have fought through centuries even within the walls of this castle, and from contact with whom we have, through the ages, received so much fruitful inspiration. Now you will realize what a shock I got when, after my address, a member of the audience came up to me and said that he could not understand why I hated the Swedes. Obviously I must have expressed myself most confusingly on that occasion, and I am afraid that also to-day I have talked in a very obscure way. Still, I hope that I have not spoken so unclearly as to give rise to any such misunderstandings of the trend of my argument.

In February 1925, Bohr, together with Harald Høffding, the great arctic explorer Knud Rasmussen, the linguist Vilhelm Thomsen and H.N. Andersen, director of the East Asiatic Company, formed a committee to prepare the establishment of a National Museum Foundation, which should in turn seek support for the building of new premises for the Danish National Museum. The initiative was successful and a new building – considerably larger than that of the earlier museum – was inaugurated in 1938. The above picture was taken in the museum's Tapestry Room on 20 June 1928 when the proceeds of the foundation were handed over to the Danish Government. Sitting in the front, from the left: bank manager C.P. Clausen and Vilhelm Slomann, director of the Copenhagen Museum of Decorative Art. Standing: Harald Høffding. Back row, from the left: Knud Rasmussen; Niels Bohr; Arthur Christensen, orientalist; Gudmund Hatt, geographer; prime minister Thomas Madsen-Mygdal; Alfred Louw, secretary of the National Museum Foundation; Frederik Graae, permanent undersecretary at the Danish Ministry of Education; education minister Jens Byskov; N.N.; Mauritz Mackeprang, director of the Danish National Museum; N.N.; and Dagmar Olrik, tapestry conservator of the museum.

III. DANISH CULTURE. SOME INTRODUCTORY REFLECTIONS

DANSK KULTUR. NOGLE INDLEDENDE BETRAGTNINGER
"Danmarks Kultur ved Aar 1940", Det Danske Forlag,
Copenhagen 1941–1943, Vol. 1, pp. 9–17

TEXT AND TRANSLATION

See Introduction to Part II, p. 220.

[251]

DANISH CULTURE. SOME INTRODUCTORY REFLECTIONS (1941)

Versions published in Danish

Dansk Kultur. Nogle indledende Betragtninger

A "Danmarks Kultur ved Aar 1940", Det Danske Forlag, Copenhagen 1941–1943, Vol. I, pp. 9–17

B Tiden: Politik, Ekonomi, Kultur **34** (1942) 415–423

C "Naturbeskrivelse og menneskelig erkendelse" (eds. J. Kalckar and E. Rüdinger), Rhodos, Copenhagen 1985, pp. 247–258

All of these versions agree with each other.

DANSK KULTUR

NOGLE INDLEDENDE BETRAGTNINGER
AF NIELS BOHR

S PØRGER vi os selv, i hvad Forstand vi kan tale om en særlig
dansk Kultur, vil Svaret ganske afhænge af, hvilke Synspunkter
der lægges til Grund. Paa den ene Side vil fælles menneskelige
Træk komme des klarere til Syne, jo dybere vi søger til Bunds i
vort Væsen; paa den anden Side vil Forskellen fra andre Samfund
føles des stærkere, jo mere vi bestræber os for i Billedet at fastholde
alle Sider af vor Tilværelse. Da Kendskabet til, hvorledes Menneskelivet kan udfolde sig under forskellige Omstændigheder, maa være
Baggrunden for enhver Bedømmelse af et Samfunds kulturelle Stade,
turde en Undersøgelse af vor Plads blandt de andre Samfund være
et naturligt Udgangspunkt for en Søgen efter at klarlægge det for os
særegne. Naar jeg trods Følelsen af mine Kundskabers Begrænsning
har fulgt Redaktionens Opfordring til at skrive nogle Ord til Indledning af dette store Værk, skyldes det, at jeg under mit Samarbejde
her hjemme og i Udlandet med Videnskabsmænd fra mange forskellige Nationer har haft megen Anledning til at beskæftige mig med vort
Forhold til Omverdenen paa Kulturens Omraade og til at tænke over
de Traditioner, der giver dansk Livsindstilling dens Særpræg.

Et Folkeslags Kultur er ofte blevet sammenlignet med en levende
Organisme, og især er det klart, at lige saa vel som Udviklingshistorie og Levevilkaar er afgørende for en Organismes Livsform,
er Kendskabet til en Kulturs Historie og Kaar Forudsætningen for
enhver dybere Forstaaelse af dens Egenart. Mere træffende vilde
en saadan Sammenligning dog være, hvis de enkelte Kulturer besad
den samme Grad af Selvstændighed som Organismerne; men om
noget er særegent for vor Tid, er det vel netop den voksende Forstaaelse af den mangfoldige Sammenhæng mellem selv tilsyneladende
vidt forskellige Kulturer. Hvor vigtig Belæring har i den sidste Menneskealder den arkæologiske og etnografiske Forskning ikke givet
os om forglemte og upaaagtede Forbindelser i ældre og nyere Tider,
der har sat sig de dybeste Spor i Udformningen af Kulturer, som
ved første Blik kun viser ringe Afhængighed af hverandre. Hvor
meget mere nødvendigt er det da ikke stadig at holde sig Spørgsmaalet
om den gensidige Vekselvirkning for Øje, naar det angaar Forholdet

[253]

mellem Kulturerne hos Folkeslag, der som de europæiske igennem Aarhundreder har opretholdt den livligste Forbindelse.

Hverken Lejligheden eller mine Forudsætninger tillader mig at give nogen indgaaende Redegørelse for vor Kulturs Historie, men jeg vil blot erindre om den Stilling, som Danmark og de andre nordiske Lande paa Grund af deres Beliggenhed ved en Udkant af vor Verdensdel gennem Tiderne har indtaget inden for Europas Kultursamfund. Kun langsomt og besværligt er mange Kulturrørelser naaet herop; men til Gengæld har vi ofte haft bedre Lejlighed til at bringe de udefra kommende Strømninger i Samklang med vor egen Indstilling. Inden for Norden har vel Landegrænser og Naturforhold i disse Henseender virket paa forskellig Vis, men Sprogenes Slægtskab og de fælles Minder har dog skabt et Broderforhold langt inderligere end det, man sædvanligvis finder mellem Nabostater. Lig-heden i Kultur mellem os og vore nordiske Brødre er jo saa stor, at hver Forskel tit forsvinder for den fremmede, og at for os selv Understregningen af Forskellighederne som oftest netop er et Middel til som i et Spejl bedre at lære det Ansigt at kende, vi vender ud mod den øvrige Verden.

I de tusind Aar der er forløbet, siden Kristendommens Indførelse i Norden aabnede nye Veje for den fredelige Forbindelse med Omverdenen, har Kultursamkvemmet mellem Danmark og andre europæiske Lande trods skiftende Former stadig fremvist visse ejendommelige Træk, betingede af, at det foregik over en Grænse, bag hvilken de nordiske Landes Sprog og Selvstændighed altid bevaredes. Ofte medbragtes Oplysning og nye Sæder af Udlændinge, der kom her til Landet uden nærmere Kendskab til vore Forhold, ja, vel som Regel ikke engang beherskede vort Sprog. Til andre Tider var det især unge Danske, der fra deres Besøg i Europas Kultur- og Lærdomscentrer hjemførte Dannelse og Kundskaber. I begge Tilfælde, om end paa forskellig Vis, frembragte Berøringen med det fremmede ikke alene ny Befrugtning, men tillige en forstærket Erkendelse af kulturelle Værdier med dybe Rødder i den hjemlige Jordbund.

Om Trangen til at værne saavel om vore Minder som om de Samfundsformer og den Livsindstilling, der havde udviklet sig her i Danmark, bringer jo allerede fra Tidsrummets første Aarhundreder Krønikerne og de danske Landskabslove det mest talende Vidnesbyrd. Ja, endog i den bildende Kunst, for hvilken Kirken aabnede helt nye Udsigter, var der, som det ses paa de første Mindesmærker fra den kristne Tid, en Hang til at bevare hjemlige Traditioner i Stil og Midler. Ogsaa i senere Brydningstider, hvor nye kulturelle Rørelser kom hertil, finder vi lignende Træk; selv under Reformationen bragtes den fremmede Forkyndelse til Folkets brede Lag gennem danske Talsmænd og gav Stødet til en dansk Salmedigtning, der fortsatte og udviklede den Digtekunst paa Modersmaalet, som de gamle Folkeviser bærer Kimen til.

Da Latinens Herredømme i Skolen, der havde betydet saa meget for at holde Mindet om den europæiske Kulturs fælles Kilder levende og reddet de mindre Nationer fra en alt for ensidig kulturel Paavirkning fra større Nabostater, efterhaanden var brudt, var Faren for det danske Sprogs Fortrængelse fra de dannedes Kreds maaske størst. Intet skulde dog bidrage mere til denne Fares Afværgelse og vort Sprogs Berigelse end den atter paa Baggrund af udenlandske Forbilleder frembragte alsidige Fornyelse af den danske Litteratur. Trods al Forskel i Tid og Opgaver bærer Holbergs Liv og Værk Træk til Skue, der minder om Absalons Skæbne og hans Bedrifter i Fædrelandets Tjeneste. I begge Tilfælde betød de fra Rejserne i det fremmede hjembragte Iagttagelser og Erfaringer en Indpodning paa dansk Grund, der satte Frugter, som med Rette regnes til vort dyrebareste nationale Eje.

Den stadige Udvikling uden større Brud har ogsaa siden været kendetegnende for danske Forhold. Fremfor alt førte den voksende Erkendelse af Menneskerettighederne, der andetsteds gav Anledning til voldsomme Samfundsomvæltninger, her hjemme til den Frigørelse af Bondestanden, der fandt Sted under saa fredelige Former og i en Forstaaelsens Aand, hvorom endnu i Dag den manende og gribende Indskrift paa Frihedsstøtten vidner. I Virkeligheden havde vor ældgamle Bondekultur trods svære Kaar bevaret en Livskraft, der satte den i Stand til, saa snart Tiderne blev bedre, igen at skyde nye, friske Skud. Det Krav om Indflydelse paa Statens Styre, som Middelalderens Tingmøder gav Udtryk for, efterlod sig jo ogsaa Traditioner, der aldrig var helt glemte og igen blev virksomme, da vi for hundrede Aar siden, i Skridt med Udviklingen i andre europæiske Lande, fik den fri Forfatning, der paa nyt Grundlag stadfæstede og udviklede vore gamle Retsidealer.

Samhørigheden med vor Fortid og Broderfølelsen inden for Norden styrkedes ved forrige Aarhundredes Begyndelse især ved den nyvakte Værdsættelse af de fælles Skatte, vi besad i den gamle Edda- og Sagadigtning. Ogsaa i denne Henseende virkede udenlandsk Indflydelse befrugtende, men, som Dønninger i Digtekunsten rundt om i Europa viser, drejede det sig her om en langt større Gensidighed i den kulturelle Vekselvirkning mellem Norden og Udlandet end nogen Sinde tidligere. Paamindelsen om, hvor meget vi havde at lære for at hævde os paa de Maader, Samtiden anviste, mindskedes dog ingenlunde og var alt for stærk til, at vi af svundne Tiders Glans kunde forledes til at glemme vor beskedne Plads mellem Nationerne. Til

Gengæld opvoksede der i de nordiske Lande en Følelse af Verdensborgerskab, der gav sig et mere harmonisk Udtryk end i mange større
Lande, hvor Fristelsen til at betragte egne Kulturer som selvstændige
organiske Enheder var saa meget stærkere.

Typiske Vidnesbyrd om, hvorledes Tanken om Forholdet mellem
Indland og Udland har beskæftiget danske Digtere, finder vi i
Grundtvigs fyndige Ord »Indfødsret for dansk alene, Gæsteret for
alt paa Jord«, og i H. C. Andersens skønne Linier »I Danmark er
jeg født, der har jeg hjemme, der har jeg Rod, derfra min Verden
gaar«. Begge disse, i Stil saavel som Stemning saa forskellige Ud-
talelser, giver Udtryk for en Indstilling til Spørgsmaalet om vort
Forhold til Verden af en Art, som man vanskeligt finder noget Sidestykke til. Kernepunktet er, at Spørgsmaalet hverken afvises eller
afgøres, men opfattes som et, der bestandigt maa stilles i stort og
smaat. Det, der vel mest af alt præger dansk Kultur, turde jo netop
være en umiddelbar Forening af Aabenhed for den Belæring, der
bringes til os udefra, eller som vi selv henter hjem, og Fastholden
ved det af vore Minder og Skæbne betingede Livssyn, der knytter os
saa fast sammen inden for den store Verden, vi uløseligt tilhører.

Om den Kilde til Inspiration, en saadan Indstilling indebærer, vidner Søren Kierkegaards berømte Slutningsord i »Stadier paa Livets
Vej«. Med Udgangspunkt i et Forsvar for det danske Sprog imod
Paastanden om de mere udbredte Kultursprogs Fortrin udtaler han
sig saa varmt og inderligt om Modersmaalets Rigdom og Skønhed
og med Anvendelse af saadan Kunst og Dybsind, at hans Betragtninger i Virkeligheden kommer til at gælde ikke blot det danske,
men ethvert menneskeligt Sprog, og i sig selv belærer os om Sprogets
uendelig forfinede Udtryks- og Virkemidler. En tilsvarende Baggrund
for Kærligheden til Hjemlandet finder man hos Poul Martin Møller,
som man undertiden har kaldt den danskeste af alle danske Digtere. Selv hvor han, som i »En dansk Students Æventyr«, ikke kommer ind paa en Sammenligning med udenlandske Forhold, rummer
den Maade, hvorpaa han hos de optrædende Personer med saa meget
Lune stiller forskellige Sider af sit eget Væsen til Skue, et bevidst
Krav om Ligevægt mellem Tilpasningen til de hjemlige Livsformer
og Indlevelsen i den fælles-menneskelige Tankeverden.

Brydningen mellem danske Traditioner og udenlandske Kultur-
strømninger spores ikke blot overalt i vor Digtning, men har tillige Gang paa Gang fundet Udtryk i aaben Diskussion. Hvor meget
bidrog ikke den af Baggesen aabnede »Pennefejde« mod Øhlenschlä-
ger til at belyse de Farer, som Romantikken, der gav vor Digtekunst
saa rig en Blomstring, kunde indebære. Hvor stærkt var ikke ogsaa
det Røre i Aandslivet her hjemme, som senere europæiske Tanke-

retninger gav Anledning til, og som kort før Aarhundredskiftet fandt saa lærerigt Udtryk i den bekendte Meningsudveksling mellem Georg Brandes og Harald Høffding. Trods al Forskel i Betoningen hos den enkelte Digter eller Forfatter var det jo et Hovedmaal for dem alle gennem Kulturforbindelsen med Omverdenen at højne vort Aandsliv uden at miste den Tilknytning til vor Fortid, som ene giver os Styrke til paa beskeden, men selvstændig Maade at bidrage til den menneskelige Kulturs Udvikling.

Følelsen af Verdensborgerskab har i lange Tider sat sit Præg paa danske Bestræbelser paa Skoleundervisningens Omraade, og Virkninger deraf møder vi ogsaa inden for den Højskolebevægelse, der, bygget paa den fra Grundtvig udgaaede folkelige Vækkelse, paa saa lykkelig Maade har hjulpet til at bringe Aandslivets Rørelser ud til større Kredse af Befolkningen, end det maaske er Tilfældet noget andet Sted. Saaledes træffer vi her en dyb Forstaaelse af, at Kilderne til en Folkeoplysning ud over den, Almenskolen kan give, foruden i vore nationale Minder maa søges i Kendskabet til det mellemfolkelige kulturelle Samarbejde. Et ejendommeligt Vidnesbyrd herom er de udmærkede Værker om Naturvidenskabernes Historie, som til Brug for Højskolens Undervisning blev skrevet af nogle af dens egne Mænd og som fik Betydning langt ud over det oprindelige Formaal. Ved Siden af den Indførelse i Filosofiens almindelige Problemer, som er forbundet med ethvert Universitetsstudium her hjemme, har Læsningen af disse Værker bidraget til at give senere Generationer af dansk studerende Ungdom en Baggrund, som man ofte kan mærke Savnet af selv ved Udlandets store Universiteter.

Hvad vort Lands Deltagelse i den videnskabelige Forsknings fællesmenneskelige Opgaver angaar, kommer vi jo ind paa et Omraade, der efter sin Art ingen nationale Skranker kender, men hvor ikke desto mindre Traditioner er af lige saa afgørende Betydning som for alt andet menneskeligt Virke. Da i Renæssancetiden Videnskaben i Europa paany blomstrede op, kunde ingen Deltagelse fra Danskes Side komme i Gang uden ved en Søgen til Udlandet, hvor ene Uddannelse og Forskningsvilkaar var at finde. Om den vaagne Interesse, der dog fra første Færd var til Stede for at være med i det gryende Værk, vidner maaske bedst den storslaaede Gavmildhed, der fra Kongemagtens Side vistes Tycho Brahe for at formaa ham til hjemme at fortsætte sit paa Rejserne i Udlandet planlagte Arbejde for Astronomiens Udvikling. Netop Tycho Brahes af hele Verden beundrede Virksomhed paa Hven og hans senere tragiske Skæbne bringer Vidnesbyrd om, hvorledes Videnskaben kan blomstre selv paa udyrket Jordbund, naar der blot er Forstaaelse for dens Værd, og hvor svært det kan være at opretholde en saadan Værdsættelse, naar Traditionerne endnu ikke er tilstrækkelig rodfæstede.

I de følgende Tider skabtes her i Landet videnskabelige Skoler med videre Uddannelsesmuligheder, om end en Deltagelse i Kappestriden, saaledes som den formede sig i Europas Videnskabscentrer, stadig var Betingelsen for at naa de største Højder. Selv om Stenos store Opdagelser udførtes i Udlandet og kun dér fandt den rette Anerkendelse, har vi dog her i Danmark Grund til at være stolte af den Aand, han medbragte hjemmefra, og som straks sikrede ham en Plads i Forskernes første Række. Hvor forskelligt Skæbnen paa de Tider kunde forme sig for danske Videnskabsmænd, der havde vundet Berømmelse i Udlandet, viser en Sammenligning mellem Steno og Ole Rømer. Medens Steno endte sit Liv i det fremmede i en her hjemme forkætret Overbevisnings Tjeneste, blev Rømer efter sin Hjemkomst betroet Løsningen af de mangfoldigste Opgaver for vort Samfunds Behov, der i den Grad lagde Beslag paa hans enestaaende Evner, at man først i vor Tid har faaet Kendskab til mange af de vigtige videnskabelige Undersøgelser, han siden naaede at udføre.

Efterhaanden formede Udviklingen i Danmark sig saaledes, at Forholdene ogsaa paa Videnskabens Omraade ikke længere adskilte sig saa meget fra dem, der herskede andetsteds; men stadig var man naturligvis her hjemme, ligesom i vore Dage, i langt højere Grad end i de større Kulturlande henvist til at søge Forbindelse med Udlandet. Til Gengæld gav den udbredte Forstaaelse af, hvor nødvendigt og vanskeligt det under de mindre Forhold var at holde Skridt med Udviklingen i Verden, sig Udtryk i de Muligheder, som den enkelte Banebryder paa Videnskabens som paa andre af Aandslivets Omraader besad for at hæve vor hele Kultur og styrke vor Samfølelse. Det store Eksempel herpaa er jo H. C. Ørsted, der ikke alene bidrog saa overordentligt til Udbredelsen af Kendskabet til Naturvidenskaben og dens Anvendelsesmuligheder, men hvis Indflydelse paa alle Kulturlivets Omraader var af en Art og et Omfang, som det i større Lande næppe vilde være tænkeligt. Hvor forskellige Udgangspunkt og Virkningsmaade end var, tør Ørsted nævnes ved Siden af Grundtvig som Folkeopdrager, medens hans Opdagelser kom til at spille en Rolle for Videnskabens Stilling her hjemme, der minder om den Maade, hvorpaa H. C. Andersens og Thorvaldsens viden om beundrede Værker bragte Følelsen for Kunstens Betydning ud til alle Kredse af vort Folk.

Om end ethvert Bidrag til den Sandhedens Erkendelse, som Videnskaben stræber efter, vel som intet andet er alment menneskeligt Eje, gælder det for Videnskaben som for Kunsten, at dens Trivsel hvert enkelt Sted er betinget af mange særlige Forhold. Saaledes er det ikke uden Forbindelse med den af hele Folket delte Interesse for vor Historie, at Udforskningen af de Oldtidsminder, hvorpaa vor Jordbund er saa rig, har faaet Betydning langt ud over vore egne Grænser ved at klarlægge fælles Udviklingstrin hos menneskelige Kulturer. Ligeledes er det næppe nogen Tilfældighed, at netop vort lille Land, med dets snævre Sprogomraade, lige siden Rasmus Rasks Dage har spillet en saa betydningsfuld Rolle i den sammenlignende Sprogvidenskabs Udvikling. Endvidere var den Omstændighed, at vore Retstraditioner vel var dybt paavirkede, men aldrig, som andetsteds, helt fortrængte af Romerretten, Grundlaget for den af Anders Sandøe Ørsted skabte danske Retsvidenskab, der har betydet saa meget for Antagelsen af fælles Grundsynspunkter for Lovgivningen i alle de nordiske Lande. De heldige Bestræbelser her hjemme for Lægevidenskabens Anvendelse til Forbedring af Sundhedstilstanden beror jo ogsaa foruden paa saadanne Opdagelser som den, til hvilken Finsens Navn er knyttet, paa den vaagne Interesse for humane og sociale Spørgsmaal, der har givet sig Udslag saavel i Statsforanstaltninger som i uafhængig Medvirken fra Befolkningens egen Side.

Ved Siden af Deltagelsen i den Uddannelse og det Samarbejde, der tilbydes os i Udlandet, er der i vor egen Tid i stadig stigende Grad skabt Vilkaar for, at ogsaa fremmede kan søge Belæring her i Landet og sammen med os deltage i Forskningen paa Felter, hvor der her er gunstige Betingelser til Stede. For saadan Deltagelse i det mellemfolkelige og kulturelle Samarbejde giver vor almindelige Indstilling os jo ogsaa særlige Forudsætninger. Denne gør det ikke alene lettere for Danske, der faar Lejlighed til at komme ud i Verden, at finde sig til Rette, men bevirker tillige, at fremmede her modtages paa en naturligere Maade end mange andre Steder. Videnskabsmænd fra de fjerneste Dele af Verden, der har deltaget i Forskningsarbejdet her hjemme, har saaledes Gang paa Gang givet Udtryk for deres Taknemmelighed over, at de af alle her, til Trods for enhver ydre Fremmedartethed, er blevet betragtet som Medmennesker, der bragte velkommen Bud om Kulturer, fra hvilke man ventede ny menneskelig Belæring. Iøvrigt danner netop denne Indstilling Baggrunden saavel for Behandlingen fra dansk Side af den indfødte Befolkning paa Grønland som for den lykkelige Maade, hvorpaa Forbindelsen med denne har beriget vor Kultur og skænket os Mænd som Knud Rasmussen, hvis straalende Opdagerfærd for altid har indskrevet Danmarks Navn i Polarforskningens Historie.

Hvor mange forskellige Sider end Digtning og Videnskab omfatter af Menneskets Stræben for at erkende sig selv og Verden, vilde uden Tone- og Billedkunsten Kulturen være langt fattigere selv paa sine

Højdepunkter. Det maatte naturligvis være begrænset, hvad et lille Land, der altid har ligget fjernt fra Kulturens Hovedveje, har kunnet bidrage til de Rigdomme, som disse Kunstformer, hvis Fremelskning kræver saa megen Sangbund, har skænket Menneskeheden. Dog har Modtageligheden for de Gaver, der bragtes os, og vor Tilbøjelighed til at søge hen, hvor Kulturlivet har staaet i rigeste Blomst, medført, at vi ogsaa paa dette Omraade har været blandt dem, der lydhørt har fulgt Kaldet. Inden for alle Kunstens Afskygninger har vi jo Traditioner, der gaar mere eller mindre langt tilbage og har givet os Kunstværker, som vi, baade hvad Indhold og Opfattelse angaar, erkender som vore egne. Endvidere maa det ikke glemmes, hvor meget Danskes Deltagelse i Kunstnerlivet ude i Verden har bidraget til stadig at bringe os Bud om de følelses- og kampombruste Stemningssvingninger, hvorpaa vi, saa at sige, mærker Kulturens Pulsslag.

Den af vort Forhold til Omverdenen betingede almindelige Livsindstilling præger ikke alene de Sider af Kulturlivet, som her er berørt, men afspejler sig ogsaa paa mange andre Maader inden for vort Samfund. Erkendelsen af, hvad enhver til fælles Gavn kan lære af menneskelige Værdier, frembragt paa de mest forskellige Arbejdsomraader, er netop Baggrunden for den fremskredne Oplysning i vort Folk, der maaske mere end noget andet kendetegner Danmarks kulturelle Stade, og som trods vore indskrænkede Livsbetingelser har muliggjort en stadig Forbedring af Befolkningens Levevilkaar. Uden Folkehøjskolen vilde jo den Andelsbevægelse, der betød saa meget for vort Landbrugs tidssvarende Udvikling og derved hjalp os over de store af udenlandske Produktionsforhold foraarsagede Vanskeligheder, næppe have været tænkelig. Naar endvidere vor Fattigdom paa Raastoffer ikke har hindret os i at følge med i Industriens Fremskridt og endda paa enkelte Felter at gaa i Spidsen, skyldes det vel allermest de Bestræbelser for at opretholde det nøjeste Samarbejde mellem Videnskab og Teknik og for at give hver enkelt Medarbejder den bedst mulige Uddannelse, der siden Ørsteds Tid har været saa levende her hjemme. Dansk Indsats i saadanne Henseender har da ogsaa fundet Anerkendelse ved den Deltagelse i Landbrugsvirksomheder og Ingeniørforetagender Verden over, der ligesom vore gamle og stadig fornyede Handels- og Søfartstraditioner har bidraget til at udvide vor Horisont.

De Opgaver og Forpligtelser, der paa det mellemfolkelige Samarbejdes Omraade frembyder sig for vort Land, har mødt Forstaaelse fra hele Befolkningens Side. Et særligt Udtryk derfor var den Beslutning af Regering og Rigsdag, ved hvilken der lige efter forrige Verdenskrig oprettedes et Fond til Fremme af videnskabelig Forbindelse mellem Danmark og Udlandet. Dette Rask-Ørsted Fond be-

gyndte sin frugtbare Virksomhed netop paa et for internationale Forbindelser kritisk Tidspunkt og har tjent som Forbillede for lignende
Institutioner, skabt i større Lande og i tilsvarende større Maalestok.
Det er dog ikke blot fra Statens Side, at der hos os ydes en i Betragtning af Forholdene meget betydelig Støtte til Kulturbestræbelser af
saadan og anden Art. Ogsaa i nyere Tid har jo den offervillige Interesse for Kulturens Højnelse, der igennem lange Tider har været
lagt for Dagen her hjemme, fundet mangfoldigt Udtryk og bevirket,
at Betingelserne saavel for kunstnerisk og videnskabelig Virksomhed
som for faglig Uddannelse paa de mest forskellige Omraader er langt
gunstigere, end det kunde ventes efter Landets Størrelse og begrænsede Rigdomskilder.

Om end, som allerførst betonet, Spørgsmaalet om, hvad vi forstaar
ved dansk Kultur, frembyder mange, med hverandre uløseligt forbundne Sider, turde det Syn paa Fællesskabet mellem Folkeslagene,
som vor hele Historie har udviklet, være det for vor Kultur mest
betegnende Træk. Netop vort Krav paa og Ansvar ved Verdens-
borgerskabet, som vi opfatter det, anviser os vore Opgaver ind-
adtil og udadtil, og uden at drage nogen Sammenligning, der vilde
ligge denne Opfattelse fjernt, har vi Lov til at føle Stolthed over den
Maade, hvorpaa vi har udnyttet vore Kaar til egen Udvikling og
til Deltagelse i Samarbejdet paa den menneskelige Kulturs Fremskridt.
Hvad Skæbnen har i Behold for os som for andre, er skjult for vort
Blik, men hvor vidtgaaende Følgerne paa alle Menneskelivets Omraader af den Krise, hvori Verden for Tiden befinder sig, end turde
blive, har vi Ret til at haabe, at vort Folk, om vi blot bevarer Frihed
til at udvikle den hos os saa dybt rodfæstede Indstilling, ogsaa i
Fremtiden med Ære vil kunne tjene Menneskehedens Sag.

TRANSLATION

DANISH CULTURE
SOME INTRODUCTORY REFLECTIONS
BY NIELS BOHR

If we ask ourselves in what respect we can speak of a specifically Danish culture, the answer will depend entirely on the adopted point of view. On the one hand, common human characteristics will come more sharply into focus the more we seek to fathom our own nature; on the other hand, the difference from other societies will be more strongly felt the more we strive to keep all facets of our existence in the overall picture. Since knowledge of how human life can flourish under different circumstances must be the basis for any evaluation of the cultural stage reached by a society, an investigation of our position among other societies would seem to be a natural point of departure for an attempt to elucidate what is special to us. I have accepted the invitation of the editors to write a few words of introduction to this large work, despite being aware of my limited knowledge, because my cooperation both at home and abroad with scientists from many different nations has often given me occasion to concern myself with our relationship within the realm of culture to the world around us and to think about the traditions that give the Danish outlook on life its special features.

The culture of a people has often been compared with a living organism, and it is obvious in particular that, just as an organism's development and living conditions are decisive for its form of life, so too is knowledge of the history and conditions of a culture a prerequisite for any fuller understanding of its character. Such a comparison would be more pertinent, however, if individual cultures possessed the same degree of independence as organisms; but if anything is characteristic for our age, it must just be the growing understanding of the multitude of interconnections even between apparently widely different cultures. Archaeological and anthropological research during the last generation has indeed given us an important lesson concerning forgotten and unnoticed links in ancient and more recent times which have had a profound influence on the development of cultures that at first glance seem to be only marginally dependent on each other. It is therefore all the more imperative always to keep the question of mutual interaction in mind with regard to the relationship between the cultures of peoples that, as in Europe, have maintained the liveliest contact for centuries.

Neither the present occasion nor my own qualifications allow me to offer any detailed account of the history of our culture, but I would like to remind the reader of the position that Denmark and the other Nordic countries have occupied within the European cultural community through the centuries because of their geographical location on the outskirts of our part of the world. Many cultural movements have only reached us slowly and with difficulty; but on the other hand, we have often had a better opportunity of bringing such external influences into harmony with our own outlook. Within the Nordic countries, national boundaries and conditions of nature may in this respect have worked in different ways, but the kinship of our languages and our common heritage have nevertheless created a much more intimate sense of brotherhood than one usually finds among neighbouring states. The similarity in culture between us and our Nordic brothers is after all so great that a foreigner often fails to see any differences at all, while for ourselves an emphasis on our differences is most often precisely a means of becoming acquainted with, as in a mirror, the face that we turn towards the outside world.

During the thousand years that have passed since the introduction of Christianity to the Nordic countries opened new paths for peaceful relations with the world around us, cultural intercourse between Denmark and other European countries has, despite changing forms, constantly displayed certain distinctive traits because it took place across a boundary behind which the languages and independence of the Nordic countries were always preserved. Enlightenment and new customs were often brought here by foreigners who came to this country without any precise knowledge of our conditions and who usually did not even have command of our language. At other times Danes, especially young Danes, returning from a stay at European centres of culture and learning, brought back education and knowledge. In both cases, although in different ways, contact with foreign cultures brought forth not only new inspiration, but also an enhanced recognition of cultural values with deep roots in the home soil.

The chronicles[1] and the Danish provincial laws as far back as the first centuries of our history testify most eloquently to the desire to safeguard our heritage as well as the forms of society and the outlook on life that had developed here in Denmark. Indeed, even in the visual arts, for which the church opened up entirely new prospects, there was, as can be seen in the earliest relics from Christian times, an inclination to conserve domestic traditions of style and technique. Also in later times of upheaval, when new cultural movements came here, we find similar traits; even during the Reformation the foreign doctrine

---

[1] [This refers to the first attempts at written history in Denmark, of which the most famous is due to Saxo (c. 1150–1220).]

was brought to the common people by Danish spokesmen and provided the impetus for a Danish hymn-writing tradition which continued and developed the art of writing in the mother tongue that has its origin in traditional folk songs.

When the supremacy of Latin in the schools, which had meant so much for keeping alive the heritage of the sources common to European culture and which had saved the smaller nations from an all too one-sided cultural influence from larger neighbouring states, was gradually broken down, the danger of the Danish language being displaced in the circle of the educated was perhaps at its greatest. And yet nothing was to contribute more to averting this danger and to enriching our language than the many-sided renewal of Danish literature brought about once again on the basis of examples from abroad. Despite the many differences in time and concerns, Holberg's[2] life and work exhibit features reminiscent of Absalon's[3] fate and his deeds in the service of his fatherland. In both cases, observations and experiences brought home from travels in foreign countries were grafted into Danish soil, and bore fruit that can rightly be considered part of our most precious national heritage.

Later, too, constant development free of any major ruptures has characterized conditions in Denmark. Above all, the growing recognition of human rights, which elsewhere gave rise to violent social upheavals, led here to the liberation of the peasantry[4], which took place in a most peaceful way and in a spirit of understanding, to which the admonishing and moving inscription on the liberty monument[5] testifies to this very day. In fact, despite harsh conditions, our ancient peasant culture had retained a vitality that enabled it, as soon as times improved, to grow new, fresh shoots. The demand for influence on the governance of the state expressed in the medieval moots also left traditions which were never completely forgotten and were reactivated when, a hundred years ago, in line with the development in other European countries, we gained

[2] [Ludvig Holberg (1684–1754), Danish author, born in Bergen, Norway. Holberg was professor of history at Copenhagen University and also taught Latin and metaphysics. Today he is known mainly for his numerous comedies which are still performed in many theatres.]

[3] [Absalon (1128–1201), Danish bishop and statesman. Founded Copenhagen in 1167.]

[4] [In 1733, a law was introduced forcing all peasants to stay for life on the estate where they were born. The abolishment of this law in 1788 marked the final liberation of the peasantry.]

[5] [The liberty monument (in Danish, "Frihedsstøtten") is an obelisk erected in 1797 at the western gateway of Copenhagen (now beside the central railway station) by Danish peasants in memory of the liberation in 1788. The inscription reads: "Kongen bød Stavnsbaandet skal ophøre, Landbolovene gives Orden og Kraft, at den frie Bonde kan vorde kjæk og oplyst, flittig og god, hæderlig Borger, lykkelig" (The king decreed that villeinage cease, the land laws come into force, so that the free peasant may be brave and enlightened, hard-working and good, honest citizen, happy).]

the free constitution[6] which consolidated and developed our old ideals of justice on a new basis.

At the beginning of the last century solidarity with our past and the feeling of brotherhood within the Nordic countries were strengthened, especially by the newly awakened appreciation of the common treasures we possessed in the Edda[7] and the old sagas. Also in this connection foreign influence proved stimulating, but, as shown by the reverberations in literary writing throughout Europe, there was now a much greater reciprocity in cultural interaction between the Nordic countries and foreign lands than ever before. Nevertheless, the awareness of how much we needed to learn in order to hold our own as the times demanded, did in no way lessen, and was much too strong to let the glory of the past ages tempt us to forget our modest position among nations. On the other hand, there arose in the Nordic countries a sense of world citizenship, which found more harmonious expression here than in many of the larger countries, where the temptation to regard their own cultures as independent organic units was so much greater.

Grundtvig's[8] telling words, "Citizenship for Danes alone, hospitality for all on earth", and Hans Christian Andersen's beautiful lines, "In Denmark I was born, there I belong, there is my home, from there my world unfolds", provide typical testimonies to how the idea of the relationship between home and abroad has engaged Danish writers. Both of these statements, so different in style as well as mood, express an attitude to the question of our relationship to the world to which one can hardly find a parallel. The crucial point is that this question is neither rejected nor resolved, but is understood as being one which must constantly be posed with regard to things great and small. What characterizes Danish culture more than anything else must surely be the immediate combination of an openness to the lesson brought to us from the outside or that we bring home ourselves, and an adherence to our outlook on life, determined by our heritage and our destiny, uniting us so closely in the big world to which we irrevocably belong.

Søren Kierkegaard's famous closing words in "Stages on Life's Way" bear

---

[6] [The Danish constitution (representative democracy) was introduced in 1849.]

[7] [The Edda is the name of two Icelandic poetic works: the "Elder Edda", which stems from the 10th century and was written down in the 13th century, is the main source of Scandinavian mythology; the "Snorra Edda", which also portrays Scandinavian mythology, was written by Snorri Sturlason (1179–1241).]

[8] [Nikolaj F.S. Grundtvig (1783–1872) was a Danish theologist and poet who had a great influence on the liberation movements within the Danish State Church and on the development of the folk high schools.]

testimony to the source of inspiration that such an outlook entails[9]. Taking as a starting point a defence of the Danish language against the claim of the merits of the more widespread languages, he expresses himself so warmly and fervently about the richness and beauty of the mother tongue, and displays such artistry and profundity, that in fact his observations become valid not only for the Danish, but for any human language, and offer a lesson about the infinitely subtle means of expression and operation contained in any language. A similar background for the love of one's native country can be found in the writings of Poul Martin Møller[10], who has sometimes been called the most Danish of all Danish writers. Even where, as in his "Adventures of a Danish Student", he does not enter into a comparison with conditions abroad, the way in which in the gallery of characters he displays with such great wit various aspects of his own nature, contains a conscious demand for a balance between adaptation to the ways of life at home and empathy with the world of ideas common to all humanity.

The struggle between Danish traditions and foreign cultural movements cannot only be traced everywhere in our literature, it has also time and time again been the subject of open discussion. Thus, Baggesen's "Polemic" against Oehlenschläger[11] did much to illuminate the dangers inherent in Romanticism, a movement which led to so rich a flowering of our literature. And the ferment in Danish intellectual life occasioned by later European schools of thought found instructive expression just before the turn of the century in the well-known exchange of opinions between Georg Brandes and Harald Høffding[12]. Despite all differences of emphasis expressed by the individual writer or author, one of the principal aims for all of them was to raise the level of our intellectual life through cultural contacts with the outside world, without losing the link to our past, which alone can give us strength to contribute, in a modest yet independent way, to the development of human culture.

For a long time the sense of world citizenship has marked Danish endeavours

---

[9] S. Kierkegaard, *Stadier paa Livets Vei*, in *Søren Kierkegaard: Samlede Værker*, Vol. 8, Gyldendal, Copenhagen 1963, pp. 277–278. English translation: *Stages on Life's Way*, Princeton University Press, Princeton 1988

[10] [See above, *General Introduction*, pp. XXXI and XLIV.]

[11] [The Danish poet Jens Baggesen (1764–1826) criticized the romanticism of the great Danish poet Adam Oehlenschläger (1779–1850). This caused a violent "Polemic" ("Pennefejde") in which many intellectuals participated in the period from 1813 to 1819.]

[12] [Georg Brandes (1842–1927), famous man of letters, was an adherent of Nietzsche's theories about aristocracy and the moral superiority of the master. Høffding opposed these views in a debate with Brandes in 1889–1890.]

with regard to education in the schools, and we also encounter the effects of this in the folk high school movement[13], which, built on the popular revival inspired by Grundtvig, has helped in such a fortunate way to bring intellectual and spiritual movements to wider circles of the population than may be the case anywhere else. Thus we meet here a deep appreciation that the sources for raising the general level of education beyond basic schooling are to be sought not only in our national heritage, but also in the knowledge of international cultural cooperation. A peculiar testimony to this are the excellent books[14] on the history of the sciences that were written for use in folk high school teaching by some of its own men and gained importance far beyond their original purpose. Alongside the introduction to general philosophical problems required for any form of university study here at home[15], the reading of these works has contributed to giving later generations of Danish students a background, the lack of which is often felt even at the great universities abroad.

As regards our country's participation in the endeavours of scientific research common to all humanity, we are of course entering an area which by its very nature knows no national boundaries, but where nevertheless traditions play just as decisive a role as in any other human activity. When science in Europe again began to flourish during the Renaissance, no Danish participation could begin without seeking abroad, for only there were education and research facilities to be found. The munificent generosity bestowed by the Crown on Tycho Brahe[16], to enable him to continue at home the work to develop astronomy that he had planned during his travels abroad, provides perhaps the best testimony of the lively interest, which after all was present from the very beginning, in being part of the dawning endeavour. Indeed, Tycho Brahe's activity on the island of Hven, admired by the whole world, and his later tragic fate, bear witness to how scientific research can thrive even on virgin soil, as long as there is an

---

[13] [Folk high schools are residential colleges for adults who wish to supplement their education with a general education without final examination. The schools receive state support and are spread all over the country. The first one was established in 1844.]

[14] [The books referred to are P. la Cour, *Historisk Matematik* (Historical Mathematics), manuscript edition, Kolding 1881, and J. Appel & P. la Cour, *Historisk Fysik I–II* (Historical Physics), Det nordiske Forlag, Copenhagen 1896–1897.]

[15] [See above, p. XLIII.]

[16] [The great Danish astronomer Tycho Brahe (1546–1601) was given the island Hven (now, Ven) in Øresund north of Copenhagen by King Frederik II and here Brahe built his observatory "Uraniborg". After a conflict with Frederik II's successor, King Christian IV, Brahe was forced to leave Denmark in 1597, and lived his last years at the court in Prague.]

understanding of its value, and of how difficult it can be to maintain such an appreciation if traditions are not yet sufficiently deep-rooted.

In the following period schools of science with opportunities for further education were established in this country, even though participation in the keen competition, as it developed in the scientific centres of Europe, was still required in order to reach the highest levels. Although Steno's great discoveries[17] were made abroad and only received proper recognition there, we have nevertheless reason to be proud here in Denmark of the spirit that he brought from home and that immediately secured him a place among the leading researchers. Just how different the fate of Danish scientists who had won recognition abroad could be at that time can be seen when comparing Steno with Ole Rømer[18]. Whereas Steno ended his days abroad adhering to a conviction denounced here as heretical, Rømer was entrusted on his return with solving a wide range of problems for the benefit of our society, which occupied so much of his unique talents that only in our time have we learned about many of the important scientific investigations he found time to carry out later.

Gradually, the situation in Denmark developed so that conditions also within the realm of science no longer differed much from those prevalent abroad; but here in Denmark we were, of course, just as in our day, still obliged to seek contact with foreign countries to a much greater extent than is the case in larger civilized countries. On the other hand, the widespread understanding of the necessity and difficulty under more modest circumstances to keep pace with the development in the world around us expressed itself in the opportunities of the individual pioneer in science or other areas of intellectual life to enhance our culture and strengthen our sense of solidarity. The outstanding example is

[17] [Nicolaus Steno (Niels Stensen) (1638–1686), Danish anatomist, founder of geology, crystallography and palaeontology. Steno converted to Catholicism in 1667 and thereby lost all possibility of obtaining a professorship at Copenhagen University. In 1677, he became bishop of Hannover and later of Münster, Hamburg and Schwerin. He discovered, among other things, the duct of the parotid gland (Stensen's duct, a principal source of saliva for the oral cavity) and the law of constancy of crystallic angles.]

[18] [While working in Paris in 1676, Ole Rømer (1644–1710) discovered that light travels with a definite velocity. After his return to Copenhagen, he worked as a professor in astronomy at the university, but at the same time also as mayor of Copenhagen, member of the supreme court, commissioner of police, chief of the fire brigade, chief of the surveying of the whole country, chief of town-planning and chief of the departments for money, weights and measures and for the calendar.]

of course H.C. Ørsted[19], who not only contributed in such an extraordinary manner to the dissemination of scientific knowledge and the possibilities for its application, but whose influence on all areas of cultural life was of a kind and of an extent scarcely imaginable in larger countries. However different their points of departure and approaches to their work, Ørsted deserves to be named alongside Grundtvig as an educator of the people, while his discoveries came to play a role for the esteem of science in this country reminiscent of the way in which Hans Christian Andersen's and Thorvaldsen's[20] widely admired works brought an appreciation of the importance of art to all circles of our people.

Although any contribution to the recognition of truth that science endeavours to attain is indeed more than anything else part of a common human heritage, it is true of both science and art that their welfare everywhere depends on many special conditions. Thus, it is not without connection to the interest in history shared by the whole population that the exploration of the prehistoric relics, with which our soil abounds, has gained significance far beyond our own borders by elucidating common stages of development in human cultures[21]. Similarly, it is scarcely a coincidence that precisely our little country, with the limited area where our language is spoken, has played such an important part in the development of comparative linguistics ever since the days of Rasmus Rask[22]. Furthermore, the circumstance that our legal traditions have indeed been deeply affected, but not as elsewhere completely supplanted, by Roman Law, constituted the basis for the Danish jurisprudence founded by Anders Sandøe Ørsted[23], which has meant so much for the adoption of common basic principles of legislation in all the Nordic countries. Besides discoveries such as the one linked to Finsen's name[24], the successful endeavours here at home to apply medical science for the improvement of people's health are due to the vigilant interest in human and social issues that has manifested itself both in government measures and in independent involvement on the part of the population.

[19] [Hans Christian Ørsted (1777–1851), Danish physicist. See Bohr's article in this volume, pp. [342]–[356] (Danish original), pp. [357]–[369] (English translation).]

[20] [Bertel Thorvaldsen (1768–1844), Danish sculptor.]

[21] [Bohr is here referring to the division of prehistoric times into Stone Age, Bronze Age and Iron Age which was introduced by the Danish archaeologist Christian Jørgensen Thomsen (1788–1865).]

[22] [Rasmus Rask (1787–1832), prominent Danish philologist and founder of Nordic philology.]

[23] [Anders Sandøe Ørsted (1778–1860) was the greatest Danish jurist of his time. A brother of H.C. Ørsted, he was Danish Prime Minister from 1853 to 1854.]

[24] [Niels Finsen (1860–1904), Danish physician who received the Nobel Prize in Physiology and Medicine in 1903 for his method of treating illnesses with concentrated light rays.]

Alongside our participation in the education and the cooperation offered us abroad, conditions are being established to a steadily increasing degree in our own age also to enable foreigners to seek instruction in this country and together with us participate in research in fields that have favourable conditions here. Thus, our general outlook provides us with special qualifications for such participation in international and cultural cooperation. Not only does this outlook make it easier for Danes who have the opportunity to go out into the world to feel at home there, but it also means that foreigners are received in this country in a more natural way than in many other places. Scientists from the most distant parts of the world who have participated in research activity here in Denmark have thus repeatedly expressed their gratitude for, despite any outward appearance of strangeness, having been considered by all as fellow human beings who brought welcome tidings about cultures from which new lessons of the human situation were expected. Incidentally, precisely this outlook is the basis for the manner in which Danes treat the indigenous population on Greenland as well as for the fortunate way in which the contact with this people has enriched our culture and given us men such as Knud Rasmussen[25], whose wonderful expeditions of exploration have for ever inscribed Denmark into the history of polar research.

However many different aspects of man's striving to know himself and the world are comprised in literature and science, culture, even at its peak, would be far poorer without musical and visual art. Naturally, it had to be limited what a small country, always remote from the main streams of culture, was able to contribute to the riches that these art forms, whose fostering requires so much responsiveness, have bestowed on mankind. Even so, the appreciation for the gifts brought to us, and our inclination to seek those places where cultural life has flourished most richly, have entailed that we, also here, have been among those who followed the call most readily. Indeed, in all kinds of art we have traditions that go far and not so far back in time and that have given us works of art which we recognize as our own, as regards both content and conception. Moreover, it should not be forgotten to what extent Danish participation in the life of artists around the world has constantly contributed to keeping us informed about the shifts of mood, laden with emotion and struggle, that, so to speak, make us feel the pulse of culture.

Our general outlook on life, determined by our relationship with the outside world, characterizes not only the aspects of cultural life mentioned here, but is also mirrored in our society in many other ways. The recognition of what

---

[25] [Knud Rasmussen (1879–1933), prominent Danish arctic explorer.]

Three of the founders of the National Museum Foundation and the education minister. From the left: Niels Bohr, Knud Rasmussen, education minister Jens Byskov and Harald Høffding. Copenhagen 1928.

each of us can learn, to the benefit of all, from human values obtained in the most varied areas of activity forms precisely the basis for the advanced stage of enlightenment that perhaps more than anything else is characteristic of the cultural stage reached by Denmark and that, despite our limited living conditions, has made possible a constant improvement of the population's standard of living. Without the folk high school, the cooperative movement[26], which meant so much in keeping the development of our agriculture up to date and thereby helped us over the great difficulties caused by conditions of production abroad, would scarcely have been conceivable. When furthermore our lack of raw materials has not prevented us from keeping up with industrial progress, and even in some fields from leading the way, it is probably most of all due to the efforts to maintain the closest possible cooperation between science and

[26] [The Danish cooperative movement started in 1866 with cooperative shops and from 1882 with dairies. It played a great role in the development of Danish agriculture and cattle breeding.]

technology and to give each individual employee the best possible education, a tradition which has been so alive here at home since Ørsted's time. As a result, Danish achievement in this regard has found recognition through our participation in agricultural enterprises and engineering concerns all over the world, which, like our old and constantly renewed traditions of commerce and shipping, has helped to broaden our horizons.

The tasks and obligations of our country in the realm of international cooperation have been met with understanding by the entire population. A special expression of this was the decision by government and parliament just after the last world war to establish a foundation for the advancement of scientific relations between Denmark and other countries. The Rask–Ørsted Foundation began its fruitful activity just at a critical point for international relations and has been a model for similar institutions established in larger countries and on a correspondingly larger scale. Even so, it is not only the state which provides, comparatively speaking, considerable support for cultural activities. Also in recent times, generous interest in the heightening of culture, which has been in evidence here at home for a long time, has found a variety of expressions and has led to far more favourable conditions for artistic and scientific activity, as well as for vocational education in many different fields, than could be expected considering the country's size and limited resources.

Even though, as stressed at the very beginning, the question of what we mean when we speak of Danish culture involves many inextricably intertwined aspects, the attitude to the fellowship among peoples that our entire history has fostered might well be the most characteristic feature of our culture. Precisely our insistence on and responsibility for world citizenship, as we understand it, define our tasks at home and abroad, and without making any comparison, which would be foreign to this viewpoint, we may feel proud of the way in which we have made use of our situation to further our own development and to participate in the common effort toward the progress of human culture. What destiny has in store for us and for others is hidden from our eyes, but no matter how far-reaching the consequences, in all areas of human life, may be of the crisis in which the world finds itself today, we are entitled to hope that our people, as long as we can retain freedom to develop the outlook that is so deeply rooted in us, will also in the future be able to serve honourably the cause of mankind.

IV. PHYSICAL SCIENCE AND THE STUDY OF RELIGIONS

Studia Orientalia Ioanni Pedersen Septuagenario A.D. VII id. Nov. Anno MCMLIII, Ejnar Munksgaard, Copenhagen 1953, pp. 385–390

Niels Bohr, the chemist Niels Bjerrum and the orientalist Johannes Pedersen.
Carlsberg Museum 1958.

PHYSICAL SCIENCE
AND THE STUDY OF RELIGIONS

BY NIELS BOHR, COPENHAGEN

Though not a scholar in the study of the history and philosophy of religions, but deeply indebted to Johannes Pedersen for inspiration from his published works and personal discussions about the many problems occupying his open and active mind, I am grateful for the opportunity to contribute to this volume, edited in honour of his 70th birthday, by a few comments on the implications of modern development of physics as regards general epistemological problems.[1]

In any attempt of clarifying man's position in existence, it is a question of the proper balance between our want for an all-embracing way of looking at life in its multifarious aspects and our power of expressing ourselves in a logically consistent manner. Essentially different starting points are here taken by physical science, aiming at the development of general methods for the comprehension of common human experience, and the religions, originating in endeavours of furthering harmony of outlook and behaviour within individual communities. Of course, in any religion, all knowledge shared by the members of the community was included in the frame of which the values and the ideals emphasized in cult and faith constituted the primary content. The inherent relation between content and frame therefore hardly called for the attention before subsequent progress of science had led to a cosmological or epistemological lesson pointing beyond common horizon at the epoch of the foundation of the religion.

The course of history offers many interesting illustrations in such

[1] Detailed account of the development of modern physics and especially of atomic theory is given in many textbooks, to which the reader may be referred. The problems with which the present article is primarily concerned are dealt with in various articles by the writer, e. g.: Natural Philosophy and Human Cultures, Nature 143, 268, 1939; On the Notions of Causality and Complementarity, Dialectica 2, 312, 1948. Suggestive considerations concerning psychological and religious problems may also be found in O. Klein, Bibeln och våra dagars naturforskning, Ord och Bild 1941, p. 471; W. Pauli, Die philosophische Bedeutung der Idee der Komplementarität, Experentia VI, 72, 1950.

25

respect, but, as a background for comments on the present situation, we shall only briefly refer to the impression on general philosophical thinking made by the great progress of physical science which followed European Renaissance and led to the mechanical conception of nature. The elucidation of the principles of mechanics did indeed not merely provide a foundation for a simple consistent description of many phenomena regarding the behaviour of the bodies surrounding us, but even the regularities exhibited by the grand display in the sky, which by innumerable generations had been witnessed with wonder and awe, received on these principles wide-going explanation.

Within its apparently unlimited scope, the mechanical conception of nature implied a clarification of the relationship between cause and effect specifically expressed by the possibility from the state of a physical system at a given time, defined by measurable quantities, to predict its state at any subsequent time. A causal—or deterministic—description of this kind, excluding all finalistic argumentation, became the foundation of the great edifice, often termed classical physics, the rapid technical exploration of which has so radically changed the conditions of practical life.

From its beginning, the whole development exerted a deep influence on European philosophy, and the adoption of the notion of causality as an ideal of rational explanation within all fields of knowledge brought about a veritable schism between science and religion. On the one hand, many phenomena, once regarded as manifestations of divine providence, appeared as consequences of general immutable laws of nature. On the other hand, the physical methods and viewpoints were far remote from the emphasis of human values and ideals essential for the religions. Common to the schools of so-called empirical and critical philosophy, an attitude therefore prevailed of a more or less vague distinction between objective knowledge and subjective belief.

By the lesson regarding our position as observers of nature, which the development of physical science in the present century has given us, a new background has, however, been created just for the use of such words as objectivity and subjectivity. From a logical standpoint, we can by an objective description only understand a communication of experience to others by means of a language which does not admit ambiguity as regards the perception of such communications. In classical physics, this goal was secured by the circumstance that, apart from unessential conventions of terminology, the description is based on pictures and ideas embodied in common language, adapted to our orientation in daily-life events. The exploration of new fields of physical experience has, however, revealed unsuspected limitations of such approach and has demanded a radical revision of the foundation for the unambiguous application of our most elementary concepts, like space and time, and cause and effect.

The first decisive step in this direction was the realization of the extent to which the finite limit of the speed of propagation of any physical signal influences the space-time coordination of events. In fact, observers moving relative to each other with velocities not negligible compared with the velocity of light will not only describe the shape of rigid bodies and the rate of clocks differently, but even events which by one observer are judged to take place simultaneously may by other observers be perceived as occurring before or after each other. The recognition of this element of subjectivity in the description permitted to combine apparently conflicting experimental evidence and proved a most powerful guidance for tracing and formulating general physical laws common to all observers. Besides lending a unity to our world picture surpassing all previous expectations, the principle of relativity has thus essentially extended the frame and scope for objective description of physical phenomena.

The next epoch in natural philosophy was inaugurated by the discovery of the universal quantum of action which disclosed a feature of wholeness in physical phenomena quite foreign to the mechanical conception of nature. Indeed, it became clear that the whole edifice of classical physics represents an idealization applicable only to phenomena in the analysis of which all actions are sufficiently large to permit the neglect of the individual quantum. While this condition is amply fulfilled in daily-life events, we meet in the exploration of the world of atoms, made possible by modern development of experimental technique, fundamental regularities which reject causal analysis and even pictorial representation. In quantum physics, we can in fact no longer uphold customary ideas of properties and behaviour of the objects under investigation as separate from the interaction between such objects and the measuring instruments, indispensable for the definition of the circumstances under which the phenomena occur.

Objectivity of communication requires, of course, that the account of the experimental arrangement and the recording of observations are given in plain language, conveniently supplemented with terminology of classical physics. Just this condition implies, however, not only the impossibility in quantum phenomena to control the interaction between the atomic objects and the measuring instruments, but even prevents the unrestricted combination of space-time coordination and the laws of conservation of momentum and energy, fundamental for the causal description of classical physics. Although equally important in the account of quantum regularities, the strict applications of these concepts are, however, mutually exclusive due to the circumstance that any unam-

25*

[277]

biguous use of the concepts of space and time refers to an experimental arrangement involving an exchange, uncontrollable in principle, of momentum and energy with the remedies like scales and clocks required for the fixation of the space-time frame.

The paradoxes with which we are confronted when trying to picture a course of a quantum phenomenon, and especially the apparent contrast exhibited by atomic phenomena observed with different experimental arrangements, obtain indeed full explanation by such analysis of the conditions for objective description in this field of experience. In particular, the essential wholeness of a quantum phenomenon finds its logical expression in the circumstance that any attempt at its subdivision would demand a change in the experimental arrangement incompatible with its appearance. Under these circumstances, phenomena described by concepts referring to mutually exclusive experimental arrangements are conveniently termed complementary in order to stress that they present a novel kind of relationship between physical experiences and constitute equally important aspects of all well defined knowledge about atomic objects.

In pointing directly to our position as regards objective description of experience, this notion of complementarity exhibits a certain resemblance with the principle of relativity. However, in the latter we have to do with the compensation for the subjective element in the description of physical evidence by observers in different states of motion, while in the former we are concerned with the relationship of evidence observed with different experimental arrangements among which every observer has a free subjective choice. Just in such respects, the lesson of physical science reminds us of our position as regards objective description in many other fields of knowledge, where relativistic or complementary relationships are apparent in the analysis and synthesis of experience.

Returning to the problems with which studies of the history and philosophy of religions are primarily concerned, we are in the first place confronted with general and specific aspects of cultures developed in more or less isolated communities. A common feature of such cultures is the strife for the harmonious display of the individuals within the social relationships equally essential in human life. The great differences exhibited by the cultures, in spite of this common background, have often been suggestively compared with the essentially different ways in which observers moving relatively to each other will judge physical experiences. Still, the very objectivity and unity of relativistic description depend on the possibility, for every observer, on general physical laws to predict how any other observer will describe experience. The divergence, fostered by history, of the traditions inherent in all cultures impedes, however, such simple comparison, and the cultures may more appropriately be regarded as complementary phenomena, each exhibiting within their limitations aspects of the richness and variety of human life.

Notwithstanding all differences in such respects, general epistemological features of objective description may be perceived in the attitude to ethical problems in the various cultures, and especially expressed in the religions. Surely, all stable human relationships demand fair play specified in judicial rules which, as regards the recognition of equal rights of individuals, remind of the principle of relativity. At the same time, however, life without attachment to kinsfolk and friends will obviously be deprived of some of its most precious values. Still, though the closest possible combination of justice and love presents a common goal in all cultures, it must be recognized that, in any situation which calls for the strict application of justice, there is no room for display of love, and that, conversely, the ultimate exigencies of a feeling of love may conflict with all ideas of justice. This situation, which in many religions is mythically illustrated by the fight between divine personifications of such concepts, is indeed one of the most striking analogies to the complementary relationship between physical phenomena described by different elementary concepts which were combined in the mechanical conception of nature, but whose strict applications in wider fields of physical experience exclude each other.

The recognition of the limited scope of the concept of causality points also directly to problems related to dogmas of religions. Not only has the deterministic description of physical events, once regarded as suggestive support of the idea of predestination, lost its unrestricted applicability by the elucidation of the conditions for the rational account of atomic phenomena, but it must even be realized that mechanistic and finalistic argumentation, each within its proper limits, present inherently complementary approaches to the objective description of the phenomena of organic life. Moreover, the problem of free will, so pertinent in the philosophy of religions, has received a new background by the recognition, in modern psychology, of the frustration of attempts to order experience regarding our own consciousness as a causal chain of events, originally suggested by the mechanical conception of nature.

Actually, ordinary language, by its use of such words as thoughts and sentiments, admits typical complementary relation between conscious experiences implying a different placing of the section line between the observing subject and the object on which attention is focussed. We are here presented with a close analogy to the relationship between atomic phenomena appearing under different experimental conditions and described by different physical concepts, according to the rôle played by the measuring instruments. In fact, the varying separation line between subject and object, characteristic of different conscious experiences, is the clue to the consistent logical use of such contrasting notions as will, conscience and aspirations, each referring to equally important aspects of the human personality.

The aim of these brief comments has, of course, not been to enlarge upon topics requiring deeper studies of the history and philosophy of religions, but merely to indicate a general attitude inspired by the novel development of physical science. This attitude may be characterized by its striving for harmonious comprehension of ever more aspects of our situation in the never ending struggle for proper relation between content and frame, recognizing that no experience is definable without a logical frame and that any disharmony apparent in such relationship can be removed only by an appropriate widening of the conceptual framework.

V. ATOMIC SCIENCE AND THE CRISIS OF HUMANITY

ATOMVIDENSKABEN OG MENNESKEHEDENS KRISE
Politiken, 20 April 1961

TEXT AND TRANSLATION

See Introduction to Part II, p. [221].

[281]

Atomvidenskaben og menneskehedens krise

SÆRTRYK AF POLITIKENS KRONIK

20. 4. 1961

[282]

JEG vil gerne udtrykke min tak-nemmelighed over den ære, som er blevet mig vist af Københavns Universitet, hvor min akademiske uddannelse netop i disse dage for halvtreds år siden afsluttedes med erhvervelsen af den filosofiske doktorgrad, og hvortil jeg siden så længe har været knyttet. Jeg føler denne ære så meget større ved tanken om de personligheder, der hidtil har modtaget Sonningprisen, og som hver på sin vis, i handlingens såvel som i tænkningens verden, har ydet så betydningsfuld en indsats, og som i vor bevidsthed står som repræsentanter for idealer, som vi alle hylder. Jeg er også dybt rørt over rektors venlige ord og over den smukke måde, hvorpå mine kolleger Christian Møller og Mogens Pihl har talt om mine bestræbelser.

✱

Udviklingen af videnskaben i vor tid hviler jo på et verdensomspændende samarbejde af enestående omfang og frugtbarhed. Ligesom på andre områder har samarbejdet inden for atomfysikken været karakterise-ret ikke blot ved det nøje fællesskab i grupper af forskere såvel ved iværksættelse af eksperimenter som ved udformningen af teoretiske synspunkter. men også ved den forbindelse mellem kolleger overalt i verden gennem diskussioner og publikationer som har været så intens, at fremgangen fra år til år groede gennem mange bidrag, der hverken med hensyn til udgangspunkt eller resultat skarpt kan skilles fra hverandre. Under sådanne forhold kan den enkelte forskers arbejde nærmest sammenlignes med det at bringe sten til en bygning, og glæden for alle er at se det hele værk rejse sig under de fælles anstrengelser. For os ældre har det været den største opmuntring at følge, hvorledes den yngre generation har formået stadig at styrke grundlaget for den voksende bygning og til denne at føje tårne og spir, som i skønhed og dristighed skulle overgå alle forventninger.

Udforskningen af atomernes verden, der skulle føre til en så uforudset teknologisk udvikling og derved rejse store menneskelige problemer, har været som er rejse i ukendt land ad ubanede veje, på hvilke vi har mødt mange overraskelser og hindringer. Gang på gang stødte vi på tilsyneladende modsigelser, hvis overvindelse kun skulle vise sig mulig ved ny og indgående undersøgelse af vor stilling som iagttagere af naturen og vore muligheder for en objektiv sammenfatning af erfaringerne – uafhængig af enhver subjektiv bedømmelse.

Sådanne spørgsmål var jo allerede bragt i forgrunden gennem udviklingen af relativitetsteorien, der netop ved erkendelsen af den forskellige måde, hvorpå iagttagere, der bevæger sig i forhold til hinanden med store hastigheder, vil ordne begivenheder i rum og tid, åbnede muligheder for at efterspore lovmæssigheder, der er fælles for alle iagttagere og derved gav vort verdensbillede en enhed, – som man tidligere næppe havde kunnet drømme om. Som ofte fremhævet indeholder afsløringen af, i hvor høj grad vor opfattelse af selv så forholdsvis simple problemer som ordningen af fysiske erfaringer afhænger af iagttagerens standpunkt, en påmindelse om nødvendigheden af, når vi beskæftiger os med mere almindelige sider af menneskelivet, at være opmærksom på den særlige baggrund, hvorfra de betragtes.

Trods den gennemgribende revision af grundlaget for den entydige anvendelse af så elementære begreber som rum og tid, kan relativitetsteo-

rien betragtes som en harmonisk afrunding af den såkaldte klassiske fysik. Derimod skulle opdagelsen af det universelle virkningskvantum afsløre et helhedstræk ved de atomare processer, som var ganske fremmed for den mekaniske naturopfattelse, der jo antager muligheden af en uindskrænket opdeling af fænomenerne og af deres beskrivelse som en kæde, hvis enkelte led ligegyldigt hvor langt underdelingen fortsættes kan sammenknyttes gennem entydig anvendelse af begreberne årsag og virkning.

På den ene side gav virkningskvantet nøglen til forståelsen af den særegne stabilitet af atomare systemer, hvorpå alle egenskaberne af de stoffer hvoraf såvel vore redskaber som vore egne legemer er bygget op, til syvende og sidst beror. På den anden side betød det nye helhedstræk, at den med enhver iagttagelse forbundne vekselvirkning mellem undersøgelsesobjekt og måleinstrumenter ikke altid, som antaget i den klassiske fysik, kan lades ude af betragtning, men tværtimod ved studiet af egentlige kvanteprocesser udgør en uadskillelig del af selve de iagttagne fænomener. Under sådanne forhold kan det ikke undre, at erfaringer vundne under forskellige forsøgsbetingelser ikke kan sammenfattes i noget enkelt billede, men i en objektiv beskrivelse må indgå som komplementære i den forstand, at de tilsammen udtømmer al oplysning om de pågældende atomare systemer, som vi kan vinde ved at stille spørgsmål til naturen i form af eksperimenter.

Dette såkaldte komplementaritetssynspunkt tillod ikke blot at opklare alle tilsyneladende paradokser, men frembød tillige en ramme, der var tilstrækkelig vid til at rumme en mangfoldighed af fysiske erfaringer, der ingen plads kunne finde i den tilvante billedlige årsagsbeskrivelse. Stillet over for denne i de eksakte videnskaber så ny situation var det en afgørende inspiration at genfinde lighedspunkter ved beskrivelsen af erfaringer på andre kundskabsområder, hvor helhedstræk altid har været iøjnefaldende. Værdien af sådanne sammenligninger beror på, at vi på disse områder møder en rigdom af erfaringer og en mangesidighed i deres sammenhæng, over for hvilken situationen i fysikken fremtræder som yderst simpel og derfor med en logisk belæring, der er så meget lettere overskuelig.

Når vi vender os til de levende organismer, stilles vi således over for en strukturel og funktionsmæssig kompleksitet af en helt anden størrelsesorden end den, vi træffer ved selv de mest sammensatte mekaniske systemer. Omend vi ikke har nogen grund til at tvivle om, at de atomare processer i organismerne i hver enkelthed beherskes af de samme love, som i så vidt omfang har tilladt at gøre rede for stoffernes fysiske og kemiske egenskaber, er det derfor ikke overraskende, at vi i organismernes opførsel møder træk, der er uden parallel i de erfaringer, der vindes ved fysiske og kemiske eksperimenter under simple reproducerbare forsøgsbetingelser. I de levende organismer har vi jo så at sige at gøre med resultater af forsøg udført af naturen selv under hele den lange organiske udviklings historie.

Når man for oversigten over de biologiske erfaringer bekvemt anvender ord, helt fremmede for de fysiske videnskaber, som selvopholdelsesdrift, drejer det sig ikke om påvisning af nogen begrænsning af de fysiske begrebers rækkevidde, – men langt snarere om en erkendelse af, at organismernes kompleksitet står hindrende i vejen for disse begrebers praktiske anvendelighed til en redegørelse for organismernes opretholdelse som helheder. Ved talen om de til livet knyttede psykiske oplevelser må det i sådan sammenhæng fremhæves, at ordet bevidsthed melder sig selv, så snart vi har at gøre med en adfærd, der er så kompliceret, at den ikke praktisk kan meddeles uden ved ord, der henviser til den pågældende organismes reaktioner på dens hele situation.

Betingelserne for beskrivelse og sammenfatning af bevidste oplevelser har altid været et vigtigt problem i filosofien. I lyset af den almindelige erkendelsesteoretiske belæring, som vi har modtaget i kvantefysikken, er det klart, at ord som tanker og følelser, der henviser til psykiske op-

[285]

levelser, som udelukker hverandre, – har været brugt på komplementær måde lige siden sprogets oprindelse. I denne forbindelse må man dog ikke glemme, at den objektive beskrivelse af atomfænomenerne netop beror på den detaljerede specifikation af de gensidigt udelukkende forsøgsbetingelser, hvorunder de komplementære erfaringer vindes, og at derfor ingen henvisning til den personlige iagttager er påkrævet. Ved meddelelsen af individuelle psykiske oplevelser, hvor vi anvender udtryk som „jeg mener" eller „jeg vil", – kommer analogien først frem derved at vi, trods anvendelsen af samme pronomen, benytter forskellige verber.

Vi møder her det gamle spørgsmål om jeg'ets art, så velkendt fra alvorlige eller spøgefulde diskussioner om enhvers mange jeg'er, som betragter hverandre og strides med hverandre. Medens naturligvis al søgen efter et sidste subjekt er i strid med objektiv beskrivelses væsen, der jo forlanger en modstilling af subjekt og objekt, er mulighederne for i menneskelivet at bevare personlighedens enhed og at opretholde forestillingen om viljens frihed netop betinget af den forskellige placering af skillelinjen mellem subjekt og objekt i situationer, som vi betegner med ord som overvejelse eller driftbetonet tilskyndelse.

Typiske komplementære træk møder vi også ved beskrivelsen af for-

holdet mellem individerne i et samfund. Særlig iøjnefaldende er jo her den begrænsede forenelighed af såkaldte etiske idealer som retfærdighed og kærlighed. Belæring om de mangfoldige veje, der frembyder sig for at søge balance inden for menneskesamfund, finder vi ikke mindst ved sammenligningen mellem forskellige nationale kulturer, der trods alle nok så dybtgående forskelle i traditioner og indstilling ved nærmere betragtning åbenbarer fælles træk i den menneskelige tilværelse. Enhver berøring med fremmedartede samfundsforhold tvinger os jo til eftertanke om oprindelsen af egne livsvaner og synspunkter og bringer os til erkendelse af fordomme, som vi næppe anede, at vi besad.

Det kunne måske synes, at man med sådanne almindelige bemærkninger har fjernet sig langt fra de umiddelbare opgaver, som vi i de fysiske videnskaber stilles overfor. – Det har imidlertid været min hensigt at understrege, hvorledes atomfysikkens udvikling, ganske uanset dens teknologiske anvendelser, ligesom det har været tilfældet ved tidligere fremskridt på naturvidenskabens område, frembyder hjælp i den stadige stræben efter dybere indblik i menneskelivets almindelige problemer.

Det forøgede herredømme over naturens kræfter, som atomfysikken har givet os, indebærer i forbindelse med de biologiske videnskabers blomstring og lægekunstens vidunderlige fremskridt de rigeste løfter for fremme af menneskelig velfærd overalt på jorden. Samtidig er alle folkeslags skæbne blevet knyttet uløseligt sammen gennem de vældige magtmidler, der er kommet menneskene i hænde, og hvis anvendelse kan true selve livets beståen på vor klode.

I denne på en gang så løfterige og farefulde situation må vi betænke, at enhver forøgelse af viden og kunnen altid har været forbundet med et større ansvar for den enkelte. Den seneste udvikling har imidlertid stillet hele vor civilisation over for den alvorligste prøve, som kun kan bestås ved et samarbejde mellem alle nationer i erkendelsen af det fælles ansvar. Ikke blot er det klart, at uoverensstemmelser mellem staterne i fremtiden må finde deres afgørelse igennem fredelig forhandling, men de nuværende forhold indebærer, at nationerne i det lange løb kun vil kunne hævde sig gennem den støtte, de kan tilbyde andre og de bidrag, de kan yde til den fælles menneskelige kultur.

I den krise, hvori menneskeheden i dag befinder sig, og som intet fortilfælde har i civilisationens historie, kan udveje kun findes ved fordomsfri undersøgelse af traditionelle synspunkters begrænsede rækkevidde. En væsentlig forudsætning herfor må være fælles adgang til fuld oplysning om samfundsforhold i ethvert land og fri diskussion over alle landegrænser om et hvilket som helst spørgsmål af betydning for samlivet mellem nationerne. Kun i en sådan åben verden vil der være grobund for den gensidige tillid mellem folkeslagene, som kræves for afkald på besiddelsen af magtmidler, der hidtil har været anset for nødvendige for forsvaret af nationale interesser.

Selvom mange vanskeligheder stiller sig i vejen for så gennemgribende ændringer i bestående forhold, er der dog opmuntring at finde ikke mindst i det stigende antal fælles menneskelige opgaver, om hvis løsning alle kan samles. I denne henseende indtager det internationale videnskabelige samarbejde en særlig stilling, idet bestræbelserne her alene er rettet mod at forøge menneskelig kundskab og erkendelse. På tværs af enhver forskel i national hjemmehørighed og politisk indstilling har samarbejdet på dette fælles mål skabt varme venskaber, der ikke alene rummer dyb glæde for den enkelte forsker, men også et håb for en lysere fremtid for hele menneskeheden.

*

At være knyttet til et universitet som vort, der efter dets formål og traditioner står som udtryk for enheden i vor viden og fællesskabet i vore bestræbelser, er et særligt privilegium og en særlig forpligtelse. Personlig har jeg følt det som en stor lykke foruden stadig at kunne finde berigende kontakt med kolleger fra vidt forskellige fagområder at være i stand til at bevare tilknytningen til den opvoksende ungdom, til hvis friske kræfter og uhildede sind vi alle må knytte vore inderligste forhåbninger, og jeg vil gerne slutte disse få ord med at udtrykke mine varmeste ønsker for Københavns Universitet.

NIELS BOHR.

Niels Bohr receiving the Sonning Prize from Carl Iversen, Vice-Chancellor of the University of Copenhagen. Centre: Leonie Sonning, far right: Margrethe Bohr. Copenhagen 1961.

TRANSLATION

ATOMIC SCIENCE AND
THE CRISIS OF HUMANITY

I should like to express my gratitude for the honour paid to me by the University of Copenhagen, where precisely at this time fifty years ago my academic education was completed with the acquisition of the degree of doctor of philosophy, and to which I since have been attached for so long. I feel that the honour is all the greater when thinking of the personalities who have hitherto received the Sonning Prize, and who, each in his own way, have made such a significant contribution in the world of action as well as in the world of thought, and who in our minds represent ideals we all cherish. I am also deeply moved by the kind words from our Rector, and by the beautiful way in which my colleagues Christian Møller and Mogens Pihl have spoken of my endeavours.

* * *

The development of science in our time is of course based on a world-wide collaboration of unique scope and fruitfulness. As in other fields, collaboration within atomic physics has been characterized not only by the close cooperation within groups of scientists in initiating experiments as well as in framing theoretical viewpoints, but also by the contact between colleagues all over the world through discussions and publications; this contact has been so intense that the progress from year to year developed through many contributions which cannot be sharply distinguished from one another with regard to either origin or result. Under such circumstances, the work of the individual scientist can best be compared to carrying bricks to a building, where everyone rejoices in seeing the whole edifice grow as a result of the common effort. To us older people it has been most encouraging to watch how the younger generation has managed to continue strengthening the foundation of the expanding building and to add turrets and spires whose beauty and boldness exceed all expectations.

The exploration of the world of atoms, which was to result in such an unforeseen technological development and thus create great problems for humanity, has been like travelling in unknown territory along unbeaten tracks, where we have encountered many surprises and obstacles. Time and time again we ran into apparent contradictions which, as it turned out, could be overcome only through a new and thorough investigation of our situation as observers of nature and our possibilities for making an objective ordering of our experiences – independent of any subjective judgement.

Such questions had of course already been brought to the fore through the development of relativity theory which, precisely through the realization of how differently observers moving at high speeds relative to each other will order events in space and time, provided possibilities for tracing regularities which are common to all observers, and thus gave our picture of the world a unity that one could scarcely have imagined before. As frequently emphasized, the disclosure of the extent to which our understanding of even such relatively simple problems as the ordering of physical experience depends on the viewpoint of the observer, involves a reminder of the need, when we deal with the more ordinary aspects of human life, to be aware of the particular background from which they are observed.

In spite of the thorough revision of the basis for the unambiguous application of such elementary concepts as space and time, the theory of relativity can be regarded as a harmonious rounding off of the so-called classical physics. The discovery of the universal quantum of action, on the other hand, was to reveal a feature of wholeness with regard to the atomic processes quite foreign to the mechanistic conception of nature, which presupposes the possibility of an unlimited division of the phenomena and of their description as a chain, the separate links of which can be tied together through an unambiguous application of the concepts of cause and effect regardless of how far the subdivision continues.

On the one hand, the quantum of action offered the key to the understanding of the characteristic stability of atomic systems on which all properties of the substances making up our tools as well as our own bodies after all depend. On the other hand, the new feature of wholeness meant that the interaction between the object of investigation and the measuring instruments connected with any observation cannot, as assumed in classical physics, always be disregarded, but constitutes on the contrary, in the study of actual quantum processes, an inseparable part of the observed phenomena themselves. Under such circumstances, it is no surprise that experiences gained under different experimental conditions cannot be summarized in any one picture, but, in an objective description, must be considered complementary in the sense that they collectively provide all information about the atomic systems in question that we may gain by posing questions to nature in the form of experiments.

This so-called complementarity viewpoint did not only permit the explanation of all apparent paradoxes, but also presented a framework wide enough to comprise a multitude of physical experiences, for which there was no room in the customary representational description of causality. Confronted with this situation, which is so new in the exact sciences, it was a decisive inspiration to rediscover points of similarity in the description of experiences in other areas of knowledge where features of wholeness have always been conspicuous. The value of such comparisons is due to the fact that in these areas we meet a richness of experiences and a many-sidedness in their relations, compared to which the situation in physics appears extremely simple and consequently holds a logical lesson that is so much easier to grasp.

When we turn to living organisms we are thus faced with a structural and functional complexity of quite another order of magnitude than that we meet in even the most complicated mechanical systems. Although we have no reason to doubt that the atomic processes in the organisms are governed in every detail by the same laws which to such great extent have permitted the explanation of the physical and chemical properties of matter, it is nevertheless not surprising that in the behaviour of the organisms we encounter traits that are without parallel in the experiences gained by physical and chemical experiments under simple, reproducible experimental conditions. In living organisms we are of course dealing, so to speak, with results of experiments made by nature itself in the course of the long history of organic development.

When, in order to describe biological observations, one finds it convenient to use words completely foreign to the physical sciences, such as instinct of self-preservation, it is not a question of pointing out a limitation in the scope of the physical concepts, but rather of recognizing that the complexity of the organisms obstructs the practical applicability of these concepts in explaining the preservation of organisms as wholes. When we speak of the psychical experiences in life, it should be stressed in this context that the word consciousness presents itself as soon as we have to do with behaviour so complex that in practice it can only be communicated with words referring to the reactions of the particular organism to its entire situation.

The conditions for describing and ordering conscious experience have always constituted an important problem in philosophy. In the light of the general epistemological lesson we have received in quantum physics, it is clear that words such as thoughts and feelings, which refer to psychical experiences that are mutually exclusive, have been used in a complementary manner ever since the origin of language. Still, in this connection one should not forget that the objective description of atomic phenomena is based precisely on the detailed specification of the mutually exclusive experimental conditions under which complementary experiences are gained, and that therefore no reference to the individual observer is required. In the communication of individual psychical experiences, where we use expressions such as "I think" or "I will", the analogy

[291]

only emerges because, although using the same pronoun, we apply different verbs.

Here we encounter the old question of the nature of the ego, which is so well known from serious or jocular discussions of anyone's many egos watching each other and fighting each other. While any search for a last subject is of course contrary to the essence of objective description, which after all demands a juxtaposition of subject and object, the possibilities in human life of preserving the unity of personality and of maintaining the notion of the freedom of the will depends precisely on the different placement of the dividing line between subject and object in situations we describe with words such as deliberation or instinctive urge.

We also encounter typically complementary features in the description of the relationship between individuals in society. Here the limited compatibility of so-called ethical ideals such as justice and love is of course particularly conspicuous. We gain a lesson about the manifold ways offering themselves in the search for balance within human society not least by comparing different national cultures which, despite all quite profound differences in traditions and attitudes, on closer inspection reveal common features of human existence. Every contact with unfamiliar social conditions forces us to reflect on the origin of our own habits and viewpoints and makes us recognize prejudices we barely knew we held.

It might perhaps seem that with such general remarks we have diverged considerably from the immediate tasks we are faced with in the physical sciences. It has, however, been my intention to emphasize how the development of atomic physics, quite irrespective of its technological applications, offers help in our constant striving for a deeper insight into the general problems of human life, just as has been the case with previous advances in the area of natural science.

The increased mastery of the forces of nature atomic physics has given us, together with the flowering of the biological sciences and the wonderful progress in medicine, holds rich promises for the advance of human welfare all over the world. At the same time, the fate of all peoples of the world has become linked inseparably through the formidable instruments of power that man now has at his disposal, and whose application can threaten the very existence of life on our planet.

In this at the same time so promising and dangerous situation we must bear in mind that any increase in knowledge and ability has always involved a greater responsibility on the part of the individual. However, the latest development has confronted our entire civilization with the most serious challenge, which can be met only through cooperation between all nations in realization of the joint responsibility. Not only is it obvious that disagreements between the states in the future must be settled through peaceful negotiation, but under present conditions nations will in the long run be able to hold their own only through the support they can give to others, and the contributions they can make to common human culture.

In the crisis confronting humanity today, which is unprecedented in the history of civilization, solutions can be found only through unbiased investigation of the limited range of traditional points of view. An essential requirement must be common access to full information about social conditions in every country and free discussion across all national borders about any question of importance for the coexistence of nations. Only such an open world can provide fertile soil for the mutual trust between the peoples of the world that is required to renounce the possession of instruments of power hitherto considered necessary for the defence of national interests.

Although many difficulties stand in the way of such sweeping changes of the established order, there is still encouragement to be found not least in the increasing number of common tasks facing mankind, for the solution of which all can unite. In this respect international scientific cooperation occupies a unique position because here efforts are directed solely towards increasing human knowledge and insight. Across any difference in national origin and political orientation, cooperation towards this common goal has created warm friendships which hold not only great happiness for the individual scientist, but also hope for a brighter future for all mankind.

* * *

It is a special privilege and a special obligation to belong to a university such as ours, which according to its purpose and traditions is a manifestation of the unity in our knowledge and the community in our efforts. Personally, I have felt great happiness not only in always being able to find rewarding contact with colleagues from widely different fields, but also in keeping the ties with the rising generations, to whose fresh strength and open minds all of us must attach our most sincere hopes, and I would like to conclude these few remarks by expressing my warmest wishes for the University of Copenhagen.

PART III

PAPERS ON HARALD HØFFDING

Harald Høffding, professor of philosophy at the University of Copenhagen, 1883–1915. Circa 1910.

INTRODUCTION

by

DAVID FAVRHOLDT

In the first two parts of this volume we have seen that Bohr was occupied with philosophical problems throughout his life, and it is natural to enquire from where this interest arose. In the literature on Bohr's philosophical background, it is often maintained that he was influenced by the Danish philosopher Harald Høffding and through Høffding possibly by other philosophers such as Søren Kierkegaard and William James[1]. One might therefore expect that Bohr's publications about Høffding – a newspaper article on the occasion of Høffding's 85th birthday and two speeches given before the Royal Danish Academy[2] – would shed light on this issue. This introduction seeks to bring these publications into a proper perspective.

Harald Høffding was born on 11 March 1843 and died on 3 July 1931. He became a master of theology (cand. theol.) in 1865 and doctor of philosophy in 1870. His doctoral thesis was on "Antiquity's perception of the human will"[3]. In 1883, he became professor of philosophy at the University of Copenhagen, a chair he held until his retirement in 1915. He won international renown, especially because of two – for the time – excellent books: "Outlines of Psychology" and "A History of Modern Philosophy"[4]. In addition, he wrote

---

[1] See, for instance, M. Jammer, *The Conceptual Development of Quantum Mechanics*, McGraw-Hill Book Company, New York 1966, pp. 173–179.

[2] N. Bohr, *Ved Harald Høffdings 85 Aars-Dag* (At Harald Høffding's 85th Birthday), Berlingske Tidende, 10 March 1928; *Mindeord over Harald Høffding* (Tribute to the Memory of Harald Høffding), Overs. Dan. Vidensk. Selsk. Virks. Juni 1931 – Maj 1932, pp. 131–136; *Harald Høffdings 100-Aars Fødselsdag* (Harald Høffding's 100th Birthday), Overs. Dan. Vidensk. Selsk. Virks. Juni 1942 – Maj 1943, pp. 57–58. All reproduced in this volume on pp. [306]–[307], [313]–[318], [324] (Danish originals), and on pp. [308]–[309], [319]–[322], [325] (English translations).

[3] *Den antike Opfattelse af Menneskets Villie*, Copenhagen University Press, Copenhagen 1870.

[4] *Psykologi i Omrids paa Grundlag af Erfaring*, Gyldendal, Copenhagen 1882 (first English edition: *Outlines of Psychology*, Macmillan, London 1891) and *Den nyere Filosofis Historie. En Fremstilling af Filosofiens Historie fra Renaissancens Slutning til vore Dage*, 2 volumes, P.G. Philipsens Forlag, Copenhagen 1894–1895 (Eng. ed.: *A History of Modern Philosophy I–II*, Macmillan, London 1900).

a number of books on, for example, Spinoza, Rousseau, Kant, Kierkegaard, ethics, philosophy of religion and epistemology[5].

In 1884, Høffding was appointed member of the Royal Danish Academy of Sciences and Letters. At that time, he knew Niels Bohr's father, whom he had consulted concerning physiological matters when he wrote his "Psychology"[6]. Christian Bohr was elected to membership in the Academy in 1888. In 1893, the two started dining together after the Academy meetings[7]. Some years later the physicist Christian Christiansen (later Niels Bohr's teacher at the university) and the philologist Vilhelm Thomsen joined them. The four thus formed a small debating group which met regularly in each other's homes until Christian Bohr's death in 1911[8]. In his brief notes on the occasion of Høffding's death reproduced below[9] (which supplements the official tribute to Høffding's memory by Vilhelm Grønbech, professor of the history of religion), Bohr describes these meetings, recalling (as Høffding does not) that the discussions often concerned psychological and biological matters.

As noted in the General Introduction, when Bohr began his studies at the University of Copenhagen in 1903, he had to attend and pass the "Filosofikum", an elementary, compulsory course in philosophy, and chose to attend Høffding's lectures. The course comprised four hours of lectures a week over two semesters followed by an oral examination. Høffding's curriculum consisted of three of his own books: the above-mentioned "Psychology" (426 pages), an abbreviated edition of Høffding's "History of Modern Philosophy" (95 pages) and a small textbook in classical logic (32 pages)[10].

As alluded to in his tribute to Høffding's memory[11], while attending the course Bohr discovered some elementary mistakes in Høffding's book on logic, and when Høffding prepared a new edition a few years later, he wrote to Bohr asking for his opinion. Unfortunately, we have only the letters from Høffding

---

[5] A complete bibliography of Høffding's work can be found in Frithiof Brandt et al., *Harald Høffding in Memoriam*, Gyldendal, Copenhagen 1932.

[6] Harald Høffding, *Erindringer* (Recollections), Gyldendalske Boghandel–Nordisk Forlag, Copenhagen 1928, pp. 171–172.

[7] See above, *General Introduction*, p. XXVIII.

[8] Høffding writes about this in *Erindringer*, ref. 6, p. 171; see quotation in Vol. 6, p. XXI.

[9] Bohr, *Mindeord*, ref. 2.

[10] The Danish editions used the year Bohr followed the course are, *Psykologi i Omrids*, ref. 4, 4th edition, Gyldendalske Boghandel–Nordisk Forlag, Copenhagen 1898; *Kort Oversigt over den nyere Filosofis Historie*, P.G. Philipsens Forlag, Copenhagen 1898; *Formel Logik til Brug ved Forelæsninger*, 4th edition, Det Nordiske Forlag Ernst Bojesen, Copenhagen 1903.

[11] Bohr, *Mindeord*, ref. 2, this volume on pp. [316]–[317].

to Bohr[12]. By comparing the two editions of Høffding's book – from 1903 (4th edition) and 1907 (5th edition) – we find the relevant alterations. In the edition from 1903 Høffding exemplifies the principle of duality (i.e. the principle of the excluded middle) and states that "a concept (B) must either contain another concept (A) or its negation (a)"[13]. Here "either–or" is used in the exclusive sense. In the 1907 edition, the principle is stated as follows: "The concept A must either be connected with B or not be connected with B". The example given is: "A vertebrate is either a mammal or not a mammal". But, Høffding adds, "In the case where B is not a necessary feature of A both AB and Ab [b being the negation of B] may occur – under different circumstances and at different times. Thus the vertebrate type may occur both in mammals and in non-mammals, for instance in the horse as well as in the eagle"[14]. Obviously, Bohr had pointed out to Høffding that a concept may comprise both another concept and that concept's negation.

In 1905, the "Ekliptika" discussion group of students at the University of Copenhagen was formed, all of whose members later gained prominence in Danish society[15]. One of the members, Peter Skov, describes the group in his

---

[12] The correspondence relating to this matter, which is deposited in the Niels Bohr Archive, is reproduced below, pp. [505] ff.

[13] H. Høffding, *Formel logik*, ref. 10, pp. 26–27.

[14] H. Høffding, *Formel logik*, 5th ed., Gyldendal, Copenhagen 1907, p. 27.

[15] Members of this club were, first, Edgar Rubin (1886–1951), later a professor of psychology; Niels and Harald Bohr; the historian Poul Nørlund (1888–1951), later the director of the Danish National Museum; and his brother, the mathematician Niels Erik Nørlund (1885–1981), later a professor of mathematics. Niels Bohr married the latter two's sister, Margrethe Nørlund, in 1912. Then there were Peter Skov (1883–1967), who was to become Danish ambassador to Moscow, Ankara, Prague and Warsaw; Vilhelm Slomann (1885–1962), an art historian and later director of the Copenhagen Museum of Decorative Art (Kunstindustrimuseet); Kaj Henriksen (1888–1940), an entomologist and later director of the Zoological Museum in Copenhagen; Einar Cohn (1885–1969), a cousin of Rubin and a political economist who would later become permanent undersecretary of a governmental department dealing with statistics; Lis Jacobsen (1882–1961), also a cousin of Rubin and a Nordic philologist; Viggo Brøndal (1887–1942), later professor in Romance philology; and, finally, Astrid Lund (1881–1935), later married to Elias Lunding. Lunding (1878–1969) was an agricultural economist and became a leading figure in Danish agriculture. He too became a member of "Ekliptika", but we do not know when. Gudmund Hatt (1884–1960), a geographer and later a professor of this subject also joined the circle at some time. Protocols or minutes of the group have not been found, which indicates that the set-up was very informal. Slomann (*Minder om samvær med Niels Bohr* [Recollections of meetings with Niels Bohr], Politiken, 7 October 1955) writes that the "Ekliptika" meetings ended when the students graduated. The Niels Bohr Archive holds a small drawing of a rosette in which the names and addresses of seven members of "Ekliptika" are inscribed; it can be dated to 1916. Of course, this does not prove that the meetings still went on regularly at that time.

recollections[16]:

> "Having finished grammar school I began to study law, partly in order to please my father, partly because I was so attracted by a number of other subjects such as literature, art and philosophy that I didn't know what to choose. My studies left me with plenty of leisure. For several years I attended Høffding's seminars. Here I found some good friends who started a small club with twelve members. Hence the name 'Ekliptika'. Among the participants were the Bohr brothers, the Nørlund brothers, Edgar Rubin, Brøndal and others, who later became professors. *Quorum minimum pars fui*. I have always mixed with people from whom I could learn something."

Many scholars in the history of science have concluded from this passage that after the "Filosofikum", Bohr and many other members of "Ekliptika" attended Høffding's seminars, i.e. his lectures given to students who had philosophy as their main subject[17]. Obviously, however, Peter Skov does not make this improbable statement. Rubin, Bohr's half-cousin and closest friend, did attend the seminars because he was a student of philosophy with psychology (which was then not taught independently at the University of Copenhagen) as his special field. We also know that Skov and Viggo Brøndal attended some of the seminars.

But "Ekliptika" existed independently of Høffding's seminars. This can be seen from the reminiscences of another member, Vilhelm Slomann[18]:

> "Starting in 1905, Edgar Rubin, later a professor of psychology, for a number of winters gathered initially twelve students of his own age for meetings of the type students usually hold – consisting of a talk, a cup of tea at a restaurant or a boarding-house and a discussion with considerable consumption of tobacco. The natural and social sciences, geography and the traditional humanistic subjects – philosophy, literature, linguistics, archae-ology and history – were all represented, and the different interests of the twelve students were manifested in the slightly pompous name of the club: Ekliptika. Rubin remained the vigilant and natural centre of this circle, most

---

[16] P. Skov, *Årenes Høst. Erindringer fra mange Lande i urolige Tider* [Harvest of the Years. Recollections from many countries in times of unrest], Munksgaard, Copenhagen 1961, p. 10.

[17] See, for instance, Jammer, *Conceptual Development*, ref. 1, p. 173 and J. Faye, *Niels Bohr: His Heritage and Legacy*, Kluwer Academic Publishers, Dordrecht 1991, pp. 25 ff.

[18] Slomann, *Minder*, ref. 15, p. 12.

of whose members became professors in due course or were in other ways involved with scholarship."

The existence of "Ekliptika", then, constitutes no argument for Bohr being influenced by Høffding, or even that he attended his seminars.

On the death in 1911 of Christian Bohr – Niels Bohr's father – Høffding wrote some commemorative words in a Danish journal[19]. The Bohr family thanked him and kept in touch with him, but we do not have any testimony of scientific contact between Høffding and Bohr in the following years until 1922. In that year, Høffding wrote a letter to Bohr[20], asking him about the use of the concept of analogy[21] in atomic physics. At the time, Høffding was preparing a book on the concept of analogy and apparently felt that there was something in Bohr's use of the concept that he had not clearly understood. Bohr's polite answer[22] is of substantial historical interest in that it contains one of his first statements about the non-visualizable character of quantum physics. But he does not deal in any detail with the question raised by Høffding.

In his extensive correspondence with the Polish–French philosopher Emile Meyerson (1859–1933), beginning in 1918, Høffding writes on 12 February 1924 that he knows Bohr[23]. Apart from the two letters from 1922, however, between 1918 and 1928 – when Bohr's main philosophical contribution germinated – there is no indication in the correspondence of either Høffding or Bohr of any scientific contact between them. For example, Høffding's correspondence includes 41 letters written between 1888 and 1931 to the German philosopher Ferdinand Tönnies (1855–1936), in which Bohr's name does not appear at all[24].

Only from 1928 did Niels and Margrethe Bohr pay several visits to Høffding. To understand what light these visits may shed on the Bohr–Høffding relationship it is necessary to trace their background. We must begin in 1922, when Høffding gave a lecture at a student meeting. There he met a young Swedish

[19] H. Høffding, *Mindetale over Christian Bohr* (Tribute to the Memory of Christian Bohr), Tilskueren 1911, 209–212.

[20] Letter from Høffding to Bohr, 20 September 1922, reproduced on pp. [510] (Danish original) and [511] (English translation).

[21] H. Høffding, *Begrebet Analogi*, Kgl. Dan. Vid. Selsk. Filos. Medd. **I**, 4, 1923.

[22] Letter from Bohr to Høffding, 22 September 1922, reproduced on pp. [512] (Danish original) and [513] (English translation).

[23] Letter from Høffding to Meyerson, 12 February 1924, *Correspondence entre Harald Høffding et Emile Meyerson* (eds. F. Brandt, H. Høffding and J. Adigard des Gautries), Ejnar Munksgaard, Copenhagen 1939, p. 70.

[24] The correspondence is deposited in the Royal Library, Copenhagen.

girl who was preoccupied with studies in the philosophy of religion. Soon after, she moved into Høffding's house, the residence of honour at the Carlsberg Brewery[25]. The bare fact that the young student lived together with Høffding, who had been a widower since 1877, created a great stir. When they married in October 1924[26], the scandal reached its peak. Høffding was then 81 years old, his wife only 24. Moreover, his wife suffered from severe psychological problems and soon became an alcoholic; she committed suicide in August 1930. Two years before, Høffding wrote to Meyerson about the situation:[27]

"Some months ago I was ill and lying in bed for several days and I believe my wife had spoken of my illness as being a very serious condition. Hence, Mr. Bohr came several evenings and sat at my bedside. He told me about his work and other interesting topics and read to me some works by his favourite poets, while Mrs. Bohr conversed with my wife. It was a great encouragement in our uniform and isolated life."

During this period, then, Høffding had become more and more isolated, and the visits by Bohr and his wife at the Carlsberg Mansion must be seen as an act of personal support rather than a sign of Bohr's intellectual dependence on the Danish philosopher.

Bohr's explanation of his work inspired Høffding to prepare his last publication, a small booklet, "Notes on the Present State of the Theory of Knowledge"[28], from which it appears that he tried to grasp Bohr's ideas – with little success, however[29]. Yet this effort seems to have made an impression on Bohr, who, as can be seen from the publications reproduced below, later always spoke of Høffding with great veneration, and always returned to the fact that

---

[25] The mansion at Carlsberg became a residence of honour in 1914. Here the founder of the Carlsberg brewery, J.C. Jacobsen, and after him his son Carl Jacobsen, had lived. In accordance with the latter's will, the mansion was taken over by the Royal Danish Academy of Sciences and Letters as an honorary residence for a distinguished member of the Academy. The first resident was Høffding who lived at the mansion from 11 May 1914 until his death on 3 July 1931. At a meeting of the Academy on 11 December 1931, it was decided that Bohr would succeed Høffding as resident.

[26] Høffding, *Erindringer*, ref. 6, p. 321.

[27] Letter from Høffding to Meyerson, 13 August 1928, *Correspondence*, ref. 23, p. 156.

[28] H. Høffding, *Bemærkninger om Erkendelsesteoriens nuværende Stilling*, Det Kongelige Danske Videnskabernes Selskab, Copenhagen 1930, 28 pages.

[29] See D. Favrholdt, *Niels Bohr's Philosophical Background*, Munksgaard, Copenhagen 1992, pp. 101 ff.

during the last years of his life, Høffding had tried to understand what quantum mechanics was about.

Bohr was elected a member of the Royal Danish Academy in 1917, and the minutes show that Bohr and Høffding attended the same meetings several times. They also met in a committee set up in 1925 to prepare the financing of a new National Museum to be set up in Copenhagen. Yet there is no further documentation showing that Bohr and Høffding discussed physics and philosophy before 1928. In his substantial "Recollections", published in 1928, Høffding provides an impressive gallery of people he has known – not only public figures such as Georg Brandes, Henrik Ibsen, Bjørnstjerne Bjørnson, Kierkegaard's family, etc., but also fellow students, colleagues, pupils and so forth. He describes his environment and philosophical development in detail. If he had had considerable scientific discussions with Bohr, he would surely have mentioned them in the "Recollections". However, Bohr is mentioned only once, in connection with Høffding's book on the concept of analogy[30] (published in 1923). Høffding writes: "But now Niels Bohr's electron physics and Hjelmslev's geometrical investigations became especially important to me"[31].

In conclusion, as expressed in Bohr's publications about Høffding, the Danish philosopher was no doubt important for Bohr's general philosophical schooling and attitude. Yet it seems that Høffding played little or no part as regards the formulation of Bohr's specific contribution to philosophy.

---

[30] Ref. 21.
[31] Høffding, *Erindringer*, ref. 6, pp. 311–312.

I. AT HARALD HØFFDING'S 85TH BIRTHDAY

VED HARALD HØFFDINGS 85 AARS-DAG
Berlingske Tidende, 10 March 1928

TEXT AND TRANSLATION

See Introduction to Part III, p. [297].

Ved
Harald Høffdings 85 Aars-Dag.

Af Professor *Niels Bohr*.

Paa Professor Høffdings 85 Aars Fødselsdag vil der fra hele den yngre Generation strømme ham de varmeste Følelser af Taknemmelighed i Møde for alt, hvad han gennem sin personlige. Belæring og sine Skrifter har betydet for os. Som staaende udenfor Fagfilosoffernes Kreds besidder jeg jo ikke Forudsætninger for at udtale mig om hans store og alsidige videnskabelige Virksomhed. Naar jeg dog gerne vil være med til at give vor Taknemmelighed Udtryk, er det frem for alt den, af Høffding selv saa ofte og stærkt fremhævede, nøje Forbindelse imellem Filosofi og Naturforskning, der ligger mig paa Sinde. Denne Forbindelse møder vi ikke alene ved Videnskabens første Begyndelse, men en vedvarende, gensidig Befrugtning har fundet Sted gennem hele Udviklingen op til de seneste Tider, hvor den store Ophobning af Erfaringsmateriale paa alle Omraader og den høje Udvikling af Metoderne til at fremskaffe og analysere dette Materiale har nødvendiggjort en vidtgaaende Faginddeling indenfor Videnskaben.

Under Studiet af Naturfænomenerne er man Gang paa Gang stødt paa Problemer, der forlangte en Revision af de til Grund for vor Opfattelse af Iagttagelserne liggende Begreber. Her har hver Gang været en af ydre Forhold betinget Nødstilstand til Stede, idet det har drejet sig om en tilsyneladende Modstrid mellem ældre og nye Erfaringer, der truede med at sætte en Stopper for den menneskelige Tankes videre Indtrængen i Naturens Hemmeligheder. Af uvurderlig Betydning har det da været, at Naturforskerne har kunnet finde Støtte og Udgangspunkter for nye Fremstød i Filosoffernes Bestræbelser for at gøre sig Grundlaget og Grænserne for menneskelig Tankevirksomhed klart. Uden at ydre Forhold har bragt dem i Nød, har Filosofferne, ene af Trang til Harmoni i Sind og Tanke, uddybet vor Erkendelse og skabt en almindelig Indstilling overfor opdukkende Vanskeligheder og en udbredt Forstaaelse af Relativiteten og Komplementariteten i alle menneskelige Begreber. Paa Grund af sin uundgaaeligt ensidige Uddannelse er Naturforskeren som oftest forhindret fra i det enkelte at gøre sig Rede for, i hvor høj Grad den Grund, vi bygger paa, foruden ved de Foregangsmænds Virksomhed, hvis Navne er knyttet til Opdagelserne indenfor Naturvidenskabens snævrere Omraade, er skabt ogsaa ved de store Tænkeres utrættelige Fordybelse i de mere almenmenneskelige Problemer; en Fordybelse, hvis Frugter vi nyder godt af allerede gennem de Orddannelser, som fra Gengivelsen af deres Tanker er gaaet over i Sproget og blevet alment Eje. Vi skylder alle Professor Høffding den største Tak for, at han, gennem den for ham egne, samtidig objektive og personligt bearbejdede, Fremstilling af Filosofiens Landvindinger i Kampen for Klargørelsen af Vilkaarene for vort Erkendelses- og Følelsesliv, har bidraget til at øge vor Forstaaelse af Grundlaget for vort Arbejde og derigennem virksomt har understøttet os i dette.

Personlig har jeg haft den Lykke fra min tidligste Ungdom at staa i nær Forbindelse med Professor Høffding, og føler mig i dyb Taknemmelighedsgæld til ham for Belæring og Opmuntring. Jeg vilde dog næppe have grebet Lejligheden til offentligt at udtrykke disse Følelser, dersom jeg ikke vidste, at jeg samtidig giver Udtryk for Tanker, der besjæler store Kredse af den yngre Generation af danske Naturforskere. Ikke alene har Høffding ført os ind i Filosofiens paa engang saa fjerne og saa nære Skønheder, men hans

stedsevarende Friskhed og den Aabenhed for ethvert nyt Fremskridt, hvormed han paa alle Omraader har fulgt Udviklingen indtil dens seneste Faser, har om muligt styrket vor Tillid og aabnet ham alle Hjerter. Naar herhjemme næppe nogen i Filosofien ser en ørkesløs Spekulation over Spørgsmaal, der ikke er egnet til at hjælpe Menneskeheden i dens Stræben efter social Udvikling og efter at gøre sig til Herre over Naturen, men vi alle hylder den som Videnskabernes Videnskab, skyldes det først og fremmest Høffdings Virksomhed og den Tradition, han har skabt. Ogsaa under Ophold i Udlandet har jeg haft rig Lejlighed til at føle, hvor højt Professor Høffding er skattet paa Grund af hans enestaaende almenvidenskabelige Indstilling. Af hans Fagfæller erkendes han overalt som en Mester, med hvem, hvad Overblik og Uhildethed angaar, næppe nogen samtidig Filosof taaler Sammenligning. Maatte han endnu længe virke med usvækket Kraft iblandt os.

TRANSLATION

AT HARALD HØFFDING'S 85TH BIRTHDAY
BY PROFESSOR NIELS BOHR

On the occasion of Professor Høffding's 85th birthday the entire younger generation will bring him the warmest feelings of gratitude for everything he has meant to us through his personal teaching and his printed works. Being outside the circle of professional philosophers, I do not have the qualifications to speak about his large and comprehensive scholarly work. When I nevertheless would like to join in giving our gratitude expression, I have at heart above all the close connection between philosophy and natural science, so often strongly emphasized by Høffding himself. We meet this connection not only in the early beginnings of science, but a continuing mutual fertilization has taken place throughout the whole development up to the most recent times, when the great accumulation of observational material in all fields, and the rapid development of methods for obtaining and analyzing this material, have necessitated extensive specialization within science.

In the study of natural phenomena we have time and time again encountered problems that demanded a revision of the concepts underlying our interpretation of observations. In every instance there has been a crisis, determined by external circumstances, in that there was an apparent discrepancy between older and newer experience, which threatened to put a stop to the further advance of human thought into the secrets of nature. Thus it has been of inestimable importance that scientists have been able to find support and points of departure for new advances in the attempts of philosophers to clarify the foundations and the limits of human thought. Without being forced by external circumstances, but only out of a desire for harmony in mind and thought, philosophers have deepened our understanding and created a general approach to emerging difficulties as well as a widespread understanding of the relativity and the complementarity in all human concepts. Because of his inevitably one-sided education, the natural scientist is most often prevented, in the specific case, from seeing to what extent the foundations on which we build have been created not only through the work of his predecessors, whose names are linked to the discoveries within the more narrow confines of science, but also through the untiring concentration of the great thinkers on more universal problems; a concentration whose fruits we enjoy already through the word-formations which by rendering their thoughts have entered our language and become common property. We all owe Professor Høffding our most sincere thanks because in his characteristic – at the same time objective and personally formed – presentation of the conquests of philosophy in the struggle for clarification of the conditions for our rational and emotional life, he has contributed to increasing our understanding of the foundations for our work and thereby actively supported this endeavour.

Personally, I have had the good fortune to be in close contact with Professor Høffding since my earliest youth, and I feel deeply indebted to him for his instruction and encouragement. I should nevertheless hardly have taken the opportunity to express these feelings in public, had I not known that, at the same time, I express thoughts which animate wide circles of the younger generation of Danish scientists. Not only has Høffding introduced us to the beauties of philosophy, at once so remote and yet so close, but his unceasing vigour and the open-mindedness to each new advance with which he has followed the development in all fields to the most recent stages have, if possible, strengthened our confidence and opened all hearts to him. When hardly anyone here at home considers philosophy to be futile speculation about questions not suited to helping humanity in its striving for social development and mastery of nature, but we all hail it as the science of sciences, it is due first and foremost to Høffding's work and the tradition he has created. Also during my travels abroad I have had plenty of opportunity to observe how highly esteemed Professor Høffding is because of his unique broadly-based scientific outlook. His colleagues everywhere acknowledge him as a master with whom, as regards breadth of view and impartiality, hardly any contemporary philosopher can bear comparison. May he work among us with undiminished energy for a long time to come.

II. TRIBUTE TO THE MEMORY OF HARALD HØFFDING

MINDEORD OVER HARALD HØFFDING

Overs. Dan. Vidensk. Selsk. Virks. Juni 1931 – Maj 1932, pp. 131–136

Communication to the Royal Danish Academy on 11 December 1931

TEXT AND TRANSLATION

See Introduction to Part III, p. [297] f.

[311]

The circle of scientists described by Bohr in his Tribute to the Memory of Harald Høffding. From the left: Vilhelm Thomsen, Harald Høffding, Christian Bohr and Christian Christiansen. Christiansen's daughter, Johanne, commissioned the drawing in 1955 as a gift to Bohr on his 70th birthday. Artist: Des Asmussen, 1955.

II.
Af N. Bohr.

Jeg er taknemmelig for, at man har opfordret mig til at sige et Par Ord i Aften, hvor Harald Høffdings Minde hædres her i Videnskabernes Selskab, der stod hans Hjerte saa nær. Ingen vil af mig vente en indgaaende Redegørelse for Høffdings menneskelige Udvikling og videnskabelige Stræben som den, vi lige med Betagelse har lyttet til; men med Udgangspunkt fra de for mig dyrebare Minder, som mit eget Forhold til Høffding rummer, vil jeg gerne forsøge ganske kort at give Udtryk for, hvad hans Personlighed og Livsværk har betydet for store Kredse af Videnskabens Dyrkere, hvis Studier kun har indirekte Forbindelse med den egentlige Filosofi.

Mine første Minder om Høffding stammer fra nogle Aftensammenkomster, beskrevet af ham selv i »Erindringer«, hvor en lille Kreds af Videnskabsmænd for omkring en Menneskealder siden mødtes regelmæssigt i hverandres Hjem og diskuterede allehaande Spørgsmaal, der havde fanget deres Interesse. De andre Medlemmer af denne Kreds var Høffdings nære Venner fra den fælles Studietid, Christian Christiansen og Vilhelm Thomsen, samt min Fader der jo var en Del yngre, men som i Aarenes Løb kom til at staa i stedse intimere Venskabsforhold til Høffding. Fra den Tid vi var gamle nok til med Udbytte at lytte til Samtalerne, og til Sammenkomsterne for vort Hjems Vedkommende afbrødes ved min Faders tidlige Død, fik vi Søskende Lov til at være tilstede, naar Møderne holdtes hos os, og vi har derfra nogle af vore tidligste og dybeste Indtryk. Un-

9*

[313]

der de ofte meget livlige Diskussioner kunde navnlig Christiansen paa sin ejendommelige lune Maade lide at drille Høffding med den almindelige Filosofis Verdensfjernhed, men ligesom enhver af de øvrige forstod han fuldt vel, i hvor høj Grad netop Høffdings Evne til Forstaaelse og Trang til almen Sammenfatning var saa at sige den nærende Jordbund, hvori de andres af deres forskellige Studier og Livssyn prægede Indfald spirede.

Den Betoning af Videnskabens Enhed, som vort Selskab staar som Symbol for, var for Høffding ikke blot en abstrakt Selvfølgelighed, men en praktisk Nødvendighed. Selv om han maaske gerne vilde karakterisere Filosofien som Videnskabernes Videnskab, laa det ham fjernt at mene, at Filosofien i snævrere Forstand skulde kunne angive de Love, som alt videnskabeligt Arbejde maatte være underkastet. Høffding var altid forberedt paa, at væsentlige Træk af det almindelige menneskelige Erkendelsesproblem kan stilles i ny og frugtbar Belysning netop gennem Studier indenfor mere begrænsede Omraader af Videnskaben, hvor særlige Lettelser frembyder sig for Overblikket mellem Erfaringernes indbyrdes Forhold. Ud fra denne Indstilling var det hans Hovedbestræbelse at frugtbargøre Synspunkter, udformede indenfor de enkelte Videnskabsgrene, til Belysning af almene Spørgsmaal. Det enestaaende Overblik over videnskabelig Tænknings Former, som han derved erhvervede sig, tillod ham til Gengæld ofte at byde Fagvidenskabsmænd en Belæring, der var saa meget mere værdifuld, som Videnskabens stadig videregaaende Forgrening gør det stedse vanskeligere for Dyrkere af forskellige Fagvidenskaber at finde umiddelbar Forstaaelse og Belæring hos hverandre.

Omend det almindelige Erkendelsesproblem i den nævnte Betydning var det centrale for Høffding og med Aarene blev det stedse mere, var det dog først og fremmest psyko-

logiske Studier, ved hvilke han udviklede den for ham
karakteristiske Arbejdsform, og hvorfra han hentede sit
Værktøj til Behandling af de abstrakte Spørgsmaal. Den Baggrund, som disse Studier afgav for hans Arbejde paa Filosofiens andre Felter, har Høffding navnlig fremhævet i sin
allersidste Afhandling, en lille Artikel med Titel »Psykologi og Autobiografi«, som han skrev for kun et Aar tilbage,
og som i nær Fremtid vil udkomme i et amerikansk Sam-
lerværk. Paa overordentlig lærerig og interessant Maade
giver Høffding her et Overblik over sin hele Forfattervirk-
somhed og den Grundstemning, hvorpaa den hviler. Det
var en uforglemmelig Oplevelse under Artiklens Tilblivelse
at høre ham tale om sit lange Arbejdsliv, og at føle hvordan han ved Erindringen om den indre Tilfredsstillelse,
dette havde givet ham, søgte Styrke under de tunge Sorger,
som Livet ikke mindst i hans sidste Aar bragte ham.

Den Fremstilling af Psykologien, som Høffding udarbejdede til Grund for sine propædeutiske Forelæsninger,
og som han betegnende kaldte »Psykologi i Omrids paa
Grundlag af Erfaring«, var jo ogsaa det Værk, hvormed
Høffding først kom i Forbindelse med en større videnskabeligt interesseret Læsekreds. Dette Værk, hvis særegne
Tiltræknings- og Overbevisningskraft først og fremmest
skyldes Forfatterens Ærbødighed overfor Emnets Storhed, har
fundet en Udbredelse og bevaret en Livskraft, som Høffding
næppe selv havde drømt om, da han udarbejdede det for
omtrent 50 Aar siden. Man vil vel deri hverken søge en
digterisk Fremstilling af Aandslivets Rørelser eller skarp-
sindige Udredninger af normale og patologiske Sindstil-
stande, men man vil finde en samtidig nøgtern og begejstret Redegørelse for en, i Ordets bedste Forstand, videnskabelig Indstilling til Foreteelserne indenfor Bevidstheden.
Fremherskende er Bestræbelsen for at bevare Ligevægten

mellem Analyse og Synthese, og det tabes aldrig af Sigte, at Helheden vel opbygges af Enkelthederne, men at disses Fremtoning igen præges af Helhedsbilledet. Denne for Høffdings Fremstilling af Psykologien særegne Objektivitet har sikkert for mange af Tilhørerne ved hans Forelæsninger og af hans Bogs endnu flere Læsere haft en Betydning, der er dybere, end hver enkelt af os let vil kunne gøre Rede for. Dette har særligt slaaet mig ved Samkvem med Studerende ved nyere Universiteter, hvor der ikke findes en Tradition som ved en gammel Højskole som vor, og hvor man kan møde en Begrænsning af den almindelige videnskabelige Indstilling, der skyldes Savnet af et saadant Indblik i de psykologiske Grundproblemer, som det Høffdings Elever rent umiddelbart modtog.

For mit eget Vedkommende var det iøvrigt gennem nogle Diskussioner vedrørende et filosofisk Omraade, der efter Høffdings hele Indstilling og Anlæg maatte ligge mere i Yderkanten af hans Interesser, nemlig den formelle Logik, at jeg for første Gang kom i direkte videnskabelig Berøring med ham. Omend Høffding næppe selv tillagde den større Vægt indenfor sit Forfatterskab, er den korte Fremstilling af Logikken, som han benyttede ved sine Forelæsninger, en lærerig Bog, hvori, betegnende for hans Tænkemaade, den levende Baggrund, selv for Diskussionerne af de logiske Dommes Systematik, stadig er de almindelige psykologiske Erfaringer. For en ung Mand, hvis Hovedinteresse var den matematiske Behandling af naturvidenskabelige Problemer, turde derimod Spørgsmaalet om Systematikkens Stringens komme stærkere i Forgrunden. Aldrig vil jeg glemme nogle Aftener for mere end 25 Aar siden, hvor jeg fik Lov til at komme op til Høffding i hans gamle Hjem i Strandgade og tale med ham om disse Spørgsmaal, og hvor han, den i Tænkningens Historie og uendeligt varierede Af-

skygninger saa bevandrede Forsker, med ubeskrivelig Ven-
lighed og Taalmodighed lyttede efter, om der i den unge
Students vel mere af Begejstring end Afklaring farvede
Betragtninger skulde være mindste Fnug af nyt for ham at
lære i videnskabelig eller pædagogisk Henseende. Denne
Høffdings enestaaende Selvforglemmelse ved saadanne Lejligheder var for en stor Del Grunden til den Indflydelse, han øvede,
og den Vejledning til selvstændig Eftertanke, som han kunde
give, næsten uden at den, der modtog den, mærkede det.

Om Høffdings Overblik over menneskelige Tankebyg-
ninger og deres Vilkaar vidner fremfor noget hans Fremstilling af Filosofiens Historie, der Verden over har vundet
saa stor Paaskønnelse paa Grund af den Uhildethed og
Taalmodighed, hvormed han har søgt at trænge til Bunds
i de Forhold, hvorunder de store Tænkeres Værker opstod,
og forstaa hvad der laa dem inderst paa Sinde. Denne
Fremstilling faar ikke mindst sit Særpræg ved den dybe
Interesse for Naturvidenskabernes Udvikling og den Forstaaelse af deres Betydning for den almene Filosofi, som
Høffding lader komme til Orde. I hans allersidste Aar fandt
denne Indstilling et naturligt Udtryk i Høffdings forstaaende
Syn paa den Revision af Fysikkens Begrebsbygning, som
Aabningen af nye Erfaringsomraader har givet Stødet til,
og hvis Forhold til Erkendelsesteorien var Emnet for hans
sidste Meddelelse her i Videnskabernes Selskab. Med en
for hans Alder forbavsende Modtagelighed og Friskhed stillede
Høffding sig fuldt sympatisk overfor Fysikernes Bestræbelser
for at udvide Rammerne for Erfaringernes Indordning, og
det glædede ham i de nye Former at genkende Træk, som
han selv for Aar tilbage havde mødt og beskrevet i sine
psykologiske Studier navnlig i Forbindelse med etiske
Spørgsmaal. Ja, mange vil maaske først paa Baggrund af
den nye Belysning, hvori Atomteoriens Udvikling har stillet

Aarsagsproblemet, tilfulde vurdere det Klarsyn og den Takt
i Valget af Udtryksmidler, som Høffding i sin Etik lægger
for Dagen ved Omtalen af den gamle Gaade, som Viljens
Frihed bereder Tænkningen.

Den stadige Udvikling og Afklaring af Høffdings filosofiske Grundsyn, som fortsattes lige indtil Døden afsluttede
hans lange Liv, hang paa det nøjeste sammen med hans
særegne Arbejdsform og hele Tænkemaade. Hver Gang
Høffding her i Selskabet gjorde Rede for sine Studier
og derfor, som det laa i Emnets Natur, ofte fik Lejlighed til at udtale sig om Spørgsmaal, som han tidligere
havde behandlet, følte den opmærksomme Lytter, hvordan
der altid indføjedes nye Træk, og hvorledes Harmonien
i hans Synspunkter stadig afrundedes og uddybedes. At
komme og besøge ham i hans sidste Aar var hver Gang
en stor og rig Oplevelse. Til Trods for det Vemod, der
til Tider kunde gribe Høffding som Følge af de Bekym-
ringer hans Nærmestes svigtende Helbred beredte ham
og den tiltagende Ensomhed hans Ungdomsvenners Bort-
gang førte med sig, var det altid med en Følelse af at
være bragt ud af det daglige og med en ny Belæring om
Dybden og Skønheden i Tilværelsens Harmoni, at man
forlod ham. Usvækket bevarede han sin Kærlighed til alt,
hvad han havde lært at skatte af værdifuldt, og talte paa
sine allersidste Dage med ungdommelig Begejstring om
den Livets Poesi, som han fandt saavel hos Platon og
Spinoza som hos Shakespeare og Goethe. Det var vel til
syvende og sidst denne Kærlighed og Trofasthed, der gjorde
Høffding til den sande Filosof, der i vide Kredse efterlader
saa stort et Savn.

TRANSLATION

II. By **N. Bohr**

I am grateful for having been asked to say a few words tonight on the occasion of the commemoration of Harald Høffding by the Academy of Sciences and Letters, which was so dear to him. Nobody will expect me to give a detailed account of Høffding's personal development and scholarly endeavour such as the one we have just listened to with admiration. Based on the precious memories arising from my own relationship with Høffding, however, I would like to express briefly what his personality and his lifework have meant for large circles of scientists, whose studies have only indirect connection to philosophy as such.

My first reminiscences about Høffding stem from some evening gatherings, described by him in his "Recollections"[1]. About a generation ago a small circle of scientists met regularly in their homes and discussed all sorts of questions that had captured their interest. The other members of this circle were Høffding's close friends from their student years, Christian Christiansen and Vilhelm Thomsen as well as my father, who was quite a bit younger, but who during the years got on more and more intimate terms of friendship with Høffding. From the time that we were old enough to benefit from listening to the conversations and until the gatherings were interrupted on our part by the early death of my father, we siblings were allowed to be present when the meetings were held in our home, and from there we have some of our earliest and deepest impressions. During the often very vivid discussions, Christiansen especially enjoyed teasing Høffding in his humorous manner about general philosophy's aloofness from the real world. Like the others, however, he appreciated quite well the full extent to which Høffding's power of comprehension and desire for a general synthesis was so to speak the nourishing soil in which the ideas of the others germinated, stamped by their different studies and outlook[2].

The emphasis on the unity of science, symbolized by our Academy, was to Høffding not only an abstract matter of course but a practical necessity. Although he probably would have liked to characterize philosophy as the science of sciences, he was far from maintaining that philosophy, in a narrower sense, should be able to set the laws to which all scientific work must submit.

---

[1] [H. Høffding, Copenhagen 1928, pp. 171–174.]
[2] [The translation of this paragraph is taken from Vol. 6, pp. xx–xxii, with minor alterations.]

Høffding was always prepared for the possibility that essential features of the general problem of human knowledge might be elucidated in a new and fruitful way precisely through studies within more limited areas of science, where one more easily may find a comprehensive view of the mutual relationships between experiences. On the basis of this approach, his main effort was to utilize viewpoints formed in the various branches of science in order to throw light on general problems. The exceptionally broad perspective on the forms of scientific thinking thus acquired enabled him, in turn, to offer scientists working in special fields a lesson, which was all the more valuable because the continuously increasing branching-out of science makes it more and more difficult for those exploring the diverse fields of science to understand and learn from each other directly.

Although the general epistemological problem, in the sense just mentioned, was central to Høffding and over the years became even more so, it was, however, first and foremost in his psychological studies that he developed his characteristic working method and he gained the tools to deal with abstract questions. These studies formed the background for his work in other fields of philosophy, as Høffding has emphasized especially in his very last paper, a short article with the title "Psychology and Autobiography", which he wrote only a year ago and is to be published in an American anthology in the near future[3]. Here Høffding provides, in an exceedingly instructive and interesting manner, a survey of his work as a writer and of the fundamental attitude on which it rests. It was an unforgettable experience to hear him, during the writing of this article, speak of his long life of work, and to sense how he, in recalling the personal satisfaction this had given him, found the strength to endure the deep sorrows life brought him, especially in his last years.

The account of psychology developed by Høffding as a basis for his introductory lectures, which he appropriately called "Outlines of Psychology on the Basis of Experience"[4], was the work that brought Høffding, for the first time, into contact with a wider circle of readers with scientific interests. With an attraction and a conviction of its own, which are first and foremost due to the author's respect for the greatness of the subject, this work has attained a distribution and retained a vitality which Høffding himself had hardly dreamt

---

[3] [H. Høffding, *Psychology and Autobiography* in *A History of Psychology in Autobiography* (ed. C. Murcheson), Vol. 2, Clark University, Worcester, Massachusetts 1932 (republished by Russell & Russell, New York 1961), pp. 197–205. The article was published posthumously in Danish (see p. [325], ref. 1).]

[4] [H. Høffding, *Psykologi i Omrids på Grundlag af Erfaring*, Gyldendal, Copenhagen 1882. Translated into English as *Outlines of Psychology*, Macmillan, London 1891.]

of when he wrote it almost 50 years ago. In it one should seek neither a literary account of intellectual currents nor penetrating explanations of normal and pathological states of mind, but one will find a sober and, at the same time, enthusiastic account of, in the best sense of the word, a scientific approach to the phenomena of consciousness. A prominent feature is the striving to maintain the balance between analysis and synthesis, and it is never forgotten that although the whole is composed of individual parts, these in their turn appear in the light of the whole. This objectivity, peculiar to Høffding's account of psychology, has certainly been more important for many listening to his lectures and among the even more numerous readers of his book than any one of us is able to explain easily. This has struck me especially in my contacts with students from more recently founded universities where there is not a tradition like that of an old institution as ours and where one can meet a limited scientific view due to the lack of insight into the fundamental problems of psychology of the kind that Høffding's pupils received directly.

Incidentally, I personally came into direct scholarly contact with Høffding for the first time through some discussions concerning a philosophical field which, considering Høffding's whole approach and aptitude, must have been at the borderline of his interests, namely formal logic. Although Høffding himself hardly considered the short exposition of logic he used in his lectures to be an important part of his writings, it is an instructive book in which, characteristic of his whole way of thinking, the vivid background is still general psychological experience, even in the discussion of the systematics of logical propositions. To a young man, whose main interest was the mathematical treatment of problems in science, however, the question of stringency of the systematics was more in the foreground. I shall never forget some evenings, more than 25 years ago, when I was allowed to visit Høffding in his old home in Strandgade and talk to him about these questions and when he, with his proficiency in the history of human thought in all its subtle nuances, listened with indescribable kindness and patience to hear whether, in the young student's considerations coloured more by enthusiasm than understanding, there should be the smallest scrap of something new to him in regard to science or learning. Høffding's unique way of forgetting himself on such occasions explains to a large extent the influence he had on others and the guidance to independent reflection he could give without the receiving person noticing it.

Above all it is Høffding's account of the history of philosophy that bears witness to his broad perspective on man's thought constructions and their preconditions. All over the world this work is held in high esteem because of the impartiality and patience with which he has sought to get to the bottom of the conditions under which the works of the great thinkers were created and to understand their deepest intentions. This account gains its distinctive mark particularly from Høffding's deep-going interest in the development of the natural sciences and his understanding of their importance to philosophy in general. In his very last years this approach found a natural expression in Høffding's sympathetic view on the revision of the conceptual framework in physics caused by the opening of new fields of experience. The relation of this revision to the theory of epistemology was the subject of his last lecture here in the Academy. With a sensitivity and freshness astonishing for his age, Høffding was fully sympathetic to the endeavours of the physicists to widen the framework for the incorporation of experience, and in the new forms he was pleased to recognize features he himself had met many years earlier and had described in his psychological studies, especially in connection with questions of ethics. Indeed, it is perhaps only in the new light shed on the causality problem by the development of atomic theory that many will fully appreciate the clarity and tact in the choice of expressions shown by Høffding in his ethics when discussing the old riddle presented to our thinking by the idea of the free will.

The continuous development and clarification of Høffding's fundamental philosophical view, which continued right until death ended his long life, was very closely connected with his characteristic working method and his whole way of thinking. Every time Høffding gave an account of his studies in this Academy and hence, as is implicit in the nature of his subject, often commented on problems he had discussed earlier, the attentive listener could feel that new features were always being added and that the harmony of his views was continuously being rounded off and deepened. To visit him during the last years of his life was a great and rich experience every time. In spite of the sadness that now and then seized Høffding because of worries over the ill-health of his nearest and dearest and growing loneliness owing to the death of the companions of his youth, one always took leave of him with a feeling of being freed from the daily grind and with a new lesson about the depth and beauty in the harmony of life. His love of everything he had learnt to value remained unabated, and in his very last days he spoke with youthful enthusiasm of the poetry of life to be found in Plato and Spinoza as well as in Shakespeare and Goethe. When all is said and done, it may be this love and faithfulness that made Høffding the true philosopher who in large circles now leaves so great a void.

III. HARALD HØFFDING'S 100TH BIRTHDAY

HARALD HØFFDINGS 100-AARS FØDSELSDAG
Overs. Dan. Vidensk. Selsk. Virks. Juni 1942 – Maj 1943, pp. 57–58

Communication to the Royal Danish Academy on 19 March 1943

TEXT AND TRANSLATION

See Introduction to Part III, p. [297].

[323]

Præsidenten forelagde en Artikel »*Psykologi og Autobiografi*«, skrevet af HARALD HØFFDING i hans sidste Aar og hidtil ikke offentliggjort paa Dansk. Trykt i Filos. Medd. Bd. II, Nr. 3.

Præsidenten udtalte ved denne Lejlighed bl. a. følgende:

»Den 11. Marts var 100 Aar forløbne, siden HARALD HØFFDING blev født, og ved denne Lejlighed blev hans store, for dansk Videnskab og Aandsliv saa betydningsfulde Gerning mindet paa forskellig Maade. Navnlig blev der ved en Højtidelighed afsløret en Mindeplade, indsat i Muren paa den Bygning, der fornylig er opført paa det Sted, hvor den gamle Høffdingske Gaard stod, og hvor Høffding fødtes og tilbragte Barndommen og sin første Ungdom, hvorunder den for hans senere Livsværk afgørende Udvikling allerede fandt Sted. Fra Videnskabernes Selskab blev der om Morgenen samme Dag henlagt en Krans ved den smukke Sten, som Carlsbergfondet har anbragt paa hans Grav paa Vestre Kirkegaard. Paa dette vort første Møde efter 100-Aarsdagen er det naturligt, at vi mindes Harald Høffdings Livsværk og alt, hvad han har betydet for vor Kreds. I de Mindetaler, der efter hans Død blev holdt her i Selskabet og andetsteds, er der dog allerede fremdraget saa meget til Belysning heraf, at det vil være vanskeligt at tilføje noget nyt.

Det træffer sig imidlertid saadan, at Høffding har efterladt et uoffentliggjort dansk Manuskript til en Artikel, som han i sit sidste Leveaar efter Opfordring skrev til et amerikansk Værk, med kortere Autobiografier af fremragende Psykologer, i hvilket Artiklen optoges i engelsk Oversættelse. Dette sidste Arbejde fra Høffdings Haand indeholder mange interessante Oplysninger og Betragtninger og vidner om den Styrke og Klarhed, som han bevarede til sine allersidste Dage. Paa Grund af de særlige Forhold ved dens Fremkomst er Artiklen kun forblevet lidet kendt, og en Offentliggørelse af det danske Manuskript i vore Publikationer turde derfor blive hilst med Glæde i større Kredse, samtidig med at det vil give Selskabet Lejlighed til paa en smuk Maade at vise sin Taknemmelighed imod Harald Høffdings Minde. Ved en Oplæsning af Manuskriptet vil vi tillige endnu en Gang faa Lejlighed til at høre Høffdings Ord, som gennem mange Aar saa ofte har lydt her i Selskabet.«

58

[324]

TRANSLATION

The *President* submitted an article, "Psykologi og Autobiografi" [Psychology and Autobiography], written by HARALD HØFFDING in his last year and hitherto not published in Danish. Printed in Filos. Medd. Vol. 2, No. 3.[1]

The *President* said on this occasion among other things the following:

"On 11 March 100 years had passed since HARALD HØFFDING was born and on this occasion his great deeds, so significant for Danish science and intellectual life, were commemorated in various ways. In particular, at a ceremony a memorial plaque was unveiled in the wall of the building recently erected on the site of the old Høffding family house, where Høffding was born and spent his childhood and his first youth, during which time the development decisive for his lifework already took place. On the same day in the morning a wreath from the Royal Academy was laid in front of the beautiful stone placed on his grave at Vestre Kirkegaard[2] by the Carlsberg Foundation. At our first meeting after the centenary it is natural that we commemorate Harald Høffding's lifework and all he has meant for our circle. In the commemorative speeches given here in the Academy and elsewhere after his death, so much, however, has been brought to our attention to elucidate his work that it would be difficult to add anything new.

However, it so happens that Høffding has left an unpublished Danish manuscript for an article which he wrote on request in his last year for an American volume of short autobiographies of eminent psychologists, where the article was included in English translation[3]. This last paper from Høffding's hand contains much interesting information and reflection and bears witness to the strength and clarity he retained to his very last days. Due to the special circumstances regarding its publication, this article has remained only little known, and the publication of the Danish manuscript in our series of journals should, therefore, be welcomed with pleasure in wider circles and will at the same time give the Academy the opportunity of showing its gratitude to the memory of Harald Høffding in a fitting way. By having the manuscript read aloud we will also once again have the opportunity of listening to Høffding's words which for many years so often have sounded here in the Academy."

---

[1] [H. Høffding *Psykologi og Autobiografi* (with a foreword by Frithiof Brandt), Kgl. Dan. Vid. Selsk. Filos. Medd. **II**, 3 (1943) 22 pp.]

[2] [Literally, "Western Churchyard".]

[3] [See p. [320], ref. 3.]

PART IV

HISTORICAL PAPERS

INTRODUCTION

by

DAVID FAVRHOLDT

All the articles reproduced in Part IV were written for special occasions and have historical content. Bohr wrote a great many obituaries and commemorative speeches on his own initiative or on request. The commemorative speeches printed here are selected on the basis both of their richness of detail and their scope. Apart from the article on H.C. Ørsted, they all contribute to elucidating Bohr's own scientific efforts and how he perceived the historical perspective in which his own work was developed. The major article in this connection is "The Rutherford Memorial Lecture 1958"[1], which, aside from drawing a portrait of Rutherford as a scientist and human being, contains autobiographical elements in describing Bohr's own development of the atomic theory. The articles about the discoveries of the Dutch physicist Pieter Zeeman[2] (1865–1943) and the Swedish physicist Johannes Robert Rydberg[3] (1854–1919) contain supplementary material. So do the articles on "The Genesis of Quantum

---

[1] N. Bohr, *The Rutherford Memorial Lecture 1958: Reminiscences of the Founder of Nuclear Science and of Some Developments Based on his Work*, Proc. Phys. Soc. **78** (1961) 1083–1115. Reproduced on pp. [383]–[415].

[2] N. Bohr, *Zeeman Effect and Theory of Atomic Constitution*, "Zeeman, Verhandelingen", Martinus Nijhoff, The Hague 1935, pp. 131–134. Reproduced on pp. [337]–[340].

[3] N. Bohr, *Rydberg's Discovery of the Spectral Laws*, Lunds Universitets Årsskrift N.F., Avd. 2, Vol. 50, no. 21 (1955) 15–21 (Proceedings of the Rydberg Centennial Conference on Atomic Spectroscopy). Reproduced on pp. [373]–[379].

Mechanics"[4] and about the Solvay Meetings[5]; these two articles present aspects of the discussion about the status of quantum mechanics which occupied Bohr so strongly from 1927 throughout his life.

Bohr was very interested in history in general, and history of science in particular. He took great care in the writing of the articles presented here. H.C. Ørsted's achievements in science and in society[6] were so to speak part of Bohr's cultural heritage, but in addition Bohr had first-hand knowledge of all of Ørsted's scientific works because already between 1914 and 1920 he had assisted Kirstine Meyer (1861–1941) – the first Danish woman to obtain a doctorate in physics, a leading educator and historian of science, and a close friend of Bohr's parents – with the publication of the large three-volume edition of Ørsted's scientific papers[7].

In preparing "The Rutherford Memorial Lecture", Bohr was assisted in particular by Léon Rosenfeld and Jørgen Kalckar, but he also gave many of the scientists who had worked with Rutherford over the years the opportunity to suggest alterations and additions to the manuscript before the final version was sent to press. When the first version of the article was finished in March 1961, Bohr sent copies to William Lawrence Bragg, James Chadwick, John D. Cockcroft, Charles G. Darwin, Jesse W.M. Dumond, George Gamow, Ernest Marsden and George Hevesy. They all replied, and Bohr conducted a lengthy correspondence with some of them[8]. This led to several minor changes in the article, especially as regards dates and names.

Of special interest is a suggestion from Darwin for a new paragraph in the article, with which Bohr did not agree. In a letter from September 1961, Darwin writes:

---

[4] N. Bohr, *The Genesis of Quantum Mechanics* in *Essays 1958–1962 on Atomic Physics and Human Knowledge*, Interscience Publishers, New York 1963, pp. 74–78. This volume is photographically reproduced as *Essays 1958–1962 on Atomic Physics and Human Knowledge, The Philosophical Writings of Niels Bohr, Vol. III*, Ox Bow Press, Woodbridge, Connecticut 1987. Bohr's article is reproduced in this volume on pp. [424]–[428].

[5] N. Bohr, *The Solvay Meetings and the Development of Quantum Physics* in *La théorie quantique des champs, Douzième Conseil de physique tenu à l'Université Libre de Bruxelles du 9 au 14 octobre 1961*, Interscience Publishers, New York/London and R. Stoops, Bruxelles 1962, pp. 13–36. Reproduced on pp. [431]–[454]. Also published in *Essays 1958–1962*, ref. 4, pp. 79–100.

[6] N. Bohr, *Hans Christian Ørsted*, Fys. Tidsskr. **49** (1951) 6–20. Reproduced on pp. [342]–[356] (Danish original) and [357]–[369] (English translation).

[7] *H.C. Ørsted Naturvidenskabelige Skrifter, Vols. I–III* (ed. K. Meyer), Andr. Fred. Høst & Søn, Copenhagen 1920. Bohr is thanked in the foreword for checking the manuscript for the introductory article in English as well as for reading the proofs.

[8] The letters, deposited in the NBA, are not yet microfilmed.

"Now I turn to the point I raised with you the other day, about drawing more emphatic attention to the start of your theory, in particular to the hydrogen spectrum. My impression of your treatment ... is that it is too unconcentrated, so to speak, with all the difficulties being brought out all the time.

Darwin to Bohr,
11 Sep 61
English
Full text on p. [463]

I am quite certain that no one can write some one else's paper for him, and I shall not be surprised or hurt if you entirely reject my suggestion, but I think that you might consider putting in a paragraph something like this, ...

'When a new theory is proposed it is natural that its discoverer should have the whole field of knowledge in his mind including all the difficulties he has to explain, and in the present case many of these difficulties will be reviewed below here. But there is usually a spark that starts the discovery – for example in the discovery of the nucleus itself it was Rutherford's idea of *single* scattering. In the present case, the spark was the hydrogen spectrum, and the discovery was that by two separate applications of the quantum principle, this spectrum might be explained.

Firstly, if the circular orbit of the electron round the hydrogen nucleus had its angular momentum a multiple of the quantum it would fix the size of the atom about right. Second, there was the idea that one must think not about the frequencies of the emitted light, but about the spectral terms of Ritz, and if these were taken as the energies of the quantised hydrogen atom the result would give the Rydberg constant. When it came out that this constant was yielded correctly in terms of already known atomic constants without any new constant having to be invoked, it was hardly possible to doubt that this must be a quite fundamental principle of nature, but it remained to see how far it might be extended.'

As I have said, nobody can write another man's paper, and I shall not be surprised if you reject this. But it does fit in with my memory of the way your theory first struck me when I heard about it from you, and I hope you will consider whether something emphatic of the kind might not help the whole paper."

Bohr answers later in the month:

"Needless to say that I have considered the matter very closely, but I find it difficult without changing the style to introduce such changes.

Bohr to Darwin,
20 Sep 61
English
Full text on p. [464]

Indeed throughout the whole lecture I have striven to use the opportunity to revive the development in a factual and detached manner without entering into details which can be found in textbooks. It has been quite a difficult task, and I have often been scared at the length of the manuscript, resulting from the many points which presented themselves to give the whole story a reasonable balance."

Bohr's reluctance to follow Darwin's suggestion is consistent with his recollection, confirmed by subsequent historical work[9], that spectroscopical data such as Balmer's formula, Ritz's formula and Rydberg's constant entered into his considerations at a late stage, when he had more or less given his own ideas of atomic structure their fully developed form. Rosenfeld relates that at some point between Bohr's two letters to Rutherford of 31 January – where he writes:

Bohr to Rutherford,
31 Jan 13
English
Full text in Vol. 2, p. [579]

"I do not at all deal with the question of calculation of the frequencies corresponding to the lines in the visible spectrum."

– and of 6 March 1913[10], Bohr's colleague Hans Marius Hansen had asked Bohr[11]:

"... how on his theory he would account for the spectral regularities. Bohr had so far not been interested in this aspect of the question, because he thought those spectra too complicated to give any clue to the structure of the atomic systems. At this stage Hansen was able to contradict him by calling his attention to the great simplicity with which the spectral series had been represented by Rydberg."

When Bohr and Hansen looked up Rydberg's work, Bohr's reaction was, according to Rosenfeld[12]: " 'As soon as I saw Balmer's formula', he told me more than once, 'the whole thing was immediately clear to me.' " In a footnote,

---

[9] See Ulrich Hoyer, *Introduction* to "Part II, Constitution of Atoms and Molecules" in Vol. 2, pp. [103]–[143], quotation on p. [110]. Hoyer refers to L. Rosenfeld, *Introduction* in N. Bohr, *On the Constitution of Atoms and Molecules. Papers of 1913 reprinted from the Philosophical Magazine with an Introduction by L. Rosenfeld*, Munksgaard Ltd., Copenhagen and W.A. Benjamin Inc., New York 1963, pp. IX–LIV.

[10] Letter from Bohr to Rutherford, 6 March 1913. Full text in Vol. 2, p. [581].

[11] Rosenfeld, *Introduction*, ref. 9. Quotation on p. XL.

[12] *Ibid.*, p. XXXIX.

Rosenfeld adds[13]: "I have this intelligence from a conversation with Bohr on the 23rd June 1954, of which I took some notes".

In "The Rutherford Memorial Lecture"[14], Bohr says: "... it struck me in the early spring of 1913 that a clue to the problem of atomic stability directly applicable to the Rutherford atom was offered by the remarkably simple laws governing the optical spectra of the elements." This came to be a strong argument for the theory, but it was not the spark that started the discovery. Bohr's answer to Darwin bears witness to the carefulness with which he worked on his historical papers.

Finally it must be noted that the last draft of the article included footnotes, which, however, Bohr chose to leave out because he felt that he did not have the expertise for treating the subject satisfactorily from a historical point of view[15], a fact which is indeed confirmed by the rather arbitrary character of the footnotes. Since the footnotes were nevertheless fully prepared, the editors have found it appropriate to include them in facsimile[16].

[13] *Ibid.*, p. XL.

[14] Bohr, *Rutherford Memorial Lecture*, ref. 1. Quotation on p. [388].

[15] Jørgen Kalckar, professor at the Niels Bohr Institute and a close collaborator of Bohr for many years, recalls Bohr's explanation to this effect.

[16] This volume on pp. [416]–[420].

I. ZEEMAN EFFECT AND THEORY OF ATOMIC CONSTITUTION

"Zeeman, Verhandelingen", Martinus Nijhoff, The Hague 1935, pp. 131–134

See Introduction to Part IV, p. [329].

Pieter Zeeman and Niels Bohr. Amsterdam 1925.

ZEEMAN EFFECT AND THEORY OF ATOMIC CONSTITUTION

by N. BOHR, Copenhagen

Z e e m a n's discovery in 1897 that the structure of the lines of emission and absorption spectra is modified when the emitting or absorbing substance is placed in a magnetic field may truly be said to have inaugurated a new epoch in the development of atomic theory. Not only did it give a most decisive confirmation of the theoretical view, which on the basis of F a r a d a y's and M a x-w e l l's work was developed especially by L o r e n t z, that the optical properties of substances originate from the motion of electric particles within the atoms, but it offered for the first time a source of direct information about the nature of these particles. Indeed the remarkable agreement between the general features of the Z e e-m a n effect and the predictions of L o r e n t z's calculations was hardly more impressive than the close coincidence of the value of the ratio between charge and mass of the intra atomic particles deduced by this theory from Z e e m a n's measurements, with the value of the corresponding ratio for the electrified corpuscles then newly discovered in experiments on cathode rays, and now known as electrons. What this coincidence meant for the recognition of the electron as a fundamental atomic constituent was always emphasized by J. J. T h o m s o n who in the subsequent years more than anyone contributed to the development of a general electron theory of matter.

No less important than the role of the Z e e m a n effect in the foundation of the electron theory of atomic constitution is the guidance it has continually offered in the stepwise progress of this theory. The further investigation of the magneto-optical phenomena which led to the discovery of Z e e m a n patterns of a more compli-cated type than could be explained by L o r e n t z's theory, should thus soon reveal the essential insufficiency of the classical fundament of electron theory in accounting for the details of spectral phenomena. Especially the remarkable relationship, pointed out by P r e s t o n, between the regularities of spectral series and the types of Z e e- m a n patterns of the series lines showed clearly the intimate connection of the origin of these patterns with the then totally obscure mechanism of emission of line spectra. At the same time, the peculiar similarity, noticed by R u n g e, between certain features in all intricate Z e e m a n patterns and the so-called normal triplets predicted by L o r e n t z's theory, as well as the gradual transformation of all „anomalous" Z e e m a n patterns into such triplets by increasing strength of the magnetic field, discovered by P a s c h e n and B a c k, held out the promise of a future solution of these riddles on the basis of electronic theory.

Any hope of attaining this aim by means of suitable assumptions regarding the nature of the intra-atomic forces had to be abandoned, however, after R u t h e r f o r d's discovery in 1911 of the atomic nucleus, which in so unsuspected way completed our picture of the atom. While the nucleus atom from the beginning has provided an unerring guidance in disentangling the wonderful phenomena of radio-activity and transmutation of elements, its deficiency as regards the spectral phenomena, when treated on the basis of the classical electron theory, was so obvious that it at once suggested the necessity of a radical departure from the ordinary ideas of electro-dynamics and led to the attempt of basing the problem of spectral emission on the non-classical element in physics disclosed by P l a n c k's discovery of the quantum of action, which in the hands of E i n s t e i n already had proved its fertility in the explanation of the photo-electric effect. While this view point offered an im-mediate interpretation of the R y d b e r g-R i t z combination principle which governs the series spectra and had resisted any classical explanation, it was for a long time in no way clear how the anomalous Z e e m a n patterns were to be understood. Thus neither the more primitive efforts in building up a quantum theory of atomic constitution relying on a limited use of classical pictures, nor the gradually established proper quantum mechanical methods, so powerful in many other respects, seemed, to begin with, to leave more room than the classical theory for the appearance of any type of Z e e m a n patterns but the normal L o r e n t z triplet. In fact, the agreement between the results of these methods regarding the

Z e e m a n effects and the consequences of ordinary electron theory embodied in the well-known theorem of L a r m o r offered an apparently unambiguous example of the so-called correspondence principle, which characterizes the way in which classical concepts, notwithstanding their limitation, are upheld in quantum theory.

Just this situation stimulated a more searching study of the Z e e-m a n patterns, which above all revealed the possibility of a complete analysis of these patterns on the basis of the general principle of combination of spectral lines, consistent with the fundamental postulates of the quantum theory of atomic constitution. The essential soundness of this analysis, especially due to S o m m e r-f e l d and L a n d é, was also most strikingly born out by the beautiful experiments of S t e r n and G e r l a c h on the deflection of molecular rays in a magnetic field, as well as by the investigation of the other remarkable magneto-mechanical effects predicted by R i c h a r d s o n, E i n s t e i n and d e H a a s, and B a r n e t t. Indeed, through all this work the ground was gradually prepared for the next fundamental departure from classical electron theory, symbolized by the idea of electron spin. This new development was initiated by the establishing of the general exclusion principle, to which P a u l i was led just by the analysis of the P a s c h e n-B a c k effect, and received a provisional completion through the introduction by U h l e n b e c k and G o u d s m i t in the theory of atomic constitution of the picturesque idea of the spinning electron with intrinsic magnetic moment, which, in the sense of the correspondence argument, offered so strikingly simple an interpretation of the essential features, not only of the anomalous Z e e m a n patterns, but also of their transition to the normal triplet with increasing field strength. A quite rational solution of the problem of the electron spin was finally given by the ingenious theory of D i r a c, which at the same time brought about a most remarkable completion of electron theory by the prediction regarding the appearance under suitable conditions of electron pairs of opposite charges, so brilliantly confirmed by later experimental discoveries.

D i r a c's theory of the electron does not rest on any explicit assumption of a proper electronic magnetism, but like the whole wealth of spectral phenomena all details of Z e e m a n patterns appear as direct consequences of the modifications, non-vizualizable by mechanical models, which the existence of the quantum of action impose on classical electron theory. In this connection, the old riddles of the Z e e m a n effect offer an especially instructive illustration of the essential limitation of the use of space-time pictures in quantum mechanics, which the formulation of the uncertainty principle by H e i s e n b e r g has illuminated so clearly. Just like E i n s t e i n's theory of general relativity, which has given such a profound harmony to classical physics, this latest development is based on an analysis of the conclusions which can be drawn from direct observations. While the relativity theory is concerned with the dependence of the interpretation of measurements on the choice of the space-time system of reference, we meet in quantum theory with an entirely new situation brought about by the inevitable interaction between objects and measuring instruments. Since according to the nature of observations this interaction is essentially uncontrollable, it implies a novel feature of mutual exclusion between the unambiguous use of space-time concepts and of dynamical conservation laws, which replaces the classical ideal of causality by the broader view point of complementarity.

Z e e m a n's discovery has thus been an invaluable guide at all stages of modern atomic theory, from the time of the first recognition of the electronic constitution of matter to the recent elucidation of the inherent limits of the methods of classical physics in their treatment of the behaviour of the electrons bound in atoms. Its importance, however, is in no way restricted to this field, but the study of the magnetic influence on the finest structure of spectral lines as well as on the most refined features of the magneto-mechanical effects has even allowed to draw important conclusions about the properties of atomic nuclei. This source of information promises indeed to be of the greatest help for our comprehension of the laws which govern the constitution of the nucleus of the atom, the exploration of which, through the ever increasing harvest of marvelous experimental discoveries, has opened in recent years quite new outlooks for physical science.

Eingegangen: 16 Feb. 1935

II. HANS CHRISTIAN ØRSTED

Fys. Tidsskr. **49** (1951) 6–20

Address on the Occasion of the Centennial of Hans Christian Ørsted's Death, given at the University of Copenhagen on 9 March 1951

TEXT AND TRANSLATION

See Introduction to Part IV, p. [330].

Hans Christian Ørsted (1777–1851). Drawing by I.V. Gertner (Lithograph).

Professor *Niels Bohrs* tale.

Det er en af de store skikkelser i videnskabens og vort samfunds historie vi mindes i dag. Ørsteds paavisning af forbindelsen mellem magnetisme og elektricitet blev en inden for fysikken skelsættende opdagelse og indledte en udvikling, der i saa overordentlig grad har fremmet vor erkendelse af sammenhængen i naturens lovmæssigheder og gjort det muligt at tage naturkræfterne i menneskets tjeneste i et omfang, der helt har forandret rammerne for vor daglige tilværelse. Inden

for det danske samfund, hvor Hans Christian Ørsted levede
og virkede, udøvede han en gerning, hvis betydning for samtiden og de følgende generationer har faa sidestykker ikke
alene her hjemme, men ogsaa hvorhen i verden vi retter vort
blik. Paa grund af emnets rigdom kan der slet ikke blive tale
om at give et fuldstændigt billede af Ørsteds virksomhed i
videnskabens og samfundets tjeneste, men jeg skal forsøge at
fremdrage nogle hovedtræk af baggrunden for hans livsværk.
og den menneskelige indstilling, hvorpaa dette hvilede.

Opdagelsen af elektromagnetismen har en tusindaarig for-
historie. Videnskabelig forskning er jo efter sit væsen at ligne
med en famlen sig frem i taage i ukendte egne. Først gennem
erfaringerne selv vinder vi holdepunkter for udviklingen af
nye forestillinger, der kan føre os til større overblik og videre
udsyn. Allerede i oldtiden kendte man magnetstene, der kunne tiltrække jern og hvis navn henviser til deres forekomst
i nærheden af byen Magnesia i Lilleasien. Tidligt havde man
ogsaa bemærket, at ravet, hvis græske navn er elektron, ved
gnidning faar evne til at tiltrække lette smaastykker af forskellige stoffer. Paa grund af den overraskende forskel fra de
livløse legemers sædvanlige opførsel blev disse fænomener
dengang opfattet som en art besjæling af magnetstenen og
ravet, der var virksom eller kunne vækkes i disse, en opfattelse, der endnu i dag genspejler sig i talemaader som »magnetisk tiltrækning« og »elektriserende virkning« mellem mennesker.

Det var først i renaissancetiden, hvor man opnaaede saa
afgørende fremskridt i oversigten over og forstaaelsen af de
mekaniske fænomener som tidligst kom til at spille en rolle
for udviklingen af menneskets værktøj, at der ogsaa gjordes
de første skridt til en nærmere undersøgelse af de elektriske
og magnetiske fænomener. Kompasnaalen havde ganske vist
længe været et af menneskenes mest nyttige hjælpemidler, men
at dens opførsel skyldes, at jordkloden selv var en kæmpe-
mæssig magnet, blev først forstaaet af Gilbert. Saavel ved
denne forskers eksperimenter med magneter som ved de berømte forsøg, som Guericke anstillede med de første større
elektrisermaskiner, vandtes mange vigtige oplysninger. Ved

8

studiet af de overraskende og undertiden uventet kraftige
virkninger, man især kunne opnaa ved udladning af den saakaldte Leidner flaske, drejede det sig dog dengang selv i
videnskabsmændenes kreds snarest om et interessant lege-
tøj, som det benyttes af børn den dag i dag.

Franklin, der baade som forsker og som samfundsborger
paa saa mange maader minder om Ørsted, var den, der ved
en indgaaende analyse af erfaringerne og gennem sine egne
sindrige forsøg bragte en klarhed, der skulle pege videre frem.
Opfindelsen af lynaflederen, Franklins berømte bedrift, hviler
paa hans redegørelse for de love, der gælder for elektriske
ladningers fordeling. Under indtryk af de forandringer hos
magnetnaale, som han iagttog ved stærke udladninger fra
Leidner flasker og i endnu større maalestok ved lynnedslag,
tumlede hans aand ogsaa med en mulig sammenhæng mellem
elektricitet og magnetisme. Men kortvarigheden af de omhandlede udladninger og deres vanskeligt kontrollerbare karakter
førte til, at hans iagttagelser kun daarligt lod sig reproducere,
saaledes at der kom til at raade tvivl om deres rigtighed i den
videnskabelige verden.

Førend tiden var moden til en klar erkendelse af forbindelsen mellem elektricitet og magnetisme, som den, der naa-
edes ved Ørsteds opdagelse, maatte der først ske en helt ny
udvikling, hvorved man kom til at raade over vedvarende og
kontrollerbare elektriske strømme. I den livløse verden, der
omgiver os, har vi rig lejlighed til at studere mange mekaniske foreteelser, medens de elektriske, der som vi nu ved, spil-
ler en afgørende rolle for alle naturfænomener, i almindelighed er skjult for vort blik. I de levende væsener forefindes
imidlertid organer, der i flere henseender udviser en lighed
med de elektriske apparater, vi nu benytter os af i teknikken,
og faktisk frembyder de nerver og muskler, som er redskaber
for livets udfoldelse, overmaade følsomme instrumenter for
at efterspore svage elektriske paavirkninger. Det var ogsaa
netop Galvanis berømte forsøg med frølaars sammentræknin-
ger, der satte Volta paa sporet efter berøringselektriciteten og
ledte ham til konstruktionen af de første kemiske strømkilder,
som han gav navnet af galvaniske elementer. Ved disses

hjælp opnaaedes i løbet af faa aar en rigdom af iagttagelser over elektrokemiske virkninger, som især i Davys haand førte til mange nye opdagelser paa kemiens omraade.

Nu aabnedes blikket for elektricitetens dybe forbindelse med mangfoldige naturfænomener, og mange tænkere ledtes til at reagere imod den fremhævelse af mekanikkens principper som grundlaget for al naturbeskrivelse, saaledes som det fandt udtryk hos forkæmpere for den saakaldte kritiske filosofi, der tog sit navn efter Kants hovedværk. I disse aar opstod den romantisk prægede naturfilosofi, hvor man i stræben efter at se en enhed i naturen gik saa vidt som Schelling, der medtog hele menneskehedens udviklingshistorie med dens etiske og æstetiske aandsrørelser. Ørsted, der i sine første studieaar var stærkt betaget af Kants filosofi, hvis konsekvenser han søgte at forfølge videre i sin disputats, blev hurtigt en begejstret tilhænger af naturfilosofien.

En af dem, der dybest har indlevet sig i Ørsteds værker og disses baggrund, var fysikeren Chr. Christiansen, hvem vi skylder en lærerig afhandling om Ørsteds naturfilosofi. Med sit ejendommelige jyske lune siger Christiansen i indledningen til denne afhandling, som forklaring paa Ørsteds beskæftigelse med vidtsvævende filosofiske tanker: »Paa en Maade var der ikke andet at gøre, naar han ikke vilde overtage en Bestilling ved et Apotek.« Dermed sigtes der jo til, at der paa den tid i Danmark ikke fandtes kemiske laboratorier uden for apotekerne og at de samlinger af fysiske instrumenter, som enkelte rigmænd havde anskaffet sig, snarere var raritetskabinetter end alment tilgængelige hjælpemidler for eksperimentel forskning.

Ørsteds dybe hang til saadan forskning fandt da ogsaa afløb saa snart han ved en midlertidig bestyrervirksomhed af Løveapoteket fik lejlighed til at anstille galvaniske forsøg. Ganske kort efter de første elektrokemiske opdagelser kunne Ørsted saaledes give et vigtigt bidrag paa dette omraade ved at vise, at de ved elektrolyse frembragte sure og alkaliske virkninger ved efterfølgende blanding fuldstændig ophævede hinanden. Ligeledes var det nære forhold, der paa Ørsteds første udenlandsrejse opstod mellem ham og Ritter, ikke alene

10

baseret paa fællesskab i filosofisk syn, men ligesaa meget paa
Ørsteds beundring for Ritters eksperimentelle arbejder. I disse
deltog han ivrigt under samværet med Ritter og paa sin videre
rejse virkede han paa energisk og uegennyttig maade for at
skaffe Ritter en paaskønnelse, som han i mange kredse havde
savnet.

I de følgende aar, hvor Ørsted efterhaanden opnaaede stedse
bedre betingelser for at udføre fysiske og kemiske forsøg, viste
han et utrætteligt og frugtbart initiativ. Maaske mest kendt
er hans undersøgelser over akustiske klangfigurer. Her viste
han sig som en maalbevidst og systematisk arbejdende eksperi-
mentator, men i sin stærke betoning af den indre sammenhæng mellem følelsen af glæde over musikens klange og den
geometriske skønhed af de fysiske fænomener vover han sig
somme tider endnu langt ud i naturfilosofiens dunkelhed. Helt
frigjort fra saadan filosofering er derimod Ørsteds kort før
elektromagnetismens opdagelse paabegyndte og senere i mange aar videreførte undersøgelser over vædskers og luftarters
sammentrykkelighed, hvor han ved en sindrigt udtænkt eks-
perimentalteknik opnaaede nye og betydningsfulde resultater.
Ligeledes maa vi mindes de rent kemiske undersøgelser, hvorved han fandt et nyt alkaloid i peberet, samt førtes til opdagelsen af aluminium, en bedrift, hvis forfølgelse han imidlertid paa grund af forholdene og sin arbejdsmaade, saaledes
som vi ogsaa i anden sammenhæng skal se eksempel paa,
overlod til andre.

Den store opdagelse, hvormed Ørsted for alle tider har indskrevet sit navn i videnskabens historie, gjordes som bekendt
i 1820. Her er vi i den heldige situation at kunne lade Ørsted
tale selv, idet han i en levnedsbeskrivelse, som udkom i 1827,
og hvori han efter tidens stil omtaler sig selv i tredie person,
har givet følgende beskrivelse af de begivenhedsrige dage:

»Aaret 1820 var det lykkeligste i Ø.'s videnskabelige Liv.
Det var i dette han opdagede Elektricitetens magnetiske Virkning. Han havde ifølge Tingenes store Sammenhæng allerede i
sine tidligste Skrifter antaget at Magnetisme og Elektricitet
frembringes ved de samme Kræfter. Denne Mening var iøvrigt

ikke nye; men tvertimod vexelviis antaget og forkastet igjennem mere end to Aarhundreder; men hidindtil var det ikke
lykket for Nogen af dem, der antoge Overeensstemmelsen, at
finde et afgjørende Beviis derfor. Man havde i alle Forsøgene
ventet at finde Magnetismen i Retningen af den elektriske
Strøm, saa at Nord- og Sydmagnetisme skulde forholde sig
enten ligefrem eller omvendt som positiv og negativ Elektricitet. Alle Forsøg havde viist at der paa denne Vei var intet
at finde. Ø. sluttede derfor, at ligesom et Legem, der gjennem-
trænges af en meget stærk elektrisk Strøm, udstraaler Lys og
Varme til alle Sider, saaledes kunde det ogsaa forholde sig
med den magnetiske Virkning, han deri forudsatte. De Erfaringer, man allerede i et Aarhundrede havde havt, at Lynild
havde forandret Polerne i Magnetnaale, den dog ei havde truf-
fen, bekræftede ham heri. Han bragtes først paa Tanken i Be-
gyndelsen af 1820, da han skulde handle om Gjenstanden i
en Række af Forelæsninger over Elektricitet, Galvanismus og
Magnetismus. Han havde opstillet Redskaberne til Forsøget
førend Forelæsningstimen, men kom ikke til at udføre det.
Under Forelæsningen oplivedes hans Overbeviisning saaledes,
at han tilbød sine Tilhørere strax at prøve Sagen. Udfaldet
svarede til Forventningen; men man erholdt ikkun en meget
svag Virkning, og ingen bestemt Lov kunde strax opdages
deri; kun saae man, at den elektriske Strøm, ligesom anden
magnetisk Virkning, gjennemtrængte Glas. Saalænge Forsøgene ikke vare mere talende, lod Erindringen om hvad der
var hendt *Franklin, Wilke, Ritter* o. fl. ham frygte, at han
ogsaa kunde være skuffet af Tilfældigheder. Bebyrdet, adskil-
lige Maaneder igjennem, med talrige Forretninger, vovede han
sig imidlertid ikke til videre Forsøg. Hvorvidt en vis Tilbøie-
lighed til at udsætte Foretagender, og benytte sine frie Øie-
blikke til at leve i Tankernes Rige, heri medvirkede, vil han
selv ikke let kunne afgjøre. I Juli Maaned tog han Forsøgene
for igjen, med en meget stor galvanisk Kjæde af Kobberkasser,
Zinkplader og fortyndet Syre, og havde til Vidner og deeltagen-
de Venner Commandeur og Navigations-Directeur *Wleugel* og
Etatsraad *Esmarch*. Man erholdt nu strax en meget stor Virkning, og prøvede denne under forskjellige Forhold; imidlertid

12

udfordredes mange Dages Forsøg før han kunde finde den
Lov, hvorefter Virkningen retter sig. Saasnart han havde fundet denne, iilte han at bekjendtgjøre sit Arbeide. Dette skete
ved et meget kort latinsk Program, paa to tættrykte Qvart-
blade, hvori han saaledes havde sammentrængt Beskrivelsen
af sine Forsøg, at man paa de mellemste to Sider omtrent vil
finde ligesaa mange Forsøg antydede, som der er Linier.«

Meddelelsen om opdagelsen gjorde overalt det dybeste indtryk, og mange steder gentog man Ørsteds eksperimenter og
optog ivrigt arbejdet paa det nye forskningsomraade, de havde
aabnet. Paa grund af afhandlingens knappe form maatte imidlertid mange af de resultater, der — som Ørsted beskriver
det — var at læse i hver af dens linier, unddrage sig læsernes
opmærksomhed. I virkeligheden er vi først for en menneskealder siden bleven i stand til fuldtud at vurdere den tanke-
rigdom og mesterskab, Ørsted havde lagt for dagen ved de
eksperimenter, hvormed han ikke alene klarlagde alle hoved-
punkter vedrørende den først opdagede virkning, men tillige
eftersporede talrige konsekvenser, der for samtiden maatte
staa som frugter af efterfølgeres arbejder. Vi skylder Kirstine
Meyers dybe historiske indsigt, og hendes evne til at efterspore
alle kilder, fremdragelsen blandt Ørsteds efterladte papirer af
indtil da ganske oversete optegnelser om de forsøgsrækker, der
danner baggrunden for meddelelsen om opdagelsen. Deri viser
Ørsted sig som en forsker, som kun kan sammenlignes med
de mest beundrede forbilleder i naturvidenskabens historie.
I optegnelserne gøres rede saavel for de tanker og forventninger, som Ørsted satte sig for at prøve, som for en fylde af iagttagelser, der peger langt udover, hvad nogen i den knappe
form uden nærmere vejledning kunne læse sig til.
I levnedsbeskrivelsen siger Ørsted videre:

»Forsøgene blev snart gjentagne i alle Lande, hvor der gives
Venner af Naturlæren. Og den største Belønning, en Opfinder
kan nyde, at see sin Opfindelse være Gjenstanden for den
flittigste Grandskning, see den udvidet og frugtbringende, blev
ham i fuldeste Maade til Deel. Tallet paa dem, som have skre-

13

vet over Elektromagnetismen, beløber sig til meget over hundrede. Ved saa Manges forenede Bestræbelser har derfor ogsaa denne Lære allerede faaet en Udstrækning, og en Rigdom paa Indhold, som overstiger hvad man skulde vente af det korte Tidsløb af 7 Aar.«

Efter en omtale af de hædersbevisninger, der straks efter opdagelsen blev ham til del fra mange lærde selskabers side tilføjer Ørsted paa en for ham saa karakteristisk maade:

»Under alle disse Bifaldsbevidnelser kunde Opfinderen ikke andet end dybt føle, at selv de Opdagelser, vi med Flid søge, dog ikke ene ere Frugten af vor Bestræbelse, men ere betingede af en Række af Begivenheder, og en Tilstand i den videnskabelige Verden, der staaer under en høiere Lov, end alle de han søger.«

Den udvikling, der sattes i gang med Ørsteds opdagelse, var rigtignok ganske enestaaende, og allerede inden aaret var omme, fremkom et stort antal meddelelser om forsøg, der bekræftede og videreførte den paa forskellig maade. Blandt de undersøgelser, der ved den skarpsindighed og mesterskab, hvormed de blev udført, med rette tiltrak sig den største opmærksomhed og beundring, kan nævnes Ampères paavisning af de indbyrdes tiltrækninger og frastødninger mellem strømledere og Biot og Savarts nøjagtige redegørelse for de magnetiske kræfter, som elektriske strømme frembringer i det omgivende rum. Dog, som det fremgaar af den afhandling Ørsted udsendte nogle maaneder efter den første meddelelse, men hvis fulde indhold vi først nu, efter fundet af hans optegnelser, rettelig kan vurdere, var Ørsted selv opmærksom paa ækvivalensen mellem sluttede strømkredse og magneter, hvis klare udredning saavel i eksperimentel som teoretisk henseende det er Ampères store fortjeneste at have givet.

Allerede en halv snes aar efter skulle imidlertid et nyt stadium i udviklingen indledes gennem Faradays opdagelse af den elektromagnetiske induktion, hvorved en endnu mere intim forbindelse mellem elektriske og magnetiske kræfter afsløredes. I nøje tilknytning til Faradays nyskabende forestil-

[349]

linger udviklede Maxwell en teori for elektromagnetiske bølgers forplantning i rummet, hvis konsekvenser fuldtud bekræftedes ved Hertz's berømte forsøg. Ikke alene blev derved grundlaget lagt for hele den moderne radioteknik, men tillige vandt man en forstaaelse af fænomenerne ved lysets udbredelse, hvorved hele det kapitel i fysikkens historie, der indledtes med Ole Rømers opdagelse af lysets hastighed, fandt saa harmonisk en afrunding. I denne forbindelse maa vi ogsaa tænke paa, at den danske fysiker Lorenz samtidigt med og uafhængigt af Maxwell udviklede en i vidt omfang overensstemmende elektromagnetisk lysteori. Ogsaa Ørsted selv gør i slutningen af meddelelsen om sin store opdagelse den vidtskuende bemærkning, at netop den fundne sammenhæng mellem retningerne af den elektriske strøm og de af denne frembragte magnetiske kræfter turde aabne en vej til at forklare lysets ejendommelige polarisationsegenskaber, som hidtil havde rummet en uløselig gaade.

Det, der maaske mest præger fysikkens senere udvikling, er den stadig klarere erkendelse af den indre sammenhæng mellem naturfænomenerne; herom har man drømt fra de ældste tider, og om noget var karakteristisk for hele Ørsteds indstilling, var det det gryende haab om denne drøms virkeliggørelse. I vore dage er et mægtigt fremskridt i denne henseende opnaaet gennem relativitetsteorien, der tog sit udgangspunkt i de overraskende erfaringer om, hvor forskelligt iagttagere, der bevæger sig i forhold til hverandre, vil beskrive de fysiske fænomener. Denne forskel gaar saa vidt, at begivenheder paa forskellige steder, der af een iagttager opfattes som samtidige, af en anden kan beskrives som følgende efter hinanden, og medfører endda, at visse virkninger, der af een iagttager vil opfattes som værende af ren elektrisk art, af en anden vil tilskrives tilstedeværelsen af baade elektriske og magnetiske kræfter. Denne erkendelse aabnede for Einstein vejen til at finde frem til en saa almindelig formulering af naturlovene, at de har gyldighed for enhver iagttager. Ved den mulighed, der saaledes opnaas for at sammenfatte træk, mellem hvilke hidtil ingen forbindelse kunne øjnes, har vort verdensbillede vundet en enhed af hidtil ukendt rækkevidde.

Denne udvikling gaar efter sin art langt ud over grænserne for den forestillingskreds, hvori fysikere og filosoffer paa Ørsteds tid bevægede sig. Forhaabningerne om at føre stoffernes saa rigt varierede egenskaber tilbage til et sammenspil mellem elektriske og magnetiske kræfter er imidlertid blevet virkeliggjort paa den mest eventyrlige maade gennem det indblik, vi i vort aarhundrede har faaet ved udforskningen af atomernes verden, som man endnu paa Ørsteds tid i almindelighed troede ville være lukket for bestandig for menneskelig erfaring. I elektronen har man lært en fælles byggesten af alle stoffers atomer at kende, og medens den elektriske strøm skyldes en bevægelse af elektroner igennem legemerne, genspejles elektronernes bindingsmaade i de enkelte atomer og molekyler i stoffernes kemiske og magnetiske egenskaber. Det er paa dette grundlag lykkedes at sammenfatte den store rigdom af erfaringer, som fysikere og kemikere i aarhundredernes løb har samlet, men løsningen af denne opgave var ikke mulig førend opdagelsen af det saakaldte virkningskvantum, hvormed Planck netop ved aarhundredeskiftet aabnede en ny epoke i naturvidenskabens historie.

Virkningskvantets eksistens er et udtryk for et træk af udelelighed i de atomare processer, der slet ingen plads kan finde inden for rammerne af den hidtidige naturbeskrivelse. Den for udvidelsen af disse rammer nødvendige omstilling har da ogsaa medført et nyt syn paa mange tidligere filosofiske anskuelser. Vi ser i dag forholdene mellem de forskellige filosofiske skoler paa en ny baggrund, hvor vi ikke fæstner os saa meget ved modsætninger, der tidligere ofte gav anledning til bitter strid, men er i stand til at værdsætte frugtbarheden af synspunkter, som fremhæver den ene eller den anden af de tilsyneladende uforenelige, men lige væsentlige sider af vor situation som iagttagere af den natur, som vi selv tilhører. Omend man nuomstunder vel vil tale paa noget anden maade end Ørsted gjorde om fornuften i naturen, kan vi netop paa den nye baggrund dybt værdsætte den for ham saa kendetegnende stræben efter at forene, hvad han kaldte det skønne og det sande.

Til billedet af Ørsted som naturforsker og tænker hører

16

uløseligt den indstilling til medmennesker, der prægede hans
virksomhed i det samfund, han tjente saa vel. Grunden for
denne indstilling blev jo allerede lagt i hans barndom, mens
han voksede op i apotekerhjemmet i Rudkøbing sammen med
sin kun et aar yngre bror, der ogsaa skulle komme til at udføre saa stor en gerning i det danske samfund. I barndoms-
byen var der dengang ingen skole, der kunne byde de bega-
vede drenge passende betingelser for uddannelse. Hans Christian og Anders Sandøe maatte derfor benytte sig af enhver
kilde til kundskabserhvervelse, og de fortalte hinanden om
alt, hvad hver især havde lært. Derved skabtes et fællesskab,
der fortsatte under deres studier i København og kom til at
vare livet ud, hvor forskelligt end deres arbejdsomraade og
virkekreds efterhaanden skulle blive.

Forholdene ved Københavns Universitet ved det nittende aarhundredes begyndelse var jo for naturvidenskabernes vedkommende yderst beskedne, og for fysik og kemi, hvor Ørsteds
interesser i første linie laa, var der hverken lærerstillinger
eller laboratorier. Alene igennem den tillid, som han i stedse
højere grad vandt og til hvilken han ved utrættelig og selv-
forglemmende indsats viste sig saa værdig, lykkedes det Ørsted trods de fattige tider efterhaanden at opbygge en videnskabelig undervisning i fysik og skabe forskningsmuligheder
for sig selv og sine medarbejdere. Blandt disse maa man i første række nævne Zeise, der ogsaa var apotekersøn og som
senere skulle deltage saa energisk i Ørsteds bestræbelser for
at højne den farmaceutiske uddannelse i Danmark. Det er
kendt, hvordan Ørsted til at begynde med kun kunne an-
bringe Zeise i et køkken i professorgaarden i Nørregade som
medhjælper ved sine kemiske arbejder, men hvordan det efterhaanden lykkedes ham at skabe et selvstændigt laboratorium,
hvori Zeise, der blev den første egentlige kemiprofessor i Danmark, kunne udføre sine navnkundige undersøgelser.

Dette er jo kun et enkelt, omend særlig lykkeligt eksempel
paa H. C. Ørsteds bestræbelser paa at skabe betingelser for
forskning og studium her hjemme. Hele historien om disse
bestræbelser fortæller om de krav, som et begejstret stræbende
menneske maa stille til sig selv for at nyskabe et millieu, og

om de evner, der fordres for at løse en saadan opgave. Naar
Ørsted kunne fængsle den tilhørerskare, der lyttede til hans
forelæsninger, laa det jo ikke alene i hans viden og fantasi,
men tillige i den foredragskunst, hvis opøvelse — som han
selv fortæller — ikke mindst var bygget paa studier og erfaringer under udenlandsrejser. Den anseelse og de venskaber,
som Ørsted vandt ude i verden, bidrog ogsaa til at skabe en
udvidelse af horisonten her hjemme. Ikke mindst laa forbindelsen paa kulturens omraade mellem de nordiske brødrefolk
Ørsted dybt paa sinde, og om hans deltagelse i de skandinaviske naturforskermøder, der begyndte paa hans tid, taler,
som ogsaa Universitetets rektor pegede paa, billedet i denne
festsal.

Det arbejde, H. C. Ørsted indlagde i sin universitetsgerning under stadig kamp med de fattige økonomiske forhold — vi er jo
allerede blevet mindet om, at de kritiske aar i hans liv faldt i
statsbankerottens tid — kalder paa vor største beundring. Aldrig svækkedes hans initiativ; tværtimod paatog han sig stadig
flere opgaver, som det næsten er ubegribeligt han kunne over-
komme. Om det overordentlige og opofrende arbejde, som Ørsted nedlagte i stiftelsen og ledelsen af Den polytekniske Lære-
anstalt, der nu under navnet Danmarks Tekniske Højskole
indtager en stadig mere betydningsfuld plads i samfundet, vil
Højskolens rektor fortælle os, men jeg vil gerne med endnu
nogle ord minde om Ørsteds virksomhed inden for Det Kongelige Danske Videnskabernes Selskab, for hvilket han i 36
aar virkede som sekretær. I denne egenskab lagde han i
Selskabets anliggender og dets forbindelse med udenlandske
selskaber et arbejde, som man maatte synes ville kræve et
enkelt menneskes fulde arbejdskraft, og læser man beretnin-
gerne om Videnskabernes Selskabs møder fra den tid, støder
man atter og atter paa Ørsteds navn saavel i forbindelse med
nyt initiativ angaaende forskellige sider af Selskabets virksomhed som med hyppige meddelelser ikke alene om resultater
fra hans egen forskergerning, men ogsaa om fremskridt paa de
mest forskellige omraader inden for naturvidenskaben, hvorom han følte det som sin opgave stadig at holde Selskabets
medlemmer underrettet.

18

Ogsaa det af Ørsted i 1824 stiftede Selskab for Naturlærens Udbredelse har en historie, der rummer gribende vidnesbyrd om, hvor vidt hans begejstring og ansvarsfølelse kunne føre ham. Paa tider, hvor man maatte synes han var fuldt optaget af sin forskning og videnskabelige undervisning, instruerede han personligt de lektorer, som Selskabet udsendte til danske provinsbyer for at bibringe haandværkere kundskaber af betydning for deres virksomhed og samtidig højne befolkningens dannelse. Selv om mange af de opgaver, som Selskabet under Ørsteds ledelse paatog sig, efterhaanden er overtaget af andre institutioner, af hvilke flere er skabt paa Ørsteds eget initiativ, lever traditionen fra hans tid videre i Selskabet. Siden H. C. Ørsteds død har Selskabet for Naturlærens Udbredelse da ogsaa betragtet det som en særlig pligt at værne om minderne om dets stifter, og allerede for en menneskealder siden, under Martin Knudsens formandskab, lagdes den første begyndelse til et H. C. Ørsted-museum, hvor man kunne samle de af hans instrumenter, der endnu er bevaret, og som omfatter selve den magnetnaal, der blev brugt ved elektromagnetismens opdagelse. Ved pietetsfølelse og gavmildhed fra Ørsteds slægt har museet ogsaa været i stand til at samle mange af Ørsteds ejendele og modtaget testamentarisk løfte om andre værdifulde minder fra hans liv, deriblandt det skrivebord, ved hvilket Ørsted har udarbejdet sine talrige afhandlinger og taler. Ved forstaaende imødekommenhed fra statsmyndighedernes side har det netop til denne mindedag været muligt paa Danmarks Tekniske Højskole at indrette et smukt lokale, der for fremtiden vil danne en værdig ramme om H. C. Ørsted-museets uerstattelige samling.

Sammenligner man H. C. Ørsted med andre store skikkelser i dansk videnskabs historie, som Tycho Brahe, Niels Steensen og Ole Rømer, falder baade ligheder og forskelle i øjnene. Hver paa sit omraade har de indledt en ny udvikling, der skulle præge naturvidenskaberne langt frem i tiden, og karakteristisk for dem alle er ogsaa den nøje forbindelse med samtidens videnskabelige verden i udlandet, hvorved de i saa høj grad bidrog til at skabe større udsyn her hjemme. Medens brydninger i nationens liv bidrog til at forme Tycho Brahes og Niels

Steensens ejendommelige, baade straalende og tragiske skæbner, inddroges under forholdsvis mere rolige tider Rømer ligesom Ørsted i en omfattende samfundsvirksomhed. Derved blev der ganske vist lagt saa stort beslag paa deres rige evner, at de i højere grad end andre af videnskabens foregangsmænd hindredes i selv at videreføre undersøgelser ad de veje, de havde banet.

At standse ved en beklagelse af saadanne forhold ville imidlertid være en alt for ensidig betragtningsmaade, naar vi tænker paa, hvor store og paatrængende opgaver for samfundets udvikling i praktiske henseender Rømer paa sin tid var i stand til at løse, og hvor tiltrængte og frugtbare Ørsteds bestræbelser var for udbygningen af grundlaget for naturvidenskabelige studier og forskning her hjemme og for spredningen af kendskab til denne forsknings resultater og dens betydning ud til de videste kredse af det danske samfund. Lader vi blikket gaa ud over verden, vil vi næppe i egentlige akademiske kredse finde et sidestykke til Ørsteds gerning. Langt snarere føres tanken til en mand som Franklin, hvis store fortjenester for videnskaben paa saa lykkelig maade hjalp ham i den virksomhed, han fra de tidligste aar udfoldede for fremme af folkeoplysningen i et nyt samfund, for hvis organisation og anseelse i verden han kom til at betyde saa meget.

Naar vi søger at danne os et billede af H. C. Ørsteds liv og virke, er det vel den dybe menneskelige samfølelse, der træder allerstærkest frem. Store byrder paatog han sig beredvilligt, naar det gjaldt samfundets tarv og medmenneskers vel, og løsningen af saadanne opgaver tillod ham en harmonisk og til hans indstilling saa vel svarende udnyttelse af hans evner. I sine bestræbelser for med sine medmennesker at dele alt, hvad der kunne styrke og berige deres aand, fandt jo Ørsted ogsaa felter langt ud over sit naturvidenskabelige oplysningsarbejde. Den indflydelse, han øvede inden for den stedse større kreds af digtere og skønaander, til hvem han var knyttet med slægtskabs- og venskabsbaand, danner et rigt kapitel i dansk aandslivs historie. Jeg skal dog ikke forsøge at gaa nøjere ind paa alt dette, hvorom vi om et øjeblik vil faa lejlighed til at høre fra kyndigste og nærmeste side.

20

Mange er vidnesbyrdene om den beundring og taknemme-lighed, hvormed H. C. Ørsted omfattedes saavel af samtiden som af eftertiden. Et mere virkningsfuldt udtryk for disse fø-lelser kan næppe anføres end de ord, J. C. Jacobsen for 75 aar siden knyttede til stiftelsen af Carlsbergfondet, hvorved der her hjemme skabtes saa langt gunstigere betingelser for viden-skabelig forskning end de, under hvilke Ørsted maatte virke:

»I levende Erkjendelse af hvormeget jeg skylder H. C. Ør-steds Lære og vækkende Indflydelse og som et Vidnesbyrd om taknemmelig Paaskjønnelse af hans Virksomhed for at ud-brede Kundskabens Lys i videre Kredse, har jeg knyttet Op-rettelsen af ovennævnte Stiftelse til denne Dag, som ved Af-sløringen af Ørsteds Monument er viet til hans Ihukommelse.«

Maatte det danske samfund i vore dage og i fremtiden vise sig den arv værdig, som Hans Christian Ørsteds skikkelse og gerning holder os saa lysende for øje.

TRANSLATION

Professor *Niels Bohr's* speech.

Today we are commemorating one of the great figures in the history of science and of our society. Ørsted's demonstration of the connection between magnetism and electricity was a milestone in the field of physics and gave rise to a development which to an extraordinary degree has furthered our understanding of the connection between the laws of nature and has made it possible for man to harness the natural forces to an extent that has completely changed the conditions of our daily existence. In the Danish society where Hans Christian Ørsted lived and worked, he pursued a task whose importance for his own time and later generations has few parallels – not only here at home but wherever in the world we choose to look. Because the topic is so rich, it is not possible to give a complete account of Ørsted's activities in the service of science and society, but I shall attempt to convey some main characteristics underlying his life's work and the humane attitude on which it was based.

The discovery of electromagnetism has a prehistory of a thousand years. By its very nature, scientific research can be compared with groping one's way forward in the fog in unknown territory. Only through experience itself do we establish a basis for the development of new conceptions which can lead us to a wider perspective and breadth of vision. Already in antiquity, magnetic stones that could attract iron were known; the name refers to their occurrence near the town of Magnesia in Asia Minor. In early times it was also noticed that amber, called electron in Greek, when rubbed acquires the capacity to attract light fragments of various substances. On account of the surprising divergence from the usual behaviour of lifeless matter, these phenomena were then seen as a kind of animation of the magnetic stone and the amber, which acted or could be awakened in them, a view reflected even today in common expressions such as "magnetic attraction" and "electrifying effect" between human beings.

It was only in the Renaissance, when decisive advances were made in the ordering and understanding of the mechanical phenomena that were first to play a role in the development of man's tools, that the first steps were also taken to examine electric and magnetic phenomena more closely. The compass needle had certainly been one of man's most useful aids for a long time, but Gilbert was the first to understand that its behaviour was due to the earth itself being a huge magnet. Much valuable information was obtained from this scientist's experiments with magnets, as well as from the famous experiments carried out by Guericke with the first large electrostatic generators. In the study of the surprising and sometimes unexpectedly powerful effects, which could be obtained in particular from the discharge of the so-called Leyden jar, it was then nevertheless, even in scientific circles, rather a question of an interesting toy, as it is used by children to this very day.

Franklin, who both as a researcher and a citizen in so many ways reminds us of Ørsted, was the one who, by means of a detailed analysis of experience and through his own ingenious experiments, brought about a clarification that would point the way ahead. The discovery of the lightning rod, Franklin's famous achievement, is based on his explanation of the laws governing the distribution of electric charges. Being impressed by the motion of magnetic needles that he observed during strong discharges from Leyden jars and, on an even greater scale, during lightning, his mind also grappled with the idea of a possible connection between electricity and magnetism. But the short duration of these discharges and the difficulty of subjecting them to control meant that his observations could only be reproduced with difficulty, a circumstance which caused doubt about their reliability in the scientific world.

Before the time was ripe for a clear recognition of the connection between electricity and magnetism, as obtained by Ørsted's discovery, a totally new development was necessary, whereby constant, controllable electric currents would become available. In the inanimate world around us we have ample opportunity to study many mechanical phenomena, whereas the electric ones, which, as we now know, play a decisive role in all natural phenomena, are usually hidden from our gaze. In living beings, however, there are organs which, in many respects, are similar to the electric devices we now use in technology, and the nerves and muscles, which are tools for the unfolding of life, provide us in fact with exceptionally sensitive instruments for tracing weak electric effects. And it was indeed precisely Galvani's famous experiment with the contractions of a frog's leg that put Volta on the track of contact electricity and led him to the construction of the first chemical sources of electric current, which he called galvanic elements. By means of these galvanic elements was obtained, in the course of a few years, a wealth of observations on electrochemical effects, which, especially in the hands of Davy, led to many new discoveries in the field of chemistry.

Now the profound connection between electricity and a whole range of natural phenomena became apparent, and many thinkers were led to react against the emphasis on mechanical principles as the basis for all description of nature, as it found expression with the advocates of the so-called critical philosophy, which took its name from Kant's major work. During these years the romantically inclined *Naturphilosophie* came into being, where, in striving to find a unity in nature, one went as far as Schelling, who included the entire history of humankind, with all its ethical and aesthetical intellectual movements. Ørsted, who during his first years of study was deeply impressed by Kant's philosophy, the consequences of which he sought to pursue further in his dissertation, quickly became an enthusiastic follower of *Naturphilosophie*.

One of those who familiarized themselves most deeply with Ørsted's works and their background was the physicist Christian Christiansen, to whom we are indebted for an instructive paper on Ørsted's *Naturphilosophie*[1]. With characteristic Jutland humour Christiansen, in the introduction to this paper, gives the following explanation of Ørsted's preoccupation with diffuse philosophical thoughts: "In a way there was nothing else to do, when he didn't want to take a post in a pharmacy". This of course refers to the fact that there were no chemical laboratories in Denmark at that time apart from the pharmacies and that the collections of physical instruments, which some wealthy men had acquired, were collections of curios rather than generally accessible tools for experimental research.

Indeed, Ørsted's deep inclination for this kind of research found its outlet as soon as he had the opportunity of carrying out galvanic experiments while working temporarily as manager of the Lion Pharmacy[2]. Quite soon after the first electrochemical discoveries, Ørsted was thus able to make an important contribution in this field by demonstrating that the acid and alkaline reactions, brought about by electrolysis, cancelled each other out completely after subsequent mixing. Similarly, the close relationship established between him and Ritter[3] during Ørsted's first journey abroad was based not only on a common philosophical view, but just as much on Ørsted's admiration for Ritter's experimental work. He eagerly took part in these experiments during the time spent with Ritter, and during the rest of his journey he made vigorous and unselfish efforts to secure Ritter the recognition he had lacked in many circles.

In the following years, when Ørsted gradually obtained ever improved conditions for carrying out physical and chemical experiments, he displayed indefatigable and productive initiative. Perhaps best known are his investigations concerning Chladni figures[4]. Here he showed himself to be a determined and systematic experimenter, but in his strong emphasis on the inner relationship

---

[1] [C. Christiansen, *H.C. Ørsted som Naturfilosof*, Overs. Dan. Vidensk. Selsk. Forh. 1903, 473–493.]

[2] [Pharmacies often bore the names of animals, linked to the alchemists' view of their various attributes. Thus the lion was a symbol of life-energy, of strength.]

[3] [Johann Wilhelm Ritter (1776–1810), German chemist.]

[4] [In 1807 Ørsted carried out a meticulous investigation of the sonorous figures discovered by the German physicist Ernst Chladni (1756–1827).]

[359]

between the feeling of joy on hearing musical sounds and the geometric beauty of the physical phenomena, he still sometimes ventures far out into the obscurity of *Naturphilosophie*. However, Ørsted's investigations of the compressibility of liquids and gases, begun shortly before the discovery of electromagnetism and continued for many years, are completely free of such philosophizing. Here he achieved new and important results by means of an ingeniously thought out experimental technique. We must also remember his purely chemical investigations which led to the discovery of a new alkaloid in pepper, as well as to the discovery of aluminium. However, because of prevailing circumstances and his own working method, which we shall also see exemplified in another connection, he left the following up of the latter achievement to others.

As is well known, the great discovery that inscribed Ørsted's name for all time in the history of science took place in 1820. Here we are in the fortunate position of being able to let Ørsted speak for himself, since, in an autobiographical account which appeared in 1827 and in which according to the style of the period he speaks of himself in the third person, he has given us the following description of those eventful days[5]:

"The year 1820 was the happiest in Ø.'s scientific life. It was in that year that he discovered the magnetic effect of electricity. In accordance with the great unity of all things he had, already in his earliest writings, assumed that magnetism and electricity are brought about by the same forces. Moreover, this opinion was not new; on the contrary, it had been alternately postulated and rejected for more than two centuries; but until then none of those who maintained such a unity had succeeded in providing irrefutable evidence. In all experiments, one had expected to find the magnetism in the direction of the electric current, so that north and south magnetism should correspond either to positive and negative electricity, or the opposite. All experiments in this direction had led to nothing. Ø. therefore concluded that, just as a body traversed by a very strong electric current radiates light and heat to all sides, this could also be the case for the magnetic effect that he assumed to be present. The observations, which had already been known for a century, that lightning changed the poles of a magnetic needle even without striking it, confirmed him in his belief. He first thought of the idea in the beginning of 1820, when he was going to speak on the topic in a series of lectures on electricity, galvanism and magnetism. He had

---

[5] [Translated from the original Danish in Hans A. Kofoed, *Conversations-Lexicon*, XXVIII, Copenhagen 1828, pp. 536–538.]

set up the tools for the experiment prior to the lecture, but did not have occasion to carry it out. During the lecture, his conviction was revived to the degree that he offered his audience to put it to the test immediately. The result met his expectations; but only a weak effect was obtained, and no particular law could be immediately discovered in it; one only saw that the electric current, just like any other magnetic effect, penetrated glass. As long as the experiments were no more telling, the recollection of what had happened to *Franklin, Wilke, Ritter* and others led him to fear that he too might have been deceived by chance. Burdened for several months with numerous duties, he did not, however, venture further experiments. Whether a certain tendency to put things off and to use his leisure time to dwell in the realm of ideas contributed to this state of affairs, he will not easily be able to decide himself. In the month of July he resumed the experiments with a very large galvanic chain of copper boxes, zinc plates and dilute acid. As witnesses and participating friends he had Commander and Navigation Director *Wleugel* and Titular Councillor of State *Esmarch*. One now immediately obtained a very great effect and tested this under various conditions; even so, it took many days of experiments before he could determine the law which governs the effect. As soon as he had found it, he hastened to make his work known. This was done in a very brief statement in Latin, consisting of two densely printed quarto sheets, into which he had compressed the description of his experiments so that on the two middle pages there are about as many experiments alluded to as there are lines."

The announcement of the discovery made a deep impression everywhere, and Ørsted's experiments were repeated in many places and work was eagerly begun in the new field of research they had opened up. Because of the condensed form of the paper, however, many of the results which – as Ørsted describes it – could be read in each line escaped the attention of the readers. In fact, not until a generation ago did we become fully able to appraise the richness of ideas and the mastery that Ørsted had displayed in the experiments, whereby he not only explained all the main points as regards the effect first discovered, but also traced numerous consequences which for his contemporaries must have seemed to be the fruits of his successors' labour. We are indebted to the deep historical insight of Kirstine Meyer[6], and her ability to trace all sources, for the retrieval among Ørsted's surviving papers of until then quite unnoticed notes about the

---

[6] [On Kirstine Meyer, née Bjerrum (1861–1941), see above, *Introduction* to Part IV, p. [330].]

series of experiments that form the basis of the discovery's announcement. In these, Ørsted appears as a researcher who only bears comparison with the most admired idols in the history of science. In the notes, Ørsted discusses the ideas and expectations he sought to test as well as a wealth of observations pointing far beyond what anyone could read out of the condensed form without further guidance.

In his autobiographical account Ørsted continues[7]:

"The experiments were soon repeated in all countries in which there are friends of science and the greatest reward an inventor can enjoy, that of seeing his invention become the object of the most industrious investigation, seeing it expanded and fructifying, was his to the fullest. The number of those who have written on electromagnetism amounts to well over a hundred. Therefore, with the united efforts of so many, this knowledge too has [already] been expanded and enriched in content far beyond what one would expect in the short period of seven years."

After a description of the honours awarded him by many learned societies immediately after the discovery, Ørsted adds, in his own characteristic way:

"In the midst of all these expressions of acclamation, the inventor could not help feeling deeply that even the discoveries that we so diligently seek are nevertheless not only the fruit of our exertion, but depend on a series of events and a state of affairs in the world of science which are subject to a higher law than all those that he seeks."

The development set in motion by Ørsted's discovery was indeed quite unique, and already before the year was over a great many reports appeared of experiments that confirmed and developed it in various ways. Among the investigations that, due to the ingenuity and mastery with which they were carried out, rightly attracted the greatest attention and admiration, one might mention Ampère's demonstration of the mutual attractions and repulsions between electric conductors and Biot and Savart's precise account of the magnetic forces that electric currents bring forth in the surrounding space. Even so, as is apparent from the paper Ørsted published some months after the first announcement, but whose full content can only be justly appreciated today after his notes have been found, Ørsted was himself aware of the equivalence

[7] [The translation of this particular passage is taken from Bern Dibner, *Oersted and the Discovery of Electromagnetism*, Burndy Library, Norwalk, Connecticut 1961, p. 18.]

between closed electric circuits and magnets, the clear exposition of which, both experimentally and theoretically, is to the great credit of Ampère.

Already a decade later, however, a new stage in the development was introduced by Faraday's discovery of electromagnetic induction, whereby an even more intimate connection between electric and magnetic forces was revealed. In close relation to Faraday's innovative ideas, Maxwell developed a theory for the propagation of electromagnetic waves through space, whose consequences were fully confirmed by Hertz's famous experiment. Not only was the foundation thereby laid for all modern radio technology, but an understanding was also gained of the phenomena associated with the propagation of light, whereby the entire chapter in the history of physics beginning with Ole Rømer's discovery of the velocity of light found such a harmonious conclusion. In this connection one must also remember that the Danish physicist Lorenz[8], simultaneously with and independently of Maxwell, developed an electromagnetic theory of light which to a great extent agreed with Maxwell's. At the end of the announcement of his great discovery, also Ørsted himself makes the far-sighted remark that precisely the discovered relationship between the directions of the electric current and of the resulting magnetic forces ought to open the way for an explanation of the characteristic polarization properties of light, which until then had posed an insoluble riddle.

What is perhaps most characteristic of the recent development in physics is the ever clearer recognition of the inner connection among natural phenomena; this has been a dream since ancient times, and if anything was typical of Ørsted's whole attitude, it was the dawning hope of this dream's realization. In our own age a tremendous advance in this regard has been achieved through the theory of relativity, which took its point of departure in the surprising experience that observers who move in relation to each other will describe physical phenomena quite differently. This difference is so great that events occurring at different places which are experienced as simultaneous by one observer, will be described by another as taking place one after the other, and implies furthermore that certain effects which are perceived by one observer as being purely electric, by another will be attributed to the presence of both electric and magnetic forces. This realization opened the way for Einstein to arrive at so general a formulation of the laws of nature that they are valid for any observer. By virtue of the possibility thus achieved for combining features

---

[8] [Ludvig Valentin Lorenz (1829–1891), whose greatest achievement was his electromagnetic theory of light. Lorenz also did other important research in optics and made original contributions to heat theory and to the understanding of the electric conductivities of metals.]

between which hitherto no connection could be seen, our world view has gained a unity of hitherto unknown scope.

By its very nature, this development goes far beyond the limits of the range of ideas within which physicists and philosophers moved in Ørsted's time. The hopes of tracing such a rich variety of properties of all matter back to an interplay between electric and magnetic forces have, however, been realized in the most marvellous way through the insight we have gained in our century by the exploration of the world of atoms, which, still in Ørsted's time, one generally considered forever closed to human experience. The electron has become recognized as a building block common to the atoms of all substances, and whereas electric current is caused by a movement of electrons through bodies, the way the electrons are bound in the individual atoms and molecules is reflected in the chemical and magnetic properties of the substances. On this basis we have succeeded in bringing together the great wealth of experience that physicists and chemists have amassed in the course of centuries, but the solution of this problem was not possible until the discovery of the so-called quantum of action, whereby Planck at the turn of the century opened a new epoch in the history of science.

The existence of the quantum of action is an expression of a feature of indivisibility in atomic processes, for which there is no room at all within the framework of the previous description of nature. Thus, the reorientation necessary for the extension of this framework has also led to a new attitude toward many earlier philosophical views. Today we perceive the relationships between the various schools of philosophy on a new background, focusing no longer so much on the differences, which previously often gave rise to bitter feud, but are able to appreciate the fruitfulness of views that emphasize the one or the other of the apparently irreconcilable but equally essential aspects of our situation as observers of the nature to which we belong ourselves. Even though we would nowadays probably speak of the reason in nature in a somewhat different way than Ørsted did[9], we can precisely on the new background appreciate deeply his characteristic striving to unite what he called the beautiful and the true.

To the picture of Ørsted as scientist and thinker belongs inextricably his attitude towards his fellow human beings, which characterized his activity in the society he served so well. The foundation for this attitude was laid already in his childhood, when he grew up in the home of a pharmacist in Rudkøbing

---

[9] [Ørsted believed that Nature was a manifestation of a divine rationality and thought that physics, in particular, constituted knowledge of a divine unity in Nature.]

together with his only one year younger brother, who was also going to perform such great work in Danish society[10]. In their native town there was at that time no school that could offer the gifted boys adequate educational conditions. Hans Christian and Anders Sandøe thus had to make use of every possible source for acquiring knowledge, and they told each other about everything each of them had learned. Thus was created a fellowship which continued during their studies in Copenhagen and was to last throughout their lives, however different their areas of work and spheres of activity were to become.

At the beginning of the nineteenth century conditions at the University of Copenhagen were extremely modest as far as the sciences were concerned, and in physics and chemistry, where Ørsted's interests in the main resided, there existed neither teaching posts nor laboratories. Only through the confidence he gained to an ever increasing degree and of which by indefatigable and self-denying efforts he proved himself so deserving, did Ørsted, despite the poverty of the time, gradually succeed in building up scientific teaching in physics and in creating research opportunities for himself and for his co-workers. Among these should be mentioned above all Zeise[11], who was also a pharmacist's son and who later was to take such an active part in Ørsted's efforts to improve pharmaceutical education in Denmark. It is well known how Ørsted was at first only able to place Zeise in a kitchen in the professorial residence in Nørregade as an assistant in his chemical experiments, but how he gradually succeeded in establishing an independent laboratory where Zeise, who became Denmark's first real professor of chemistry, could carry out his renowned investigations.

This is of course only one, if particularly fortunate, example of H.C. Ørsted's efforts to create conditions for research and study in this country. The whole story of these efforts tells us something of the demands that an enthusiastic striving human being must make on himself in order to create a new environment and of the talents required for accomplishing such a task. Ørsted was able to captivate the audience listening to his lectures not only by his knowledge and imagination, but also by his skill at lecturing, the development of which – as he himself explains – was based not least on studies and experiences during his travels abroad. The esteem and the friendships Ørsted gained around the world contributed also to creating a broadening of the horizon in this country. Not least, the cultural links between the Nordic sister nations were close to Ørsted's

---

[10] [On A.S. Ørsted (1778–1860), see p. [269], ref. 23.]
[11] [William Zeise (1789–1847), Danish chemist.]

The picture of the Scandinavian Meeting of Natural Scientists in 1847 mentioned in ref. 14. The original painting is part of the decoration of the University of Copenhagen's ceremonial hall, where Bohr gave his speech in commemoration of Ørsted. The picture shows H.C. Ørsted on the platform. Seated at the table farthest to the left is Swedish chemist Jöns Jacob Berzelius. Grundtvig (whom Bohr mentions in his article "Danish Culture" – see ref. 8 on p. [265]) can be seen between the two men seated at the middle of the table. Standing at the front of the picture to the right is Christopher Hansteen, Norwegian physicist and astronomer.
Painted by Erik Henningsen, 1896 (photograph by Kaj Lergaard).

heart, and his participation in the Scandinavian meetings of natural scientists[12], which came into being in his time, is illustrated, as the Vice-Chancellor of the university pointed out[13] by the painting in this ceremonial hall[14].

The effort that H.C. Ørsted invested in his university duties, while constantly struggling with the poor economic conditions – we have already been reminded that the critical years of his life coincided with the time of state bankruptcy – calls for our greatest admiration. His initiative never faltered; on the contrary, he assumed an ever increasing number of responsibilities, and it is almost unbelievable that he could cope with them. The Vice-Chancellor of the Technical University is going to tell us about the exceptional and self-denying work Ørsted devoted to the foundation and the leadership of the Polytechnic College, which now, under the name of Denmark's Technical University, occupies an ever more important place in society, but I would like to remind you with a few more words of Ørsted's activity in the Royal Danish Academy of Sciences and Letters, for which he served as secretary for thirty-six years. In this capacity he put an effort into the affairs of the Academy and its relations with foreign academies that one would think required a single individual's full working capacity, and reading the minutes of the Academy's meetings from that period, we encounter time and time again Ørsted's name in connection with new initiatives concerning various aspects of the Academy's activity as well as in frequent announcements not only of results from his own research, but also of progress in the most diverse areas of science, about which he felt it his task to keep the Academy's members constantly informed.

The Society for the Dissemination of Natural Science, founded by Ørsted in 1824, also has a history that bears poignant witness to how far his enthusiasm and sense of responsibility could lead him. At times when one would think that he was fully occupied with his research and science teaching, he personally instructed the lecturers the Society sent out to Danish provincial towns to give craftsmen knowledge of importance for their craft, and at the same time to raise the population's general level of education. Although many of the tasks the Society undertook under Ørsted's leadership have gradually been taken over by other institutions, several of which were created on Ørsted's own initiative, the tradition of his time lives on in the Society. Thus, since H.C. Ørsted's

[12] [From about 1830, a close cooperation in the arts and the sciences developed among Norway, Sweden and Denmark. The first Scandinavian Meeting of Natural Scientists was held in Gothenburg in 1839.]

[13] [Bohr refers to the talk, preceding his own, by H.M. Hansen (1886–1956).]

[14] [The painting depicts Ørsted addressing the 1847 Scandinavian Meeting of Natural Scientists. Bohr had a print of this picture in his office at the Institute of Theoretical Physics.]

[367]

death, the Society for the Dissemination of Natural Science has considered it to be its special duty to cherish the memory of its founder, and already a generation ago, under the chairmanship of Martin Knudsen[15], the first steps were taken to set up an H.C. Ørsted museum, where those of his instruments that have been kept, including the very magnetic needle that was used in the discovery of electromagnetism, could be gathered. Thanks to the sense of piety and the generosity of Ørsted's family, the museum has also been able to collect many of Ørsted's belongings and has received testamentary promise of other valuable mementoes of his life, including the writing-desk where Ørsted worked on his numerous papers and talks. Sympathetic cooperation on the part of the authorities has made it possible, in preparation for just this commemoration day, to arrange a beautiful room in Denmark's Technical University, which will henceforth provide a proper setting for the irreplaceable H.C. Ørsted collection[16].

If one compares H.C. Ørsted with other great figures in the history of Danish science, such as Tycho Brahe, Niels Stensen and Ole Rømer, both similarities and differences become apparent. Each of them, within his own field, initiated a new development that was to leave its mark on the sciences far into the future, and characteristic for them all is also the close connection with the contemporary scientific world abroad, whereby they contributed so greatly to creating a broader outlook here at home. Whereas ferment in the life of the nation contributed to the moulding of the remarkable, at the same time brilliant and tragic, destinies of Tycho Brahe and Niels Steensen, Rømer and Ørsted were drawn into comprehensive activity for society in relatively more tranquil times. Thereby, however, such great demands were made on their rich talents that they, to a higher degree than other scientific pioneers, were themselves prevented from continuing research on the paths they had opened.

To end by deploring such a state of affairs would be, however, a much too one-sided viewpoint in light of the great and urgent tasks for the development of society in practical respects that Rømer in his time was able to accomplish, and in light of how badly needed and fruitful Ørsted's strivings were for the extension of the basis for scientific studies and research here at home and for the dissemination of knowledge about the results of this research and its importance to all circles of Danish society. If we look around the world, we can hardly find a parallel to Ørsted's deeds in academic circles proper. We are much rather led to think of a man such as Franklin, whose great services to

---

[15] [See Vol. 1, p. [109].]

[16] [The collection was moved to the Danish Technical Museum, Elsinore, in 1984.]

science helped him in such a fortunate way in his efforts, from an early age, to promote general education in a new society, for whose organization and esteem in the world he came to mean so much.

When we try to form a picture of H.C. Ørsted's life and work, his deep human sympathy is probably the most striking characteristic. He readily accepted great burdens in matters concerning the needs of society and the well-being of fellow human beings, and the accomplishment of such tasks allowed him to make use of his talents in a harmonious way which agreed so well with his outlook. In his endeavours to share with his fellow human beings everything that could strengthen and enrich their minds, Ørsted also discovered fields which lay far beyond his work for scientific enlightenment. The influence he exerted within an ever increasing circle of writers and beaux esprits, to whom he was linked through ties of family and friendship, makes up a rich chapter in the history of Danish intellectual life. However, I shall not attempt to go into all this in more detail, as in a moment we shall have the opportunity of hearing about it from the most competent and closest source[17].

There is ample testimony of the admiration and gratitude felt for H.C. Ørsted by his contemporaries as well as by following generations. One can hardly cite a more effective expression of such feelings than the words J.C. Jacobsen spoke 75 years ago on the establishment of the Carlsberg Foundation, which was to create conditions here at home for scientific research far more favourable than those under which Ørsted had had to work[18]:

"In deep recognition of the debt I owe H.C. Ørsted's teachings and inspiring influence and as a testimony of the grateful appreciation of his work in disseminating the light of knowledge to wider circles, I have linked the establishment of the above-mentioned foundation to this day, which, with the unveiling of Ørsted's monument, is dedicated to his memory."

May Danish society of today and tomorrow prove itself worthy of the heritage that Hans Christian Ørsted's character and work provide so clearly for us to see.

[17] [Bohr refers to the talk, following upon his own, by the Norwegian literary historian Francis Bull (1887–1974).]

[18] [The quotation is from the brewer J.C. Jacobsen's introductory speech in the Royal Danish Academy of Sciences and Letters at the presentation of the articles for the Carlsberg Foundation on 29 September 1876. The speech is printed in *Det Kongelige Danske Videnskabernes Selskab 1742–1942, Samlinger til Selskabets Historie, I*, Munksgaard, Copenhagen 1942, pp. 632–633.]

III. RYDBERG'S DISCOVERY OF THE SPECTRAL LAWS

Proceedings of the Rydberg Centennial Conference on Atomic Spectroscopy, Lunds Universitets Årsskrift. N.F. Avd. 2. Bd. 50. Nr 21 (1955) 15–21

See Introduction to Part IV, p. [329].

Swedish physicist Johannes Robert Rydberg (1854–1919).

[372]

Rydberg's discovery of the spectral laws

by

NIELS BOHR

(Universitetets Institut for Teoretisk Fysik, Copenhagen)

It is a great pleasure to me to accept the invitation at this Rydberg Centennial Conference, where the present stand of our knowledge about spectroscopy will be reviewed by so many experts from different parts of the world, to remind about Rydberg's pioneering work in this field. Especially I shall stress the direct consequences of his great discoveries for the development of our ideas of atomic constitution and, in this connection, recall some personal remembrances.

As is well known, Rydberg's discovery of the spectral regularities was the outcome of his intense interest in the problem of the relationship between the chemical elements which, in the latter part of the former century, was brought into the foreground especially through the work of Mendelejew. The remarkable periodicity in the physical and chemical properties of the elements when arranged in order of increasing atomic weights took the fancy of the searching spirit of Rydberg who, with his disposition for numerical calculations, became especially interested in the optical spectra, where the great refinement of measurements permits the establishment of arithmetical relationships of high accuracy.

For Rydberg's great achievements in this respect, it was a happy intuition from the beginning to look for relations not between the directly measured wavelengths of the spectral lines, but between the reciprocal figures expressing the number of waves per unit length, now known as wave numbers. To this choice he was led through the constant differences between wave numbers occurring in so-called doublet and triplet lines. From Rydberg's side, the tracing of these equal spacings was an original discovery, but when his work was already far advanced, he became, as he modestly and honestly remarks, aware that the occurrence of such relations in complex lines had been pointed out by Hartley a few years before. Rydberg, however, went himself far deeper into the matter and extensively used wave number differences as a main tool for his disentangling of spectral regularities.

To this purpose, further directives were offered by the study of the so-called line series which, in the previous decade, were discovered in many spectra by Liveing and Dewar, and are characterized by similarity in appearance (sharp, diffuse, etc.) as well as by gradual harmonious decrease in intensity and separation. Rydberg now found that all the series in the spectra analyzed by him, when described by wave numbers and suitably arranged by displacement of the wave number scale, showed such close relationship that he was led to represent the wave numbers for the lines in each series by the difference between a constant term and a term which, in a common manner, decreased when progressing through the series. This relation he expressed by the formula

$$\sigma = a - \varphi(n + a),\qquad(1)$$

where n is an integer serving as ordinal index of the series lines, and φ a universal function, while a and a are constants specific for the individual series.

As a first attempt to determine this function φ, which obviously had to converge to zero for increasing n, Rydberg tried the expression

$$\varphi(n+a) = \frac{C}{n+a},\qquad(2)$$

but obtained neither a satisfactory agreement for any longer series nor the required constancy of C for all series. As a better choice, Rydberg next tried

$$\varphi(n+a) = \frac{R}{(n+a)^2},\qquad(3)$$

and he tells in his famous article in the Swedish Academy from 1889 that just when he was occupied by testing this formula he learned about Balmer's discovery of the simple law

$$\lambda = B \cdot \frac{n^2}{n^2 - 4}\qquad(4)$$

which with such extraordinary accuracy represents the wavelengths of the well-known series of hydrogen lines. Substituting wavelength by the corresponding wave numbers, Rydberg wrote the Balmer formula in the form

$$\sigma = \frac{R}{2^2} - \frac{R}{n^2},\qquad(5)$$

representing a special case of his own formula. Guided in this way to an accurate fixation of the suspected universal constant R, now known as the Rydberg constant, he also soon found it possible not only to prove the far-reaching validity of formulae (1) and (3), but could with its help with considerable precision determine the constants a and a for any series.

This great advance made it possible for Rydberg to trace even more intimate connections between the different series composing the spectrum of an element. In fact, he found not only that certain series with different values of a exhibited the same value of a, but that quite generally the value of the constant term a in any series coincided with a member of the sequence of the variable terms in some other series of the element. In particular, Rydberg found that the difference between the limit for the principal series and the common limit for the diffuse and sharp series was just equal to the wave number of the first member of the principal series, a result which, as is well known, was later independently obtained by Schuster. In his original paper, Rydberg thus proposed as a comprehensive formula for every spectral line of an element

$$\sigma = \frac{R}{(n_1+a_1)^2} - \frac{R}{(n_2+a_2)^2},\tag{6}$$

according to which each series corresponds to a constant value of n_1 and a sequence of values of n_2. In this scheme, the characteristic complexity of many series lines is directly accounted for by a multiplicity of the a-values.

In arguing about the scope of his final formula, Rydberg expressed himself with remarkable caution and subtility. On the one hand, he was quite aware that the special form of the two combining terms in formula (6) could not rigorously fit the observations. On the other hand, he stressed that his formula in essentials fulfilled the conditions of universality demanded of fundamental laws of nature. This attitude finds special expression in the discussion, in the last part of Rydberg's paper, of the type of series formulae just at that time used by Kayser and Runge to represent their extensive refined measurements of spectral lines. While fully appreciating the great accuracy of such formulae, he points out that it had not been his main intention to develop a proper interpolation formula for each individual series, but rather, by using the smallest number of specific constants in his computations, to trace universal relationships.

The search for a mechanism which might explain the spectral regularities, however, was at that time confronted with apparently insurmountable difficulties. Especially we may recall the pertinent remark of Rayleigh that any analysis of the normal modes of vibrations of a stable mechanical system leads to relations between the squares of the frequencies, and not between the frequencies themselves. It is true that Ritz, inspired by Lorentz' explanation of the Zeeman effect, attempted to account for the spectral laws by introducing the idea of atomic magnetic fields, whose effects on the electric constituents of the atom — in contrast to ordinary mechanical forces — depend intrinsically on the velocities. In spite of all ingenuity of these endeavours, it was not found possible, however, on such lines to arrive at an explanation of the spectral laws, consistent with the interpretation of other atomic properties.

Still, Ritz' penetrating inquiry into spectral problems, furthered especially by his close collaboration with Paschen, led him to various refinements of the numerical formulae for spectral series and to the prediction of new series which essentially completed the analysis of various line spectra. In connection with this work, in which Rydberg's discoveries and original conceptions proved of decisive importance, Ritz in 1908 enunciated a general law, now known as the Rydberg-Ritz combination principle, according to which the wave number of any line of a spectrum can be rigorously expressed as

$$\sigma = T_1 - T_2,\tag{7}$$

where T_1 and T_2 are two members of the set of terms characteristic of the element.

A new epoch in the development of our ideas of atomic structure was soon after

2

initiated by Rutherford's discovery, in 1911, of the atomic nucleus. This discovery led to a remarkably simple picture of the atom as a system of electrons moving round a central charge of minute extension, in which practically the whole mass of the atom is concentrated. Indeed, it became at once clear that all physical and chemical properties of an element, depending on the binding of the atomic electrons, were widely regulated by the total charge of the nucleus determining the number of electrons in the neutral atom. This so-called atomic number was obviously to be identified with the ordinal index of the element in the periodic table, which Rydberg had so clearly recognized as the principal factor governing the relationships between the properties of the elements. As is well known, this view found, a few years later, decisive confirmation by Moseley's fundamental researches on the characteristic X-ray spectra of the elements, and it is interesting to recall that his fixation of the atomic number of the whole sequence of chemical elements in several respects confirmed Rydberg's expectations of the lengths of the periods in Mendelejew's table.

An immediate consequence of Rutherford's discovery was, however, the stressing of the difficulties in accounting for the spectral regularities on the basis of classical physics. In fact, on the ideas of ordinary mechanics and electrodynamics, a system of point charges possesses no inherent stability which could account for the constancy of the specific properties of the elements so impressively revealed by their line spectra. In particular, the radiation from the motion of the electrons would give rise to a continuous dissipation of energy, accompanied by gradual alteration of frequencies of motion and shrinking of orbital dimensions until all electrons would be amalgamated with the nucleus within a minute neutral system.

A clue to the problem of atomic stability and to the origin of line spectra was, however, offered by the discovery of the universal quantum of action, to which Planck, in the first years of this century, was led by his ingenious analysis of the phenomena of thermal radiation. As is well known, Einstein a few years later pointed out not only that Planck's formula $E = nh\omega$ for the possible energy values of a harmonic oscillator of proper frequency ω permitted to explain observed anomalies of specific heats of various substances at low temperatures, but that also characteristic features of atomic photo effects demanded that exchange of the energy by radiation of frequency $\nu = c\sigma$ should take place in so-called light quanta or photons of energy $h\nu$. Notwithstanding the impossibility of a closer analysis of these phenomena on accustomed lines, conspicuous especially in the dilemma as regards the constitution of radiation implied in the concept of the photon, it was clear that we were here dealing with an essential feature of wholeness of atomic processes, quite foreign to the classical ideas of physics.

On this foundation, the idea suggested itself that, in any change of the energy of an atom, we have to do with a process consisting in a complete transition between two stationary quantum states, and that any radiation involved in such transition processes is exchanged in the form of a photon. In fact, this so-called quantum postulate offered an immediate interpretation of the combination principle by identifying the numerical value of each spectral term multiplied by hc with the energy of a possible stationary state of the atom. Moreover, a solution was indicated of the puzzle of the apparently capricious occurrence of selective absorption of radiation by atoms. Under ordinary conditions, the atom will be in its normal state of lowest energy, corresponding to the largest spectral term, given by the limit of the principal series. We therefore understand that only this series appears in selective absorption and, in particular, that continuous absorption sets in at its limit, evidently corresponding to the removal of an electron from the atom. Soon after, these conclusions were also directly borne out by the famous experiments of Franck and Hertz on the excitation of spectral lines by impact of electrons. The experiments showed that any possible exchange of energy between electron and atom corresponds to a transition from the normal state of the atom to a higher stationary state, and that the minimum energy for producing ionization of the atom is just equal to hc times the wave number of the limit of the principal series.

Remembering the vivid discussions of those years, it may perhaps be of interest to recall a conversation between Einstein and Hevesy, to whom as a co-pupil of Rutherford I had early communicated the new views and prospects. When asked about his attitude to such ideas, Einstein answered that they were not completely foreign to his mind, but humorously added that he felt that, if they were to be taken seriously, it would mean the end of physics. Looking back, one will admit the pertinence of this utterance; surely, we have had to revise our ideas of what to understand by physical explanation. Meanwhile it proved possible step by step to make ever more extensive use of the spectral evidence to advance our knowledge of atomic constitution. The attainment of this goal should, as we know, demand the development of an appropriate mathematical formalism departing radically from that of classical physics. To begin with, however, one had to approach the problems in a tentative manner by more primitive methods. Guidance was mainly afforded by so-called correspondence considerations characterized by the endeavour to make use of ordinary physical concepts in all considerations not directly opposed to the quantum postulate.

The first step was the establishment of a relation

$$R = \frac{2\pi^2 e^4 m}{ch^3} \tag{8}$$

expressing Rydberg's constant in terms of the mass m and charge e of the electron, and the fundamental constants c and h. In fact, this relation could be shown to be a necessary condition for the asymptotic approach of the frequencies of the hydrogen spectrum and those of the motion of an electron in a Keplerian orbit around a heavy nucleus of unit charge. Such considerations also offered a simple explanation of the appearance of the Rydberg constant in the spectra of the other elements by assuming that the series in question originate from transition between stationary states in which one of the atomic electrons is bound more loosely to the nucleus than the others and that therefore the forces exerted on it by the residual ion, at any rate at large distances, closely resemble the forces to which the electron in the hydrogen atom is subjected.

A special problem arose, however, as regards the origin of the series of lines first observed in 1899 by Pickering in stellar spectra and expressed with great accuracy by the formula

$$\sigma = R\left(\frac{1}{2^2} - \frac{1}{(n+1/2)^2}\right). \tag{9}$$

Due to its close relation with the Balmer series, the Pickering series was attributed to hydrogen, and this assignment was apparently strongly supported by considerations of Rydberg who compared the relationship between the Balmer and the Pickering series with that of the diffuse and sharp series in other spectra, and in this connection predicted the existence of a further hydrogen series

$$\sigma = R\left(\frac{1}{(3/2)^2} - \frac{1}{n^2}\right), \tag{10}$$

corresponding to an ordinary principal series.

Just in 1912, not only the Pickering lines, but also a series of lines represented by (10) as well as a series of lines given by

$$\sigma = R\left(\frac{1}{(3/2)^2} - \frac{1}{(n+1/2)^2}\right) \tag{11}$$

were observed by Fowler in intense discharges through mixtures of hydrogen and helium gases. The assignment of all these lines to the hydrogen atom was, however, irreconcilable with correspondence considerations which, on the contrary, suggested that the Pickering series as well as the series (10) and (11) had to be attributed to the helium ion consisting of an electron bound to a nucleus of two unit charges. In fact, such a system would just be expected to give a spectrum of the same type as that of the hydrogen atom, but in which R was replaced by $4R$.

These ideas were to begin with contested by leading spectroscopists like Fowler and Runge. I especially recall the warning, given by the latter at a colloquium in Göttingen, against such apparently arbitrary use of spectral evidence by theoreticians who did not seem properly to appreciate the beauty and harmony of the general pattern of series spectra, revealed above all by the ingenuity of Rydberg. The dispute was, however, rapidly settled to the general satisfaction. Not only were the Pickering and Fowler lines soon after observed by Evans in highly purified helium, showing no trace of hydrogen lines, but it could even be shown that the slight deviations of the lines measured by Fowler from the formulae suggested by Rydberg corresponded exactly to a small correction in the Rydberg constant deduced from theoretical arguments when taking the actual masses of the atomic nuclei into account.

An important outcome of the whole discussion was the recognition that certain series of the magnesium spectrum observed by Fowler in intense spark discharges could be united into a simpler series scheme just by the replacement of the Rydberg constant by $4R$. Such series systems, to the discovery of which in many elements Fowler as well as Paschen contributed so largely in the following years, are now known as spark spectra. In contrast to the ordinary arc spectra which originate from neutral atoms, these spectra are assigned to ions of unit charge, in which a loosely bound electron is exposed to conditions resembling those of the electron in the helium ion. The expectation that ions of still higher charge, Ne, would give rise to spectra of a generalized Rydberg scheme, in which the common constant is given by N^2R, has also been widely fulfilled. I hardly need to enlarge on this point in this beautiful institute directed by Edlén who, to the admiration of all physicists, has succeeded by such skill and perseverance through the years in producing and analyzing a multitude of spectra corresponding to a high degree of electron stripping of the atoms.

Within the compass of this short address, I have had to confine myself to Rydberg's pioneer work and to a few aspects of his discoveries which came to play such a decisive part in the initial stage of a development by which spectroscopic evidence gave us an ever deeper insight into the problem of atomic constitution, and in particular led to a classification of the states of electron binding in the shell structure of the atom, which in every detail accounts for the periodic relationships of the properties of the elements. A veritable culmination of the semi-empirical approach characteristic of this first tentative period was the enunciation by Pauli of the exclusion principle, which should subsequently find such an appropriate incorporation in the rational methods of quantum theory. These methods, though discarding accustomed pictorial representations, rival classical mechanics and electrodynamics in consistency and completeness, and provide the firm foundation for the exploitation of the inexhaustible wealth of spectroscopic evidence.

It has truly been a great adventure for our generation to witness this whole development which sometimes proceeded with almost tumultuous rapidity. Thus, I remember especially the successful conference in 1919 here in Lund, when a new stage was just being initiated by the work of Sommerfeld and his school, and the prospects were discussed with great enthusiasm and mutual benefit. We were assembled in the old Fysikum, where the rich traditions were happily upheld by Rydberg's young followers, among whom Siegbahn with such experimental mastery brilliantly pursued Moseley's work, and Heurlinger made so important contributions to the theoretical interpretation of band spectra. Although Rydberg himself by illness was prevented from attending the conference, we all felt most vividly his guiding spirit among us, just as we do today at this memorial meeting.

[379]

Ernest Rutherford photographed at the Carlsberg honorary mansion when visiting Bohr in 1932.

IV. THE RUTHERFORD MEMORIAL LECTURE 1958: REMINISCENCES OF THE FOUNDER OF NUCLEAR SCIENCE AND OF SOME DEVELOPMENTS BASED ON HIS WORK

Proc. Phys. Soc. **78** (1961) 1083–1115

Elaborated version, completed in 1961, of a lecture given at a meeting of the Physical Society of London, 28 November 1958

References (unpublished)

See Introduction to Part IV, p. [329] ff.

THE RUTHERFORD MEMORIAL LECTURE 1958: REMINISCENCES OF THE FOUNDER OF NUCLEAR SCIENCE AND OF SOME DEVELOPMENTS BASED ON HIS WORK (1958)

Versions published in English, Danish and German

English: The Rutherford Memorial Lecture 1958: Reminiscences of the Founder of Nuclear Science and of Some Developments Based on his Work
A   Proc. Phys. Soc. **78** (1961) 1083–1115
B   "Rutherford at Manchester" (ed. J.B. Birks), London 1962, pp. 114–167
C   "Essays 1958–1962 on Atomic Physics and Human Knowledge", Interscience Publishers, New York 1963, pp. 30–73 (reprinted in: "Essays 1958–1962 on Atomic Physics and Human Knowledge, The Philosophical Writings of Niels Bohr, Vol. III", Ox Bow Press, Woodbridge, Connecticut 1987, pp. 30–73)

Danish: Rutherford mindeforelæsning 1958. Erindringer om grundlæggeren af kernefysikken og om den udvikling, der bygger på hans værk
D   "Atomfysik og menneskelig erkendelse II", J.H. Schultz Forlag, Copenhagen 1964, pp. 43–94
E   "Naturbeskrivelse og menneskelig erkendelse" (eds. J. Kalckar and E. Rüdinger), Rhodos, Copenhagen 1985, pp. 167–228

German: Rutherford-Gedenkvorlesung 1958: Erinnerungen an den Begründer der Kernphysik und an die von seinem Werk ausgehende Entwicklung
F   "Atomphysik und menschliche Erkenntnis II", Friedr. Vieweg & Sohn, Braunschweig 1966, pp. 30–74

All of these versions agree with each other, with the exception that the next to last paragraph is omitted in *B*.

[382]

THE RUTHERFORD MEMORIAL LECTURE 1958†

Reminiscences of the Founder of Nuclear Science and of Some Developments Based on his Work

By NIELS BOHR

I T has been a pleasure for me to accept the invitation from the Physical Society to contribute to the series of Rutherford Memorial Lectures in which, through the years, several of Rutherford's closest collaborators have commented on his fundamental scientific achievements and communicated reminiscences about his great human personality. As one who in early youth had the good fortune to join the group of physicists working under Rutherford's inspiration, and owes so much to his warm friendship through the many succeeding years, I welcome the task of recalling some of my most treasured remembrances. Since it is impossible, of course, in a single lecture to attempt a survey of the immense and many-sided life-work of Ernest Rutherford and its far reaching consequences, I shall confine myself to periods of which I have personal recollections and to developments I have followed at close hand.

I

The first time I had the great experience of seeing and listening to Rutherford was in the autumn of 1911 when, after my university studies in Copenhagen, I was working in Cambridge with J. J. Thomson, and Rutherford came down from Manchester to speak at the annual Cavendish Dinner. Although I did not on this occasion come into personal contact with Rutherford I received a deep impression of the charm and power of his personality by which he had been able to achieve almost the incredible wherever he worked. The dinner took place in a most humorous atmosphere and gave the opportunity for several of Rutherford's colleagues to recall some of the many anecdotes which already then were attached to his name. Among various illustrations of how intensely he was absorbed in his researches, a laboratory assistant in the Cavendish was reported to have noted that, of all the eager young physicists who through the years had worked in the famous laboratory, Rutherford was the one who could swear at his apparatus most forcefully.

From Rutherford's own address I especially remember the warmth with which he greeted the latest success of his old friend C.T.R. Wilson who by the ingenious cloud chamber method had just then obtained the first photographs of tracks of α-rays exhibiting clear cases of sharp bends in their usual remarkably straight path. Of course, Rutherford was thoroughly acquainted with the phenomenon which only a few months before had led him to his epoch-making discovery of the atomic nucleus, but that such details of the life history of α-rays could now be witnessed

† The present text is an elaborated version, completed in 1961, of the lecture delivered without a prepared manuscript at a meeting of The Physical Society of London at the Imperial College of Science and Technology on 28th November 1958.

[383]

directly by our eyes, he admitted to be a surprise, causing him extreme pleasure. In this connection Rutherford spoke most admiringly of the persistence with which Wilson already during their comradeship in the Cavendish had pursued his researches on cloud formation with ever more refined apparatus. As Wilson later told me, his interest in these beautiful phenomena had been awakened when as a youth he was watching the appearance and disappearance of fogs as air currents ascended the Scottish mountain ridges and again descended in the valleys.

A few weeks after the Cavendish Dinner I went up to Manchester to visit one of my recently deceased father's colleagues who was also a close friend of Rutherford. There, I again had the opportunity to see Rutherford who in the meantime had attended the inaugural meeting of the Solvay Council in Brussels, where he had met Planck and Einstein for the first time. During the conversation, in which Rutherford spoke with characteristic enthusiasm about the many new prospects in physical science, he kindly assented to my wish to join the group working in his laboratory when, in the early spring of 1912, I should have finished my studies in Cambridge where I had been deeply interested in J. J. Thomson's original ideas on the electronic constitution of atoms.

In those days, many young physicists from various countries had gathered around Rutherford, attracted by his genius as a physicist and by his unique gifts as a leader of scientific cooperation. Although Rutherford was always intensely occupied with the progress of his own work, he had the patience to listen to every young man, when he felt he had any idea, however modest, on his mind. At the same time, with his whole independent attitude, he had only little respect for authority and could not stand what he called 'pompous talk'. On such occasions he could even sometimes speak in a boyish way about venerable colleagues, but he never permitted himself to enter into personal controversies, and he used to say: '' There is only one person who can take away one's good name, and that is oneself!''

Naturally, to trace in every direction the consequences of the discovery of the atomic nucleus was the centre of interest of the whole Manchester group. In the first weeks of my stay in the laboratory, I followed, on Rutherford's advice, an introductory course on the experimental methods of radioactive research which under the experienced instruction of Geiger, Makower and Marsden was arranged for the benefit of students and new visitors. However, I rapidly became absorbed in the general theoretical implications of the new atomic model and especially in the possibility it offered of a sharp distinction as regards the physical and chemical properties of matter, between those directly originating in the atomic nucleus itself and those primarily depending on the distribution of the electrons bound to it at distances very large compared with nuclear dimensions.

While the explanation of the radioactive disintegrations had to be sought in the intrinsic constitution of the nucleus, it was evident that the ordinary physical and chemical characteristics of the elements manifested properties of the surrounding electron system. It was even clear that, owing to the large mass of the nucleus and its small extension compared with that of the whole atom, the constitution of the electron system would depend almost exclusively on the total electric charge of the nucleus. Such considerations at once suggested the prospect of basing the account of the physical and chemical properties of every element on a single integer, now generally known as the atomic number, expressing the nuclear charge as a multiple of the elementary unit of electricity.

In the development of such views, I was encouraged not least by discussions with George Hevesy who distinguished himself among the Manchester group by his uncommonly broad chemical knowledge. In particular, as early as 1911, he had conceived the ingenious tracer method which has since become so powerful a tool in chemical and biological research. As Hevesy has himself humorously described, he was led to this method by the negative results of elaborate work undertaken as a response to a challenge by Rutherford who had told him that " if he was worth his salt " he ought to be helpful by separating the valuable radium D from the large amount of lead chloride extracted from pitchblende and presented to Rutherford by the Austrian government.

My views took more definite shape in conversations with Hevesy about the wonderful adventure of those Montreal and Manchester years, in which Rutherford and his collaborators, after the discoveries of Becquerel and Madame Curie, had built up the science of radioactivity by progressively disentangling the succession and interconnections of radioactive disintegrations. Thus, when I learned that the number of stable and decaying elements already identified exceeded the available places in the famous table of Mendeleev, it struck me that such chemically inseparable substances, to the existence of which Soddy had early called attention and which later by him were termed 'isotopes', possessed the same nuclear charge and differed only in the mass and intrinsic structure of the nucleus. The immediate conclusion was that by radioactive decay the element, quite independently of any change in its atomic weight, would shift its place in the periodic table by two steps down or one step up, corresponding to the decrease or increase in the nuclear charge accompanying the emission of α- or β-rays, respectively.

When I turned to Rutherford to learn his reaction to such ideas, he expressed, as always, alert interest in any promising simplicity but warned with characteristic caution against overstressing the bearing of the atomic model and extrapolating from comparatively meagre experimental evidence. Still, such views, probably originating from many sides, were at that time lively discussed within the Manchester group and evidence in their support was rapidly forthcoming, especially through chemical investigations by Hevesy as well as by Russell.

In particular, a strong support for the idea of the atomic number as determining the general physical properties of the elements was obtained from spectroscopic investigations by Russell and Rossi of mixtures of ionium and thorium, which pointed to the identity of the optical spectra of these two substances in spite of their different radioactive properties and atomic weights. On the basis of an analysis of the whole evidence then available, the general relationship between the specified radioactive processes and the resulting change of the atomic number of the element was indicated by Russell in a lecture to the Chemical Society in the late autumn of 1912.

In this connection it is interesting that, when after further research especially by Fleck, the radioactive displacement law in its complete form was enunciated a few months later by Soddy working in Glasgow, as well as by Fajans in Karlsruhe, these authors did not recognize its close relation to the fundamental features of Rutherford's atomic model, and Fajans even regarded the change in chemical properties evidently connected with the electron constitution of the atoms as a strong argument against a model according to which the α- as well as the β-rays.

had their origin in the nucleus. About the same time, the idea of the atomic number was independently introduced by van den Broek in Amsterdam, but in his classification of the elements a different nuclear charge was still ascribed to every stable or radioactive substance.

So far, the primary object of the discussions within the Manchester group was the immediate consequences of the discovery of the atomic nucleus. The general programme of interpreting the accumulated experience about the ordinary physical and chemical properties of matter on the basis of the Rutherford model of the atom presented, however, more intricate problems which were to be clarified gradually in the succeeding years. Thus in 1912, there could only be question of a preliminary orientation as to the general features of the situation.

From the outset it was evident that, on the basis of the Rutherford model, the typical stability of atomic systems could by no means be reconciled with classical principles of mechanics and electrodynamics. Indeed, on Newtonian mechanics, no system of point charges admits of a stable static equilibrium, and any motion of the electrons around the nucleus would, according to Maxwell's electrodynamics, give rise to a dissipation of energy through radiation accompanied by a steady contraction of the system, resulting in the close combination of the nucleus and the electrons within a region of extension far smaller than atomic dimensions.

Still, this situation was not too surprising, since an essential limitation of classical physical theories had already been revealed by Planck's discovery in 1900 of the universal quantum of action which, especially in the hands of Einstein, had found such promising application in the account of specific heats and photochemical reactions. Quite independent of the new experimental evidence as regards the structure of the atom, there was therefore a widespread expectation that quantum concepts might have a decisive bearing on the whole problem of the atomic constitution of matter.

Thus, as I later learned, A. Haas had in 1910 attempted, on the basis of Thomson's atomic model, to fix dimensions and periods of electronic motions by means of Planck's relation between the energy and the frequency of a harmonic oscillator. Further, J. Nicholson had in 1912 made use of quantized angular momenta in his search for the origin of certain lines in the spectra of stellar nebulae and the solar corona. Above all, however, it deserves mention that, following early ideas of Nernst about quantized rotations of molecules, N. Bjerrum already in 1912 predicted the band structure of infra-red absorption lines in diatomic gases, and thereby made a first step towards the detailed analysis of molecular spectra eventually achieved on the basis of the subsequent interpretation, by quantum theory, of the general spectral combination law.

Early in my stay in Manchester in the spring of 1912 I became convinced that the electronic constitution of the Rutherford atom was governed throughout by the quantum of action. A support for this view was found not only in the fact that Planck's relation appeared approximately applicable to the more loosely bound electrons involved in the chemical and optical properties of the elements, but especially in the tracing of similar relationships as regards the most firmly bound electrons in the atom revealed by the characteristic penetrating radiation discovered by Barkla. Thus, measurements of the energy necessary to produce the Barkla radiation by electron bombardment of various elements, performed by Whiddington at the time when I was staying in Cambridge, exhibited simple regularities of the kind to be expected from an estimate of the firmest binding energy of an electron rotating in a Planck orbit round a nucleus with a charge given by the atomic number. From Lawrence Bragg's recently published Rutherford Lecture I have been very interested to learn that William Bragg, then in Leeds, in his first investigation of x-ray spectra, based on Laue's discovery in 1912, was fully aware of the bearing of Whiddington's results on the connection between the Barkla radiation and the ordering of the elements in Mendeleev's table, a problem which through Moseley's work in Manchester soon was to receive such complete elucidation.

During the last month of my stay in Manchester I was mainly occupied with a theoretical investigation of the stopping power of matter for α- and β-rays. This problem, which originally was discussed by J. J. Thomson from the point of view of his own atomic model, had just been re-examined by Darwin on the basis of the Rutherford model. In connection with the considerations mentioned above regarding the frequencies involved in the electron binding in the atom, it occurred to me that the transfer of energy from the particles to the electrons could be simply treated in analogy to the dispersion and absorption of radiation. In this way, it proved possible to interpret the results of the stopping power measurements as additional support for ascribing to hydrogen and helium the atomic numbers 1 and 2 in conformity with general chemical evidence, and in particular with Rutherford and Royds' demonstration of the formation of helium gas by the collection of α-particles escaping from thin-walled emanation tubes. Also for the more complex case of heavier substances, approximate agreement was ascertained with the expected atomic numbers and the estimated values for the binding energies of the electrons, but the theoretical methods were much too primitive to yield more accurate results. As is well known, an appropriate treatment of the problem by modern methods of quantum mechanics was first achieved in 1930 by H. Bethe.

Although Rutherford just at that time was concentrating on the preparation of his great book, *Radioactive Substances and Their Radiations*, he nevertheless followed my work with a constant interest, which gave me the opportunity to learn about the care which he always took in the publications of his pupils. After my return to Denmark I was married in mid-summer 1912 and, on our wedding trip in August to England and Scotland, my wife and I passed through Manchester to visit Rutherford and deliver the completed manuscript of my paper on the stopping problems. Both Rutherford and his wife received us with a cordiality which laid the foundation of the intimate friendship that through the many years connected the families.

II

After settling down in Copenhagen, I remained in close contact with Rutherford, to whom I regularly reported about the development of the work on general atomic problems, which I had started in Manchester. Common to Rutherford's answers, which were always very encouraging, was the spontaneity and joy with which he told about the work in his laboratory. It was indeed the beginning of a long correspondence which lasted over 25 years and which revives, every time I look into it, my memories of Rutherford's enthusiasm for the progress in the field he had opened up and the warm interest he took in the endeavours of every one trying to contribute to it.

My letters to Rutherford in the autumn of 1912 concerned the continued endeavours to trace the role of the quantum of action for the electronic

[387]

constitution of the Rutherford atom, including problems of molecular bindings and radiative and magnetic effects. Still, the stability question presented in all such considerations intricate difficulties stimulating the search for a firmer hold. However, after various attempts to apply quantum ideas in a more consistent manner, it struck me in the early spring of 1913 that a clue to the problem of atomic stability directly applicable to the Rutherford atom was offered by the remarkably simple laws governing the optical spectra of the elements.

On the basis of the extremely accurate measurements of the wavelengths of spectral lines by Rowland and others, and after contributions by Balmer and by Schuster, Rutherford's predecessor in the Manchester Chair, the general spectral laws were most ingeniously clarified by Rydberg. The principal result of the thorough analysis of the conspicuous series in the line spectra and their mutual relationship was the recognition that the frequency ν of every line in the spectrum of a given element could be represented with unparalleled accuracy as $\nu = T' - T''$, where T' and T'' are two among a multitude of spectral terms T characteristic of the element.

This fundamental combination law obviously defied ordinary mechanical interpretation, and it is interesting to recall how in this connection Lord Rayleigh had pertinently stressed that any general relationship between the frequencies of the normal modes of vibration of a mechanical model would be quadratic and not linear in these frequencies. For the Rutherford atom we should not even expect a line spectrum since, according to ordinary electrodynamics, the frequencies of radiation accompanying the electronic motion would change continuously with the energy emitted. It was therefore natural to attempt to base the explanation of spectra directly on the combination law.

In fact, accepting Einstein's idea of light quanta or photons with energy $h\nu$, where h is Planck's constant, one was led to assume that any emission or absorption of radiation by the atom is an individual process accompanied by an energy transfer $h(T' - T'')$, and to interpret hT as the binding energy of the electrons in some stable, or so-called stationary, state of the atom. In particular, this assumption offered an immediate explanation of the apparently capricious appearance of emission and absorption lines in series spectra. Thus, in emission processes we witness the transition of the atom from a higher to a lower energy level, whereas in the absorption processes we have in general to do with a transfer of the atom from the ground state, with the lowest energy, to one of its excited states.

In the simplest case of the hydrogen spectrum, the terms are with great accuracy given by $T_n = R/n^2$, where n is an integer and R the Rydberg constant. Thus, the interpretation indicated led to a sequence of decreasing values for the binding energy of the electron in the hydrogen atom, pointing to a steplike process by which the electron, originally at a large distance from the nucleus, passes by radiative transitions to stationary states of firmer and firmer binding, characterized by lower and lower n-values, until the ground state, specified by $n = 1$, is reached. Moreover, a comparison of the binding energy in this state with that of an electron moving in a Keplerian orbit around the nucleus yielded orbital dimensions of the same order as the atomic sizes derived from the properties of gases.

On the basis of the Rutherford atomic model, this view also immediately suggested an explanation of the appearance of the Rydberg constant in the more complex spectra of other elements. Thus it was concluded that we were here faced with transition processes involving excited states of the atom, in which one of the electrons had been brought outside the region occupied by the other electrons bound to the nucleus, and therefore exposed to a field of force resembling that surrounding a unit charge.

The tracing of a closer relation between the Rutherford atomic model and the spectral evidence obviously presented intricate problems. On the one hand, the very definition of the charge and mass of the electron and the nucleus rested entirely on an analysis of physical phenomena in terms of the principles of classical mechanics and electromagnetism. On the other hand, the so-called quantum postulate, stating that any change of the intrinsic energy of the atom consists in a complete transition between two stationary states, excluded the possibility of accounting on classical principles for the radiative processes or any other reaction involving the stability of the atom.

As we know today, the solution of such problems demanded the development of a mathematical formalism, the proper interpretation of which implied a radical revision of the foundation for the unambiguous use of elementary physical concepts and the recognition of complementary relationships between phenomena observed under different experimental conditions. Still, at that time, some progress could be made by utilizing classical physical pictures for the classification of stationary states based on Planck's original assumptions on the energy states of a harmonic oscillator. In particular, a starting point was offered by the closer comparison between an oscillator of given frequency and the Keplerian motion of an electron around a nucleus, with a frequency of revolution depending on the binding energy.

In fact, just as in the case of a harmonic oscillator, a simple calculation showed that, for each of the stationary states of the hydrogen atom, the action integrated over an orbital period of the electron could be identified with nh, a condition which in the case of circular orbits is equivalent to a quantization of the angular momentum in units $h/2\pi$. Such identification involved a fixation of the Rydberg constant in terms of the charge e and mass m of the electron and Planck's constant, according to the formula

$$R = \frac{2\pi^2 m e^4}{h^3}$$

which was found to agree with the empirical value within the accuracy of the available measurements of e, m and h.

Although this agreement offered an indication of the scope for the use of mechanical models in picturing stationary states, of course the difficulties involved in any combination of quantum ideas and the principles of ordinary mechanics remained. It was therefore most reassuring to find that the whole approach to the spectral problems fulfilled the obvious demand of embracing the classical physical description in the limit where the action involved is sufficiently large to permit the neglect of the individual quantum. Such considerations presented indeed the first indication of the so-called correspondence principle expressing the aim of representing the essentially statistical account of quantum physics as a rational generalization of the classical physical description.

Thus, in ordinary electrodynamics, the composition of the radiation emitted from an electron system should be determined by the frequencies and amplitudes of the harmonic oscillations into which the motion of the system can be resolved. Of course no such simple relation holds between the Keplerian motion of an electron around a heavy nucleus and the radiation emitted by transitions between the stationary states of the system.   However, in the limiting case of transitions between states for which the values of the quantum number *n* are large compared with their difference, it could be shown that the frequencies of the components of the radiation, appearing as the result of the random individual transition processes, coincide asymptotically with those of the harmonic components of the electron motion.   Moreover, the fact that in a Keplerian orbit, in contrast to a simple harmonic oscillation, there appears not only the frequency of revolution but also higher harmonics, offered the possibility of tracing a classical analogy as regards the unrestricted combination of the terms in the hydrogen spectrum.

Still, the unambiguous demonstration of the relation between the Rutherford atomic model and the spectral evidence was for a time hindered by a peculiar circumstance.   Already twenty years before, Pickering had observed in the spectra of distant stars a series of lines with wavelengths exhibiting a close numerical relationship with the ordinary hydrogen spectrum.   These lines were therefore generally ascribed to hydrogen and were even thought by Rydberg to remove the apparent contrast between the simplicity of the hydrogen spectrum and the complexity of the spectra of other elements including those of the alkalis, whose structure comes nearest to the hydrogen spectrum.   This view was also upheld by the eminent spectroscopist A. Fowler, who just at that time in laboratory experiments with discharges through a mixture of hydrogen and helium gas had observed the Pickering lines and new related spectral series.

However, the Pickering and Fowler lines could not be included in the Rydberg formula for the hydrogen spectrum, unless the number *n* in the expression for the spectral terms were allowed to take half integrals as well as integral values; but this assumption would evidently destroy the asymptotic approach to the classical relationship between energy and spectral frequencies.   On the other hand, such correspondence would hold for the spectrum of a system consisting of an electron bound to a nucleus of charge Ze, whose stationary states are determined by the same value nh of the action integral.   Indeed, the spectral terms for such a system would be given by Z^2R/n^2, which for $Z=2$ yields the same result as the introduction of half-integral values of n in the Rydberg formula.   Thus, it was natural to ascribe the Pickering and Fowler lines to helium ionized by the high thermal agitation in the stars and in the strong discharges used by Fowler. Indeed, if this conclusion were confirmed, a first step would have been made towards the establishment of quantitative relationships between the properties of different elements on the basis of the Rutherford model.

III

When in March 1913 I wrote to Rutherford, enclosing a draft of my first paper on the quantum theory of atomic constitution, I stressed the importance of settling the question of the origin of the Pickering lines and took the opportunity of asking whether experiments to that purpose could be performed in his laboratory, where from Schuster's days appropriate spectroscopic apparatus were available.   I received a prompt answer, so characteristic of Rutherford's acute scientific judgment and helpful human attitude, that I shall quote it in full:

March 20, 1913.

Dear Dr. Bohr,

I have received your paper safely and read it with great interest, but I want to look over it again carefully when I have more leisure. Your ideas as to the mode of origin of spectrum and hydrogen are very ingenious and seem to work out well; but the mixture of Planck's ideas with the old mechanics make it very difficult to form a physical idea of what is the basis of it. There appears to me one grave difficulty in your hypothesis, which I have no doubt you fully realise, namely, how does an electron decide what frequency it is going to vibrate at when it passes from one stationary state to the other ? It seems to me that you would have to assume that the electron knows beforehand where it is going to stop.

There is one criticism of minor character which I would make in the arrangement of the paper. I think in your endeavour to be clear you have a tendency to make your papers much too long, and a tendency to repeat your statements in different parts of the paper. I think that your paper really ought to be cut down, and I think this could be done without sacrificing anything to clearness. I do not know if you appreciate the fact that long papers have a way of frightening readers, who feel that they have not time to dip into them.

I will go over your paper very carefully and let you know what I think about the details. I shall be quite pleased to send it to the *Phil. Mag.* but I would be happier if its volume could be cut down to a fair amount. In any case I will make any corrections in English that are necessary.

I shall be very pleased to see your later papers, but please take to heart my advice, and try to make them as brief as possible consistent with clearness. I am glad to hear that you are coming over to England later and we shall be very glad to see you when you come to Manchester.

By the way, I was much interested in your speculations in regard to Fowler's spectrum. I mentioned the matter to Evans here, who told me that he was much interested in it, and I think it quite possible that he may try some experiments on the matter when he comes back next term. General work goes well, but I am held up momentarily by finding that the mass of the α-particle comes out rather bigger than it ought to be. If correct it is such an important conclusion that I cannot publish it until I am certain of my accuracy at every point. The experiments take a good deal of time and have to be done with great accuracy.

Yours very sincerely,

E. RUTHERFORD.

P.S.   I suppose you have no objection to my using my judgment to cut out any matter I may consider unnecessary in your paper ? Please reply.

Rutherford's first remark was certainly very pertinent, touching on a point which was to become a central issue in the subsequent prolonged discussions. My own views at that time, as expressed in a lecture at a meeting of the Danish Physical Society in October 1913, were that just the radical departure from the accustomed demands on physical explanation involved in the quantum postulate should of itself leave sufficient scope for the possibility of achieving in due course the incorporation of the new assumptions in a logically consistent scheme. In connection with Rutherford's remark, it is of special interest to recall that Einstein, in his famous paper of 1917 on the derivation of Planck's formula for temperature radiation, took the same starting point as regards the origin of spectra, and pointed to the analogy between the statistical laws governing the occurrence of spontaneous radiation processes and the fundamental law of radioactive decay, formulated by Rutherford and Soddy already in 1903. Indeed,

this law, which allowed them at one stroke to disentangle the multifarious phenomena of natural radioactivity then known, also proved the clue to the understanding of the later observed peculiar branching in spontaneous decay processes.

The second point raised with such emphasis in Rutherford's letter brought me
into a quite embarrassing situation. In fact, a few days before receiving his answer,
I had sent Rutherford a considerably extended version of the earlier manuscript,
the additions especially concerning the relation between emission and absorption
spectra and the asymptotic correspondence with the classical physical theories. I
therefore felt the only way to straighten matters was to go at once to Manchester
and talk it all over with Rutherford himself. Although Rutherford was as busy
as ever, he showed an almost angelic patience with me, and after discussions
through several long evenings during which he declared he had never thought I
should prove so obstinate, he consented to leave all the old and new points in the
final paper. Surely, both style and language were essentially improved by
Rutherford's help and advice, and I have often had occasion to think how right
he was in objecting to the rather complicated presentation and especially to the
many repetitions caused by reference to previous literature. This Rutherford
Memorial Lecture has therefore offered a welcome opportunity to give a more
concise account of the actual development of the arguments in those years.

During the following months, the discussion about the origin of the spectral
lines ascribed to helium ions took a dramatic turn. In the first place, Evans was
able to produce the Fowler lines in discharges through helium of extreme purity,
not showing any trace of the ordinary hydrogen lines. Still, Fowler was not yet
convinced and stressed the spurious manner in which spectra may appear in gas
mixtures. Above all he noted that his accurate measurements of the wavelengths
of the Pickering lines did not exactly coincide with those calculated from my
formula with $Z = 2$. An answer to the last point was, however, easily found, since
it was evident that the mass m in the expression for the Rydberg constant had to be
taken not as the mass of a free electron but as the so-called reduced mass
$mM(m+M)^{-1}$, where M is the mass of the nucleus. Indeed, taking this correction into account, the predicted relationship between the spectra of hydrogen and
ionized helium was in complete agreement with all the measurements. This
result was at once welcomed by Fowler who took the opportunity of pointing out
that also in the spectra of other elements series were observed in which the ordinary
Rydberg constant had to be multiplied by a number close to four. Such series
spectra, which are generally referred to as spark spectra, could now be recognized
as originating from excited ions in contrast to the so-called arc spectra due to
excited neutral atoms.

Continued spectroscopical investigations were in the following years to reveal
many spectra of atoms, from which not only one but even several electrons were
removed. In particular, the well-known investigations of Bowen led to the
recognition that the origin of the nebular spectra discussed by Nicholson had to
be sought not in new hypothetical elements, but in atoms of oxygen and nitrogen
in a highly ionized state. Eventually, the prospect arose of arriving, by analysis
of the processes by which the electrons one by one are bound to the nucleus, at a
survey of the binding of every electron in the ground state of the Rutherford
atom. In 1913, of course, the experimental evidence was still far too scarce, and
the theoretical methods for classification of stationary states were not yet
sufficiently developed to cope with so ambitious a task.

IV

In the meantime, the work on the electronic constitution of the atom gradually proceeded, and soon again I permitted myself to ask Rutherford for help and advice. Thus, in June 1913 I went to Manchester with a second paper which, besides a continued discussion of the radioactive displacement law and the origin of the Barkla radiation, dealt with the ground state of atoms containing several electrons. As regards this problem, I tried tentatively to arrange the electron orbits in closed rings resembling the shell structure originally introduced by J. J. Thomson in his early attempt to account by his atomic model for the periodicity features in Mendeleev's table of the elements.

In Rutherford's laboratory, I met on that occasion Hevesy and Paneth, who told me of the success of the first systematic investigations by the tracer method of the solubility of lead sulphide and chromate, which at the beginning of that year they had carried out together in Vienna. In every way, these repeated visits to Manchester were a great stimulation and gave me the welcome opportunity to keep abreast of the work in the laboratory . At that time, assisted by Robinson, Rutherford was busily engaged in the analysis of β-ray emission and in cooperation with Andrade studied γ-ray spectra. Moreover, Darwin and Moseley were then intensely occupied with refined theoretical and experimental investigations on the diffraction of x-rays in crystals.

A special opportunity to see Rutherford again soon arose in connection with the meeting of the British Association for the Advancement of Science in Birmingham in September 1913. At the meeting, attended by Madame Curie, there was in particular a general discussion about the problem of radiation with the participation of such authorities as Rayleigh, Larmor and Lorentz, and especially Jeans who gave an introductory survey of the application of quantum theory to the problem of atomic constitution. His lucid exposition was, in fact, the first public expression of serious interest in considerations which outside the Manchester group were generally received with much scepticism.

An incident which amused Rutherford and us all was the remark of Lord Rayleigh in response to a solemn request by Sir Joseph Larmor to express his opinion on the latest developments. The prompt reply from the great veteran, who in earlier years had contributed so decisively to the elucidation of radiation problems, was : " In my young days I took many views very strongly and among them that a man who has passed his sixtieth year ought not to express himself about modern ideas. Although I must confess that today I do not take this view quite so strongly, I keep it strongly enough not to take part in this discussion !"

On my visit to Manchester in June I had discussed with Darwin and Moseley the question of the proper sequence for the arrangement of the elements according to their atomic number, and learned then for the first time about Moseley's plans to settle this problem by systematic measurements of the high-frequency spectra of the elements by the Laue–Bragg method. With Moseley's extraordinary energy and gifts of purposeful experimentation, his work developed astonishingly quickly, and already in November 1913 I received a most interesting letter from him with an account of his important results and with some questions regarding their interpretation on the lines which had proved applicable to the optical spectra.

In modern history of physics and chemistry, few events have from the outset attracted such general interest as Moseley's discovery of the simple laws allowing an unambiguous assignment of the atomic number to any element from its high-frequency spectrum.   Not only was the decisive support of the Rutherford atomic model immediately recognized, but also the intuition which had led Mendeleev at certain places in his table to depart from the sequence of increasing atomic weights was strikingly brought out.   In particular, it was evident that Moseley's laws offered an unerring guide in the search for as yet undiscovered elements fitting into vacant places in the series of atomic numbers.

Also as regards the problem of the configuration of the electrons in the atom, Moseley's work was to initiate important progress.   Certainly, the predominance, in the innermost part of the atom, of the attraction exerted by the nucleus on the individual electrons over their mutual repulsion afforded the basis for an understanding of the striking similarity between Moseley's spectra and those to be expected for a system consisting of a single electron bound to the bare nucleus. The closer comparison, however, brought new information pertaining to the shell structure of the electronic constitution of the atoms.

An important contribution to this problem was soon after given by Kossel who, as the origin of the K, L and M types of Barkla radiation, pointed to the removal of an electron from one of the sequence of rings or shells surrounding the nucleus.   In particular he ascribed the $K\alpha$ and $K\beta$ components of Moseley's spectra to individual transition processes by which the electron lacking in the K-shell is replaced by one of the electrons in the L- and M-shells, respectively. Proceeding in this way, Kossel was able to trace further relationships between Moseley's measurements of the spectral frequencies which permitted him to represent the whole high-frequency spectrum of an element as a combination scheme in which the product of any of the terms and Planck's constant was to be identified with the energy required to remove an electron from a shell in the atom to a distance from the nucleus beyond all the shells.

In addition, Kossel's views offered an explanation of the fact that the absorption of penetrating radiation of increasing wavelength practically begins with an absorption edge representing the complete removal in one step of an electron of the respective shell.   The absence of intermediate excited states was assumed to be due to the full occupation of all shells in the ground state of the atom.   As is well known, this view eventually found its final expression through Pauli's formulation in 1924 of the general exclusion principle for electron binding states, inspired by Stoner's derivation of finer details of the shell structure of the Rutherford atom from an analysis of the regularities of the optical spectra.

V

In the autumn of 1913, another stir among physicists was created by Stark's discovery of the surprisingly large effect of electric fields on the structure of the lines in the hydrogen spectrum.   With his vigilant attention to all progress in physical science, Rutherford, when he had received Stark's paper from the Prussian Academy, at once wrote to me: "I think it is rather up to you at the present time to write something on the Zeeman and electric effects, if it is possible to reconcile them with your theory."   Responding to Rutherford's challenge, I tried to look into the matter, and it was soon clear to me that in the effects of electric and magnetic fields we had to do with two very different problems.

The essence of Lorentz' and Larmor's interpretations of Zeeman's famous discovery in 1896 was that it pointed directly to electron motions as the origin of line spectra in a way largely independent of special assumptions about the binding mechanism of the electrons in the atom. Even if the origin of the spectra is assigned to individual transitions between stationary states, the correspondence principle thus led one, in view of Larmor's general theorem, to expect a normal Zeeman effect for all spectral lines emitted by electrons bound in a field of central symmetry, as in the Rutherford atom. Rather did the appearance of so-called anomalous Zeeman effects present new puzzles which could only be overcome more than ten years later when the complex structure of the lines in series spectra was traced to an intrinsic electron spin. A most interesting historical account of this development, to which important contributions were given from various sides, can be found in the well-known volume recently published in memory of Pauli.

In the case of an electric field, however, no effect proportional to its intensity was to be expected for the radiation emitted by a harmonic oscillator, and Stark's discovery therefore definitely excluded the conventional idea of elastic vibrations of electrons as the origin of line spectra. Still, for a Keplerian motion of the electron around the nucleus even a comparatively weak external electric field will through secular perturbations produce considerable changes in the shape and orientation of the orbit. By the study of particular cases in which the orbit remains purely periodic in the external field it was possible, by arguments of the same type as those applied to the stationary states of the undisturbed hydrogen atom, to deduce the order of magnitude of the Stark effect and especially to explain its rapid rise from line to line within the hydrogen spectral series. Yet, these considerations clearly showed that, for an explanation of the finer details of the phenomenon, the methods for a classification of stationary states of atomic systems were not sufficiently developed.

In just this respect a great advance was achieved in the following years by the introduction of quantum numbers specifying components of angular momenta and other action integrals. Such methods were first suggested by W. Wilson in 1915 who applied them to electron orbits in the hydrogen atom. However, owing to the circumstance that on Newtonian mechanics every orbit in this case is simply periodic with a frequency of revolution depending only on the total energy of the system, no physical effects were disclosed. Still, the velocity dependence of the electron mass, predicted by the new mechanics of Einstein, removes the degeneracy of the motion and introduces a second period in its harmonic components through a continual slow progression of the aphelion of the Keplerian orbit. In fact, as was shown in Sommerfeld's famous paper of 1916, the separate quantization of the angular momentum and of the action in the radial motion permitted a detailed interpretation of the observed fine structure of the lines in the spectra of the hydrogen atom and helium ion.

Moreover, the effect of magnetic and electric fields on the hydrogen spectrum was treated by Sommerfeld and Epstein who by a masterly application of the methods for quantization of multiperiodic systems were able, in complete accordance with observations, to derive the spectral terms by the combination of which the resolution of the hydrogen lines appears. The compatibility of such methods with the principle of adiabatic invariance of stationary states, which Ehrenfest had formulated in 1914 in order to meet thermodynamical requirements, was secured by the circumstance that the action integrals to which the quantum numbers refer according to classical mechanics are not modified by a variation of the external field slow compared with the characteristic periods of the system.

Further evidence of the fruitfulness of the approach was derived from the application of the correspondence principle to the radiation emitted by multiperiodic systems, permitting qualitative conclusions regarding the relative probabilities for the different transition processes. These considerations were not least confirmed by Kramers' explanation of the apparently capricious variations in the intensities of the Stark effect components of the hydrogen lines. It was even found possible to account by the correspondence argument for the absence of certain types of transitions in other atoms, beyond those which, as pointed out by Rubinowicz, could be excluded by the conservation laws for energy and angular momentum applied to the reaction between the atom and the radiation.

With the help of the rapidly increasing experimental evidence about the structure of complicated optical spectra, as well as the methodical search for finer regularities in the high-frequency spectra by Siegbahn and his collaborators, the classification of the binding states in atoms containing several electrons continually advanced. In particular, the study of the way in which the ground states of the atoms could be built up by the successive bindings of the electrons to the nucleus led to a gradual elucidation of the shell structure of the electronic configuration in the atom. Thus, although such essential elements of the explanation as the electron spin were still unknown, it became in fact possible within about ten years of Rutherford's discovery of the atomic nucleus to achieve a summary interpretation of many of the most striking periodicity features of Mendeleev's table.

The whole approach, however, was still of largely semi-empirical character, and it was soon to become clear that, for an exhaustive account of the physical and chemical properties of the elements, a radically new departure from classical mechanics was needed in order to incorporate the quantum postulate in a logically consistent scheme. To this well-known development we shall have occasion to return, but I shall first proceed with the account of my reminiscences of Rutherford.

VI

The outbreak of the first world war brought about an almost complete dissolution of the Manchester group, but I was lucky to remain in close contact with Rutherford who in the spring of 1914 had invited me to succeed Darwin in the Schuster Readership of Mathematical Physics. On our arrival in Manchester in early autumn, after a stormy voyage round Scotland, my wife and I were most kindly received by the few of our old friends who remained in the laboratory after the departure of colleagues from abroad and the participation in military duties by most of the British. Rutherford and his wife were at that time still in America on their way back from a visit to their relatives in New Zealand, and it goes without saying that their safe return to Manchester some weeks later was greeted by all of us with great relief and joy.

Rutherford was himself soon drawn into military projects, especially concerning the development of methods of sound tracing of submarines, and teaching the students was almost entirely left to Evans, Makower and me. Still, Rutherford found time to continue his own pioneer work, which already before the end of the war was to give such great results, and showed the same warm interest as ever in the endeavours of his collaborators. As regards the problem of atomic constitution, a new impulse was given by the publication in 1914 of the famous experiments by Franck and Hertz on the excitation of atoms by electron impact.

On the one hand, these experiments, carried out with mercury vapour, gave most conspicuous evidence of the stepwise energy transfer in atomic processes; on the other hand, the value of the ionization energy of mercury atoms apparently indicated by the experiments was less than half of that to be expected from the interpretation of the mercury spectrum. One was therefore led to suspect that the ionization observed was not directly related to the electronic collisions but was due to an accompanying photoeffect on the electrodes, produced by the radiation emitted by the mercury atoms on their return from the first excited state to the ground state. Encouraged by Rutherford, Makower and I planned experiments to investigate this point, and an intricate quartz apparatus with various electrodes and grids was constructed with the help of the competent German glass blower in the laboratory, who in the earlier days had made the fine α-ray tubes for Rutherford's investigations on the formation of helium.

With his liberal human attitude, Rutherford had tried to obtain permission for the glass blower to continue his work in England in the war time, but the man's temper, not uncommon for artisans in his field, and releasing itself in violent super-patriotic utterances, eventually led to his internment by the British authorities. Thus when our fine apparatus was ruined by an accident in which its support caught fire, there was no help to reconstruct it, and when also Makower shortly afterwards volunteered for military service, the experiments were given up. I need hardly add that the problem was solved with the expected result quite independently by the brilliant investigations of Davis and Gauthier in New York in 1918, and I have only mentioned our fruitless attempts as an indication of the kind of difficulties with which work in the Manchester laboratory was faced in those days, and which were very similar to those the ladies had to cope with in their households.

Still, Rutherford's never failing optimism exerted a most encouraging influence on his surroundings, and I remember how at the time of a serious set-back in the war, Rutherford quoted the old utterance ascribed to Napoleon about the impossibility of fighting the British because they were too stupid to understand when they had lost. To me, it was also a most pleasant and enlightening experience to be admitted to the monthly discussions among a group of Rutherford's personal friends including Alexander, the philosopher, the historian Tout, the anthropologist Elliot Smith, and Chaim Weizmann, the chemist who thirty years later was to become the first president of Israel and for whose distinctive personality Rutherford had great esteem.

A terrible shock to us all was the tragic message in 1915 of Moseley's untimely death in the Gallipoli campaign, deplored so deeply by the community of physicists all over the world, and which not least Rutherford, who had endeavoured to get Moseley transferred from the front to less dangerous duties, took much to heart.

In the summer of 1916 my wife and I left Manchester and returned to Denmark where I had been appointed to the newly created professorship of theoretical physics in the University of Copenhagen. Notwithstanding the ever increasing difficulties of postal communication, a steady correspondence with Rutherford was kept up. From my side, I reported about the progress with the work on a more general representation of the quantum theory of atomic constitution which

[397]

at that time was further stimulated by the development as regards the classification of stationary states, already referred to. In that connection, Rutherford took an interest in what news I could give from the continent and in particular of my first personal contact with Sommerfeld and Ehrenfest. In his own letters, Rutherford also gave a vivid description of how, in spite of the increasing difficulties and the pressure of other obligations, he strove to continue his investigations in various directions. Thus, in the autumn of 1916, Rutherford wrote about his intense interest in some surprising results regarding the absorption of hard γ-rays produced by high voltage tubes which had just then become available.

In the next years Rutherford was more and more occupied with the possibilities of producing nuclear disintegrations by means of fast α-rays and already in a letter on 9th December 1917 he writes: "I occasionally find an odd half day to try a few of my own experiments and have got I think results that will ultimately prove of great importance. I wish you were here to talk matters over with. I am detecting and counting the lighter atoms set in motion by α-particles, and the results, I think, throw a good deal of light on the character and distribution of forces near the nucleus. I am also trying to break up the atom by this method. In one case, the results look promising but a great deal of work will be required to make sure. Kay helps me and is now an expert counter." A year later, 17th November 1918, Rutherford in his characteristic manner announced further progress: "I wish I had you here to discuss the meaning of some of my results in collision of nuclei. I have got some rather startling results, I think, but it is a heavy and long business getting *certain* proofs of my deductions. Counting weak scintillations is hard on old eyes, but still with the aid of Kay I have got through a good deal of work at odd times the past four years."

In Rutherford's famous papers in the *Philosophical Magazine*, 1919, containing the account of his fundamental discovery of controlled nuclear disintegrations, he refers to the visit to Manchester, in November 1918, of his old collaborator Ernest Marsden who at the Armistice had got leave from military service in France. With his great experience of scintillation experiments from the old Manchester days when, in collaboration with Geiger, he performed the experiments which led Rutherford to his discovery of the atomic nucleus, Marsden helped him to clear up some apparent anomalies in the statistical distribution of the high-speed protons released by the bombardment of nitrogen with α-rays. From Manchester, Marsden returned to New Zealand to take up his own university duties, but kept in close contact with Rutherford through the years.

In July 1919, when after the Armistice travelling was again possible, I went to Manchester to see Rutherford and learned in more detail about his great new discovery of controlled, or so-called artificial, nuclear transmutations, by which he gave birth to what he liked to call 'modern Alchemy', and which in the course of time was to give rise to such tremendous consequences as regards man's mastery of the forces of nature. Rutherford was at that time almost alone in the laboratory, and as told in his letters, the only help in his fundamental researches, apart from Marsden's short visit, was his faithful assistant William Kay, who by his kindness and helpfulness through the years had endeared himself to everyone in the laboratory. During my visit Rutherford also spoke about the great decision he had had to make in response to the offer of the Cavendish professorship in Cambridge left vacant by the retirement of J. J. Thomson. Certainly, it had not been easy for Rutherford to decide to leave Manchester after the many rich years there, but of course he had to follow the call to succeed the unique series of Cavendish professors.

VII

From the beginning, Rutherford gathered around him in the Cavendish Laboratory a large and brilliant group of research workers. A most notable figure was Aston who through many years had worked with J. J. Thomson and already during the wartime had started the development of mass spectroscopic methods which was to lead to the demonstration of the existence of isotopes of almost every element. This discovery, which gave such a convincing confirmation of Rutherford's atomic model, was not entirely unexpected. Already in the early Manchester days, it was understood that the apparent irregularities in the sequence of the atomic weights of the elements when they were ordered according to their chemical properties suggested that, even for the stable elements, the nuclear charge could not be expected to have a unique relation to the nuclear mass. In letters to me in January and February 1920, Rutherford expressed his joy in Aston's work, particularly about the chlorine isotopes which so clearly illustrated the statistical character of the deviations of chemical atomic weights from integral values. He also commented humorously on the lively disputes in the Cavendish Laboratory about the relative merits of different atomic models to which Aston's discovery gave rise.

It was a great help in the continuation of Rutherford's own pioneering work on the constitution and disintegrations of atomic nuclei as well as in the management of the great laboratory, that from the very beginning he was joined by James Chadwick from the old Manchester group, who returned from a long detention in Germany where at the outbreak of the war he had been working in Berlin with Geiger. Among Rutherford's collaborators in the early Cambridge years were also Blackett and Ellis, both coming from a career in the defence services, Ellis having been initiated to physics by Chadwick during their comradeship under German imprisonment. A further asset to the group at the Cavendish was the arrival, a few years later, of Kapitza, who brought with him ingenious projects, in particular for the production of magnetic fields of hitherto unheard-of intensities. In this work he was from the start assisted by John Cockcroft, who with his singular combination of scientific and technological insight was to become such a prominent collaborator of Rutherford.

At the beginning, Charles Darwin, whose mathematical insight had been so helpful in the Manchester years, shared with Ralph Fowler responsibility for the theoretical part of the activities at the Cavendish. In collaboration, they made at that time important contributions to statistical thermodynamics and its application to astrophysical problems. After Darwin's departure for Edinburgh, the principal theoretical adviser and teacher in Cambridge right up to the second world war was Fowler, who had become Rutherford's son-in-law. Not only did Fowler with enthusiastic vigour participate in the work at the Cavendish, but he also soon found numerous gifted pupils who benefitted from his inspiration. Foremost among these were Lennard-Jones and Hartree who both contributed, each along his own line, to the development of atomic and molecular physics, and especially Dirac, who from his early youth distinguished himself by his unique logical power.

[399]

Ever since I left Manchester in 1916, I had of course tried to use the experience gained in Rutherford's laboratory and it is with gratitude that I recall how Rutherford from the very outset most kindly and effectively supported my endeavours in Copenhagen to create an institute to promote intimate collaboration between theoretical and experimental physicists. It was a special encouragement that already in the autumn of 1920, when the Institute building was nearing completion, Rutherford found time to visit us in Copenhagen. As a token of appreciation, the University conferred upon him an honorary degree, and on that occasion he gave a most stimulating and humorous address which was long remembered by all present.

For the work in the new Institute it was of great benefit that we were joined shortly after the war by my old friend from the Manchester days, George Hevesy, who during the more than twenty years he worked in Copenhagen carried out many of his famous physico-chemical and biological researches, based on the isotopic tracer method. A special event, in which Rutherford took great interest, was the application of Moseley's method by Coster and Hevesy in 1922 to the successful search for the missing element now called hafnium, the properties of which gave strong additional support to the interpretation of the periodic system of the elements. An auspicious start was given to the general experimental work by a visit, at the opening of the laboratory, of James Franck, who during the following months most kindly instructed the Danish collaborators in the refined technique of excitation of atomic spectra by electron bombardment which he had so ingeniously developed together with Gustav Hertz. The first among the many distinguished theoretical physicists who stayed with us for a longer period was Hans Kramers who as a quite young man came to Copenhagen during the war and proved to be such an invaluable asset to our group during the ten years he worked with us until, in 1926, he left his position as lecturer in the Institute to take over a professorship in Utrecht. Shortly after Kramers' arrival in Copenhagen came two promising young men, Oscar Klein from Sweden and Svein Rosseland from Norway, who already in 1920 made their names known by pointing to the so-called collisions of the second kind, in which atoms are transferred by electron bombardment from a higher to a lower stationary state with gain of velocity for the electron. Indeed, the occurrence of such processes is decisive for ensuring thermal equilibrium in a way analogous to the induced radiative transitions which played an essential role in Einstein's derivation of Planck's formula for temperature radiation. The consideration of collisions of the second kind proved particularly important for the elucidation of the radiative properties of stellar atmospheres, to which at that time Saha, working in Cambridge with Fowler, made such fundamental contributions.

The group at the Copenhagen Institute was joined in 1922 by Pauli, and two years later by Heisenberg, both pupils of Sommerfeld, and who, young as they were, had already accomplished most brilliant work. I had made their acquaintance and formed a deep impression of their extraordinary talent in the summer of 1922 during a lecturing visit to Göttingen, which initiated a long and fruitful cooperation between the group working there under the leadership of Born and Franck, and the Copenhagen group. From the early days our close connection with the great centre in Cambridge was maintained especially by longer visits to Copenhagen of Darwin, Dirac, Fowler, Hartree, Mott, and others.

VIII

Those years, when a unique cooperation of a whole generation of theoretical physicists from many countries created step by step a logically consistent generalization of classical mechanics and electromagnetism, have sometimes been described as the 'heroic' era in quantum physics.   To everyone following this development, it was an unforgettable experience to witness how, through the combination of different lines of approach and the introduction of appropriate mathematical methods, a new outlook emerged regarding the comprehension of physical experience.   Many obstacles had to be overcome before this goal was reached, and time and again decisive progress was achieved by some of the youngest among us.

The common starting point was the recognition that, notwithstanding the great help which the use of mechanical pictures had temporarily offered for the classification of stationary states of atoms in isolation or exposed to constant external forces, it was clear, as already mentioned, that a fundamentally new departure was needed.   Not only was the difficulty of picturing the electronic constitution of chemical compounds on the basis of the Rutherford atomic model more and more evident, but insurmountable difficulties also arose in any attempts to account in detail for the complexity of atomic spectra, especially conspicuous in the peculiar duplex character of the arc spectrum of helium.

The first step to a more general formulation of the correspondence principle was offered by the problem of optical dispersion.   Indeed, the close relation between the atomic dispersion and the selective absorption of spectral lines so beautifully illustrated by the ingenious experiments of R. W. Wood and P. V. Bevan on the absorption and dispersion in alkali vapours, suggested from the very beginning a correspondence approach.   On the basis of Einstein's formulation of the statistical laws for the occurrence of radiation-induced transitions between stationary states of an atomic system, Kramers in 1924 succeeded in establishing a general dispersion formula, involving only the energies of these states and the probabilities of spontaneous transitions between them.   This theory, further developed by Kramers and Heisenberg, included even new dispersion effects connected with the appearance, under the influence of the radiation, of possibilities for transitions not present in the unperturbed atom, and an analogue to which is the Raman effect in molecular spectra.

Shortly afterwards, an advance of fundamental significance was achieved by Heisenberg who in 1925 introduced a most ingenious formalism, in which all use of orbital pictures beyond the general asymptotic correspondence was avoided.   In this bold conception, the canonical equations of mechanics are retained in their Hamiltonian form, but the conjugate variables are replaced by operators subject to a non-commutative algorism involving Planck's constant as well as the symbol $\sqrt{-1}$.   In fact, by representing the mechanical quantities by hermitian matrices with elements referring to all possible transition processes between stationary states, it proved possible without any arbitrariness to deduce the energies of these states and the probabilities of the associated transition processes.   This so-called quantum mechanics, to the elaboration of which Born and Jordan as well as Dirac from the outset made important contributions, opened the way to a consistent statistical treatment of many atomic problems which hitherto were only amenable to a semi-empirical approach.

For the completion of this great task the emphasis on the formal analogy between mechanics and optics, originally stressed by Hamilton, proved most helpful and instructive. Thus, pointing to the similar roles played by the quantum numbers in the classification of stationary states by means of mechanical pictures, and by the numbers of nodes in characterizing the possible standing waves in elastic media, L. de Broglie had already in 1924 been led to a comparison between the behaviour of free material particles and the properties of photons. Especially illuminating was his demonstration of the identity of the particle velocity with the group velocity of a wave-packet built up of components with wavelengths confined to a small interval, and each related to a value of the momentum by Einstein's equation between the momentum of a photon and the corresponding wavelength of radiation. As is well known, the pertinence of this comparison soon received a decisive confirmation with the discoveries by Davisson and Germer and by George Thomson of selective scattering of electrons in crystals.

The culminating event of this period was Schrödinger's establishment in 1926 of a more comprehensive wave mechanics in which the stationary states are conceived as proper solutions of a fundamental wave equation, obtained by regarding the Hamiltonian of a system of electric particles as a differential operator acting upon a function of the coordinates which define the configuration of the system. In the case of the hydrogen atom, not only did this method lead to a remarkably simple determination of the energies of the stationary states, but Schrödinger also showed that the superposition of any two proper solutions corresponded to a distribution of electric charge and current in the atom which on classical electrodynamics would give rise to the emission and resonance absorption of a monochromatic radiation of a frequency coinciding with some line of the hydrogen spectrum.

Similarly, Schrödinger was able to explain essential features of the dispersion of radiation by atoms by representing the charge and current distribution of the atom perturbed by the incident radiation as the effect of a superposition of the proper functions defining the manifold of possible stationary states of the unperturbed system. Particularly suggestive was the derivation on such lines of the laws of the Compton effect which, in spite of the striking support it gave to Einstein's original photon idea, at first presented obvious difficulties for a correspondence treatment, attempting to combine conservation of energy and momentum with a division of the process in two separate steps, consisting in an absorption and an emission of radiation resembling radiative transitions between the stationary states of an atomic system.

This recognition of the wide scope of arguments implying the use of a superposition principle similar to that of classical electromagnetic field theory, which was only implicitly contained in the matrix formulation of quantum mechanics, meant a great advance in the treatment of atomic problems. Still, it was from the beginning obvious that wave mechanics did not point to any less radical modification of the classical physical approach than the statistical description envisaged by the correspondence principle. Thus, I remember how, on a visit of Schrödinger to Copenhagen in 1926, when he gave us a most impressive account of his wonderful work, we argued with him that any procedure disregarding the individual character of the quantum processes would never account for Planck's fundamental formula of thermal radiation.

Notwithstanding the remarkable analogy between essential features of atomic processes and classical resonance problems, it must indeed be taken into account that in wave mechanics we are dealing with functions which do not generally take real values, but demand the essential use of the symbol $\sqrt{-1}$ just as the matrices of quantum mechanics. Moreover, when dealing with the constitution of atoms with more than one electron, or collisions between atoms and free electric particles, the state functions are not represented in ordinary space but in a configuration space of as many dimensions as there are degrees of freedom in the total system. The essentially statistical character of the physical deductions from wave mechanics was eventually clarified by Born's brilliant treatment of general collision problems.

The equivalence of the physical contents of the two different mathematical formalisms was completely elucidated by the transformation theory formulated independently by Dirac in Copenhagen and Jordan in Göttingen, which introduced in quantum physics possibilities for the change of variables similar to those offered by the symmetrical character of the equations of motion in classical dynamics in the canonical form given by Hamilton. An analogous situation is met with in the formulation of a quantum electrodynamics incorporating the photon concept. This aim was first achieved in Dirac's quantum theory of radiation treating phases and amplitudes of the harmonic components of the fields as non-commuting variables. After further ingenious contributions by Jordan, Klein and Wigner, this formalism found, as is well known, essential completion in the work of Heisenberg and Pauli.

A special illustration of the power and scope of the mathematical methods of quantum physics is presented by the peculiar quantum statistics pertaining to systems of identical particles where we have to do with a feature as foreign to classical physics as the quantum of action itself. Indeed, any problem which calls for relevant application of Bose–Einstein or Fermi–Dirac statistics in principle excludes pictorial illustration. In particular, this situation left room for the proper formulation of the Pauli exclusion principle, which not only gave the final elucidation of the periodicity relations in Mendeleev's table, but in the following years proved fertile for the understanding of most of the varied aspects of the atomic constitution of matter.

A fundamental contribution to the clarification of the principles of quantum statistics was afforded by Heisenberg's ingenious explanation in 1926 of the duplicity of the helium spectrum. In fact, as he showed, the set of stationary states of atoms with two electrons consists of two non-combining groups corresponding to symmetric and antisymmetric spatial wave functions, respectively associated with opposite and parallel orientations of the electron spins. Shortly afterwards, Heitler and London succeeded on the same lines in explaining the binding mechanism in the hydrogen molecule and thereby opened the way for the understanding of homopolar chemical bonds. Even Rutherford's famous formula for the scattering of charged particles by atomic nuclei had, as was shown by Mott, to be essentially modified when applied to collisions between identical particles like protons and hydrogen nuclei or α-rays and helium nuclei. However, in the actual experiments of large angle scattering of fast α-rays by heavy nuclei, from which Rutherford drew his fundamental conclusions, we are well within the range of validity of classical mechanics.

The increasing use of more and more refined mathematical abstractions to ensure consistency in the account of atomic phenomena found in 1928 a temporary climax in Dirac's relativistic quantum theory of the electron.   Thus the concept of electron spin, to the treatment of which Darwin and Pauli had made important contributions, was harmoniously incorporated in Dirac's spinor analysis.   Above all, however, in connection with the discovery of the positron by Anderson and Blackett, Dirac's theory prepared the recognition of the existence of antiparticles of equal mass but opposite electric charges and opposite orientations of the magnetic moment relative to the spin axis.   As is well known, we have here to do with a development which in a novel manner has restored and enlarged that isotropy in space and reversibility in time which has been one of the basic ideas of the classical physical approach.

The wonderful progress of our knowledge of the atomic constitution of matter and of the methods by which such knowledge can be acquired and interrelated has indeed carried us far beyond the scope of the deterministic pictorial description brought to such perfection by Newton and Maxwell.   Following this development at close hand, I have often had occasion to think of the dominating influence of Rutherford's original discovery of the atomic nucleus, which at every stage presented us with so forceful a challenge.

IX

In all the long and rich years during which Rutherford worked with untiring vigour in the Cavendish, I often came to Cambridge, where on Rutherford's invitation I gave several courses of lectures on theoretical problems including the epistemological implications of the development of quantum theory.   On such occasions it was always a great encouragement to feel the open mind and intense interest with which Rutherford followed the progress in the field of research which he had himself so largely initiated and the growth of which should carry us so far beyond the horizon which limited the outlook at the early stages.

Indeed, the extensive use of abstract mathematical methods to cope with the rapidly increasing evidence about atomic phenomena brought the whole observational problem more and more to the foreground.   In its roots this problem is as old as physical science itself.   Thus the philosophers in ancient Greece, who based the explanation of the specific properties of substances on the limited divisibility of all matter, took it for granted that the coarseness of our sense organs would forever prevent the direct observation of individual atoms.   In such respect, the situation has been radically changed in our days by the construction of amplification devices like cloud chambers and the counter mechanisms originally developed by Rutherford and Geiger in connection with their measurements of the numbers and charges of α-particles.   Still, the exploration of the world of atoms was, as we have seen, to reveal inherent limitations in the mode of description embodied in common language developed for the orientation in our surroundings and the account of events of daily life.

In words conforming with Rutherford's whole attitude, one may say that the aim of experimentation is to put questions to nature, and of course Rutherford owed his success in this task to his intuition in shaping such questions so as to permit the most useful answers.   In order that the enquiry may augment common knowledge it is an obvious demand that the recording of observations as well as the construction and handling of the apparatus, necessary for the definition of the experimental conditions, be described in plain language. In actual physical research, this demand is amply satisfied with the specification of the experimental arrangement through the use of bodies like diaphragms and photographic plates, so large and heavy that their manipulation can be accounted for in terms of classical physics, although of course the properties of the materials of which the instruments, as well as our own bodies, are built up depend essentially on the constitution and stability of the component atomic systems defying such account.

The description of ordinary experience presupposes the unrestricted divisibility of the course of the phenomena in space and time and the linking of all steps in an unbroken chain in terms of cause and effect. Ultimately, this viewpoint rests on the fineness of our senses which for perception demands an interaction with the objects under investigation so small that in ordinary circumstances it is without appreciable influence on the course of events. In the edifice of classical physics, this situation finds its idealized expression in the assumption that the interaction between the object and the tools of observation can be neglected or, at any rate, compensated for.

The element of wholeness, symbolized by the quantum of action and completely foreign to classical physical principles has, however, the consequence that in the study of quantum processes any experimental enquiry implies an interaction between the atomic object and the measuring tools which, although essential for the characterization of the phenomena, evades a separate account if the experiment is to serve its purpose of yielding unambiguous answers to our questions. It is indeed the recognition of this situation which makes the recourse to a statistical mode of description imperative as regards the expectations of the occurrence of individual quantum effects in one and the same experimental arrangement, and which removes any apparent contradiction between phenomena observed under mutually exclusive experimental conditions. However contrasting such phenomena may at first sight appear, it must be realized that they are complementary in the sense that taken together they exhaust all information about the atomic object, which can be expressed in common language without ambiguity.

The notion of complementarity does not imply any renunciation of detailed analysis limiting the scope of our enquiry, but simply stresses the character of objective description, independent of subjective judgment, in any field of experience where unambiguous communication essentially involves regard to the circumstances in which evidence is obtained. In logical respect, such a situation is well known from discussions about psychological and social problems where many words have been used in a complementary manner since the very origin of language. Of course we are here often dealing with qualities unsuited to the quantitative analysis characteristic of so-called exact sciences, whose task, according to the programme of Galileo, is to base all description on well-defined measurements.

Notwithstanding the help which mathematics has always offered for such a task, it must be realized that the very definition of mathematical symbols and operations rests on simple logical use of common language. Indeed, mathematics is not to be regarded as a special branch of knowledge based on the accumulation of experience, but rather as a refinement of general language, supplementing it with appropriate tools to represent relations for which ordinary verbal communication is imprecise or too cumbersome. Strictly speaking, the mathematical formalism of quantum mechanics and electrodynamics merely offers rules of calculation for the deduction of expectations about observations obtained under well-defined experimental conditions specified by classical physical concepts. The exhaustive character of this description depends not only on the freedom, offered by the formalism, of choosing these conditions in any conceivable manner, but equally on the fact that the very definition of the phenomena under consideration for their completion implies an element of irreversibility in the observational process emphasizing the fundamentally irreversible character of the concept of observation itself.

Of course all contradictions in the complementary account in quantum physics were beforehand excluded by the logical consistency of the mathematical scheme upholding every demand of correspondence. Still, the recognition of the reciprocal latitude for the fixation of any two canonically conjugate variables, expressed in the principle of indeterminacy formulated by Heisenberg in 1927, was a decisive step towards the elucidation of the measuring problem in quantum mechanics. Indeed it became evident that the formal representation of physical quantities by non-commuting operators directly reflects the relationship of mutual exclusion between the operations by which the respective physical quantities are defined and measured.

To gain familiarity with this situation, the detailed treatment of a great variety of examples of such arguments was needed. Notwithstanding the generalized significance of the superposition principle in quantum physics, an important guide for the closer study of observational problems was repeatedly found in Rayleigh's classic analysis of the inverse relation between the accuracy of image-forming by microscopes and the resolving power of spectroscopic instruments. In this connection not least Darwin's mastery of the methods of mathematical physics often proved helpful.

With all appreciation of Planck's happy choice of words when introducing the concept of a universal 'quantum of action', or the suggestive value of the idea of 'intrinsic spin', it must be realized that such notions merely refer to relationships between well-defined experimental evidence which cannot be comprehended by the classical mode of description. Indeed, the numbers expressing the values of the quantum or spin in ordinary physical units do not concern direct measurements of classically defined actions or angular momenta, but are logically interpretable only by consistent use of the mathematical formalism of quantum theory. In particular the much discussed impossibility of measuring the magnetic moment of a free electron by ordinary magnetometers is directly evident from the fact that in Dirac's theory the spin and magnetic moment do not result from any alteration in the basic Hamiltonian equation of motion, but appear as consequences of the peculiar non-commutative character of the operator calculus.

The question of the proper interpretation of the notions of complementarity and indeterminacy was not settled without lively disputes, in particular at the Solvay meetings of 1927 and 1930. On these occasions, Einstein challenged us with his subtle criticism which especially gave the inspiration to a closer analysis of the role of the instruments in the measuring process. A crucial point, irrevocably excluding the possibility of reverting to causal pictorial description, was the recognition that the scope of unambiguous application of the general conservation laws of momentum and energy is inherently limited by the circumstance that any experimental arrangement, allowing the location of atomic objects in space and time, implies a transfer, uncontrollable in principle, of momentum and energy to the fixed scales and regulated clocks indispensable for the definition of the reference frame.   The physical interpretation of the relativistic formulation of quantum theory ultimately rests on the possibility of fulfilling all relativity exigencies in the account of the handling of the macroscopic measuring apparatus.

This circumstance was especially elucidated in the discussion of the measurability of electromagnetic field components raised by Landau and Peierls as a serious argument against the consistency of quantum field theory.   Indeed, a detailed investigation in collaboration with Rosenfeld showed that all the predictions of the theory in this respect could be fulfilled when due regard was taken to the mutual exclusiveness of the fixation of the values of electric and magnetic intensities and the specification of the photon composition of the field.   An analogous situation is met with in positron theory where any arrangement suited for measurements of the charge distribution in space necessarily implies uncontrollable creation of electron pairs.

The typical quantum features of electromagnetic fields do not depend on scale, since the two fundamental constants—the velocity of light c and the quantum of action h—do not allow of any fixation of quantities of dimensions of a length or time interval.   Relativistic electron theory, however, involves the charge e and mass m of the electron, and essential characteristics of the phenomena are limited to spatial extensions of the order h/mc.   The fact that this length is still large compared with the 'electron radius' e^2/mc^2, which limits the unambiguous application of the concepts of classical electromagnetic theory, suggests, however, that there is still a wide scope for the validity of quantum electrodynamics, even though many of its consequences cannot be tested by practical experimental arrangements involving measuring instruments sufficiently large to permit the neglect of the statistical element in their construction and handling.   Such difficulties would of course also prevent any direct enquiry into the close interactions of the fundamental constituents of matter, whose number has been so largely increased by recent discoveries, and in the exploration of their relationships we must therefore be prepared for a new approach transcending the scope of present quantum theory.

It need hardly be stressed that such problems do not arise in the account of the ordinary physical and chemical properties of matter, based on the Rutherford atomic model, in the analysis of which use is only made of well-defined characteristics of the constituent particles.   Here, the complementary description offers indeed the adequate approach to the problem of atomic stability with which we were faced from the very beginning.   Thus, the interpretation of spectral regularities and chemical bonds refers to experimental conditions mutually exclusive of those which permit exact control of the position and displacement of the individual electrons in the atomic systems.

In this connection, it is of decisive importance to realize that the fruitful application of structural formulae in chemistry rests solely on the fact that the atomic nuclei are so much heavier than the electrons that, in comparison with molecular dimensions, the indeterminacy in the position of the nuclei can be largely neglected.   When we look back on the whole development we recognize indeed that the discovery of the concentration of the mass of the atom within a region so small compared with its extension has been the clue to the understanding

[407]

of an immense field of experience embracing the crystalline structure of solids as well as the complex molecular systems which carry the genetic characters of living organisms.

As is well known, the methods of quantum theory have also proved decisive for the clarification of many problems regarding the constitution and stability of the atomic nuclei themselves. To some early disclosed aspects of such problems I shall have occasion to refer in continuing the account of my reminiscences of Rutherford, but it would be beyond the scope of this Memorial Lecture to attempt a detailed account of the rapidly increasing insight in the intrinsic nuclear constitution, brought about by the work of the present generation of experimental and theoretical physicists. This development reminds indeed the elders among us of the gradual clarification of the electronic constitution of the atom in the first decades after Rutherford's fundamental discovery.

X

Every physicist is of course acquainted with the imposing series of brilliant investigations with which Rutherford to the very end of his life augmented our insight into the properties and constitution of atomic nuclei. I shall therefore here mention only a few of my remembrances from those years when I often had occasion to follow the work in the Cavendish Laboratory and learned in talks with Rutherford about the trend of his views and the problems occupying him and his collaborators.

With his penetrating intuition, Rutherford was early aware of the strange and novel problems presented by the existence and stability of composite nuclei. Indeed, already in the Manchester time he had pointed out that any approach to these problems demanded the assumption of forces of short range between the nuclear constitutents, of a kind essentially different from the electric forces acting between charged particles. With the intention of throwing more light on the specific nuclear forces, Rutherford and Chadwick, in the first years in Cambridge, performed thorough investigations of anomalous scattering of α-rays in close nuclear collisions.

Although much important new evidence was obtained in these investigations, it was more and more felt that, for a broader attack on nuclear problems, the natural α-ray sources were not sufficient and that it was desirable to have available intense beams of high-energy particles produced by artificial acceleration of ions. In spite of Chadwick's urge to start the construction of an appropriate accelerator, Rutherford was during several years reluctant to embark upon such a great and expensive enterprise in his laboratory. This attitude is quite understandable when one considers the wonderful progress which Rutherford hitherto had achieved with the help of very modest experimental equipment. The task of competing with natural, radioactive sources must also have appeared quite formidable at that time. The outlook, however, was changed by the development of quantum theory and its first application to nuclear problems.

Rutherford himself had as early as 1920 in his second Bakerian Lecture clearly pointed out the difficulties of interpreting α-ray emission from nuclei on the basis of the simple mechanical ideas, which had proved so helpful in explaining the scattering of α-particles by nuclei, since the velocity of the ejected particles was not large enough to allow them by reversal to re-enter the nuclei against the electric repulsion. However, the possibilities of penetration of particles through potential barriers was soon recognized as a consequence of wave mechanics, and in 1928 Gamow, working in Göttingen, as well as Condon and Gurney in Princeton, gave on this basis a general explanation of α-decay and even a detailed account of the relationship between the lifetime of the nucleus and the kinetic energy of the emitted α-particles, in conformity with the empirical regularities found by Geiger and Nuttall in the early Manchester days.

When, in the summer of 1928, Gamow joined us in Copenhagen, he was investigating the penetration of charged particles into nuclei by a reverse tunnel effect. He had started this work in Göttingen and discussed it with Houtermans and Atkinson, with the result that the latter were led to suggest that the source of solar energy might be traced to nuclear transmutations induced by impact of protons with the great thermal velocities which according to Eddington's ideas were to be expected in the interior of the sun.

During a brief visit to Cambridge in October 1928, Gamow discussed the experimental prospects arising from his theoretical considerations with Cockcroft, who by more detailed estimates convinced himself of the possibility of obtaining observable effects by bombardment of light nuclei with protons of an energy far smaller than that of α-particles from natural radioactive sources. As the result appeared promising, Rutherford accepted Cockcroft's proposal to build a high voltage accelerator for such experiments. Work on the construction of the apparatus was started by Cockcroft at the end of 1928 and was continued during the following year with the collaboration of Walton. The first experiments they made with accelerated protons in March 1930, in which they looked for gamma rays emitted as a result of the interaction of the protons with the target nuclei, gave no result. The apparatus then had to be rebuilt owing to a change of laboratory and, as is well known, production of high-speed α-particles by proton impact on lithium nuclei was obtained in March 1932.

These experiments initiated a new stage of most important progress, during which both our knowledge of nuclear reactions and the mastery of accelerator techniques rapidly increased from year to year. Already Cockcroft and Walton's first experiments gave results of great significance in several respects. Not only did they confirm in all details the predictions of quantum theory as regards the dependence of the reaction cross section on the energy of the protons, but it was also possible to connect the kinetic energy of the emitted α-rays with the masses of the reacting particles which were at that time known with sufficient accuracy thanks to Aston's ingenious development of mass spectroscopy. Indeed, this comparison offered the first experimental test of Einstein's famous relation between energy and mass, to which he had been led many years before by relativity arguments. It need hardly be recalled how fundamental this relation was to prove in the further development of nuclear research.

The story of Chadwick's discovery of the neutron presents similar dramatic features. It is characteristic of the broadness of Rutherford's views that he early anticipated the presence in nuclei of a heavy neutral constituent of a mass closely coinciding with that of the proton. As gradually became clear, this idea would indeed explain Aston's discoveries of isotopes of nearly all elements with atomic masses closely approximated by multiples of the atomic weight of hydrogen. In connection with their studies of many types of α-ray induced nuclear disintegrations Rutherford and Chadwick made an extensive search for evidence concerning the existence of such a particle. However, the problem came to a climax through the observation by Bothe and the Joliot–Curies of a penetrating radiation resulting from the bombardment of beryllium by α-particles. At first this radiation was assumed to be of γ-ray type, but with Chadwick's thorough familiarity with the multifarious aspects of radiative phenomena he clearly perceived that the experimental evidence was not compatible with this view.

Indeed, from a masterly investigation, in which a number of new features of the phenomenon were revealed, Chadwick was able to prove that one was faced with momentum and energy exchanges through a neutral particle, the mass of which he determined as differing from that of the proton by less than one part in a thousand. On account of the ease with which neutrons, compared with charged particles, can pass through matter without transfer of energy to the electrons and penetrate into atomic nuclei, Chadwick's discovery opened great possibilities of producing new types of nuclear transmutations. Some most interesting cases of such new effects were immediately demonstrated in the Cavendish by Feather, who obtained cloud chamber pictures showing nitrogen nuclei disintegrating under α-particle release by neutron bombardment. As is well known, continued studies in many laboratories along such lines were rapidly to increase our knowledge of nuclear constitution and transmutation processes.

In the spring of 1932, at one of our yearly conferences in the Copenhagen Institute, where as always we were happy to see many of our former collaborators, one of the main topics of discussion was of course the implications of the discovery of the neutron, and a special point raised was the apparently strange circumstance that in Dee's beautiful cloud chamber pictures no interaction whatever was observed between the neutrons and the electrons bound in the atoms. In relation to this point, it was argued that, owing to the dependence in quantum physics of the scattering cross section on the reduced mass of the colliding particles, this fact would not be inconsistent even with the assumption of short range interaction between the neutron and an electron of strength similar to that between the neutron and a proton. A few days later, I got a letter from Rutherford touching incidentally on this point, and which I cannot resist quoting in full:

April 21, 1932.

My dear Bohr,

I was very glad to hear about you all from Fowler when he returned to Cambridge and to know what an excellent meeting of old friends you had. I was interested to hear about your theory of the Neutron. I saw it described very nicely by the scientific correspondent of the Manchester Guardian, Crowther, who is quite intelligent in these matters. I am very pleased to hear that you regard the Neutron with favour. I think the evidence in its support, obtained by Chadwick and others, is now complete in the main essentials. It is still a moot point how much ionization is, or should be, produced to account for the absorption, disregarding the collisions with nuclei.

It never rains but it pours, and I have another interesting development to tell you about of which a short account should appear in *Nature* next week. You know that we have a High Tension Laboratory where steady D.C. voltages can be readily obtained up to 600,000 volts or more. They have recently been examining the effects of a bombardment of light elements by protons. The protons fall on a surface of the material inclined at 45° to the axis of the tube, and the effects produced were observed at the side by the scintillation method, the zinc sulphide screen being covered with sufficient mica to stop the protons. In the case of lithium brilliant scintillations are observed, beginning at about 125,000 volts and mounting up very rapidly with voltage when many hundreds per minute can be obtained with a protonic current of a few milliamperes. The α-particles apparently had a definite range, practically independent of voltage, of 8 cm in air. The simplest assumption to make is that the lithium 7 captures a proton breaking up with the emission of two ordinary α-particles. On this view the total energy liberated is about 16 million volts and this is of the right order for the changes in mass involved, assuming the Conservation of Energy.

Later special experiments will be made to test the nature of the particles but from the brightness of the scintillations and the trail in a Wilson chamber it seems probable they are α-particles. In experiments in the last few days similar effects have been observed in Boron and Fluorine but the ranges of the particles are smaller although they look like α-particles. It may be, Boron 11 captures a proton and breaks up into three alphas, while fluorine breaks up into oxygen and an alpha. The energy changes are in approximate accord with these conclusions. I am sure you will be much interested in these new results which we hope to extend in the near future.

It is clear that the α-particle, neutron and proton will probably give rise to different types of disintegration and it may be significant that so far results have only been observed in $4n + 3$ elements. It looks as if the addition of the 4th proton leads at once to the formation of an α-particle and the consequent disintegration. I suppose, however, the whole question should be regarded as the result of one process rather than of steps.

I am very pleased that the energy and expense in getting high potentials has been rewarded by definite and interesting results. Actually they ought to have observed the effect a year or so ago but did not try it in the right way. You can easily appreciate that these results may open up a wide line of research in transmutation generally.

We are all very well at home and I start lectures tomorrow. With best wishes to you and Mrs. Bohr.

Yours ever

RUTHERFORD

Beryllium shows some queer effects—still to be made definite.
I shall possibly refer to these experiments in the Royal Society discussion on nuclei on Thursday April 25.

Of course, in reading this letter, it must be borne in mind that my previous visits to Cambridge had kept me acquainted with the work in progress in the Cavendish Laboratory, so that Rutherford had no need to specify the individual contributions of his collaborators. The letter is indeed a spontaneous expression of his exuberant joy in the great achievements of those years and his eagerness in pursuing their consequences.

XI

As a true pioneer, Rutherford never relied merely on intuition, however far it carried him, but was always on the look-out for new sources of knowledge which could possibly lead to unexpected progress. Thus, also in Cambridge, Rutherford and his collaborators continued with great vigour and steadily refined apparatus the investigations of the radioactive processes of α- and β-decay. The important work of Rutherford and Ellis on β-ray spectra revealed the possibility of a clear distinction between intranuclear effects and the interaction of the β-particle with the outer electron system and led to the clarification of the mechanism of internal conversion.

[411]

Moreover, Ellis' demonstration of the continuous spectral distribution of the electrons directly emitted from the nucleus raised a puzzling question about energy conservation, which was eventually answered by Pauli's bold hypothesis of the simultaneous emission of a neutrino, affording the basis for Fermi's ingenious theory of β-decay.

By the great improvement of accuracy in measurements of α-ray spectra by Rutherford, Wynn-Williams and others, much new light was thrown on the fine structure of these spectra and their relation to the energy levels of the residual nucleus resulting from the α-decay. A special adventure at an earlier stage was the discovery of the capture of electrons by α-rays which, after the first observation of the phenomenon in 1922 by Henderson, was explored by Rutherford in one of his most masterly researches. As is well known, this work, which brought so much information about the process of electron capture, was to attract new attention a few years after Rutherford's death when, with the discovery of the fission processes of heavy nuclei by neutron impact, the study of the penetration of highly charged nuclear fragments through matter, where electron capture is the dominating feature, came into the foreground.

Great progress both as regards general outlook and experimental technique was initiated in 1933 by the discovery by Frédéric Joliot and Irène Curie of so-called artificial β-radioactivity produced by nuclear transmutations initiated by α-ray bombardment. I need hardly here remind how, by Enrico Fermi's brilliant systematic investigations of neutron induced nuclear transmutations, radioactive isotopes of a great number of elements were discovered and much information gained about nuclear processes initiated by capture of slow neutrons. Especially the continued study of such processes revealed most remarkable resonance effects of a sharpness far surpassing that of the peaks in the cross section of α-ray induced reactions first observed by Pose and to Gurney's explanation of which, on the basis of the potential well model, Gamow at once drew Rutherford's attention.

Already Blackett's observations with his ingenious automatic cloud chamber technique had shown that, in the very process investigated in Rutherford's original experiments on artificial nuclear disintegrations, the incident α-particle remained incorporated in the residual nucleus left after proton escape. It now became clear that all types of nuclear transmutations within a large energy region take place in two well-separated steps. Of these the first is the formation of a relatively long-lived compound nucleus, while the second is the release of its excitation energy as a result of a competition between the various possible modes of disintegration and radiative processes. Such views, in which Rutherford took a vivid interest, were the theme for the last course of lectures which on Rutherford's invitation I gave in 1936 in the Cavendish Laboratory.

Less than two years after Rutherford's death in 1937, a new and dramatic development was initiated by the discovery of the fission processes of the heaviest elements by his old friend and collaborator in Montreal, Otto Hahn, working in Berlin with Fritz Strassmann. Immediately after this discovery, Lise Meitner and Otto Frisch, then working in Stockholm and Copenhagen, and now both in Cambridge, made an important contribution to the understanding of the phenomenon by pointing out that the critical decrease in stability of nuclei of high charge was a simple consequence of the balancing of cohesive forces between the nuclear constituents and the electrostatic repulsion. A closer investigation of the fission process in collaboration with Wheeler showed that many of its characteristic features could be accounted for in terms of the mechanism of nuclear reactions involving as a first step the formation of a compound nucleus.

In Rutherford's last years he found in Marcus Oliphant a collaborator and friend whose general attitude and working power reminds us so much of his own. At that time new possibilities of research were opened by Urey's discovery of the heavy hydrogen isotope ^{2}H or deuterium, and by the construction of the cyclotron by Lawrence, who in his first investigations on nuclear disintegrations by deuteron beams obtained a number of new striking effects. In the classical experiments of Rutherford and Oliphant, in which by bombardment of separated lithium isotopes with protons and deuterons they were led to the discovery of ^{3}H, or tritium, and ^{3}He, the foundation was indeed created for the vigorous modern attempt to apply thermonuclear reactions to the realization of the full promises of atomic energy sources.

From the very beginning of his radioactive researches, Rutherford was acutely aware of the wide perspectives they opened in several directions. In particular, he early took deep interest in the possibility of arriving at an estimate of the age of the earth and of understanding the thermal equilibrium in the crust of our planet. Even if the liberation of nuclear energy for technological purposes was still to come, it must have been a great satisfaction for Rutherford that the explanation of the hitherto completely unknown source of solar energy as a result of the development he had initiated had come within the horizon in his lifetime.

XII

When we look back on Rutherford's life we perceive it, of course, against the unique background of his epoch-making scientific achievements, but our memories will always remain irradiated by the enchantment of his personality. In earlier Memorial Lectures, several of Rutherford's closest co-workers have recalled the inspiration which emanated from his vigour and enthusiasm and the charm of his impulsive ways. Indeed, in spite of the large and rapidly expanding scope of Rutherford's scientific and administrative activities, the same spirit reigned in the Cavendish as we all had enjoyed so much in the early Manchester days.

A faithful account of Rutherford's eventful life from childhood till his last days has been written by his old friend from the Montreal period, A. S. Eve. Especially the many quotations in Eve's book from Rutherford's astonishingly large correspondence give a vivid impression of his relations with colleagues and pupils all over the world. Eve also does not fail to report some of the humorous stories which constantly grew around Rutherford, and to which I alluded in a speech, reproduced in his book, when Rutherford for the second and last time visited us in Copenhagen in 1932.

Characteristic of Rutherford's whole attitude was the warm interest he took in any one of the many young physicists with whom he came into contact for shorter or longer periods. Thus I vividly remember the circumstances of my first meeting in Rutherford's office in the Cavendish with the young Robert Oppenheimer, with whom I was later to come into such close friendship. Indeed, before Oppenheimer entered the office, Rutherford, with his keen appreciation of talents, had described the rich gifts of the young man, which in the course of time were to create for him his eminent position in scientific life in the United States.

As is well known, Oppenheimer, shortly after his visit to Cambridge, during his studies in Göttingen was among the first who called attention to the phenomenon of particle penetration through potential barriers, which should prove basic for the ingenious explanation of α-decay by Gamow and others. After his stay in Copenhagen, Gamow came in 1929 to Cambridge, where his steady contributions to the interpretation of nuclear phenomena were highly appreciated by Rutherford, who also greatly enjoyed the bizarre and subtle humour which Gamow unfolded in daily intercourse and to which he later gave so abundant expression in his well-known popular books.

Of the many young physicists from abroad working in the Cavendish Laboratory in those years, one of the most colourful personalities was Kapitza, whose power of imagination and talent as a physical engineer Rutherford greatly admired. The relationship between Rutherford and Kapitza was very characteristic of them both and was, notwithstanding inevitable emotional encounters, marked from first to last by a deep mutual affection. Such sentiments were also behind Rutherford's efforts to support Kapitza's work after his return to Russia in 1934, and were from Kapitza's side most movingly expressed in a letter which I received from him after Rutherford's death.

When, in the beginning of the nineteen thirties, as an extension to the Cavendish, the Mond Laboratory was created on Rutherford's initiative for the promotion of Kapitza's promising projects, Kapitza wanted in its decoration to give expression for his joy in Rutherford's friendship. Still, the carving of a crocodile on the outer wall caused comments which could only be appeased by reference to special Russian folklore about animal life. Above all, however, the relief of Rutherford, in Eric Gill's artistic interpretation, placed in the entrance hall, deeply shocked many of Rutherford's friends. On a visit to Cambridge I confessed that I could not share this indignation, and this remark was so welcomed that Kapitza and Dirac presented me with a replica of the relief; installed above the fireplace in my office at the Copenhagen Institute it has since given me daily enjoyment.

When, in recognition of his position in science Rutherford was given a British peerage, he took a keen interest in his new responsibilities as a member of the House of Lords, but there was certainly no change in the directness and simplicity of his behaviour. Thus I do not remember any more severe utterance of his to me than, when at a Royal Society Club dinner in a conversation with some of his friends I had referred to him in the third person as Lord Rutherford, he furiously turned on me with the words: "Do you lord me?".

In the nearly twenty years during which Rutherford, right up to his death, worked with undiminished energy in Cambridge, my wife and I kept in close touch with him and his family. Almost every year, we were hospitably received in their beautiful home in Newnham Cottage at the backs of the old colleges, with the lovely garden in which Rutherford found relaxation and the upkeep of which gave Mary Rutherford much enjoyable work. I remember many peaceful evening hours in Rutherford's study spent discussing not merely new prospects of physical science but also topics from many other fields of human interest. In such conversation one was never tempted to overrate the interest of one's own contributions since Rutherford after a long day's work was apt to fall asleep as soon as the discourse seemed pointless to him. One then just had to wait until he woke up and resumed the conversation with usual vigour as if nothing had happened.

On Sundays, Rutherford regularly played golf in the morning with some close friends and dined in the evening in Trinity College where he met many eminent scholars and enjoyed discussions on the most different subjects.   With his insatiable curiosity for all aspects of life, Rutherford had great esteem for his learned colleagues; however, I remember how he once remarked, on our way back from Trinity, that to his mind so-called humanists went a bit too far when expressing pride in their complete ignorance of what happened in between the pressing of a button at their front door and the sounding of a bell in the kitchen.

Some of Rutherford's utterances have led to the misunderstanding that he did not fully appreciate the value of mathematical formalisms for the progress of physical science.   On the contrary, as the whole branch of physics, created so largely by himself, rapidly developed, Rutherford often expressed admiration for the new theoretical methods, and even took interest in questions of the philosophical implications of quantum theory.   I remember especially how, at my last stay with him a few weeks before his death, he was fascinated by the complementary approach to biological and social problems and how eagerly he discussed the possibility of obtaining experimental evidence on the origin of national traditions and prejudices by such unconventional procedures as the interchange of newborn children between nations.

A few weeks later, at the Centenary Celebrations for Galvani in Bologna, we learned with sorrow and consternation of Rutherford's death, and I went at once to England to attend his funeral.   Having been with them both so shortly before and found Rutherford in full vigour and in the same high spirits as always, it was under tragic circumstances, indeed, that I met Mary Rutherford again.   We talked about Ernest's great life in which from their early youth she had been so faithful a companion, and how to me he had almost been as a second father.   On one of the following days, Rutherford was buried in Westminster Abbey, close to the sarcophagus of Newton.

Rutherford did not live to see the great technological revolution which was to ensue from his discovery of the atomic nucleus and his subsequent fundamental researches.   However, he was always aware of the responsibility connected with any increase in our knowledge and abilities.   We are now confronted with a most serious challenge to our whole civilization, to see to it that disastrous use of the formidable powers which have come into the hands of man be prevented, and that the great progress be turned into promoting the welfare of all humanity.   Some of us, who were called to take part in the war projects, often thought of Rutherford and modestly strove to act in the way which we imagined he himself would have taken.

The memory which Rutherford has left us remains to everyone who had the good fortune to know and come close to him a rich source of encouragement and fortitude.   The generations who in coming years pursue the exploration of the world of atoms will continue to draw inspiration from the work and life of the great pioneer.

THE RUTHERFORD MEMORIAL LECTURE 1958 †

Reminiscences of the Founder of Nuclear Science
and of Some Developments Based on his Work

By NIELS BOHR

Footnotes to an article published in the
PROCEEDINGS OF THE PHYSICAL SOCIETY
Vol. LXXVIII, p.1083, 1961

[416]

No.	Page	Line from above	below	Reference
1)	1083		last	E. Rutherford: Phil.Mag. 21, 669 (1911).
2)	1085	10		G. Hevesy: Perspectives in Biology and Medicine 1, 345 (1958); cf. also Naturwiss. 27, 604 (1923).
3)	1085		5	F. Soddy: Chem.News 107, 97 (1913).
4)	1085		3	K. Fajans: Verh.deutsch. Phys.Ges. 15, 240 (1913).
5)	1086	2		A. v.d.Broek: Phys.Z. 14, 32 (1913).
6)	1086	28		A. Haas: Sitz.Ber.Wiener Akad.d.Wiss., mat.nat.Kl. Abt. IIa (1910).
7)	1086		19	J. Nicholson: Month.Not. Roy.Astr.Soc. 72, 679 (1912).
8)	1086		16	N. Bjerrum: Nernst-Festschrift, p. 90 (Halle 1912); cf. also Verh.deutsch.Phys. Ges. 16, 640, 737 (1914).
9)	1087	5		R. Whiddington: Proc.Roy. Soc.A 85, 323 (1911).
10)	1087	21		E. Rutherford and T. Royds: Phil.Mag. 17, 281 (1909).
11)	1087	25		N. Bohr: Phil.Mag. 25, 10 (1913). For the proper treatment of this problem by modern methods of quantum mechanics, cf. H.Bethe: Ann.d.Phys. (5) 5, 325 (1930).
12)	1087	28		E. Rutherford: Radioactive Substances and their Radiations. Cambridge University Press (1913).

No.	Page	Line from above	below	Reference
13)	1088	11		cf. N. Bohr: Rydberg's Discovery of the Spectral Law. Proc. Rydberg Centennial Conf. on Atomic Spectroscopy (Lund 1955).
14)	1090	24		A. Fowler: Month.Not. Roy.Astr.Soc. (Dec.1912).
15)	1091		7	cf. N. Bohr: Fysisk Tidsskr. $\underline{12}$, 3 (1913); translated in The Theory of Spectra and Atomic Constitution. Cambridge University Press (1922).
16)	1091		4	A. Einstein: Phys.Z. $\underline{18}$, 121 (1917).
17)	1092	23		E.J. Evans: Nature $\underline{92}$, 5 (1913).
18)	1092	23		A. Fowler: Nature $\underline{92}$, 95 (1913).
19)	1092		18	N. Bohr: Nature $\underline{92}$, 231 (1913).
20)	1092		17	A. Fowler: Nature $\underline{92}$, 232 (1913).
21)	1093	6		N. Bohr: Phil.Mag. $\underline{26}$, 476 (1913).
22)	1094	2		H.G.J. Moseley: Phil.Mag. $\underline{26}$, 1024 (1913).
23)	1094	16		W. Kossel: Verh.deutsch. Phys.Ges. $\underline{16}$, 898, 953 (1914).
24)	1094	27		N. Bohr: Phil. Mag. $\underline{30}$, 394 (1915).
25)	1095	9		E. Uhlenbeck and S. Goudsmit: Naturwiss. $\underline{13}$, 953 (1925) and Nature $\underline{117}$, 264 (1926); cf. also The Pauli Memorial Volume, ed. by M. Fierz and V. Weisskopf (Interscience Publishers, New York, London, 1960), where the history of the concept of electron spin is treated in various articles.

No.	Page	Line from above	Line from below	Reference
26)	1095	22		N. Bohr: Phil.Mag. 27, 506 (1914).
27)	1095		21	W. Wilson: Phil.Mag. 29, 795 (1915)
28)	1095		13	A. Sommerfeld: Ann.d. Phys. (4) 51, 1 (1916).
29)	1096	4		cf. N. Bohr: The Quantum Theory of Line Spectra I-III. Det Kgl.Dan.Vid. Selsk.Skr. (1918-22), and H.A. Kramers: Intensities of Spectral Lines, ibid. (1919).
30)	1096	10		A. Rubinowicz: Phys.Z. 19, 441, 465 (1918).
31)	1096	22		cf. N. Bohr: Nature 107, 104 (1921) and Z.Phys. 9, 1 (1922); N. Bohr and D. Coster: Z.Phys. 12, 342 (1923), and in particular E. Stoner: Phil. Mag. 48, 79 (1924).
32)	1097	11		cf. N. Bohr: Phil.Mag. 30, 394 (1915).
33)	1102		12	N. Bohr, H.A. Kramers and J.C. Slater: Phil.Mag. 47, 785 (1924).
34)	1106	14		W. Heisenberg: Z.Phys. 43, 172 (1927); cf. also N. Bohr: Nature 121, 580 (1928) and C.G. Darwin: Proc.Roy.Soc.A 130, 632 (1931).
35)	1106		13	cf., e.g., W. Pauli: Institut Solvay, 6me Conseil de Physique, p. 175 (1930) and N. Bohr: Atti del Congresso di Fisica Nucleare (Roma 1932).

No.	Page	Line from above	below	Reference
36)	1106		5	cf. N. Bohr: Atomic Physics and Human Knowledge. (John Wiley and Sons, New York 1958).
37)	1107	16		cf. N. Bohr and L. Rosenfeld: Det Kgl.Dan.Vid. Selsk.,Mat.-fys.Medd. $\underline{12}$, no.8 (1933) and Phys.Rev. $\underline{78}$, 794 (1950).
38)	1108	last		cf. E. Rutherford: Proc. Roy.Soc.A $\underline{97}$, 374 (1920).
38a)	1109		6	cf. 38).
39)	1112	12		E. Rutherford: Phil.Mag. (6) $\underline{47}$, 277 (1924).
40)	1112	17		cf. N. Bohr: Det Kgl.Dan. Vid.Selsk.Mat.-fys.Medd. $\underline{18}$, no.8 (1948).
41)	1112		12	N. Bohr: Nature $\underline{137}$, 344 (1936); N. Bohr and F. Kalckar: Det Kgl.Dan.Vid. Selsk.Mat.-fys.Medd. $\underline{14}$, no.10 (1937).
42)	1113	3		N. Bohr and J.A. Wheeler: Phys.Rev. $\underline{56}$, 426 (1939).
43)	1113		15	A.S. Eve: Rutherford. Cambridge University Press (1939).

V. THE GENESIS OF QUANTUM MECHANICS

"Essays 1958–1962 on Atomic Physics and Human Knowledge", Interscience Publishers, New York 1963, pp. 74–78

See Introduction to Part IV, p. [330].

Niels Bohr, Werner Heisenberg and P.A.M. Dirac in discussion. Lindau 1962.

THE GENESIS OF QUANTUM MECHANICS (1961)

Versions published in German, English and Danish

German: Die Entstehung der Quantenmechanik
A "Werner Heisenberg und die Physik unserer Zeit", Friedr. Vieweg & Sohn, Braunschweig 1961, pp. IX–XII
B "Atomphysik und menschliche Erkenntnis II", Friedr. Vieweg & Sohn, Braunschweig 1966, pp. 75–79

English: The Genesis of Quantum Mechanics
C "Essays 1958–1962 on Atomic Physics and Human Knowledge", Interscience Publishers, New York 1963, pp. 74–78 (reprinted in: "Essays 1958–1962 on Atomic Physics and Human Knowledge, The Philosophical Writings of Niels Bohr, Vol. III", Ox Bow Press, Woodbridge, Connecticut 1987, pp. 74–78)

Danish: Kvantemekanikkens tilblivelse
D "Atomfysik og menneskelig erkendelse II", J.H. Schultz Forlag, Copenhagen 1964, pp. 95–100

All of these versions agree with each other, with the exception of a few minor improvements of formulation from A to B.

The Genesis of Quantum Mechanics[1]

1962

The sixtieth birthday of Werner Heisenberg gives me a welcome opportunity to recount some of my memories from the time when he worked with us in Copenhagen and with such genius created the foundations of quantum mechanics.

I met the young student Heisenberg for the first time almost forty years ago, in the spring of 1922. It was in Göttingen, where I was invited to give a series of lectures on the state of the quantum theory of atomic constitution. In spite of the great progress achieved by Sommerfeld and his school with their supreme mastery of the methods, developed by Hamilton and Jacobi, for treating mechanical systems in terms of invariant action quantities, the problem of incorporating the quantum of action in a consistent generalization of classical physics still contained deep-lying difficulties. The divergent attitudes to this problem gave rise to lively discussions, and I remember with pleasure the interest with which especially the younger listeners responded to my emphasis on the correspondence principle as a guide for the further development.

On this occasion we discussed the possibility that two of Sommerfeld's youngest pupils, of whom he had the greatest expectations, should come to Copenhagen. While Pauli joined our group already in the same year, Heisenberg stayed for another year in Munich at Sommerfeld's advice to complete his doctoral thesis. Before Heisenberg came

[1] Translated from the German article *Die Entstehung der Quantenmechanik.*

74

to Copenhagen for a longer stay in the fall of 1924, we had the pleasure of seeing him here briefly already the previous spring. The Göttingen discussions were then continued in the Institute as well as on long walks, and I gained an even stronger impression of Heisenberg's rare gifts.

Our conversations touched upon many problems in physics and philosophy, and the requirement of unambiguous definition of the concepts in question was particularly emphasized. The discussions of problems in atomic physics were concerned above all with the strange character of the quantum of action in relation to the concepts employed in the description of all experimental results, and in this connection we also talked about the possibility that mathematical abstractions here, as in relativity theory, might prove to be useful. At that time no such perspectives were yet at hand, but the development of the physical ideas had just entered a new stage.

An attempt to encompass individual atomic reactions within the framework of classical radiation theory had been made in collaboration with Kramers and Slater. Although at first we encountered difficulties regarding the strict conservation of energy and momentum, these investigations led to further development of the notion of virtual oscillators as the connecting link between atoms and radiation fields. Soon after, a great step forward was achieved by the dispersion theory of Kramers, developed on correspondence lines, which established a direct connection with Einstein's general probability rules for processes of absorption and spontaneous and induced emission.

Heisenberg and Kramers immediately took up a close collaboration which resulted in an extension of the dispersion theory. In particular, they investigated a novel type of atomic reaction connected with perturbations caused by radiation fields. However, the treatment remained semi-empirical in the sense that there was still no self-contained basis for the derivation of the spectral terms of atoms or their reaction probabilities. There was then only a vague hope that the connection just mentioned between dispersion and perturbation effects could be exploited for a gradual reformulation of the theory, by which, step by step, every inappropriate use of classical ideas would be eliminated. Impressed with the difficulties of such a programme, we therefore all felt the greatest admiration when the 23-year-old Heisenberg found out how the goal could be reached with one stroke.

With his ingenious representation of kinematical and dynamical quantities through non-commutable symbols the foundation had indeed been laid on which the further development was to rest. The formal completion of the new quantum mechanics was soon achieved in close cooperation with Born and Jordan. In this connection I would like to mention that Heisenberg, on receiving a letter from Jordan, described his feelings in roughly the following words: "Now the learned Göttingen mathematicians talk so much about Hermitian matrices, but I do not even know what a matrix is." Soon afterwards Dirac, whom Heisenberg had told about his new ideas on a visit to Cambridge, gave another brilliant example of a young physicist able to create by himself the mathematical tools suited for his work.

Although decisive progress in the consistent representation of quantum problems had evidently been achieved by the new formalism, it appeared for a time as though the correspondence requirements had not yet all been fulfilled. Thus, I recall how Pauli, whose treatment of the energy states of the hydrogen atom was one of the first fruitful applications of Heisenberg's views, expressed his dissatisfaction with the situation. He stressed that it ought to be obvious that the position of the moon in its orbit around the earth can be determined, whereas, according to matrix mechanics, every state of a two-body system with well-defined energy allows only statistical expectations concerning the kinematical quantities in question.

Just in this respect new light was to come from the analogy between the motion of material particles and the wave propagation of photons, to which de Broglie had referred already in 1924. On this basis Schrödinger succeeded in 1926 by the establishment of his famous wave equation in applying with brilliant success the powerful means of function theory to the treatment of many atomic problems. In regard to the correspondence problem, it was above all essential that every solution of the Schrödinger equation could be represented as a superposition of harmonic eigenfunctions, thus making it possible to follow in detail how the motion of particles can be compared to the propagation of wave packets.

In the beginning, however, a certain lack of clarity remained concerning the mutual relationship between the apparently so different mathematical treatments of the quantum problems. As an example of the discussions from those days, I might mention how a doubt, expressed by Heisenberg regarding the possibility of explaining the Stern–Gerlach effect in terms of wave propagation, was dissipated by Oskar Klein. The latter, who was particularly familiar with the analogy between mechanics and optics, pointed out by Hamilton, and had himself come upon the track of the wave equation, could just refer to Huygens' old explanation of the double refraction in crystals. Schrödinger's visit to Copenhagen in the fall of 1926 afforded a special opportunity for lively exchange of views. On this occasion, Heisenberg and I tried to convince him that his beautiful treatment of dispersion phenomena could not be brought into conformity with Planck's law of black-body radiation without expressly taking into account the individual character of the absorption and emission processes.

The statistical interpretation of Schrödinger's wave mechanics was soon clarified by Born's investigation of collision problems. The complete equivalence of the different methods was also established already in 1926 by the transformation theory of Dirac and Jordan. In this connection I remember how, in an Institute colloquium, Heisenberg pointed out that matrix mechanics permits the determination not only of the expectation value of a physical quantity but also of the expectation values of every power of this quantity, and how in the subsequent discussion Dirac stated that this remark had given him the clue to general transformations.

In the winter of 1925–26 Heisenberg worked in Göttingen, where I also came for a few days. We talked especially about the discovery of the electron spin, whose dramatic history has recently been illuminated from many sides in the Pauli Memorial Volume. It was a great pleasure for the group in Copenhagen that during this visit Heisenberg agreed to take over the lectureship at our Institute after Kramers, who had accepted the chair of theoretical physics in Utrecht. His lectures in the following academic year were appreciated by the students not only for their content but also for Heisenberg's perfect command of the Danish language.

This year was an exceedingly fruitful one for the continuation of Heisenberg's fundamental scientific work. An outstanding achievement was the clarification of the duplicity of the helium spectrum, long considered one of the greatest difficulties in the quantum theory of atomic constitution. Through Heisenberg's treatment of the electron spin in connection with the symmetry properties of the wave functions, the Pauli principle appeared in a much clearer light, and this was at once to bring about most important consequences. Heisenberg himself was led directly to an understanding of ferromagnetism, and soon came the clarification of the homopolar chemical bonds by Heitler and London, as well as Dennison's solution of the old riddle of the specific heat of hydrogen.

In connection with the rapid development of atomic physics in those years the interest was increasingly focused on the question of the logical ordering of the wealth of empirical data. Heisenberg's deep-going investigation of these problems found expression in the famous paper "The Visualizable Content of the Kinematics and Mechanics of Quantum Theory", which appeared towards the end of his stay in

Copenhagen and in which the indeterminacy relations were formulated for the first time. From the beginning, the attitude towards the apparent paradoxes in quantum theory was characterized by the emphasis on the features of wholeness in the elementary processes, connected with the quantum of action. While so far it had been clear that energy content and other invariant quantities could be strictly defined only for isolated systems, Heisenberg's analysis revealed the extent to which the state of an atomic system is influenced during any observation by the unavoidable interaction with the measuring tools.

The emphasis on observational problems again brought to the fore-ground the questions Heisenberg and I had talked about on his first visit to Copenhagen and gave rise to further discussions about general epistemological problems. Just the requirement that it be possible to communicate experimental findings in an unambiguous manner implies that the experimental arrangement and the results of the observation must be expressed in the common language adapted to our orientation in the environment. Thus, the description of quantum phenomena requires a distinction in principle between the objects under investigation and the measuring apparatus by means of which the experimental conditions are defined. In particular, the contrasts met with here, hitherto so unfamiliar in physics, emphasize the necessity, well known in other domains of experience, to take into consideration the conditions under which the experience has been obtained.

In rendering some of my recollections from the old days, it has above all been on my mind to stress how the close collaboration among a whole generation of physicists from many countries succeeded step by step in creating order in a vast new domain of knowledge. In this period of development of physical science, in which it was a wonderful adventure to participate, Werner Heisenberg occupied an outstanding position.

VI. THE SOLVAY MEETINGS AND THE DEVELOPMENT OF QUANTUM PHYSICS

"La théorie quantique des champs", Douzième Conseil de physique tenu à l'Université Libre de Bruxelles du 9 au 14 octobre 1961, Interscience Publishers, New York 1962, pp. 13–36

Address at the Twelfth Solvay Meeting in Brussels, October 1961

See Introduction to Part IV, p. [330].

[429]

THE SOLVAY MEETINGS AND THE DEVELOPMENT OF QUANTUM PHYSICS (1961)

Versions published in English, Danish and German

English: The Solvay Meetings and the Development of Quantum Physics

A "La théorie quantique des champs", Douzième Conseil de physique, Brusselles, 9–14 October 1961, Interscience Publishers, New York 1962, pp. 13–36

B "Essays 1958–1962 on Atomic Physics and Human Knowledge", Interscience Publishers, New York 1963, pp. 79–100 (reprinted in: "Essays 1958–1962 on Atomic Physics and Human Knowledge, The Philosophical Writings of Niels Bohr, Vol. III", Ox Bow Press, Woodbridge, Connecticut 1987, pp. 79–100)

Danish: Solvay-møderne og kvantefysikkens udvikling

C "Atomfysik og menneskelig erkendelse II", J.H. Schultz Forlag, Copenhagen 1964, pp. 101–126

German: Die Solvay-Konferenzen und die Entwicklung der Atomphysik

D "Atomphysik und menschliche Erkenntnis II", Friedr. Vieweg & Sohn, Braunschweig 1966, pp. 80–102

All of these versions agree with each other, with the exception of a few minor improvements of formulation from *A* to *B*.

[430]

THE SOLVAY MEETINGS AND THE DEVELOPMENT OF QUANTUM PHYSICS

by N. BOHR

The series of conferences originally convened, just fifty years ago, at the far-sighted initiative of Ernest Solvay and continued under the auspices of the International Institute of Physics founded by him, have been unique occasions for physicists to discuss the fundamental problems which were at the centre of interest at the different periods, and have thereby in many ways stimulated modern development of physical science.

The careful recording of the reports and of the subsequent discussions at each of these meetings will in the future be a most valuable source of information for students of the history of science wishing to gain an impression of the grappling with the new problems raised in the beginning of our century. Indeed, the gradual clarification of these problems through the combined effort of a whole generation of physicists was in the following decades not only so largely to augment our insight in the atomic constitution of matter, but even to lead to a new outlook as regards the comprehension of physical experience.

As one of those who in the course of time have attended several of the Solvay conferences and have had personal contact with many of the participants in the earliest of these meetings, I have welcomed the invitation on this occasion to recall some of my reminiscences of the part played by the discussions for the elucidation of the problems confronting us. In approaching this task I shall endeavour to present these discussions against the background of the many-sided development which atomic physics has undergone in the last fifty years.

13

The very theme of the first Solvay conference in 1911, Radiation Theory and Quanta, indicates the background for the discussions in those days. The most important advances in physics in the former century were perhaps the development of Maxwell's electromagnetic theory, which offered so far-reaching an explanation of radiative phenomena, and the statistical interpretation of the thermo-dynamical principles culminating in Boltzmann's recognition of the relation between the entropy and probability of the state of a complex mechanical system. Still, the account of the spectral distribution of cavity radiation in thermal equilibrium with the enclosing walls presented unsuspected difficulties, especially brought out by Rayleigh's masterly analysis.

A turning point in the development was reached by Planck's discovery, in the first year of our century, of the universal quantum of action revealing a feature of wholeness in atomic processes completely foreign to classical physical ideas and even transcending the ancient doctrine of the limited divisibility of matter. On this new background the apparent paradoxes involved in any attempt at a detailed description of the interaction between radiation and matter were early stressed by Einstein, who did not only call attention to the support for Planck's ideas offered by investigations of the specific heat of solids at low temperature but, in connection with his original treatment of the photoelectric effect, also introduced the idea of light quanta or photons as carriers of energy and momentum in elementary radiative processes.

Indeed the introduction of the photon concept meant a revival of the old dilemma from Newton's and Huygens' days of the corpuscular or undulatory constitution of light, which had seemed resolved in favour of the latter by the establishment of the electromagnetic theory of radiation. The situation was most peculiar since the very definition of the energy or momentum of the photon, given by the product of Planck's constant and the frequency or wave number of the radiation, directly refers to the characteristics of a wave picture. We were thus confronted with a novel kind of complementary relationship between the applications of different fundamental concepts of classical physics, the study of which in the course of time was to make the limited scope of deterministic description evident and to call for an essentially statistical account of even the most elementary atomic processes.

1 4

The discussions at the meeting were initiated by a brilliant exposition by Lorentz of the argumentation based on classical ideas leading to the principle of equipartition of energy between the various degrees of freedom of a physical system, including not only the motion of its constituent material particles but also the normal modes of vibration of the electromagnetic field associated with the electric charge of the particles. This argumentation, following the lines of Rayleigh's analysis of thermal radiative equilibrium led, however, to the well known paradoxical result that no temperature equilibrium was possible, since the whole energy of the system would be gradually transferred to electromagnetic vibrations of steadily increasing frequencies.

Apparently the only way to reconcile radiation theory with the principles of ordinary statistical mechanics was the suggestion by Jeans that under the experimental conditions one did not have to do with a true equilibrium but with a quasi-stationary state, in which the production of high frequency radiation escaped notice. A testimony to the acuteness with which the difficulties in radiation theory were felt was a letter from Lord Rayleigh, read at the conference, in which he admonishes to take Jeans' suggestion into careful consideration. Still, by closer examination it was soon to become evident that Jeans' argument could not be upheld.

In many respects the reports and discussions at the conference were most illuminating. Thus, after reports by Warburg and Rubens of the experimental evidence supporting Planck's law of temperature radiation, Planck himself gave an exposition of the arguments which had led him to the discovery of the quantum of action. In commenting on the difficulties of harmonizing this new feature with the conceptual framework of classical physics, he stressed that the essential point was not the introduction of a new hypothesis of energy quanta, but rather a remoulding of the very concept of action, and expressed the conviction that the principle of least action, which was also upheld in relativity theory, would prove a guidance for the further development of quantum theory.

In the last report at the conference, Einstein summarized many applications of the quantum concept and dealt in particular with the fundamental arguments used in his explanation of the anomalies of specific heats at low temperatures. The discussions of these phenomena had been introduced at the meeting in a report by Nernst

15

on the application of quantum theory to different problems of physics and chemistry, in which he especially considered the properties of matter at very low temperatures. It is of great interest to read how Nernst in his report remarked that the well known theorem regarding the entropy at absolute zero, of which since 1906 he had made important applications, now appeared as a special case of a more general law derived from the theory of quanta. Still, the phenomenon of the superconductivity of certain metals at extremely low temperatures, on the discovery of which Kamerlingh Onnes reported, presented a great puzzle, which should first many years later find its explanation.

A new feature, commented upon from various sides, was Nernst's idea of quantized rotations of gas molecules, which was eventually to receive such beautiful confirmation in the measurements of the fine structure of infra-red absorption lines. Similar use of quantum theory was suggested in the report by Langevin on his successful theory of the variation of the magnetic properties of matter with temperature, in which he made special reference to the idea of the magneton, introduced by Weiss to explain the remarkable numerical relations between the strength of the elementary magnetic moments of atoms deduced from the analysis of his measurements. Indeed, as Langevin showed, the value of the magneton could at any rate be approximately derived on the assumption that the electrons in atoms were rotating with angular momenta corresponding to a Planck quantum.

Other spirited and heuristic attempts at exploring quantum features in many properties of matter were described by Sommerfeld, who especially discussed the production of x-rays by high speed electrons as well as problems involving the ionization of atoms in the photo-effect and by electronic impact. In commenting upon the latter problem, Sommerfeld called attention to the resemblance of some of his considerations with those exposed in a recent paper by Haas, who in an attempt at applying quantum ideas to the electron binding in an atomic model like that suggested by J.J. Thomson, involving a sphere of uniform positive electrification, had obtained rotational frequencies of the same order of magnitude as the frequencies in optical spectra. As regards his own attitude, Sommerfeld added that instead of trying from such considerations to deduce Planck's constant, he would rather take the existence of the quantum of action as the fundament for any approach to questions of the constitution of atoms

16

and molecules. On the background of the most recent trend of the development this utterance has indeed an almost prophetic character.

Although at the time of the meeting there could, of course, be no question of a comprehensive treatment of the problems raised by Planck's discovery, there was a general understanding that great new prospects had arisen for physical science. Still, notwithstanding the radical revision of the foundation for the unambiguous application of elementary physical concepts, which was here needed, it was an encouragement to all that the firmness of the building ground was just in those years so strikingly illustrated by new triumphs for the classical approach in dealing with the properties of rarefied gases and the use of statistical fluctuations for the counting of atoms. Most appropriately, detailed reports on these advances were in the course of the conference given by Martin Knudsen and Jean Perrin.

A vivid account of the discussions at the first Solvay meeting I got from Rutherford, when I met him in Manchester in 1911, shortly after his return from Brussels. On that occasion, however, Rutherford did not tell me, what I only realized some months ago by looking through the report of the meeting, that no mention was made during the discussions at the conference of a recent event which was to influence the following development so deeply, namely his own discovery of the atomic nucleus. Indeed, by completing in such unsuspected manner the evidence about the structure of the atom, interpretable by simple mechanical concepts, and at the same time revealing the inadequacy of such concepts for any problem related to the stability of atomic systems, Rutherford's discovery should not only serve as a guidance, but also remain a challenge at many later stages of the development of quantum physics.

II

By the time of the next Solvay conference in 1913, the subject of which was the Structure of Matter, most important new information had been obtained by Laue's discovery in 1912 of the diffraction of Röntgen rays in crystals. The discovery removed indeed all doubts about the necessity of ascribing wave-properties to this penetrating radiation, the corpuscular features of which in its interaction with matter, as especially stressed by William Bragg, had been so strikingly illustrated by Wilson's cloud chamber pictures showing the tracks of

17

high speed electrons liberated by the absorption of the radiation in gases. As is well known, Laue's discovery was the direct incentive to the brilliant explorations of crystalline structures by William and Lawrence Bragg, who by analyzing the reflection of monochromatic radiation from the various sequences of parallel plane configurations of atoms in crystal lattices were able both to determine the wave length of the radiation and deduce the type of symmetry of the lattice.

The discussion of these developments, which formed the main topic of the conference, was preceded by a report by J.J. Thomson about the ingenious conceptions regarding the electronic constitution of atoms, by which without departing from classical physical principles he had been able, at least in a qualitative way, to explore many general properties of matter. It is illuminating for the understanding of the general attitude of physicists at that time that the uniqueness of the fundament for such exploration given by Rutherford's discovery of the atomic nucleus was not yet generally appreciated. The only reference to this discovery was made by Rutherford himself, who in the discussion following Thomson's report insisted on the abundance and accuracy of the experimental evidence underlying the nuclear model of the atom.

Actually, a few months before the conference my first paper on the quantum theory of atomic constitution had been published, in which initial steps had been taken to use the Rutherford atomic model for the explanation of specific properties of the elements, depending on the binding of the electrons surrounding the nucleus. As already indicated, this question presented unsurmountable difficulties when treated on ordinary ideas of mechanics and electrodynamics, according to which no system of point charges admits of stable static equilibrium, and any motion of the electrons around the nucleus would give rise to a dissipation of energy through electromagnetic radiation accompanied by a rapid contraction of the electron orbits into a neutral system far smaller than the size of atoms derived from general physical and chemical experience. This situation therefore suggested that the treatment of the stability problems be based directly on the individual character of the atomic processes demonstrated by the discovery of the quantum of action.

A starting point was offered by the empirical regularities exhibited by the optical spectra of the elements, which, as first recognized by Rydberg, could be expressed by the combination principle, according

1 8

to which the frequency of any spectral line was represented with extreme accuracy as the difference between two members of a set of terms characteristic for the element. Leaning directly on Einstein's treatment of the photo-effect, it was in fact possible to interpret the combination law as evidence of elementary processes in which the atom under emission or absorption of monochromatic radiation was transferred from one to another of the so-called stationary states of the atom. This view, which permitted the product of Planck's constant and any of the spectral terms to be identified with the binding energy of the electrons in the corresponding stationary state, also offered a simple explanation of the apparently capricious relationship between emission and absorption lines in series spectra, since in the former we are confronted with transitions from an excited state of the atom to some state of lower energy, while in the latter we generally have to do with a transition process from the ground state with the lowest energy to one of the excited states.

Provisionally picturing such states of the electron system as planetary motions obeying Keplerian laws, it was found possible to deduce the Rydberg constant by suitable comparison with Planck's original expression for the energy states of a harmonic oscillator. The intimate relation with Rutherford's atomic model appeared not least in the simple relationship between the spectrum of the hydrogen atom and that of the helium ion, in which one has to do with systems consisting of an electron bound to a nucleus of minute extension and carrying one and two elementary electric charges, respectively. In this connection it is of interest to recall that at the very time of the conference, Moseley was studying the high frequency spectra of the elements by the Laue-Bragg method, and had already discovered the remarkably simple laws which not only allowed the identification of the nuclear charge of any element, but even were to give the first direct indication of the shell-structure of the electronic configuration in the atom responsible for the peculiar periodicity exhibited in Mendeleev's famous table.

III

Owing to the upsetting of international scientific collaboration by the first world war, the Solvay meetings were not resumed until the spring of 1921. The conference, entitled Atoms and Electrons, was opened by Lorentz with a lucid survey of the principles of classical electron theory, which in particular had offered the explanation of

19

essential features of the Zeeman effect, pointing so directly to electron motions in the atom as the origin of spectra.

As the next speaker, Rutherford gave a detailed account of the numerous phenomena which in the meantime had received such convincing interpretation by his atomic model. Apart from the immediate understanding of essential features of radioactive transformations and of the existence of isotopes which the model provided, the application of quantum theory to the electron binding in the atom had then made considerable progress. Especially the more complete classification of stationary quantum states by the use of invariant action integrals had, in the hands of Sommerfeld and his school, led to an explanation of many details in the structure of spectra and especially of the Stark effect, the discovery of which had so definitely excluded the possibility of tracing the appearance of line spectra to harmonic vibrations of the electrons in the atom.

In the next following years it should indeed be possible through the continued study of high frequency and optical spectra by Siegbahn, Catalan and others, to arrive at a detailed picture of the shell-structure of the electron distribution in the ground state of the atom, which clearly reflected the periodicity features of Mendeleev's table. Such advances implied the clarification of several significant points, like the Pauli principle of mutual exclusion of equivalent quantum states, and the discovery of the intrinsic electron spin involving a departure from central symmetry in the states of electron binding necessary to account for the anomalous Zeeman effect on the basis of the Rutherford atomic model.

While such developments of theoretical conceptions were still to come, reports were given at the conference of recent experimental progress regarding characteristic features of the interaction between radiation and matter. Thus Maurice de Broglie discussed some most interesting effects encountered in his experiments with x-rays, which in particular revealed a relationship between absorption and emission processes reminding of that exhibited by spectra in the optical region. Moreover, Millikan reported about the continuation of his systematic investigations on the photo-electric effect which, as is well known, led to such improvement in the accuracy of the empirical determination of Planck's constant.

A contribution of fundamental importance to the foundation of quantum theory was already during the war given by Einstein, who

20

showed how the Planck formula of radiation could be simply derived by the same assumptions that had proved fruitful for the explanation of spectral regularities, and had found such striking support in the famous investigations by Franck and Hertz on the excitation of atoms by electron bombardment. Indeed, Einstein's ingenious formulation of general probability laws for the occurrence of the spontaneous radiative transitions between stationary states as well as of radiation induced transitions, and not least his analysis of the conservation of energy and momentum in the emission and absorption processes, was to prove basic for future developments.

At the time of the conference, preliminary progress had been made by the utilization of general arguments to ensure the upholding of thermodynamical principles and the asymptotic approach of the description of the classical physical theories in the limit where the action involved is sufficiently large to permit the neglect of the individual quantum. In the first respect, Ehrenfest had introduced the principle of adiabatic invariance of stationary states. The latter demand had come to expression through the formulation of the so-called correspondence principle, which from the beginning had offered guidance for a qualitative exploration of many different atomic phenomena, and the aim of which was to let a statistical account of the individual quantum processes appear as a rational generalization of the deterministic description of classical physics.

For the occasion I was invited to give a general survey of these recent developments of quantum theory, but as I was prevented by illness from taking part in the conference, Ehrenfest kindly undertook the task of presenting my paper, to which he added a very clear summary of the essential points of the correspondence argument. Through the acute awareness of deficiencies and warm enthusiasm for any even modest advance, characteristic of Ehrenfest's whole attitude, his exposition faithfully reflects the state of flux of our ideas at that time, as well as the feeling of expectation of approaching decisive progress.

IV

How much remained to be done before appropriate methods could be developed for a more comprehensive description of the properties of matter was illustrated by the discussions at the next Solvay conference in 1924, devoted to the problem of metallic conduction. A survey of the procedures by which this problem could be treated on

21

the principles of classical physics was given by Lorentz, who in a series of famous papers had traced the consequences of the assumption that the electrons in metals behaved like a gas obeying the Maxwell velocity distribution law. In spite of the initial success of such considerations, serious doubts about the adequacy of the underlying assumptions had, however, gradually arisen. These difficulties were further stressed during the discussions at the conference, at which reports on the experimental progress were given by experts as Bridgman, Kamerlingh Onnes, Rosenhain and Hall, and the theoretical aspects of the situation were commented upon especially by Richardson, who also tentatively applied quantum theory on the lines utilized in atomic problems.

Still, at the time of the conference it had become more and more evident that even such limited use of mechanical pictures as was so far retained in the correspondence approach could not be upheld when dealing with more complicated problems. Looking back on those days, it is indeed interesting to recall that various progress, which should be of great importance for the subsequent development, was already initiated. Thus Arthur Compton had in 1923 discovered the change in frequency of x-rays by scattering from free electrons and had himself, as well as Debye, stressed the support which this discovery gave for Einstein's conception of the photon, notwithstanding the increased difficulties of picturing the correlation between the processes of absorption and emission of photons by the electron in the simple manner used for the interpretation of atomic spectra.

Within a year such problems were, however, brought in a new light by Louis de Broglie's pertinent comparison of particle motion and wave propagation, which soon was to find striking confirmation in the experiments by Davisson and Germer and George Thomson on the diffraction of electrons in crystals. I need not at this place remind in detail how de Broglie's original idea in the hands of Schrödinger should prove basic for the establishment of a general wave equation, which by a novel application of the highly developed methods of mathematical physics was to afford such a powerful tool for the elucidation of multifarious atomic problems.

As everyone knows another approach to the fundamental problem of quantum physics had been initiated in 1924 by Kramers, who a month before the conference had succeeded in developing a general theory of dispersion of radiation by atomic systems. The treatment of dispersion had from the beginning been an essential part of the

22

classical approach to radiation problems, and it is interesting to recall that Lorentz had himself repeatedly called attention to the lack of such guidance in quantum theory. Leaning on correspondence arguments Kramers showed, however, how the dispersion effects could be brought in direct connection with the laws formulated by Einstein for the probabilities of spontaneous and induced individual radiative processes.

It was in fact in the dispersion theory, further developed by Kramers and Heisenberg to include new effects originating in the perturbation of the states of atomic systems produced by electromagnetic fields, that Heisenberg should find a stepping stone for the development of a formalism of quantum mechanics, from which all reference to classical pictures beyond the asymptotic correspondence was completely eliminated. Through the work of Born, Heisenberg and Jordan as well as Dirac this bold and ingenious conception was soon given a general formulation in which the classical kinematic and dynamical variables are replaced by symbolic operators obeying a non-commutative algebra involving Planck's constant.

The relationship between Heisenberg's and Schrödinger's approaches to the problems of quantum theory and the full scope of the interpretation of the formalisms were shortly after most instructively elucidated by Dirac and Jordan with the help of canonical transformations of variables on the lines of Hamilton's original treatment of classical mechanical problems. In particular, such considerations served to clarify the apparent contrast between the superposition principle in wave mechanics and the postulate of the individuality of the elementary quantum processes. Dirac even succeeded in applying such considerations to the problems of electromagnetic fields and, by using as conjugate variables the amplitudes and phases of the constituent harmonic components, developed a quantum theory of radiation, in which Einstein's original photon concept was consistently incorporated. This whole revolutionary development should form the background for the next conference, which was the first of the Solvay meetings I was able to attend.

V

The conference of 1927, the theme of which was Electrons and Photons, was opened by reports by Lawrence Bragg and Arthur Compton about the rich new experimental evidence regarding scat-

2 3

tering of high frequency radiation by electrons exhibiting widely different features when firmly bound in crystalline structures of heavy substances and when practically free in atoms of light gases. These reports were followed by most instructive expositions by Louis de Broglie, Born and Heisenberg as well as by Schrödinger about the great advances as regards the consistent formulation of quantum theory, to which I have already alluded.

A main theme for the discussion was the renunciation of pictorial deterministic description implied in the new methods. A particular point was the question, to what extent the wave mechanics indicated possibilities of a less radical departure from ordinary physical description than hitherto envisaged in all attempts at solving the paradoxes to which the discovery of the quantum of action had from the beginning given rise. Still, the essentially statistical character of the interpretation of physical experience by wave pictures was not only evident from Born's successful treatment of collision problems, but the symbolic character of the whole conception appeared perhaps most strikingly in the necessity of replacing ordinary three-dimensional space coordination by a representation of the state of a system containing several particles as a wave function in a configuration space with as many coordinates as the total number of degrees of freedom of the system.

In the course of the discussions the last point was in particular stressed in connection with the great progress already achieved as regards the treatment of systems involving particles of the same mass, charge and spin, revealing in the case of such " identical " particles a limitation of the individuality implied in classical corpuscular concepts. Indications of such novel features as regards electrons were already contained in Pauli's formulation of the exclusion principle, and in connection with the particle concept of radiation quanta Bose had at an even earlier stage called attention to a simple possibility of deriving Planck's formula for temperature radiation by the application of a statistics involving a departure from the way followed by Bolzmann in the counting of complexions of a many-particle system, which had proved so adequate for numerous applications of classical statistical mechanics.

Already in 1926 a decisive contribution to the treatment of atoms with more than one electron had been made by Heisenberg's explanation of the peculiar duplexity of the helium spectrum, which

24

through many years had remained one of the main obstacles for the quantum theory of atomic constitution. By exploring the symmetry properties of the wave function in configuration space, considerations independently taken up by Dirac and subsequently pursued by Fermi, Heisenberg succeeded in showing that the stationary states of the helium atom fall into two classes, corresponding to two non-combining sets of spectral terms and represented by symmetrical and antisymmetrical spatial wave functions associated with opposite and parallel electron spins, respectively.

I need hardly recall how this remarkable achievement initiated a true avalanche of further progress, and how within a year Heitler and London's analogous treatment of the electronic constitution of the hydrogen molecule gave the first clue to the understanding of non-polar chemical bonds. Moreover, similar considerations of the proton wave function of the rotating hydrogen molecule led to the assignment of a spin to the proton and thereby to an understanding of the separation between ortho and para states, which, as shown by Dennison, supplied an explanation of the hitherto mysterious anomalies in the specific heat of hydrogen gas at low temperature.

This whole development culminated in the recognition of the existence of two families of particles, now referred to as fermions and bosons. Thus, any state of a system composed of particles with half-integral spin like electrons or protons is to be represented by a wave function which is antisymmetrical in the sense that it changes its sign, when the coordinates of two particles of the same kind are interchanged. Conversely, only symmetrical wave functions come into consideration for photons, to which according to Dirac's theory of radiation the spin one has to be ascribed, and for entities like α-particles without spin.

This situation was soon beautifully illustrated by Mott's explanation of the marked deviations from Rutherford's famous scattering formula, in the case of collisions between identical particles like α-particles and helium nuclei, or protons and hydrogen nuclei. With such applications of the formalism we are indeed not only faced with the inadequacy of orbital pictures, but even with a renunciation of the distinction between the particles involved. Indeed, whenever customary ideas of the individuality of the particles can be upheld by ascertaining their location in separate spatial domains, all application

25

of Fermi-Dirac and Bose-Einstein statistics is irrelevant in the sense that they lead to the same expression for the probability density of the particles.

Only a few months before the conference Heisenberg had made a most significant contribution to the elucidation of the physical content of quantum mechanics by the formulation of the so-called indeterminacy principle, expressing the reciprocal limitation of the fixation of canonically conjugate variables. This limitation appears not only as an immediate consequence of the commutation relations between such variables, but also directly reflects the interaction between the system under observation and the tools of measurement. The full recognition of the last crucial point involves, however, the question of the scope of unambiguous application of classical physical concepts in accounting for atomic phenomena.

To introduce the discussion on such points, I was asked at the conference to give a report on the epistemological problems confronting us in quantum physics, and took the opportunity to enter upon the question of an appropriate terminology and to stress the viewpoint of complementarity. The main argument was that unambiguous communication of physical evidence demands that the experimental arrangement as well as the recording of the observations be expressed in common language, suitably refined by the vocabulary of classical physics. In all actual experimentation this demand is fulfilled by using as measuring instruments bodies like diaphragms, lenses and photographic plates so large and heavy that, notwithstanding the decisive role of the quantum of action for the stability and properties of such bodies, all quantum effects can be disregarded in the account of their position and motion.

While within the scope of classical physics we are dealing with an idealization, according to which all phenomena can be arbitrarily subdivided, and the interaction between the measuring instruments and the object under observation neglected, or at any rate compensated for it was stressed that such interaction represents in quantum physics an integral part of the phenomena, for which no separate account can be given if the instruments shall serve the purpose of defining the conditions under which the observations are obtained. In this connection it must also be remembered that recording of observations ultimately rests on the production of permanent marks on the measur-

26

ing instruments, like the spot produced on a photographic plate by impact of a photon or an electron. That such recording involves essentially irreversible physical and chemical processes does not introduce any special intricacy, but rather stresses the element of irreversibility implied in the very concept of observation. The characteristic new feature in quantum physics is merely the restricted divisibility of the phenomena, which for unambiguous description demands a specification of all significant parts of the experimental arrangement.

Since in one and the same arrangement several different individual effects will in general be observed, the recourse to statistics in quantum physics is therefore in principle unavoidable. Moreover, evidence obtained under different conditions and rejecting comprehension in a single picture must, notwithstanding any apparent contrast, be regarded as complementary in the sense that together they exhaust all well defined information about the atomic object. From this point of view, the whole purpose of the formalism of quantum theory is to derive expectations for observations obtained under given experimental conditions. In this connection it was emphasized that the elimination of all contradictions is secured by the mathematical consistency of the formalism, and the exhaustive character of the description within its scope indicated by its adaptability to any imaginable experimental arrangement.

In the very lively discussions on such points, which Lorentz with his openness of mind and balanced attitude managed to conduct in fruitful directions, ambiguities of terminology presented great difficulties for agreement regarding the epistemological problems. This situation was humorously expressed by Ehrenfest who wrote on the blackboard the sentence from the Bible, describing the confusion of languages that disturbed the building of the Babel tower.

The exchanges of views started at the sessions were eagerly continued within smaller groups during the evenings, and to me the opportunity of longer talks with Einstein and Ehrenfest was a most welcome experience. Reluctance to renounce deterministic description in principle was especially expressed by Einstein, who challenged us with arguments suggesting the possibility of taking the interaction between the atomic objects and the measuring instruments more explicitly into account. Although our answers regarding the futility

27

of this prospect did not convince Einstein, who returned to the problems at the next conference, the discussions were an inspiration further to explore the situation as regards analysis and synthesis in quantum physics and its analogies in other fields of human knowledge, where customary terminology implies attention to the conditions under which experience is gained.

VI

At the meeting of 1930, Langevin presided for the first time, after the demise of Lorentz, and spoke of the loss sustained by the Solvay Institute through the death of Ernest Solvay, by whose initiative and generosity the Institute was created. The President also dwelled on the unique way in which Lorentz had assumed the leading of all previous Solvay meetings and on the vigour with which he had continued his brilliant scientific researches until his last days. The subject of the meeting was the Magnetic Properties of Matter, to the understanding of which Langevin himself had given such important contributions, and the experimental knowledge of which had been so much augmented in those years, especially through the studies of Weiss and his school.

The conference was opened by a report by Sommerfeld on magnetism and spectroscopy, in which he in particular discussed the knowledge of the angular momenta and magnetic moments, which had been derived from the investigations of the electron constitution of atoms, resulting in the explanation of the periodic table. As to the interesting point of the peculiar variation of the magnetic moments within the family of rare earths, van Vleck reported about the latest results and their theoretical interpretation. A report was also given by Fermi on the magnetic moments of atomic nuclei, in which, as first pointed out by Pauli, the origin of the so-called hyperfine structure of spectral lines was to be found.

General surveys of the rapidly increasing experimental evidence about the magnetic properties of matter were given in reports by Cabrera and Weiss, who discussed the equation of state of ferromagnetic materials, comprising the abrupt changes of the properties of such substances at definite temperatures like the Curie point. In spite of earlier attempts at correlating such effects, especially by Weiss' introduction of an interior magnetic field associated with the

28

ferromagnetic state, a clue to the understanding of the phenomena had first recently been found by Heisenberg's original comparison of the alignment of the electron spins in ferromagnetic substances with the quantum statistics governing the symmetry properties of the wave functions responsible for the chemical bonds in Heitler and London's theory of molecular formation.

At the conference a comprehensive exposition of the theoretical treatment of magnetic phenomena was given in a report by Pauli. With characteristic clearness and emphasis on essentials he also discussed the problems raised by Dirac's ingenious quantum theory of the electron, in which the relativistic wave equation proposed by Klein and Gordon was replaced by a set of first order equations allowing the harmonious incorporation of the intrinsic spin and magnetic moment of the electron. A special point discussed in this connection was the question, how far one can regard such quantities as measurable in the same sense as the electron mass and charge whose definition rests on the analysis of phenomena which can be entirely accounted for in classical terms. Any consistent use of the concept of spin, just as that of the quantum of action itself, refers, however, to phenomena resisting such analysis, and in particular the spin concept is an abstraction permitting a generalized formulation of the conservation of angular momentum. This situation is borne out by the impossibility, discussed in detail in Pauli's report, of measuring the magnetic moment of a free electron.

The prospects which recent development of experimental technique opened for further investigations of magnetic phenomena were at the meeting reported upon by Cotton and Kapitza. While by Kapitza's bold constructions it had become possible to produce magnetic fields of unsurpassed strength within limited spatial extensions and time intervals, the ingenious design by Cotton of huge permanent magnets permitted to obtain fields of a constancy and extension greater than hitherto available. In a complement to Cotton's report, Madame Curie drew special attention to the use of such magnets for the investigations of radioactive processes, which especially through Rosenblum's work should give important new results as regards the fine structure of α-ray spectra.

While the principal theme of the meeting was the phenomena of magnetism, it is interesting to recall that at that time great advances

29

had also been made in the treatment of other aspects of the properties of matter. Thus many of the difficulties hampering the understanding of electric conduction in metals, so acutely felt in the discussions at the conference in 1924, had in the meantime been overcome. Already in 1928 Sommerfeld had, by replacing the Maxwell velocity distribution of the electrons by a Fermi distribution, obtained most promising results in the elucidation of this problem. As is well known, Bloch succeeded on this basis by appropriate use of wave mechanics in developing a detailed theory of metallic conduction explaining many features, especially regarding the temperature dependence of the phenomena. Still, the theory failed in accounting for the superconductivity, to the understanding of which a clue has been found only in the last years by the development of refined methods for treating interactions in many-body systems. Such methods also seem suitable to account for the remarkable evidence recently obtained about the quantized character of the supercurrents.

A special reminiscence, however, from the meeting in 1930 is connected with the opportunity it gave to resume the discussion of the epistemological problems debated at the conference in 1927. At the occasion Einstein brought up new arguments, by which he tried to circumvent the indeterminacy principle by utilizing the equivalence of energy and mass derived from relativity theory. Thus he suggested that it should be possible to determine the energy of a timed pulse of radiation with unlimited accuracy by the weighing of an apparatus containing a clock connected with a shutter releasing the pulse. However, by closer consideration the apparent paradox found its solution in the influence of a gravitational field on the timing of a clock, by which Einstein himself had early predicted the red-shift in the spectral distribution of light emitted by heavy celestial systems. Still the problem, which most instructively emphasized the necessity in quantum physics of the sharp distinction between objects and measuring instruments, remained for several years a matter of lively controversy, especially in philosophical circles.

It was the last meeting which Einstein attended, before the political developments in Germany forced him to emigrate to the United States. Shortly before the following meeting in 1933 we were all shocked by the news of the untimely death of Ehrenfest, of whose inspiring personality Langevin spoke in moving terms when we were again assembled.

30

VII

The conference of 1933, especially devoted to the Structure and Properties of Atomic Nuclei, took place at a time when this subject was in a stage of most rapid and eventful development. The meeting was opened by a report by Cockcroft, in which, after briefly referring to the rich evidence about nuclear disintegrations by impact of α-particles obtained in the preceding years by Rutherford and his co-workers, he described in detail the important new results obtained by bombardment of nuclei with protons accelerated to great velocities with appropriate high voltage equipment.

As is well known, Cockcroft and Walton's initial experiments on the production of high speed α-particles by the impact of protons on lithium nuclei gave the first direct verification of Einstein's formula for the general relation between energy and mass which in the following years afforded constant guidance in nuclear research. Moreover, Cockcroft described how closely the measurements of the variations of the cross section for the process with proton velocity confirmed the predictions of wave mechanics, to which Gamow was led in connection with the theory of spontaneous α-decay developed by himself and others. In the report comprising the whole evidence available at that time as regards so-called artificial nuclear disintegrations, Cockcroft also compared the results of the experiments in Cambridge with proton bombardment with those just obtained in Berkeley with deuterons accelerated in the cyclotron newly constructed by Lawrence.

The following discussion was opened by Rutherford who, after giving expression for the great pleasure that the recent development of what he used to call modern alchemy had given him, told about some most interesting new results, which he and Oliphant had just obtained by the bombardment of lithium with protons and deuterons. Indeed, these experiments yielded evidence about the existence of hitherto unknown isotopes of hydrogen and helium with atomic mass 3, the properties of which have in recent years attracted so much attention. Also Lawrence, who in more detail described his cyclotron construction, gave an account of the latest investigations of the Berkeley group.

Another progress of the utmost consequence was Chadwick's discovery of the neutron, which represented so dramatic a development, resulting in the confirmation of Rutherford's anticipation of

31

a heavy neutral constituent of atomic nuclei. Chadwick's report, beginning with a description of the purposeful search in Cambridge for anomalies in α-ray scattering, ended up by some most pertinent considerations of the part played by the neutron in nuclear structure, as well as of its important role in inducing nuclear transmutations. Before the theoretical aspects of this development were discussed at the conference, the participants had been told about another decisive progress, namely the discovery of so-called artificial radioactivity, produced by controlled nuclear disintegrations.

An account of this discovery, which was made only a few months before the conference, was included in a report by Frédéric Joliot and Irène Curie, containing a survey of many aspects of their fruitful researches, in which processes of β-ray decay with emission of positive as well as negative electrons were ascertained. In the discussion following this report, Blackett told the story of the discovery of the positron by Anderson and himself in cosmic ray researches and its interpretation in terms of Dirac's relativistic electron theory. One was indeed here confronted with the beginning of a new stage in the development of quantum physics, concerned with the creation and annihilation of material particles analogous to the processes of emission and absorption of radiation in which photons are formed and disappear.

As is well known, the starting point for Dirac was his recognition that his relativistically invariant formulation of quantum mechanics applied to electrons included, besides the probabilities of transition processes between ordinary physical states, also expectations of transitions from such states to states of neagtive energy. To avoid such undesired consequences he introduced the ingenious idea of the so-called Dirac sea, in which all states of negative energy are filled up to the full extent reconcilable with the exclusion principle of equivalent stationary states. In this picture the creation of electrons takes place in pairs, of which the one with usual charge is simply lifted out of the sea, while the other with opposite charge is represented by a hole in the sea. This conception was, as is well known, to prepare the idea of antiparticles with opposite charge and reversed magnetic moment relative to the spin axis, proving to be a fundamental property of matter.

At the conference, many features of radioactive processes were discussed, and a most instructive report was given by Gamow on the

32

interpretation of γ-ray spectra, based on his theory of spontaneous and induced α-ray and proton emission and their relation to the fine structure in α-ray spectra. A special point, which was eagerly discussed, was the problem of continuous β-ray spectra. Especially Ellis' investigations of the thermal effects produced by absorption of the emitted electrons seemed irreconcilable with detailed energy and momentum balance in the β-decay process. Moreover, evidence on the spins of the nuclei involved in the process seemed contradictory to the conservation of angular momentum. It was, in fact, to evade such difficulties that Pauli introduced the bold idea, which should be most fruitful for the later development, that a very penetrating radiation, consisting of particles with vanishing rest mass and spin one-half, the so-called neutrinos, were emitted in β-decay together with the electrons.

The whole question of the structure and stability of atomic nuclei was dealt with in a most weighty report by Heisenberg. From the point of view of the uncertainty principle he had acutely felt the difficulties of assuming the presence of particles as light as electrons within the small spatial extensions of atomic nuclei. He therefore grasped the discovery of the neutron as foundation for the view of considering only neutrons and protons as proper nuclear constituents, and on this basis developed explanations of many properties of nuclei. In particular Heisenberg's conception implied that the phenomenon of β-ray decay be considered as evidence of the creation of positive or negative electrons and neutrinos under release of energy in the accompanying change of a neutron to a proton, or vice versa. In fact, great progress in this direction was soon after the conference achieved by Fermi who on this basis developed a consistent theory of β-decay, which in subsequent developments should prove a most important guidance.

Rutherford, who with usual vigour took part in many of the discussions, was of course a central figure at the Solvay meeting in 1933, which should be the last he had the opportunity to attend before his death in 1937 ended a life-work of a richness with few counterparts in the history of physical science.

VIII

The political events leading to the second world war interrupted for many years the regular succession of the Solvay meetings, which

33

were only resumed in 1948. In those troubled years, the progress of nuclear physics had not relented and had even resulted in the realization of the possibilities of liberation of the immense energy stored in atomic nuclei. Though the serious implications of this development were in everybody's mind, no mention of them was made at the conference, which dealt with the problem of Elementary Particles, a domain in which new prospects had been opened by the discovery of particles with rest mass between that of the electron and the nucleons. As is well known, the existence of such mesons was already before their detection in cosmic radiation by Anderson in 1937 anticipated by Yukawa as quanta for the short range force fields between the nucleons, differing so essentially from the electromagnetic fields studied in the first approach to quantum physics.

The richness of these new aspects of the particle problem had just before the conference been revealed by the systematic investigations by Powell and his collaborators in Bristol of the tracks in photographic plates exposed to cosmic radiation, and by the study of the effects of high energy nucleon collisions first produced in the giant cyclotron in Berkeley. In fact, it had become clear that such collisions lead directly to the creation of so-called π-mesons which subsequently decay under neutrino emission into μ-mesons. In contrast to the π-mesons, the μ-mesons were found to exhibit no strong coupling to the nucleons and to decay, themselves, into electrons under emission of two neutrinos. At the conference, detailed reports on the new experimental evidence were followed by most interesting comments from many sides on its theoretical interpretation. In spite of promising advances in various directions there was, however, a general understanding that one stood before the beginning of a development where new theoretical viewpoints were needed.

A special point discussed was how to overcome the difficulties connected with the appearance of divergencies in quantum electrodynamics, not least conspicuous in the question of the self-energy of charged particles. Attempts at solving the problem by a reformulation of classical electron theory, fundamental for the correspondence treatment, were clearly frustrated by the dependence of the strength of the singularities on the kind of quantum statistics obeyed by the particle in question. In fact, as first pointed out by Weisskopf, the singularities in quantum electrodynamics were largely reduced in the case of fermions, whereas in the case of bosons the self-energy diverges

34

even more strongly than in classical electrodynamics, within the frame of which, as was already stressed in the discussions at the conference in 1927, all distinction between different quantum statistics is excluded.

Notwithstanding the radical departure from deterministic pictorial description, with which we are here concerned, basic features of customary ideas of causality are upheld in the correspondence approach by referring the competing individual processes to a simple superposition of wave functions defined within a common space-time-extension. The possibility of such treatment rests, however, as was stressed during the discussions, on the comparatively weak coupling between the particles and the fields expressed by the smallness of the non-dimensional constant $\alpha = e^2/\hbar c$, which permits a distinction with high degree of approximation between the state of a system of electrons and its radiative reaction with an electromagnetic field. As regards quantum electrodynamics, great progress was just at that time initiated by the work of Schwinger and Tomonaga, leading to the so-called renormalization procedure involving corrections of the same order as α, especially conspicuous in the discovery of the Lamb-effect.

The strong coupling between the nucleons and the pion fields prevented, however, adequate application of simple correspondence arguments, and especially the study of collision processes, in which a large number of pions are created, indicated the necessity of a departure from linearity in the fundamental equations and even, as suggested by Heisenberg, the introduction of an elementary length representing the ultimate limit of space-time-coordination itself. From the observational point of view such limitations might be closely related to the restrictions imposed on space-time measurements by the atomic constitution of all apparatus. Of course, far from conflicting with the argument of the impossibility in any well-defined description of physical experience of taking the interaction between the atomic objects under investigation and the tools of observation explicitly into account, such a situation would only give this argumentation sufficient scope for the logical comprehension of further regularities.

The realization of prospects involving, as condition of the consistency of the whole approach, the possibility of the fixation of the constant α, as well as the derivation of other non-dimensional rela-

35

tions between the masses of elementary particles and coupling constants, was at the time of the conference hardly yet attempted. Meanwhile, however, a way to progress was sought in the study of symmetry relations, and has since been brought to the fore by the rapid succession of discoveries of a manifold of particles exhibiting a behaviour so unexpected that it was even characterized by various degrees of " strangeness ". Thinking of the very latest developments, a great advance has, as is well known, been initiated by the bold suggestion by Lee and Yang in 1957 of the limited scope of the conservation of parity, verified by the beautiful experiments by Mrs Wu and her collaborators. The demonstration of the helicity of the neutrino was indeed anew to raise the old question of a distinction between right and left in the description of natural phenomena. Still, the avoidance of an epistemological paradox in such respect was achieved by the recognition of the relationship between reflection symmetry in space and time and the symmetries between particles and antiparticles.

Of course it is not my intention with such cursory remarks in any way to anticipate problems which will form the main theme for the discussions at the present conference, taking place at a time of new momentous empirical and theoretical advances, about which we are all eager to learn from the participants of the younger generation. Yet we shall often miss the assistance of our deceased colleagues and friends, like Kramers, Pauli and Schrödinger, who all took part in the conference of 1948, which was the last one I, so far, attended. Likewise we deplore the illness that has prevented the presence of Max Born among us.

In concluding, I want to express the hope that this review of some features of the historical development may have given an indication of the debt which the community of physicists owe to the Solvay Institute, and of the expectations which we all share for its future activity.

36

INSTITUT INTERNATIONAL DE PHYSIQUE SOLVAY

douzième Conseil de Physique — Bruxelles, 9–14 octobre 1961

S. MANDELSTAM G. CHEW M.L. GOLDBERGER G.C. WICK M. GELL-MANN G. KÄLLEN E.P. WIGNER G. WENTZEL J. SCHWINGER M. CINI

A.S. WIGHTMAN

I. PRIGOGINE A. PAIS A. SALAM W. HEISENBERG F.J. DYSON R.P. FEYNMAN L. ROSENFELD P.A.M. DIRAC L. VAN HOVE O. KLEIN

S. TOMONAGA W. HEITLER Y. NAMBU N. BOHR F. PERRIN J.R. OPPENHEIMER Sir W. LAWRENCE BRAGG C. MØLLER C.J. GORTER H. YUKAWA R.E. PEIERLS H.A. BETHE

PART V

SELECTED CORRESPONDENCE

INTRODUCTION

Letters to and from Bohr cited or quoted in the Introductions to Parts I through IV are reproduced here in the original language, followed by a translation whenever the original letter was not written in English. The letters are arranged in alphabetical order according to correspondent.

The list preceding the text of the letters supplies page numbers in the Introductions where the letters are referred to so that the reader can readily find the context in which a particular letter is introduced. A page number in parentheses indicates that the letter is referred to but not quoted in the Introductions. Unless otherwise noted on the list, the letters are found in the BSC.

In the reproduction of the letters the editors have attempted to make the layout of letterheads etc. correspond as closely as possible to that of the original letters. "Trivial" mistakes, e.g. in spelling and punctuation, have been tacitly corrected whereas what can be considered "characteristic" mistakes have been retained. In this regard, Bohr's letters in German, in particular those to Jordan, constitute a special case. Here, obvious grammatical mistakes, such as a noun being given the wrong gender, have been corrected without comment. In these letters, as well as in a very few other instances, a word may have been added in square brackets by the editors to clarify the meaning.

Footnotes originally in a letter are marked with asterisks and placed at the end of the letter. Editorial footnotes are numbered sequentially throughout the correspondence; in translated letters, any editorial footnote is placed with the original letter, with the footnote number repeated in the appropriate place in the translation. The editorial footnotes serve two main functions. First, they introduce persons mentioned in the letters who may be unfamilier to the reader. Second, they specify publications and manuscripts mentioned, but not fully described, in the text.

CORRESPONDENCE INCLUDED

	Reproduced p.	Translation p.	Quoted p.
HARALD BOHR			
Niels to Harald Bohr, 26 June 1910[1]	Vol. 1, p. [510]	Vol. 1, p. [511]	XXIX
CHARLES G. DARWIN			
Darwin to Bohr, 11 September 1961[1]	463	–	331
Bohr to Darwin, 20 September 1961[1]	464	–	331

MAX DELBRÜCK

		Reproduced p.	Translation p.	Quoted p.
Delbrück to Bohr,	30 November 1934[2]	465	466	(20)
Bohr to Delbrück,	8 December 1934	469	470	(20)
Delbrück to Bohr,	5 April 1935	471	472	(22)
Bohr to Delbrück,	10 August 1935	472	473	(23)
Delbrück to Bohr,	14 April 1953	474	–	24
Bohr to Delbrück,	2 May 1953	475	–	(24)
Delbrück to Bohr,	1 December 1954	475	–	(24)
Delbrück to Bohr,	30 June 1959	478	–	(XLIII)
Delbrück to Bohr,	30 June 1959	479	–	(XLIII)
Bohr to Delbrück,	25 July 1959	479	481	(XLIII)
Delbrück to Bohr,	3 August 1959	482	–	(XLIII)
Bohr to Delbrück,	19 November 1959	484	486	(XLIII)
Bohr to Delbrück,	17 March 1960	487	–	(25)
Delbrück to Bohr,	15 April 1962	488	–	25
Bohr to Delbrück,	27 April 1962	489	490	(25)
Delbrück to Bohr,	6 May 1962	491	–	(25)
Kalckar to Rosenfeld,	19 May 1962	492	492	(25)
Rosenfeld to Kalckar,	21 May 1962	493	493	(25)
Bohr to Delbrück,	21 May 1962	494	494	(25)

PAUL A.M. DIRAC

	Reproduced p.	Translation p.	Quoted p.
Bohr to Dirac, 24 March 1928	495	–	XXXVI

---

[1] NBA, not microfilmed.
[2] With enclosure.

	Reproduced p.	Translation p.	Quoted p.
WALTER M. ELSASSER			
Bohr to Elsasser, 19 November 1959	497	–	(14)
Elsasser to Bohr, 18 December 1959	498	–	(14)
Bohr to Elsasser, 29 December 1959	500	–	(14)
H.P.E. HANSEN			
Bohr to Hansen, 8 September 1938[1]	501	503	(XXXIV)
HARALD HØFFDING			
Høffding to Bohr, 22 November 1906	505	505	(298)
Høffding to Bohr, 1 December 1906	505	506	(298)
Høffding to Bohr, 4 December 1906	506	507	(298)
Høffding to Bohr, 6 December 1906	507	508	(299)
Høffding to Bohr, 1 January 1922	509	509	(299)
Høffding to Bohr, 20 September 1922	510	511	(301)
Bohr to Høffding, 22 September 1922	512	513	(301)
PASCUAL JORDAN			
Bohr to Jordan, 25 January 1930	514	515	(16)
Jordan to Bohr, 20 May 1931	516	517	(17)
Bohr to Jordan, 5 June 1931	517	520	17
Jordan to Bohr, 22 June 1931	523	525	(18)
Bohr to Jordan, 23 June 1931	527	528	(18)
Jordan to Bohr, 26 November 1932	529	530	(18)
Bohr to Jordan, 27 December 1932	530	532	18
OSKAR KLEIN			
Bohr to Klein, 19 January 1933	533	534	(28)
Bohr to Klein, 6 March 1940	535	537	(173)
OTTO MEYERHOF			
Bohr to Meyerhof, 5 September 1936	538	541	(20)

	Reproduced p.	Translation p.	Quoted p.
CARL W. OSEEN			
Bohr to Oseen, 5 November 1928	Vol. 6, p. [430]	Vol. 6, p. [189]	XXXI
WOLFGANG PAULI			
Bohr to Pauli, 31 December 1953	543	547	XLI, 13, (173)
Pauli to Delbrück, 16 February 1954	551	–	(13)
Pauli to Bohr, 19 February 1954	553	–	(13)
Pauli to Bohr, 26 March 1954[1]	557	–	(13)
Bohr to Pauli, 6 April 1954	558	560	(13)
Bohr to Pauli, 7 February 1955	561	562	(13)
Pauli to Bohr, 15 February 1955	563	–	(XXXVI)
Bohr to Pauli, 2 March 1955	567	–	(XXXVI)
Pauli to Bohr, 11 March 1955	569	–	(XXXVI)
Bohr to Pauli, 25 March 1955	572	–	(XXXVI)
EDGAR RUBIN			
Bohr to Rubin, 20 May 1912[1]	575	576	(XXX)
ERNEST RUTHERFORD			
Bohr to Rutherford, 31 January 1913	Vol. 2, p. [579]	–	332
Bohr to Rutherford, 6 March 1913	Vol. 2, p. [581]	–	332

CHARLES G. DARWIN

DARWIN TO BOHR, 11 September 1961
[Typewritten]

NEWNHAM GRANGE,
CAMBRIDGE.
11th September, 1961.

Dear Niels,

I have now gone through your paper[3]. There were a number of trivial suggestions I thought of as I went along – some of them merely grammatical – and it seemed therefore simplest to return the whole paper to you with these marked in the margin.

You asked me if I had any suggestions about your treatment of Moseley, and I would say that I think you have done it very well, and I have no suggestions for its improvement.

Now I turn to the point I raised with you the other day, about drawing more emphatic attention to the start of your theory, in particular to the hydrogen spectrum. My impression of your treatment between about pages 7–12 is that it is too unconcentrated, so to speak, with all the difficulties being brought out all the time.

I am quite certain that no one can write some one else's paper for him, and I shall not be surprised or hurt if you entirely reject my suggestion, but I think that you might consider putting in a paragraph something like this, perhaps a little below the head of page 7 or perhaps at the start of page 9.

"When a new theory is proposed it is natural that its discoverer should have the whole field of knowledge in his mind including all the difficulties he has to explain, and in the present case many of these difficulties will be reviewed below here. But there is usually a spark that starts the discovery – for example in the discovery of the nucleus itself it was Rutherford's idea of *single* scattering. In the present case, the spark was the hydrogen spectrum, and the discovery was that by two separate applications of the quantum principle, this spectrum might be explained.

---

[3] N. Bohr, Manuscript, *Rutherford Memorial Lecture*. Bohr MSS, microfilm no. 25. Published as *The Rutherford Memorial Lecture 1958: Reminiscences of the Founder of Nuclear Science and of Some Developments Based on his Work*, Proc. Phys. Soc. **78** (1961) 1083–1115. Reproduced in this volume on pp. [383]–[415].

Firstly, if the circular orbit of the electron round the hydrogen nucleus had its angular momentum a multiple of the quantum it would fix the size of the atom about right. Second, there was the idea that one must think not about the frequencies of the emitted light, but about the spectral terms of Ritz, and if these were taken as the energies of the quantised hydrogen atom the result would give the Rydberg constant. When it came out that this constant was yielded correctly in terms of already known atomic constants without any new constant having to be invoked, it was hardly possible to doubt that this must be a quite fundamental principle of nature, but it remained to see how far it might be extended."

As I have said, nobody can write another man's paper, and I shall not be surprised if you reject this. But it does fit in with my memory of the way your theory first struck me when I heard about it from you, and I hope you will consider whether something emphatic of the kind might not help the whole paper. If you agree, I am afraid it might mean a little further revision which I have not attempted. For example, it might be well to shift R from the foot of page 11 back to just below my draft paragraph and also you might mention that the radii are $\hbar^2 n^2/me^2$, which I do not think you have put in anywhere in detail, though you indicate it at the foot of page 10.

However whether you adopt this suggestion or not, it seems to me that what you have written will constitute a most important permanent contribution to the history of the subject.

It was very pleasant seeing you and Margrethe in Manchester, even though the meetings did keep us so busy.

Yours ever,
Charles Darwin

BOHR TO DARWIN, 20 September, 1961.
[Carbon copy]

[København,] September 20, 1961.

Dear Charles,

It was a very great pleasure to Margrethe and me to meet you and Kathrine in Manchester, and I enjoyed very much our talks about the old days. I am also very grateful for the care with which you have read the manuscript of my Rutherford Lecture and appreciate the great interest in the history of the subject expressed in your proposals of some changes in the second chapter. Needless to say that I have considered the matter very closely, but I find it difficult without changing the style to introduce such changes.

Indeed, throughout the whole lecture I have striven to use the opportunity to revive the development in a factual and detached manner without entering into details which can be found in textbooks. It has been quite a difficult task, and I have often been scared at the length of the manuscript, resulting from the many points which presented themselves to give the whole story a reasonable balance. In this situation it has been a great encouragement to learn that my old friends find the whole representation not quite out of place, and not least I have been happy for your own attitude.

From Manchester Margrethe and I went for some days to Wales to stay with the Chadwicks, and my talks with James led to some further improvements in the later chapters of the manuscript, which I now think has got the final shape, and I shall be glad when it soon will be published.

With kindest regards to Kathrine and yourself from Margrethe and me,

Yours ever,
[Niels Bohr]

MAX DELBRÜCK

DELBRÜCK TO BOHR, 30 November 1934
[Handwritten]

KAISER WILHELM-INSTITUT FÜR CHEMIE
Abteilungen HAHN und MEITNER

BERLIN-DAHLEM, DEN 30. Nov. 1934
THIEL-ALLEE 63

Lieber Professor Bohr,

vor ein paar Tagen hat Jordan hier einen Vortrag in der Gesellschaft für empirische Philosophie über Quantenmechanik und Biologie gehalten, zu dem die Biologen vollzählig erschienen waren. Der Vortrag war im physikalischen Teil sehr dürftig, er schildert etwa den Stand der Diskussion im Jahre 1927. Im biologischen Teil, der noch dürftiger war, verdrehte er jedes Ihrer Argumente vollständig, soweit er sie erwähnte. In der nachfolgenden Diskussion schimpfte der Biologe Hartmann[4] ganz fürchterlich über die Verwirrung, die durch Ihre und Jordans Aufsätze in der biologischen Litteratur hervorgerufen sei. Dabei verdrehte er noch diejenigen Ihrer Sätze die Jordan nicht erwähnt hatte. Das Ergebnis war, dass hinterher alle Biologen auf alle Physiker schimpften. Ich habe mir nun erlaubt, die beiliegende Zusammenstellung von dem, was wir* behaupten und was wir nicht behaupten, Herrn Hartmann zu überreichen; ich würde gerne wissen, wie weit Sie die Formulierung billigen. Es kam dabei darauf an, sehr knapp zu formulieren, denn die Biologen sind gewöhnt, längere Arbeiten sehr flüchtig zu lesen. Dabei können sie niemals einen neuen subtilen Gedanken richtig erfassen, sie müssen ihn immer in das bereitstehende Schema ihrer Begriffe hineinzwängen. Aus meiner Zusammenstellung von dem was wir *nicht* behaupten, können Sie die Art der hervorgerufenen Missverständnisse entnehmen. Ich glaube, es wäre sehr nützlich, wenn Sie eine kurze Erklärung dieser Art veröffentlichen würden.

Ich habe mir in diesem Semester ein Privatseminar mit Biologen, Biochemikern und Physikern eingerichtet, wo recht tüchtige Leute hinkommen, so dass wir alle viel lernen.

Mit besten Gruss

Ihr

Max Delbrück

* d.h. das, wovon ich glaube dass Sie es behaupten.

Translation

Berlin-Dahlem, November 30, 1934

Dear Professor Bohr,

a few days ago, Jordan gave a lecture here before the Society of Empirical Philosophy and spoke about quantum mechanics and biology, to which all the biologists showed up. In the part dealing with physics, the lecture was very poor, it described the state of the discussion in the year 1927. In the biological part, which was even poorer, he completely distorted each of your arguments, to the extent that he mentioned them. In the discussion that followed, the biologist Hartmann[4] complained bitterly about the confusion in the biological literature caused by your and Jordan's papers. Thereby he distorted also those of your arguments that Jordan had not even mentioned. The result was that

---

[4] Max Hartmann (1876–1962), German biologist.

subsequently all the biologists scolded all the physicists. I have now taken the liberty of giving Mr. Hartmann the enclosed survey of what we* assert and what we do not assert. I should like to know to what extent you approve of the formulation of the survey. It is important to express oneself very briefly because biologists are wont to read long papers very superficially. Therefore they can never comprehend a new subtle thought correctly, they must always force it into the ready-made scheme of their concepts. From my survey of what we do *not* assert you can gather the kind of misunderstandings that arose. I believe it would be very useful if you were to publish a short explanation of this kind.

This semester I have arranged a private seminar with biologists, biochemists and physicists in which rather able people participate, so that all of us learn a lot.

With best wishes

yours

Max Delbrück

* i.e. what I believe you assert.

Enclosure, DELBRÜCK TO BOHR, 30 November 1934
[Typewritten]

Behauptung: Die zur kausalen Ordnung der biologischen Phänomene zu machenden Annahmen dürfen zum Teil in formalen Widerspruch stehen zu den Gesetzen der Physik und Chemie, weil die Experimente an lebenden Wesen mit *Sicherheit* komplementär sind zu solchen, die die physikalischen und chemischen Vorgänge mit *atomarer* Genauigkeit festlegen.

Erläuterung:

1) Es wird *nicht* behauptet, dass die Gesetze der Atomtheorie *spezifische* Lebenserscheinungen erklären können. Im Gegenteil

a) die Gesetze der *Atomtheorie* sind die gemeinsame Wurzel von *Physik* und *Chemie*. Früher glaubte man, dass die Chemie sich auf klassische Physik müsse zurückführen lassen. Es hat sich aber gezeigt, dass man in der Physik und Chemie einander formal widersprechende Annahmen einführen darf und muss, weil die gemeinsame Wurzel gerade im atomaren liegt und deshalb die Experimente dieser Forschungszweige z.T. komplementärer Natur sind. Z.B. ist das Experiment der Herstellung einer chemischen Verbindung (makroskopisches Experiment!!) komplementär zu dem Experiment: Messung der Bahnen der die Bindung erzeugenden Elektronen. Die nachträgliche Rechtfertigung zu den formalen Widersprüchen (Quantensprünge) ergab sich aus der Erkenntnis, dass eine *atomar* genaue Beschreibung sich nur auf atomare Experimente beziehen kann, bei solchen aber das Experiment, wie man sagt, den Vorgang stört, genauer, bei solchen sich Objekt und Beobachtung nicht eindeutig trennen lassen. Dadurch entfällt die Möglichkeit zu kausaler Beschreibung.

b) Gerade *weil* im lebenden Organismus physikalische und chemische Erscheinungen bis ins atomare Gebiet hinein verwoben sind, *muss* die gemeinsame Wurzel von Biologie *und* Physik *und* Chemie im *atomaren* liegen. Gerade deshalb *kann* aber eine *kausale* Zusammenhangsbeschreibung *nicht* mit *physikalischen und chemischen Begriffen allein* arbeiten. Denn im atomaren lassen *Physik* und *Chemie* keine gemeinsame kausale Beschreibung zu.

2) Es wird *nicht* behauptet, dass der Biologe bei seinen Experimenten das Leben tötet, oder gar töten müsse. Im Gegenteil: Für die Genetik *und* die Entwicklungsmechanik *und* die Physiologie *und* die Biochemie *und* die Biophysik ist charakteristisch und wesentlich, dass sie die Vorgänge am *lebenden* Organismus untersuchen. Eben deshalb können diese Forschungsmethoden nicht zur Erforschung der *individuellen atomaren* Elementarprozesse vordringen, wovon sie auch weit entfernt sind, worüber volle Einigkeit besteht. Sie müssen auch weit von diesem Gebiet entfernt bleiben, wenn ihre Beschreibungen *streng* kausal bleiben sollen. Und eben deshalb sind diese Gebiete nicht kausal aufeinander reduzierbar, wie Physik und Chemie nicht kausal aufeinander reduzierbar sind.

Translation

Assertion: The assumptions having to do with the causal order of biological phenomena may in part stand in formal contradiction to the laws of physics and chemistry, because experiments on living organisms are *certainly* complementary to experiments establishing physical and chemical processes with *atomic* precision.

Explanation:

1) It is *not* asserted that the laws of the atomic theory can explain *specific* life phenomena. On the contrary

a) the laws of the *atomic theory* are the common root of *physics* and *chemistry*. In earlier times, one believed that chemistry must be based on classical physics. However, it turned out that one may and must introduce assumptions in physics and chemistry that formally contradict one another, because the common root is precisely in the atomic realm, and experiments in this field of research are thus partly of a complementary nature. For instance, an experiment for preparing a chemical compound (a macroscopic experiment!!) is complementary to an experiment for measuring the orbits of the electrons generating the bond. The after-the-fact justification of the formal contradictions (quantum jumps) resulted from the recognition that a precise description on the *atomic* level can only refer to atomic experiments where the experiment – as one says – disturbs the process, or more precisely, where object and observation cannot be clearly distinguished from each other. Thereby the possibility for a causal description disappears.

b) Precisely *because* in a living organism physical and chemical phenomena are interwoven far into the atomic domain, the common root of biology *and* physics *and* chemistry *must* be found in the *atomic* domain. Just for this reason, however, a *causal* description of the relationship *cannot* be based on *physical and chemical concepts alone*. For in the atomic domain, *physics* and *chemistry* allow no common causal description.

2) It is *not* asserted that the biologist kills, or even must kill, in his experiments. On the contrary: For genetics *and* developmental mechanics *and* physiology *and* biochemistry *and* biophysics, it is characteristic and essential that they study processes in the *living* organism. Just for this reason, these methods of research cannot penetrate into the exploration of *individual atomic* elementary processes from which they are indeed far removed; there is complete agreement on this point. They must indeed remain far removed from this domain if their descriptions are to remain *strictly* causal. And just for this reason these domains are not causally reducible to one another, just as physics and chemistry are not causally reducible to one another.

BOHR TO DELBRÜCK, 8 December 1934
[Carbon copy]

[København,] 8. Dezember [19]34.

Lieber Delbrück,

Eben zurückgekommen von einer kleinen Reise nach Schweden, wo ich einen Vortrag gehalten habe und einige schönen Tage bei Klein verbracht habe, finde ich Ihren freundlichen Brief mit der Zusammenstellung der Gesichtspunkte über das Verhalten zwischen biologischen und physikalischen Begriffsbildungen, die Sie Herrn Hartmann überreicht haben, und ich beeile mich, Ihnen mitzuteilen, dass ich mit der Formulierung ganz einverstanden bin. Ich verstehe ja, dass es sich hier nicht um eine allseitige Darstellung der Gesichtspunkte handelt, sondern nur um eine Berichtigung der Missverständnisse, die leider unter den Biologen verbreitet sind. Ich kenne ja nicht die genaue Diskussion, die in Berlin stattgefunden hat, aber soweit ich von Ihrem Brief sowie von den verschiedenen neulich veröffentlichten Bemerkungen von Seiten einiger Biologen beurteilen kann, glaube ich, dass Ihre Formulierung eine sehr zweckmässige ist. Wegen der genannten Veröffentlichungen hatte ich schon diesen Sommer einen kleinen Artikel vorbereitet, aber bisher liegen lassen müssen besonders wegen der vielen praktischen Probleme, mit denen sich Wissenschaftler in diesen Jahren beschäftigen müssen. Besonders durch Ihren Brief aufgemuntert, beabsichtige ich aber, dies jetzt fertig zu machen, und Ihnen dann sofort zu schicken, um Ihre Meinung darüber zu hören.

Mit vielen Grüssen, auch an alle gemeinsamen Freunde,

Ihr

[Niels Bohr]

Translation

[Copenhagen,] December 8, [19]34

Dear Delbrück,

Having just returned from a short trip to Sweden where I gave a lecture and spent a few wonderful days with Klein, I find your kind letter with the survey of the viewpoints concerning the relation between the biological and physical definitions of concepts that you gave to Mr Hartmann, and I hasten to inform you that I quite agree with the formulation. I understand, of course, that this is not a comprehensive account of the viewpoints but only a correction of the misunderstandings that are unfortunately widespread among the biologists. I do not, of course, know the exact discussion that took place in Berlin, but as far as I can judge from your letter as well as from the various recently published remarks on the part of some biologists, I think that your formulation is a very useful one. As regards the publication you mention, I prepared a short paper already this summer, but I had to put it aside in particular because of the many practical problems that scientists must deal with these years. Very much encouraged by your letter, I intend, however, to finish it now and to send it to you immediately, in order to hear your opinion of it.

With kindest regards, also to all mutual friends,

Yours

[Niels Bohr]

DELBRÜCK TO BOHR, 5 April 1935
[Handwritten]

KAISER WILHELM-INSTITUT FÜR CHEMIE
Abteilungen HAHN und MEITNER

BERLIN-DAHLEM, DEN 5. April 1935
THIEL-ALLEE 63

Lieber Professor Bohr,

ich erlaube mir, Ihnen das Manuskript einer leider sehr umfangreichen Arbeit über Mutationen zu schicken, die ich gemeinsam mit einem hiesigen Spezialisten für Röntgenmutationen (Timoféeff) und einem Strahlenphysiologen (Zimmer) geschrieben habe, und die in den Göttinger Nachrichten erscheinen wird[5].

Die Arbeit enthält keinerlei Komplementaritätsargumente. Im Gegenteil, es zeigt sich, dass man eine einheitliche atomphysikalische Theorie der Mutation und der Stabilität der Moleküle aufstellen kann. Das liegt daran, dass man über die nähere Wirkungsweise der Gene in der Entwicklung gar nichts zu wissen braucht; die Merkmalsdifferenzen, die als Folge einer Mutation resultieren, sind nur die *Indikatoren* dessen, was im Gen passiert.

Ich wäre sehr dankbar, wenn ich in der nächsten Zeit einmal nach Kopenhagen kommen dürfte, um mit Ihnen und Franck über diese Fragen sprechen zu können. Insbesondere wäre ich froh zu kommen, solange Franck noch da ist, da ich von ihm gern viel lernen möchte über biochemische Fragen. Bei unsren Diskussionen werden wir immer mehr in dieses Gebiet hineingezogen.

Bezüglich der kohärenten Streuung bin ich noch immer nicht zu greifbaren Resultaten gekommen.

Ich weiss nicht, ob Sie schon gehört haben, dass der neue Rektor unserer Universität ein junger Tierarzt ist und der Dekan unserer Fakultät Herr Bieberbach[6].

Mit den besten Grüssen Ihnen und Ihrer Familie.

Ihr

Max Delbrück

---

[5] N.W. Timoféeff-Ressovsky, K.G. Zimmer and M. Delbrück, *Über die Natur der Genmutation und der Genstruktur*, Nach. Ges. Wiss. Göttingen, Math–Phys. Kl., Fachgruppe VI: Biologie 1 (1935) 189–245.

[6] Wilhelm Krüger (1898–1977) and the German mathematician Ludwig Bieberbach (1886–1982). Delbrück probably provided this information on account of Bohr's concern for the difficult situation of scientists in Nazi Germany.

Translation

Berlin-Dahlem, April 5, 1935

Dear Professor Bohr,

I allow myself to send you the manuscript of an unfortunately very volu-
minous paper about mutations, which I have written together with a specialist
in X-ray mutations (Timoféeff) and a radiation physiologist (Zimmer) – both
living here – and which is to appear in the Göttinger Nachrichten[5].

The paper contains no complementarity arguments at all. On the contrary, it
turns out that one can formulate a unified atomic-physical theory of mutation
and molecular stability. This is due to the fact that we do not need to know
anything at all about the the precise way in which genes act in the developmental process; the differences in characteristics resulting as a consequence of
a mutation are only the *indicators* of what happened in the gene.

I should be most grateful if I could come to Copenhagen in the near future in
order to talk with you and Franck about these questions. I should be especially
happy to come while Franck is still there because I want to learn a lot from
him about biochemical questions. In our discussions, we are drawn more and
more into this field.

As regards coherent scattering, I have not yet reached any tangible results.

I do not know whether you have heard that the new president of our university is a young veterinary and the dean of our faculty is Mr Bieberbach[6].

With best regards to you and your family

Yours
Max Delbrück

BOHR TO DELBRÜCK, 10 August 1935
[Carbon copy]

[København,] 10 August [19]35

Kære Delbrück.

Mange Tak for Deres venlige Brev med den store Afhandling, til hvis
Fuldendelse, jeg endnu en Gang ønsker Dem hjertelig Til Lykke. Jeg længes
meget efter en Gang i Ro, at tale rigtigt med Dem om de biologiske Problemer,
og jeg er derfor meget ked af [at] skrive, at vi af forskellige Grunde har
besluttet at opsætte vores Konferens til Foraaret: Dels har vi indset, at de temmelig betydelige Udvidelser af Instituttet i Efteraaret vil gøre Arrangementet af
Konferensen lidt mere besværlig end vi havde tænkt, samt at det ogsaa vilde

være morsommere at vise Instituttet frem i dets nye Skikkelse til Foraaret, hvor vi da foruden Kærneproblemerne vilde gøre de biologiske Problemer til et Hovedemne for Konferensen; Dels har jeg i de sidste Maaneder lidt en Del af Reumatisme i den højre Skulder, der har forstyrret mig ikke saa lidt i Arbejdet, og Lægerne har nu indstændigt raadet mig til i September at gennemgaa en rigtig Badekur for at blive helt arbejdsdygtig igen.

Saa snart, jeg er færdig med Kuren, skal jeg skrive til Dem igen for at træffe nærmere Aftale om Mulighederne for at se Dem heroppe og helst for længere Tid. Som vi talte om sidst vil Rockefeller-Fondet sikkert gerne bevillige et Stipendium for Deres Ophold her et halvt Aar, og maaske vilde netop det kommende Foraars-Semester være den allerbedste Tid for alle Parter.

Med mange venlige Hilsner fra os alle

Deres
[Niels Bohr]

Translation

[Copenhagen,] August 10, [19]35

Dear Delbrück,

Many thanks for your kind letter with the big paper, for the completion of which I once again congratulate you heartily. I long very much for the opportunity to really discuss biological problems with you at length, and I am thus very sad to have to write that for various reasons we have decided to postpone our conference until the spring: firstly, we have realized that the fairly large extension of the Institute in the autumn would make the arrangement of the conference slightly more difficult than we had thought, plus the fact that it would also be more fun to present the Institute in its new shape in the spring, when besides nuclear problems we will make biological problems a main topic of the conference; secondly, in recent months I have been much troubled by rheumatism in my right shoulder which has interfered greatly with my work, the doctors have earnestly advised me to undergo a proper cure at a spa in September in order to become completely fit again.

As soon as I have finished the cure I'll write again to make more detailed arrangements about the possibility of seeing you here, preferably for a longer period. As we discussed last time, the Rockefeller Foundation would probably be willing to give a grant for your stay here for six months, and perhaps just the next spring term would be the best period for all parties.

With many kind regards from all of us

Yours

[Niels Bohr]

DELBRÜCK TO BOHR, 14 April 1953
[Typewritten]

CALIFORNIA INSTITUTE OF TECHNOLOGY
PASADENA

DIVISION OF BIOLOGY
KERCKHOFF LABORATORIES OF BIOLOGY

April 14, 1953

Dear Bohr:

The "News Report" of the Natl. Acad. Sci.[7] carried a notice saying that the "Symposium on Concepts of Complementarity and of Individuality in Biology and Sociology" has been postponed till 1954. I suppose this is the meeting you and Fraser[8] were talking about in 1951. If it really does take place in '54 I would like to come, if I still may. I have tentative plans to be in Göttingen April–July, 1954, bringing the whole family. I would still be reluctant to make a speech at this meeting but certainly anxious to listen. I wonder, in this connection, whether the Edinburgh lectures are out[9]. Nobody here seems to know.

Very remarkable things are happening in biology. I think that Jim Watson has made a discovery which may rival that of Rutherford in 1911.

With kind regards,

Sincerely,

Max Delbrück

---

[7] National Academy of Sciences, Washington, D.C.

[8] Ronald Fraser, ICSU Liaison Officer. Later editor of ICSU Review.

[9] The reference is to the 1949 Gifford Lectures, which were never published. See Bohr's manuscript *Summary of the Gifford Lectures*, this volume on pp. [174]–[181].

BOHR TO DELBRÜCK, 2 May 1953[10]
[Typewritten]

UNIVERSITETETS INSTITUT
FOR
TEORETISK FYSIK

BLEGDAMSVEJ 15–17, KØBENHAVN Ø.
DEN May 2nd 1953.

Dear Delbrück,

I was very happy from your letter to learn that you have plans to be in Göttingen April–July 1954 and that you might consider at that time to take part in Symposium on Complementarity, an eventuality of which we talked with Fraser in 1951, but which has not yet been definitely arranged. I am just now going to England for a week where among other things this matter will be discussed, and I shall let you know how it stands when I come back. I am ashamed to confess that my Edinburgh lectures are not yet ready for publication, but I have just in the last years thought very much of the general problems and think at various points that some progress has been achieved. Remarkable things are surely happening in biology and I need not say that I should be most interested to learn about the discovery Watson has made, which you mentioned in your letter.

With kindest regards from home to home,

yours,

[Niels Bohr]

DELBRÜCK TO BOHR, 1 December 1954
[Typewritten with handwritten postscript]

CALIFORNIA INSTITUTE OF TECHNOLOGY
PASADENA

DIVISION OF BIOLOGY
KERCKHOFF LABORATORIES OF BIOLOGY

December 1, 1954

Dear Bohr:

I would like to tell you about some supplementary thoughts in connection

[10] The letter was not sent.

with our Princeton conversations. I would like to make clearer the connection between three things. First, why I work on phycomyces, second why I am not enthusiastic about discussing with you at the present stage of development the problems concerned with replication and crossing over and the like, and third, why I am really desirous that you, or Aage and you, should write up the thoughts about complementarity in physics in greater detail.

First, about phycomyces. I am afraid during the seminar talk I failed to make clear my real ulterior motive. I talked about this system as something analogous to a gadget of physics, and explained at some length why it seemed more hopeful to me to analyze this gadget in great detail, rather than the many other biological gadgets which have been the subject of conventional research for many years. What I failed to stress was my suspicion, you might almost say hope, that when this analysis is carried sufficiently far, it will run into a paradoxical situation analogous to that into which classical physics ran in its attempts to analyze atomic phenomena. This, of course, has been my ulterior motive in biology from the beginning. What I have in mind is an application of the complementarity principle not in a form which is just vaguely analogous to the way it is used in physics, as having something to do with a shift in the dividing line between observer and object, but something much more closely related to the physics situation, springing directly from the individuality and indivisibility of the quantum processes.

Now, the point I would like to see brought out more clearly is the role that the interaction between observer and observed plays in forcing us to adopt a multi-sided view of any group of phenomena for which the shift of the dividing line plays a crucial role. In physics we have a measure for this interaction, and it seems clear that the whole argument depends on the ugly fact that we cannot get around its finiteness. One should therefore suspect that in all the other situations where complementarity is supposed to play a decisive role in our arguments, it should also be possible to introduce a measure for the interaction, or to make it clear to ourselves in some other way that the interaction cannot be disregarded. This seems to me quite essential. Without this point the whole argument loses its force. This must be so, since we always refer to classical physics as the case in which we can neglect the interaction and, therefore, do not need to have recourse to complementarity arguments. Now it seems to me that in biology we should be able to reach a point where we can see more clearly how the interaction between the observer and the object disturbs the phenomenon in a manner which is essentially unaccountable. It seems to me likely that this interaction that we are talking about is really the one symbolized by the quantum of action. In the old days you used to lay great stress on the fact that the organization of living material is of such fineness that one may well expect the quantum of action to play a decisive role in exactly the same manner as in physics. In more recent years I notice that you omit reference to this point, and that instead you discuss complementarity arguments for situations in which we have no hint as to how one could possibly measure the interaction or convince oneself otherwise of its finiteness. I wonder, therefore, whether you consider that one can arrive at the conclusion that a class of phenomena must be described in a complementary manner without being able to introduce quantities. If this is so, then this should also be possible in physics, and I would like to see the whole physics argument gone over again from this more general point of view. Perhaps you will say this is very foolish. It is certainly on account of the fact that in physics we can express everything in centimeters and seconds, that we can state our case so clearly. Still, I am not convinced that this is a proper procedure. If the argument is to have general validity, it should be possible to state it in more general terms, divested of the accidental features of the physical situation. So I think what ought to be in your book is a double statement of the complementarity principle, once in terms of physics, as in your paper on your discussions with Einstein[11], and once in a more general and abstract form, which should enable the student to see for himself whether indeed the same argument ought to have applications in other fields.

Now I have talked about phycomyces and about your book, perhaps about both in a very presumptuous and foolish manner, but perhaps not so foolish that you will not even be inclined to feel the necessity of commenting on it, and I do hope the comments will be in The Book[12].

As to replication, etc., I simply feel that the experimental situation is in a tremendous state of flux and does not at this moment need the guidance of a

---

[11] N. Bohr, *Discussion with Einstein on Epistemological Problems in Atomic Physics* in *Albert Einstein: Philosopher–Scientist* (ed. P.A. Schilpp), Library of Living Philosophers, Vol. VII, The Library of Living Philosophers, Evanston, Illinois 1949, pp. 201–241. Reproduced in Vol. 7, pp. [341]–[381]. Also published in N. Bohr, *Atomic Physics and Human Knowledge*, John Wiley & Sons, Inc., New York 1958, pp. 32–66. The latter volume is photographically reproduced as *Essays 1933–1957 on Atomic Physics and Human Knowledge, The Philosophical Writings of Niels Bohr, Vol. II*, Ox Bow Press, Woodbridge, Connecticut 1987.

[12] From at least 1949, Bohr wanted to publish a comprehensive presentation of his complementarity viewpoint as a book. See S. Rozental, *Schicksalsjahre mit Niels Bohr: Erinnerungen an den Begründer der modernen Atomtheorie*, Deutsche Verlags-Anstalt, Stuttgart 1991, p. 100. More extensive bibliographic information on Rozental's book is provided in the *Introduction* to Part II, p. [220], ref. 3.

new general point of view. It will in a couple of years, I am sure, and then I would like to take it up again.

With kindest regards,
Max Delbrück

I told the people of the 8th International Congress of Radiology that I had nothing to contribute to a Symposium on Physics, but might come as a listener.

DELBRÜCK TO BOHR, 30 June 1959
[Handwritten]

30. Juni 1959

Dear Bohr:

it bothers me that I expressed myself so poorly today about the idea of the immortality of the soul, that I was not able to make my point. It bothers me all the more because Pauli, too, did not react to it at all in the way I had expected, that is, he reacted very strongly but, to my mind, talked, and wrote, about different things. The same with Feynman. To use your phrase, none of you saw the challenge of the argument. So I will try again.

Perhaps the clearest way to put it is by overstating it paradoxically: it is impossible to imagine the world after your own extinction, because, of course, you cannot imagine the world except from a vantage point, i.e. by implying yourself viewing it, i.e. yourself *not* extinct. To my mind this is the *only* root, and an inescapable one, of the idea of the persistence of yourself as a spectator. You cannot conceive of a world except as a spectator. Of course the argument is as fallacious as the proof that a cat has three tails, and yet, it is to me inescapable. Can you tell me, not why it should not be inescapable, but why it *is*, to *me*, inescapable, and why to me, and apparently not to my best friends?

The answer can wait until I see you next time. My uncle Harnack told us this story, when we were small: Two monks who had lived many years together in a monastery, and had often had converse with each other about the life hereafter, and what it was like, and had vowed each other that whoever dies first should appear to the other in his dream and tell him what it was like. So, when one died, he dutifully appeared to the surviving one in his dream, who eagerly asked him: "Taliter qualiter?" (Is it so as we had thought). And the answer came: "Totaliter aliter!"

Now that Pauli cannot write to you any more, perhaps I should take up the formula

Dein treuer alter

Max

DELBRÜCK TO BOHR, 30 June 1959
[Handwritten]

June 30, 1959

Dear Bohr:

here is still another echo of our visit today. In the book of Chuang Tzu (a Taoist) there is a story of a wheelwright who does not believe in books recording the words of the sages who are dead. He explains that in his own craft the essential thing, namely the art of driving in the nails with just the right kind of stroke, can not be taught in words. "That is why it is impossible for me to let my son take over my work, and here I am, at the age of seventy, still making wheels. In my opinion it must have been the same with the wise men of old. All that was worth handing on, died with them; the rest, they put into their books."

M.

BOHR TO DELBRÜCK, 25 July 1959
[Carbon copy]

Tisvilde, den 25. juli 1959.

Kære Delbrück.

Det var en meget stor glæde for Margrethe og mig at se dig og Manni og børnene i Pasadena, og jeg nød vore samtaler både om de biologiske og de mere almindelige menneskelige spørgsmål. Også dine rare breve med de morsomme anekdoter var meget velkomne bidrag til vore diskussioner.

Jeg forstår imidlertid stadig ikke dine vanskeligheder ved at tænke os selv som ikke længere existerende. Naturligvis rummer enhver brug af ordet "tænke" med dets implicite henvisning til et bevidst subjekt en tvetydighed, så længe man ikke klargør sig sprogets formål, at meddele sig på eentydig eller måske rettere objektiv måde til vore medmennesker. Som vi talte om, har man jo allerede i matematikken gang på gang mødt vanskeligheder i argumentationer, der benyttede ordet tænkelig, og som først kunne overkommes ved udviklingen af en veldefineret formalisme. Jeg behøver vist heller ikke at erindre om den humoristiske historie om Licentiaten i En Dansk Students Eventyr[13], som jeg fremdrog ved mit foredrag i Pasadena til belysning af den komplementære sprogbrug i psykologien. Pointen er jo her, at selvom enhver eentydig meddelelse kræver distinktion mellem et subjekt og et objekt, kan det subjekt, der underforstås i en given situation, helt eller delvis indgå i det objektive indhold af en meddelelse om en anden situation. Vel kommer digteren ikke ind på spørgsmålet, hvorvidt vi kan tænke os, hvordan verden vil være, når vi selv er døde, men det forekommer mig, at dine vanskeligheder minder om dem, som Licentiaten mødte i det praktiske liv. Vi vil vel begge sige, at vi meget vel kan tænke på, hvordan andre vil bedømme os, når vi er døde, og måske forfængeligt håbe, at de i erindringen om os på en eller anden måde kan finde støtte for deres egne bestræbelser, men det som jeg gerne vil påstå er, at en sådan sætning ikke rummer større farer for misbrug af ord end de der lurer på os ved enhver gensidig meddelelse. Jeg er selvfølgelig forberedt på, at du ikke er tilfreds med en sådan påstand, og glæder mig til at høre om din reaktion, som jeg håber du vil give ligeså uforbeholdent udtryk, som jeg var vant til fra Pauli's side.

Med de venligste hilsner fra os alle, også Aage Petersen[14], med hvem jeg har diskuteret indholdet af dette brev,

Din gamle
[Niels Bohr]

P.S. Indlagt sender jeg særtrykket af min lille artikel om kvantefysik og epistemologi[15].

Iøvrigt lovede jeg Elsasser at skrive til ham, så snart jeg har kunnet danne mig en nærmere forestilling om indholdet af hans bog[16]. Jeg skal naturligvis også skrive til dig derom.

---

[13] P.M. Møller, *En dansk Students Eventyr* (The Adventures of a Danish Student) in *P.M. Møller: Efterladte Skrifter, Vol. 3*, Reitzel, Copenhagen 1843. See above, *General Introduction*, pp. XXXI f, XLIV.

[14] Aage Petersen (b. 1927) was Bohr's assistant during the last ten years of Bohr's life, and later moved to the United States.

[15] Most likely, N. Bohr, *Quantum Physics and Philosophy – Causality and Complementarity* in *Philosophy in the Mid-Century, A Survey* (ed. R. Klibansky), La nuova Italia editrice, Firenze 1958, pp. 308–314. Reproduced in Vol. 7, pp. [388]–[394]. The article is also published in N. Bohr, *Essays 1958–1962 on Atomic Physics and Human Knowledge*, Interscience Publishers, New York 1963, pp. 1–7. The latter volume is photographically reproduced as *Essays 1958–1962 on Atomic Physics and Human Knowledge, The Philosophical Writings of Niels Bohr, Vol. III*, Ox Bow Press, Woodbridge, Connecticut 1987.

[16] W.M. Elsasser, *The Physical Foundation of Biology: An Analytical Study*, Pergamon Press, New York 1958.

Translation

Tisvilde, July 25, 1959

Dear Delbrück,

It was a great pleasure for Margrethe and me to see you and Manni and the children in Pasadena, and I enjoyed our conversations on both the biological and the more general human topics. Your nice letters with the amusing anecdotes were also a very welcome contribution to our discussions.

However, I still do not understand your difficulties in thinking of ourselves as not existing any more. Naturally, any use of the the word "think" with its implicit reference to a conscious subject contains an ambiguity, as long as one does not realize the purpose of language: to communicate in an unambiguous or, perhaps more correctly, objective way with our fellow human beings. As we discussed, already in mathematics one has met with difficulties time and time again in arguments using the word: conceivable, and these could only be solved by the development of a well-defined formalism. Nor do I need bring to mind the amusing story about the licentiate in "The Adventures of a Danish Student"[13], which I related at my talk in Pasadena to elucidate the complementary use of terminology in psychology. The point here is, of course, that even though every unambiguous communication requires distinction between a subject and an object, the subject implied in a given situation can wholly or partially be included in the objective content of a communication about another situation. Even though the author does not discuss the question as to whether we can imagine what the world will be like when we ourselves are dead, it appears to me that your difficulties are similar to those the licentiate met in practical life. We would probably both say that we can very well imagine how others will judge us when we are dead, and perhaps conceitedly hope that they, in remembrance of us, in one way or another may find support for their own endeavours, but what I want to state is: such a sentence does not hold greater dangers for misusing words than those lying in wait for us in any mutual communication. I am of course prepared for the fact that you are not satisfied with such a statement, and I look forward to hearing your reaction, which I hope you will express with the same candidness as I was used to from Pauli.

[481]

With kindest regards from us all, including Aage Petersen[14], with whom I have discussed the contents of this letter,

> Your old friend
> [Niels Bohr]

P.S. I enclose the reprint of my small article about quantum physics and epistemology[15].

Moreover, I promised Elsasser that I would write to him as soon as I was able to form a more detailed impression of the contents of his book[16]. I will of course also write to you about it.

DELBRÜCK TO BOHR, 3 August 1959
[Typewritten]

CALIFORNIA INSTITUTE OF TECHNOLOGY
PASADENA, CALIFORNIA

DIVISION OF BIOLOGY

August 3, 1959

Dear Bohr:

I'm afraid the only thing that your letter demonstrated to me with respect to what I had said about death is that I had been quite unable to express what was in my mind. Perhaps what is in my mind is not a thought at all, properly speaking, but something akin to what was in the mind of a certain man who, every time he went to the dentist, and was given an anaesthetic, as he came out of the anaesthesia, in his state of half-consciousness, had a marvelous but extraordinarily elusive revelation, which, as it seemed to him, explained the whole meaning of life. Finally, he arranged with the dentist to have a paper and a pencil ready and put into his hand so he could scribble down at once his revelation which, on every previous occasion, had completely vanished at the time he reached full consciousness, except for the feeling of euphoria, and the curiosity to take a closer look at the thing. On this final occasion, then, he did indeed have the same revelation, and scribbled it down hastily, but was much disappointed when, a minute later, he read what he had written: "Petroleum is oily". Nevertheless, I will try again.

Of course I do not claim that there is any logical or epistemological difficulty in the notion of one's own death. I think the difficulty lies in a mixture of the logical and the emotional aspect. Suppose one tries to imagine the situation of one's wife after your death. Unless you are emotionally detached from her at this very moment, you will find it very hard to imagine a situation in which you would have the most urgent wish to comfort her, but the essence of which is the meaninglessness of any further communication, in either direction. In this sense, in fact, the death of another person is as difficult to imagine as your own, and my supposition was that the notions about the persistence of the soul in the various religions derive from this *incompatibility of being both involved and not involved* emotionally. Logically, of course, there is no difficulty: "The guy is dead, so what?" as Feynman might say (though even he confessed that he had talked to his wife after her death). Similarly, nobody will have difficulty in agreeing with his lawyer, who tells you that you must use clear and unambiguous language in your last Will and Testament, as you have no chance to speak again and correct yourself. Yet, in trying to foresee how Aunt Jenny, to whom you bequeath the grand piano, or your friend Karl, to whom you give the Arabian Nights Tales, will react to your Will, it is very difficult to avoid the attitude that you may yet have a chance later on to explain the meaning of your Will to them, and to smooth their ruffled feelings.

You may say that this is a very trivial psychological problem, and that it has nothing to do with the epistemological ones in which you are interested, and yet it seems to me that there is a connection. It is not merely an oddity that our language is so structured that every object has a subject. The subject which views the world, not only views it passively, but is also acting, and it is *this* subject–object relation which gives rise to the strange constructions such as immortality, and which make it in practice impossible to conceive of the death of any person to which you are close, and especially of yourself. It has always seemed to me that your epistemology side-stepped this issue, and it has puzzled me that it did so. This is all I seem to be able to say on the subject. Incidentally, who wrote "En Dansk Students Eventyr"[17]? I would like to read it, to keep up my practice of the Danish language.

As to the meeting in Baltimore at the end of March, the McCollum Pratt Conference on Photobiology, where you are billed on the tentative program with an opening lecture on "Light and Life"[18], you may be interested to hear

[17] Møller, *En dansk Students Eventyr*, ref. 13.
[18] N. Bohr, *Physical Models and Living Organisms* in *Light and Life* (eds. W.D. McElroy and B. Glass), The Johns Hopkins Press, Baltimore 1961, pp. 1–3. Reproduced on pp. [135]–[137].

some details. These are very good conferences covering facts and theories in broad fields of biology. They publish a volume of papers and discussions *within four months after the meeting*. Attendance is only by invitation, though in the end generally about 300 people manage to get in, mostly bio-chemists. I expect to go there, but have refused to give a paper this time, since the papers they expect are rather elaborate review papers.

We have been in Europe meanwhile, and one of the objects of this trip was further negotiations about the proposition in Cologne. The upshot is that it now seems that we will be moving to Cologne in April, 1961, for two years. Then, we will be neighbors, and this is one of the things we look forward to.

With warm regards to you and Margrethe, and to Aage Petersen,

Yours,

Max D.

BOHR TO DELBRÜCK, 19 November 1959
[Carbon copy]

[København,] November 19, 1959.

Kære Max,

Jeg er ked af at der er gået så lang tid før jeg har svaret på dit brev af 3. august, men det har været en tid meget optaget med videnskabelige diskussioner og Instituttets anliggenheder. Imidlertid har jeg tænkt meget på forskellige spørgsmål af fælles interesse, og har blandt andet sendt nogle få sider, som jeg vedlægger, med indledende bemærkninger til diskussionerne ved en kongres i Bristol om brugen af modeller i biologiske betragtninger[19]. Bemærkningerne vil blive trykt[20] sammen med uddrag af den artikel[21] om fysik og filosofi, som jeg gav dig i sommer, og hvori kongressens ledere havde været interesseret. I bemærkningerne har jeg efter nøje overlæg ikke benyttet betegnelsen komplementaritet, fordi alle kvantemekanikkens konsekvenser jo er inddraget i den kemiske kinetik. I mit gamle foredrag om lys og liv[22] drejede det sig om en nøgtern sproglig udtryksmåde, der pegede på væsentlige punkter i

[19] Manuscript, *Models in Biology* (manuscript for Symposium at the University of Bristol, 7 September 1959). Bohr MSS, microfilm no. 23.
[20] N. Bohr, *Quantum Physics and Biology* in *Symposion of the Society for Experimental Biology*, Number XIV: *Models and Analogues in Biology*, Cambridge 1960, pp. 1–5. Reproduced on pp. [126]–[131].
[21] Bohr, *Quantum Physics and Philosophy*, ref. 15.
[22] *Light and Life*. The various publications of this lecture are listed on p. [28].

beskrivelsen af situationen i biologien og fra hvilken jeg naturligvis ikke tænker på at tage afstand, men som jo så let kan misforstås. Det var også en stor glæde for mig at deltage i det genetiske symposium her i København og navnlig atter at møde og tale med Watson. Jeg er, som du også vil se af indlagte kopi af et brev til Elsasser[23], for tiden meget begejstret over det synspunkt, som også Watson såvidt jeg forstod helt tilsluttede sig, at betragte alle organiske processer, inklusive DNA-kædernes reproduktion og deres dirigerende indflydelse på udviklingen i cellerne, som udtryk for en stadig tiltagende stabilitet hos organismerne. Det er måske ikke meget nyt for nogen, men har i det mindste givet mig selv fornyet inspiration til indlevelse i de mest forskellige manifestationer af livets udfoldelse. Med hensyn til kongressen i Baltimore i marts er det endnu ikke helt klart om det vil være muligt for mig at deltage, men jeg håber meget at jeg kan komme og det skal jo blive en stor glæde om vi kunne mødes der. Hvad de almindelige erkendelsesteoretiske spørgsmål angår, tænker jeg stadig meget over komplementaritetsbegrebets virkelige nytte og fristende misbrug, og jeg må tilstå at det endnu ikke har været muligt for mig at gøre mig klart, hvorhen du vil med at understrege vanskeligheden i at forestille os vor egen død. Jeg kan blot gentage, at bestræbelsen for den logiske analyse af bevidsthedslivets beskrivelse for mig ikke mindst har været at pege på, hvordan de paradokser som knytter sig til forestillingen om viljens frihed og sjælens udødelighed netop opløser sig ved nærmere undersøgelse af subjekt–objekt skillelinjens stadige forskydning. Naturligvis ophører subjektet, ligesom hele indholdet af bevidstheden, at eksistere når vi ikke længere lever, men dette kan jo ingenlunde hindre at vi så længe vi lever kan tænke ikke alene på fortidige, men også på fremtidige hændelser.

Jeg var glad for din interesse for Poul Martin Møllers humor og sender som en lille hilsen fra os alle et eksemplar af en ny udgave af hans bog.

Med de varmeste ønsker fra hjem til hjem

altid din

[Niels Bohr]

[23] Letter from Bohr to Elsasser, 19 November 1959. Reproduced on p. [497].

Translation

[Copenhagen,] November 19, 1959

Dear Max,

I am sorry that so much time has passed before I answer your letter of 3rd August, but it has been a period very occupied by scientific discussions and Institute matters. However, I have been thinking about various questions of mutual interest, and have, among other things, sent a few pages, which I enclose, with introductory remarks to the discussions at a conference in Bristol about the use of models in biology[19]. The remarks will be published[20] together with an excerpt of the article[21] on physics and philosophy which I gave you this summer, and in which the organizers of the congress had been interested. After careful consideration I have not used the term complementarity in the remarks, because all the consequences of quantum mechanics are included in chemical kinetics. In my old lecture about light and life[22] a sober way of expression as regards language was involved, calling attention to important points in the description of the situation in biology, which I naturally do not intend to repudiate, but which can so easily be misunderstood. It was also a great pleasure for me to take part in the genetics symposium here in Copenhagen, and especially to meet and talk with Watson again. As you will see from the enclosed copy of a letter to Elsasser[23], I am at present very enthusiastic about the point of view, with which to my knowledge Watson is also in total agreement, to consider all organic processes, including the reproduction of DNA chains and their directing influence on the development in cells, as an expression of a steadily increasing stability in organisms. This is perhaps not very new to some people, but has at least given me renewed inspiration for understanding the most various manifestations of the unfolding of life. As regards the congress in Baltimore in March, it is still not quite certain whether I shall be able to attend, but I hope that I can come and it would of course be delightful if we could meet there. Regarding general epistemological problems, I still give a good deal of thought to the real usefulness and the tempting misuse of the concept of complementarity, and I must admit that I am still unable to understand what you want to achieve by emphasizing the difficulty involved in imagining our own death. I will just say again that the endeavour to make a logical analysis of the description of consciousness has been for me, not least, to point out how the paradoxes linked to the idea of a free will and an immortal soul are solved precisely on closer examination of the continual shifting of the dividing line between subject and object. Naturally, the subject expires, as does the content of the consciousness, when we are no longer alive, but as long as we live this

[486]

cannot in any way prevent us from thinking not only about past events but also about events in the future.

I appreciated your interest in Poul Martin Møller's humour, and I am sending a new edition of his book as a small greeting from us all.

With warmest regards from home to home,

Yours ever
[Niels Bohr]

BOHR TO DELBRÜCK, 17 March 1960
[Carbon copy]

[København,] March 17, 1960.

Dear Max,

I am sorry not to have written you for so long, but ever since Margrethe and I returned from a visit in January to India, which was a wonderful experience of which I hope to tell you some time, I have been constantly occupied with accumulated pressing duties. To my regret, I have also found it impossible to come to the Symposium in Baltimore, and I am sorry to miss this occasion to meet you. As you may have understood from my previous letters, I have not lost my interest in the biological problems, rather it has recently been activated through regular seminars in this Institute on microbiology, to which Westergaard[24] and Aage have taken the initiative. We have at these seminars heard splendid lectures by Maaløe[25], which have revived many reminiscences of our talks and I need not say how much we all hope for renewed contact with you when you come to Europe.

With kindest regards and best wishes from home to home,

Yours ever,
[Niels Bohr]

---

[24] Mogens Westergaard (1912–1975), professor of genetics at the University of Copenhagen 1949–1962.

[25] Ole Urban Maaløe (1914–1988), professor of microbiology at the University of Copenhagen from 1958.

DELBRÜCK TO BOHR, 15 April 1962
[Handwritten]

Inst. f. GENETIK   Albergo de la Posta
WEYERTAL 117   Ronco S. Ascona,
KÖLN-LINDENTHAL   April 15, 1962

Dear Bohr

On June 22 we are planning to dedicate the Institut für GENETIK der Universität zu KÖLN, and I would like to invite you to give the principal lecture on this occasion.

To me, this would seem wonderfully fitting. It was you who inspired me 30 years ago to go into biology and I believe I am the only one of your disciples who has made his way in this direction. Also, the dedication of this Institute, even though I shall be associated with it for only two years, may be looked upon as some kind of culmination in my career, and perhaps it will be the only opportunity I shall have to invite you to give such an address. Also, it has seemed to me, when I saw you last, a little over a year ago, that you were as interested as ever in the question of the relevance of the complementarity argument to biology and might want to take such an occasion as this to elaborate on what you said 30 years ago, in the light of the new developments that have taken place. Even if I was impatient on that occasion (a year ago, as well as 30 years ago!), you know, probably better than I, what is behind this impatience. Indeed, there is no complex of scientific questions that affects me more deeply and it has, through the years, provided the sole motivation for my work.

The Institut für Genetik, as I may have told you, represents an attempt to establish a center of modern biology, within a University, somewhat similar to the group which was to be Maaløe, Kalckar, Westergaard, and Ussing, and which has had such difficult times. Our Institute is now in full operation. My own association with it has been in its planning during the last five years, and now, for a period of two years, its Direction. We are now approaching the middle of this two year period. I believe it will not be too difficult to find a reasonable arrangement for the continuance of the Institute after we go back to Pasadena. I feel, though, that much will depend on the spirit in which this question is approached, i.e., the spirit with which one looks upon the meaning of such an Institute. To see to it that this spirit is the right one I can think of no stronger move than to have you at the center of the dedication.

We are planning to separate the official dedication, with the government people, from the scientific one. The official one shall be in the morning, the scientific one in the late afternoon. In this way there will be freedom of choice for those who prefer not to attend one or the other of these functions. It will be quite allright for you to speak in English, since many of our speakers do. In such cases we always have somebody give brief summaries in German, from time to time, interspersed into the lecture, but I doubt whether this would work in your case.

At the moment we are spending a brief vacation at the Lago Maggiore, above Ascona, a place where I was last with Pauli, Gamow, and Rossi, in 1932. Tonight we will be joined by the Weisskopfs, who will be bringing with them our older boy Jonathan who has been skiing in the WALLIS. Yesterday we had spring weather as it should be, but today it snows! So we are room-bound, and can concentrate on writing letters to our friends.

With all good wishes, in which Manny joins me, to you and Margrethe.

Yours

Max D.

P.S. I am planning *to come up to Copenhagen* some time this summer, and, if you like, could arrange to do so before June 22, to bring you into the mood of biology.

BOHR TO DELBRÜCK, 27 April 1962
[Carbon copy]

[København,] Den 27. april 1962.

Kære Delbrück,

Hjertelig tak for dit venlige brev med indbydelsen til at komme til stede og tale ved indvielsen af det nye genetiske institut i Köln. Jeg var rørt over alt hvad du skrev, og det ville glæde mig meget, hvis det var muligt for mig at overvære den store begivenhed og at holde et foredrag af den art, som du ønsker. Når jeg dog endnu ikke kan give et definitivt svar på din indbydelse, ligger det i, at jeg netop i øjeblikket er travlt beskæftiget med forskellige sager, der må gøres færdige. Imidlertid har jeg, så snart jeg fik dit brev, begyndt at tænke på plan og indhold af et eventuelt foredrag, i hvilket jeg ville prøve på at give en så klar fremstilling som muligt af de erkendelsesteoretiske spørgsmål, som i dag melder sig for en fysikers indstilling til biologiens almindelige problemer. Om nogle få uger, når jeg har tænkt nærmere over indholdet af foredraget, skal jeg skrive igen, hvorledes det står, og forhåbentlig kunne meddele at jeg kommer og er forberedt på at tale ved instituttets indvielse.

Med de venligste hilsner fra os alle,

Din hengivne
[Niels Bohr]

Translation

[Copenhagen,] April 27, 1962

Dear Delbrück,

Many thanks for your kind letter with the invitation to come and give a talk at the inauguration of the new Institute of Genetics at Cologne. I was touched by all that you wrote, and it would give me great pleasure if it should prove possible for me to be present at the great occasion and to give a talk of the kind you wish. I am very busy at the moment with various matters which must be completed and thus I cannot give a definite reply to your invitation. However, as soon as I received your letter I began to think of the plan and content of an eventual talk, in which I will try to give as clear an exposition as possible of the epistemological problems arising today as regards a physicist's attitude towards the general problems of biology. In a few weeks, when I have considered the content of the talk in greater detail, I will write again as to how things are going, and hopefully be able to say that I am coming and that I am prepared to give a talk at the inauguration of the Institute.

With kindest regards from all of us,

Yours affectionately
[Niels Bohr]

DELBRÜCK TO BOHR, 6 May 1962
[Handwritten]

Institut für Genetik   Köln-Lindenthal
der Universität zu Köln   Weyertal 115

                                                    Sunday May 6, 1962

Dear Bohr

Nu skulde jeg ogsaa skrive paa Dansk og benytte den "Du" som Du offer-
erest mig mange Aar siden og som vi aldrig har fundet Lejligheden for. Der
er jo ikke Rum for denne fortrolige Distinktion i vor Engelsk Konversation
og Korrespondance[26]. Thus there has been, all these years, a complementary
relationship with respect to intelligibility of communication and genuineness
in the expression of feelings. So much for complementarity and cultures.

Otherwise I just want to say that I am exceedingly happy at the thought that
you are willing to come, and do hope that practical circumstances will permit
you to do it. From our side it will be sufficient if we have a definite answer by
the 20th of May, so we can complete our plans and send out the invitations.

I looked at the English version (April 1, 1933, NATURE) and the German
version (March 30, 1933, Naturwissenschaften) of "Light and Life". Would you
like to call the new lecture: "Light and Life, revisited"??

                                            With *all* good wishes
                                                    yours
                                                Max Delbrück

We saw Franca Pauli on our way back from Switzerland, in her home in
Zollikon. She is busying herself very much with Pauli's correspondence, but
she also seemed much at a loss what to do with it.

---

[26] English translation: "Now I should also write in Danish and use the 'Thou' that you offered me
many years ago and which we have never found the opportunity to use. There is of course not
room for this intimate distinction in our English conversation and correspondence."

KALCKAR TO ROSENFELD, 19 May 1962
[Carbon copy]

[København,] Den 19. maj 1962.

Kære Professor Rosenfeld.

I forbindelse med udarbejdelsen af det foredrag, som professor Bohr skal holde ved indvielsen af det genetiske institut i Köln i juni, vil vi være meget glade for, om De kunne give os nogle henvisninger til litteraturen om termodynamikken af ikke-reversible processer (specielt Onsagers[27] og Prigogine's[28] arbejder). Jeg tror, at de spørgsmål, som professor Bohr særligt har i tankerne er dem, som Catalski[29] nævnte i sit foredrag på instituttet. I betragtning af, at tidspunktet jo allerede er ret fremskredet, ville vi naturligvis være meget taknemmelige, om De så snart som muligt kunne sende os Deres råd.

Med mange venlige hilsener

Deres hengivne

[Jørgen Kalckar]

Translation

[Copenhagen,] May 19, 1962

Dear Professor Rosenfeld,

In connection with the preparation of the talk which Professor Bohr is to give at the inauguration of the Institute of Genetics in Cologne in July, we would be very grateful if you could give us some references to literature about the thermodynamics of non-reversible processes (especially Onsager's[27] and Prigogine's[28] papers). I think the questions Professor Bohr particularly has in mind are those that Catalski[29] mentioned in his talk at the Institute. Since time

---

[27] Lars Onsager (1903–1976), Norwegian-born chemist who moved to the United States in 1928.

[28] Ilya Prigogine (b. 1917), Russian-born physicist, at that time director of the International Institutes of Physics and Chemistry (Solvay Institutes), Brussels. Prigogine moved to the United States in 1967.

[29] Probably Polish-born Aharon Katzir-Katchalsky (1913–1972), professor of physics at the Weizmann Institute in Rehovoth, Israel.

is already quite short, we would naturally be very pleased if you could send your advice as soon as possible.

With many kind greetings

Yours sincerely

[Jørgen Kalckar]

ROSENFELD TO KALCKAR, 21 May 1962
[Handwritten]

I N S T I T U U T - L O R E N T Z   LEIDEN, 21 maj 1962

VOOR THEORETISCHE NATUURKUNDE

STEENSCHUUR, LEIDEN

Kære Kalckar,

Hele termodynamikkens udvidelse til ikke reversible processer har nu fået en udmærket fremstilling i en bog ved de Groot og Mazur, Non-equilibrium thermodynamics, udgivet af North-Holland Publishing Co, Amsterdam. Jeg skal netop i dag være i Amsterdam og skal sørge for, at et eksemplar sendes til Carlsberg.

Desuden sender jeg særskilt et par særtryk ved Mazur og de Groot, deriblandt et på hollandsk (som De uden vanskelighed vil være i stand til at læse) som netop handler om Entropi og liv.

Med de hjerteligste hilsner til professoren og hans trofaste hjælpere.

Deres

L. Rosenfeld

Translation

Leiden, May 21, 1962

Dear Kalckar,

The whole expansion of thermodynamics to non-reversible processes has now been given an excellent presentation in a book by de Groot and Mazur, Non-equilibrium thermodynamics, published by North-Holland Publishing Co, Amsterdam. I shall in fact be in Amsterdam today and I'll make sure that a copy is sent to Carlsberg.

In addition, I am sending under separate cover a couple of reprints by Mazur and de Groot, including one in Dutch (which you will be able to read without difficulty) which deals with precisely entropy and life.

With the warmest greetings to the professor and his faithful helpers.

Yours

L. Rosenfeld

[493]

BOHR TO DELBRÜCK, 21 May 1962
[Carbon copy]

[København,] Den 21. maj 1962.

Kære Max,

Jeg tror, du kender mig så godt, at du ikke er blevet overrasket over, at jeg ikke nåede at skrive igen før den 20. maj. Forhåbentlig er det dog tidsnok i dag at fortælle, at jeg har ordnet det sådan, at jeg kan komme til indvielsen af dit institut den 22. juni. Jeg har også i disse uger tænkt meget over, hvad jeg kunne sige ved denne lejlighed, og skal være forberedt på at holde et sådant foredrag, som du har ønsket. Dit forslag til titlen har faktisk været en inspiration for mig, men indtil jeg har fået foredraget færdigt, vil jeg gerne vente med at tage endelig beslutning herom, og jeg håber ikke, at dette er ubekvemt for din planlæggelse af programmet.

Margrethe og jeg glæder os til at se jer igen og sender vore venligste hilsner,

Din gamle
[Niels Bohr]

Translation

[Copenhagen,] May 21, 1962

Dear Max,

I think you know me so well that it is no surprise to you that I have not managed to write again before 20th May. Hopefully, though, it is not too late to tell you today that I have organized things so that I can come to the inauguration of your institute on 22nd June. In these weeks I have also thought a lot about what I could say on this occasion, and I shall be prepared to give a talk of the kind you have asked for. Your proposal for the title has in fact been an inspiration for me, but I would like to wait with a final decision about this until I have finished the talk, and I hope this is not inconvenient for your planning of the programme.

Margrethe and I are looking forward to seeing you again and we send our warmest greetings,

Your old,
[Niels Bohr]

PAUL A.M. DIRAC

BOHR TO DIRAC, 24 March 1928[30]
[Carbon copy]

[København,] March 24, [19]28.

Dear Dirac,

Klein, who was very happy for his visit to Cambridge and for all the kindness with which he was received, tells me that we may look forward to a short visit of you in a few weeks' time on your way to Leiden. It shall be a great pleasure to us all indeed to see you here again and to hear about the latest progress of your work. I hope you will stay in the institute in the rooms which Heisenberg occupied when you were here last and which have been left as guest rooms. As you will have understood already, all expenses connected with your journey and your stay here will be covered by funds which are at disposal of the institute for such purpose. I shall be glad soon to hear from you what date we may expect you here, and for how long time you can stay in Copenhagen.

I was very thankful for your kind help with the proof of my article[31]. From our discussions in Cambridge and from what Klein told me I do not know, however, whether you are quite in sympathy with the point of view, from which I have tried to represent the paradoxes of the quantum theory. Although of course I realise the tentative character of the formulation, I still believe that the point of view of complementarity is suited to describe the situation. I think, we can not too strongly emphasize the inadequacy of our ordinary perception when dealing with quantum problems. Of course I quite appreciate your remarks that in dealing with observations we always witness through some permanent effects a choice of nature between the different possibilities. However, it appears to me that the permanency of results of measurements is inherent in the very idea of observation; whether we have to do with marks on a photographic plate or with direct sensations the possibility of some kind of remembrance is of course the necessary condition for making any use of observational results. It appears to me that the permanency of

---

[30] This letter is also reproduced in Vol. 6, p. [44].

[31] N. Bohr, *The Quantum Postulate and the Recent Development of Atomic Theory*, Nature (Suppl.) **121** (1928) 580–590. Reproduced in Vol. 6, pp. [148]–[158].

such results is the very essence of the ordinary causal space–time description. This seems to me so clear that I have not made a special point of it in my article. What has been in my mind above all was the endeavour to represent the statistical quantum theoretical description as a natural generalisation of the ordinary causal description and to analyze the reasons why such phrases like a choice of nature present themselves in the description of the actual situation. In this respect it appears to me that the emphasis on the subjective character of the idea of observation is essential. Indeed I believe that the contrast between this idea and the classical idea of isolated objects is decisive for the limitation which characterises the use of all classical concepts in the quantum theory. Especially in relation with the transformation theory the situation may, I think, be described by saying that any such concepts can be used unaltered if only due regard is taken to the unavoidable feature of complementarity.

A point not directly referred to in the article but in which I have been very interested lately is the question of the uni-direction of time. For an isolated system this question has of course no sense. In considering observations, however, it is essential that the light travels towards our eye or the photographic plate. I believe that a closer analysis offers the proper answer to such paradoxes regarding the nature of light as brought forward especially by Lewis[32]. I have been considering to send a short note to Nature concerning these paradoxes, and in this connection I should like also to discuss some of the remarks brought forward at the Solvay conference by Einstein and by yourself. Before a publication of course I should wish to discuss the questions in detail with you, but I hope that your visit here will offer a good opportunity also in this respect.

With kindest regards from us all,

Yours,
[Niels Bohr]

[32] Gilbert Newton Lewis (1875–1946). American physical chemist who devoted the last years of his career to the study of photochemistry. As shown in numerous articles reproduced in Vol. 4, Lewis was frequently referred to in Bohr's work in the early 1920s to explain the periodic system on the basis of his atomic model.

WALTER M. ELSASSER

BOHR TO ELSASSER, 19 November 1959.
[Carbon copy]

[København,] November 19, 1959.

Dear Dr. Elsasser,

It was a great pleasure meeting you this summer in La Jolla and I have in the meantime with deep interest read your book about the physical foundation of biology[33]. With all appreciation of the width of your knowledge and the lucidity of your presentation I feel, however, unable to share your general views. The whole question of the peculiar stability of living organisms and their ability to store and utilize information seems to me to be unapproachable by the study of mechanical models and to be centered entirely on chemical kinetics which has found a proper basis in quantum mechanics. Indeed, it seems to be the stability of the molecular structures which is basic for the storage of the information and its use in the organisms. In this respect, the distinction between "soft" and "rigid" tissues, so tempting in a comparison with mechanical models, seems hardly relevant. This fall, I have had various occasions to discuss these problems with biochemists and, for a congress in Bristol on the use of models in biology, I have prepared some short remarks of which I enclose a copy[34]. I hope that they convey some idea of my general attitude, and I need not say that I shall be deeply interested to learn about your reaction.

With kindest regards, also to common friends in La Jolla,

Yours,
[Niels Bohr]

---

[33] Elsasser, *Physical Foundation of Biology*, ref. 16.
[34] Manuscript, *Models in Biology*, ref. 19.

ELSASSER TO BOHR, 18 December 1959
[Typewritten]

UNIVERSITY OF CALIFORNIA

SCRIPPS INSTITUTION OF OCEANOGRAPHY   18 December 1959
LA JOLLA, CALIFORNIA

Dear Professor Bohr:

Thank you for your kind note. It reminds me somewhat of a situation long since past which you will perhaps forgive me for recalling. In 1934–35 I was in Paris working on the shell structure of the nucleus. The experimental evidence seemed patent and incontrovertible. Every two or three months somebody came through Paris who had just been in Copenhagen and who informed me that I must be plain wrong. Niels Bohr had indeed just shown that the nucleus was a homogeneous fluid droplet subject to hardly any structural and dynamical limitations other than those derived from an appreciable surface energy. So, how could the nucleus possibly have shells? May I suggest that perhaps a similar, happy relationship of unsuspected complementarity could exist in the present problem!

Now to this problem: I wish to make two points. The first concerns stability. All stability, *both* in quantum and in classical mechanics may be reduced to the quantum-mechanical stability of the ground state or lowest group of states of a system not subject to disturbances so large as to produce transitions into highly excited states. Thus a "hard" body such as a crystalline substance is one in which the geometrical configuration of the ground state (or, rather, group of lowest states) is not appreciably altered by a moderate perturbation, e.g. an external macroscopic shear. On the other hand, a "soft" body is one in which the set of lowest states is degenerate in such a way that a near-linear transformation of these states occurs when such a shear is applied, leading to a geometrical rearrangement. Similarly, all constraints (in the Lagrangian sense) of classical dynamics can ultimately be reduced to quantum-mechanical stability of sets of lowest states with regard to geometrical rearrangement. But this is not confined to mechanics. The same no doubt holds for constraints of an electrical or of a chemical nature. Thus *all constraints* that can possibly be subsumed under the most generalized classical Lagrangian scheme (whether the corresponding dynamical variables may be interpreted mechanically or otherwise) are essentially expressions of the quantum-mechanical stability of sets of low-lying states. For sufficiently highly excited states there can be, in the average, no similar stability, as is clear from the correspondence principle.

Now my idea of a computer is not that of some specialized macroscopic, mechanical or electrical device; it is meant to represent the most general dynamical system which has the ability to process information for long periods of time without loss by noise. To do this, the system must have *stable constraints*. Again, it makes little difference whether on a conceptual approach one calls these constraints mechanical, electrochemical, or anything else. What I mean to say is that in order to qualify a system as a computer (that is, information-processing device) it is necessary to demonstrate that the system has definite stable constraints which prevent the information from being dissipated into noise. Now the empirical evidence seems to be rather clear on this point. The DNA molecules in the chromosomes are extremely stable chemically over long periods of time. But they only *store* information, they do not *process* it. The processing is done mainly by protein molecules (with their associated enzymatic functions), and the empirical evidence here is rather distinctly that *there are no long-term constraints*. There is constant turnover of amino acids in the proteins. There are of course powerful temporary constraints, namely, the so-called prosthetic groups which catalyze highly specific reactions, but all the evidence seems to be against the existence of long-term constraints outside of the chromosomes, of either mechanical, electrical or chemical nature; and without these it is most difficult to understand how stable processing of information can take place for more than a short time. I feel that any proposed theory of organic stability based entirely on quantum-mechanical arguments must demonstrate the nature, at least the qualitative nature, of the long-term constraints on which the long-term conservation of information during the time this information is processed, could be based. No such effort exists to my knowledge.

I come now to my second point, which is of a quite different nature. It has to do with what I have called the "principle of finite classes" (early part of my Chapter 4) as follows: It can readily be shown that the laws of quantum mechanics, in spite of their statistical characters, exclude the co-existence of any other laws of Nature (that is, laws which are not deductions from quantum mechanics). Let the system be represented by a psi-function or more generally by a rho-matrix. Any physical statement about the system whatever can be expressed as the expectation value of some suitable operator, if "projection" operators are included (von Neumann). But if the system is given at $t = 0$, it follows that all expectation values are determined for all future times by virtue of the equations of motion. Q.E.D. (abbreviated proof). Now my principle of finite classes shows, on an essentially empirical basis, that this exclusiveness of the laws of quantum mechanics is not operationally verifiable for systems as immensely complex as organisms. The reason is that their phase space has an immensely complicated structure and that image points of actual systems are immensely rare occurrences in this phase space. I have concluded from this that biotonic laws are possible for classes of systems whose membership is *finite* and suitably limited. I would like to say that I think of this as the most important contribution of my book. *This fact creates an entirely novel epistemological situation*, and one which has no counterpart in traditional physics. This, I believe, requires the closest attention of theoretical physicists, since it is clearly of a type requiring more abstract arguments than most biologists or biochemists are dealing with. I hope that you and your associates can some time give a little thought to this particular problem. Also, some day I hope to write a paper on it. The question is clearly independent of any assumptions as to how organisms function in detail.

With the seasons best wishes,

Very sincerely yours,
Walter M. Elsasser

BOHR TO ELSASSER, 29 December 1959
[Carbon copy]

[København,] December 29, 1959.

Dear Dr. Elsasser,

Thank you for your kind letter of December 18 which I read with great interest. I must confess, however, that the comparison between complementary relationships in quantum physics and in biology, which you suggest, does not seem to me to be quite relevant. I believe we agree that the use of mutually exclusive classical pictures in atomic physics bears a certain resemblance to the difference of the terminology used in molecular biology and in practical physiology. Still, it appears to me that such apparently contrasting approaches as were provisionally applied, e.g. in the analysis of nuclear constitution and reactions, and which now have been to a large extent unified on the basis of elementary quantum principles, cannot be compared with the introduction of ideas like biotonics which, as far as I understand, suppose an inherent limit of the quantum-physical approach, beyond such problems of elementary particles which still await elucidation, but which hardly play any role in biology.

In molecular biology, all reference to quantum physics is embodied in chemical kinetics based on statistical thermodynamics. In fact, this affords not only the definition of molecular structures, but also implies the possibility through competition between the different possible reactions to ensure the peculiar stability of the organism. This circumstance stresses the difference between "soft" and "solid" structure; indeed, the latter may possess maximal stability, but not the flexibility demanded for the life processes. Thus, your arguments against the existence of long-time constraints appear to me to be at variance with general evidence of chemistry and with a proper quantum-mechanical treatment of the reactions between chemical compounds.

While trying to represent the divergence between our views quite frankly, I am of course prepared that I may have misunderstood your argumentation at some point, and I need not say that I am looking forward to see the new article on which you are working.

With kind regards also to common friends in La Jolla and best wishes for the New Year,

Yours very sincerely

[Niels Bohr]

H.P.E. HANSEN

BOHR TO HANSEN[35], 8 September 1938.
[Carbon copy]

Gl. Carlsberg, Valby, 8. Sept. 1938.

Kære Hr. Forfatter H.P.E. Hansen,

Paa Grund af Bortrejse har jeg først for nogle Dage siden modtaget det sidste af Deres venlige Breve[36], hvori De saa indgaaende gør Rede for Deres Opfattelse af det erkendelsesteoretiske Standpunkt, jeg har beskrevet i mine to smaa Artikler i "Naturens Verden"[37]. Uden at fornægte min Beundring for Deres intuitive Forstaaelse og den smukke Maade, hvorpaa De finder Udtryk derfor, er jeg imidlertid ikke helt sikker paa, at Artiklerne, der trods den populære Form er opstaaet som Indlæg i en fagvidenskabelig Diskussion, har været egnede til at give Dem et helt rigtigt Indtryk af Komplementaritetssynspunktet.

Navnlig ligger det mig paa Sinde at betone, at det ikke drejer sig om nogen Grænse for videnskabelig Erkendelse, men tværtimod om en Fjernelse af tilsyneladende Modsigelser gennem Erkendelsen af, at de Former, der

---

[35] Hans Peter Erhard Hansen (1888–1946). Danish author of novels and literary essays.
[36] Deposited at NBA.
[37] N. Bohr, *Lys og Liv*, Naturens Verden **17** (1933) 49–59; N. Bohr, *Kausalitet og Komplementaritet*, Naturens Verden **21** (1937) 113–122. Reproduced in English on pp. [29]–[35] and [39]–[48].

sædvanligvis anvendes ved Naturfænomenernes Indordning, er Idealisationer med begrænset Formaalstjenelighed.

Hvad Forholdet mellem Matematik og Naturvidenskab angaar, ser jeg ligesom De i Stuart Mills Bemærkning om, at to og to for Beboerne af en anden Klode kunde være fem, ikke andet end et trivielt Nomenklaturspørgsmaal; ved Logiken selv maa vi jo, saa længe vi vil være ærlige, ubetinget holde fast. Forskellen mellem Matematikken og de egentlige Naturvidenskaber beror alene paa, at vi i den første, hvor vi selv bestemmer Spillets Regler, kan angive Forudsætningerne for Definitioner og Begreber, medens vi i de sidste altid maa være forberedt paa af nye Erfaringer at lære, at enhver Beskrivelsesforms Formaalstjenlighed kan være betinget af hidtil upaaagtede Forudsætninger.

Uden at undervurdere Deres interessante Skelnen mellem forskellige psykologiske Typer kan jeg heller ingen principiel Forskel se mellem en Logiker og en Mystiker. Dersom Mystik ikke simpelthen skal betyde Begrebsforvirring, maa den jo netop bunde i en mere eller mindre bevidst logisk Analyse af vore Konventioners Begrænsning. Det største Eksempel i denne Henseende er vel den Afvisning af al sædvanlig Tale om Tilværelsens Mening eller Meningsløshed, som Buddha førtes til gennem sin Erkendelse af Ordet "Mening"s relative Karakter, der medfører, at den eneste "Mening", vi kan tillægge Tilværelsen, er, at det er principielt meningsløst at sige, at Tilværelsen er uden "Mening".

Kun en saadan Indstilling, hvor sofistisk den end i første Øjeblik kunde forekomme, giver jo Plads for Menneskelivets Rigdom med dets Glæder og Sorger paa ganske samme Maade, som det kun har været muligt gennem Erkendelsen af Kausalitetsidealets principielle Begrænsning at opnaa en logisk Indordning af Atomfænomenernes Mangfoldighed.

Jeg er bange for, at det heller ikke ved disse Bemærkninger vil lykkes mig at give Dem mere end en Antydning af min Indstilling, men jeg haaber, at Foredraget ved Antropologkongressen i Sommer, som jeg endnu ikke har haft Tid til at bringe i trykfærdig Stand[38], maaske vil være mere egnet til at give Dem et fyldigere Indtryk deraf.

> Med venlige Hilsener
> Deres Hengivne
> [Niels Bohr]

---

[38] N. Bohr, *Natural Philosophy and Human Cultures*, Congrès international des sciences anthropologiques et ethnologiques, compte rendu de la deuxième session, Copenhague 1938, Ejnar Munksgaard, Copenhagen 1939, pp. 86–95. Reproduced on pp. [240]–[249]. Also published in Bohr, *Atomic Physics and Human Knowledge*, ref. 11, pp. 23–31.

P.S. Jeg sender ogsaa denne Gang et Par Særtryk. Det første er af mere teknisk–videnskabelig Natur, og det er ikke Meningen at bebyrde Dem med et Studium deraf, men jeg tænkte, at det maaske kunde interessere Dem ved at kigge deri at faa et Indtryk af, hvor vanskeligt det er faldet selv en Forsker med saa frit et Blik og dyb en Indsigt som Einstein at bekvemme sig til at tage Revisionen af Forudsætningen for selve Realitetsbegrebet op for Alvor[39]. Den anden handler om det samme Emne som min lille Artikel om Liv og Lys, men Anledningen til dens Fremkomst har ført med sig, at den historiske Baggrund for Fysikkens og Biologiens Udvikling er lidt nærmere udmalet[40].

Translation

Gl. Carlsberg, Valby, September 8, 1938

Dear Mr H.P.E. Hansen, author,

Due to my absence on a journey, I received only a few days ago the last of your friendly letters[36] in which you so carefully explain your interpretation of the epistemological point of view that I have described in my two short articles in "Naturens Verden"[37]. Without denying my admiration for your intuitive understanding and the beautiful way in which you express it, I am, however, not quite sure that the articles, which despite their popular style were written as contributions to a scholarly discussion, have been suitable to provide you with an entirely correct impression of the complementarity viewpoint.

I feel in particular that I must emphasize that it is not a question of any limit to scientific knowledge, but on the contrary of the removal of apparent contradictions by recognizing that the frames normally used for the ordering of natural phenomena are idealizations of limited usefulness.

Concerning the relation between mathematics and natural science I, like you, regard Stuart Mill's remark that for the inhabitants of another planet two plus two could equal five as no more than a trivial question of nomenclature; if we are to be honest we must, indeed, adhere to logic unconditionally. The difference between mathematics and the natural sciences proper rests alone on the fact that in the former, where we define the rules of the game ourselves, we can state the preconditions for definitions and concepts, whilst in the latter we must always be prepared to learn from new experiences that the usefulness

---

[39] N. Bohr, *Can Quantum-Mechanical Description of Physical Reality be Considered Complete?* Phys. Rev. **48** (1935) 696–702. Reproduced in Vol. 7, pp. [292]–[298].

[40] N. Bohr, *Biologi og Atomfysik*, Naturens Verden **22** (1938) 433–443. Reproduced in English on pp. [52]–[62].

[503]

of any form of description may be contingent upon previously unnoticed pre-conditions.

Without underestimating your interesting distinction between different psychological types, I cannot see any fundamental difference between a logician and a mystic. If mysticism is not simply to mean conceptual confusion, it must indeed be rooted precisely in a more or less conscious logical analysis of the limitation of our conventions. The most striking example in this connection is probably the rejection of all the usual talk about the meaning or the meaninglessness of life, to which Buddha was led through his recognition of the relative character of the word "meaning", which implies that the only "meaning" we can give life is that it is in principle meaningless to say that life is without "meaning".

Only such an attitude, however sophistic it might appear at first glance, allows for the richness of human life with its joys and sorrows in precisely the same way that it has been possible to achieve a logical ordering of the variety of atomic phenomena only through the recognition of the fundamental limitation of the ideal of causality.

I fear that I will not succeed with these remarks either in giving you more than an inkling of my outlook, but I hope that the address at the Anthropology Conference this summer, which I have not yet had time to prepare for printing[38], will perhaps be better suited to give you a more complete impression.

> With kind regards
> Yours sincerely
> [Niels Bohr]

P.S. This time, too, I enclose a couple of reprints. The first one is of a more technical–scientific nature, and I do not mean to burden you with studying it, but I thought that it might be of interest for you, by looking through it, to get an impression of how difficult it has been, for even a scientist with so open an outlook and so deep an understanding as Einstein, to bring himself to seriously reconsider the basis for the very concept of reality[39]. The second one deals with the same topic as my short article on life and light, but the occasion for its appearance has given rise to a more detailed description of the historical background for the development of physics and biology[40].

HARALD HØFFDING

HØFFDING TO BOHR, 22 November 1906
[Handwritten]

Strandgade 26. C   22/11 [19]06

Kære Student Bohr!

Herved sender jeg Dem første Ark af den nye Udgave af "Formel Logik" og beder Dem gennemlæse det med Deres vante Kritik. Deres eventuelle Bemærkninger bedes skrevne paa et Stykke Papir og vedlagt – Jeg haaber ikke, at jeg gør Dem for megen Ulejlighed hermed.

Deres hengivne
Harald Høffding

Translation

Strandgade 26. C, November 22, [19]06

Dear Student Bohr,

I hereby send you the first sheet of the new edition of "Formel Logik" asking you to read it through in your usual critical fashion. Please write any comments on a piece of paper and enclose it – I hope that I do not thus cause you too much inconvenience.

Yours sincerely,
Harald Høffding

HØFFDING TO BOHR, 1 December 1906
[Handwritten]

1/12 [19]06

Kære Student Bohr!

Hermed andet Ark. Vil De sende mig det i en Konvolut, naar De har læst det, og give mig Deres Anmærkninger paa et hoslagt Papir. Vil De især lægge Mærke til Dualitetsprincipet (p. 27).

Deres hengivne
Harald Høffding

Translation

December 1, [19]06

Dear Student Bohr,

Herewith the second sheet. Please send it to me in an envelope when you have read it, and give me your remarks on an enclosed piece of paper. Please note in particular the principle of duality (p. 27).

Yours sincerely,
Harald Høffding

HØFFDING TO BOHR, 4 December 1906
[Handwritten]

Strandgade 26. C
d. 4. decbr 1906

Kære Student Bohr!

Jeg har stadigt nogle Skrupler angaaende Deres Forslag om Dualitetsprincipets Formulering.

Efter Deres Forslag skulde det formuleres saaledes

$$CA = CAB$$

$$cA = cAb.$$

Men hvorfor tage vi kun de to Tilfælde CA og cA? Forudsættes her ikke netop den gamle Formulering ($A = AC + Ac$)? Vil ikke enhver Forklaring af, hvorfor vi anvende denne Formulering, fremkalde det Spørgsmaal, hvorfor vi kun tale om CA og cA?

Meget tydeligt træder dette frem, naar vi benytte Deres Formulering ved Beviset for Kontraposition:

Haves $A = AB$; kunne vi sætte $Cb = CbA$ $cb = cba$, og da $Cb = CbAB = 0$, maa $cb = cba$ gælde. Her antage vi uden videre at cb gælder, naar Cb er ugyldig. Er det ikke den gamle Vanskelighed om igen? Jeg ser ikke Andet, end at vi her komme ind i en regressus in adfinitum [*sic*].

Afset herfra tror jeg i hvert Tilfælde, at det vil være det Simpleste at beholde den gamle Formulering – med det Forbehold, De har paavist Nødvendigheden af, og som naturligvis stadigt maa haves in mente.

Jeg er Dem meget taknemmelig for Deres gode Hjælp, og det interesserer mig meget at drøfte disse Ting med Dem. Jeg haaber, at jeg ikke tager for meget af Deres Tid ved at bede Dem sige mig Deres Mening om de anførte Betænkeligheder.

Deres hengivne
Harald Høffding

Translation

Strandgade 26. C, December 4, 1906

Dear Student Bohr,

I still have some scruples regarding your suggestion as to how to formulate the principle of duality.

According to your suggestion it should be formulated thus:

$$CA = CAB$$

$$cA = cAb.$$

But why should we consider only the two cases CA and cA? Are we not just presupposing the old formulation ($A = AC + Ac$) here? Will not any explanation of why we use this formulation give rise to the question as to why we speak only of CA and cA?

This becomes particularly clear when we apply your formulation in connection with the proof of contraposition:

If we assume $A = AB$; we could then put $Cb = CbA$ $cb = cba$, and since $Cb = CbAB = 0$, then $cb = cba$ must hold. Here we assume as a matter of course that cb holds when Cb does not hold. Is this not the old difficulty once more? I cannot see anything other than that we here get into a regressus in adfinitum [*sic*].

Apart from this I believe that in any case it would be simplest to keep the old formulation – with the reservation you have proved necessary, and which we must, naturally, still keep in mind.

I am very grateful to you for your good help, and I find great interest in discussing these matters with you. I hope that I do not take too much of your time by asking you to tell me your opinion about the misgivings mentioned.

Yours sincerely,

Harald Høffding

HØFFDING TO BOHR, 6 December 1906
[Handwritten]

Den 6. Decbr 1906

Kære Student Bohr!

Hermed følger det sidste Ark, som jeg beder Dem gennemse. Det vigtigste af det er p. 37. De vil af mine Rettelser se, hvilken Formulering af Dualitets-princippet jeg nu har bestemt mig for. Idet Hovedvægten maa lægges paa, at

Bogstaverne betyde Begrebs*indhold*, og at Dualitetsprincipets Mening er, at der er bundet Valg mellem de to *Kombinationer AB* og *Ab*, opgiver jeg at formulere det som en Ligning og skriver blot, at "det kan formuleres saaledes: $AB\cdot \mid \cdot Ab$ (idet Tegnet $\cdot \mid \cdot$ betyder et gensidigt Udelukkelsesforhold …)". I det Følgende kommer der saa, at Dualitetsprincipet udtrykker et Forhold mellem to Kombinationer (ikke, som der før stod, mellem to Domme). Nu behøves der intet Forbehold. Ti de Kendemærker, der gælder for Hvirveldyrstypen, maa enten forekomme sammen med de Kendemærker, der gælder for Pattedyrstypen, eller med saadanne, som ikke gælder for denne Type. Formuleringen passer nu baade paa højere og lavere Begreber.

Med Hensyn til den Brug, der gøres af Dualitetsprincipet, passer den nye Formulering godt til Opstilling af Kombinationer ved indirekte Slutning. Og ogsaa Beviserne for umiddelbar Slutning (p. 37) kunne fremstilles tydeligt, da det jo her blot gælder om at skaffe en af de to tænkelige Kombinationer bort.

Deres Brev ryddede ikke mine Betænkeligheder bort. Hvis man i det hele gør Forskel mellem Kontradiktions- og Dualitetsprincipet, *forudsættes* unægtelig dette sidste naar man gaar ud fra, at man ikke behøver at tage Hensyn til andre Kombinationer end *AC* og *Ac*. Nu vil jeg gerne, ved Tilbagesendelsen af det hoslagte Ark, høre, om De har Indvendinger at gøre om den nye Formulering.

Deres hengivne
Harald Høffding

Translation

December 6, 1906

Dear Student Bohr,

I enclose herewith the last sheet, which I ask you to look through. The most important part of it is p. 37. From my corrections you will see which formulation of the principle of duality I have now decided upon. Since the main emphasis must be on the fact that the letters represent conceptual *content* and that the meaning of the principle of duality is that there is a limited choice between the two *combinations*, *AB* and *Ab*, I refrain from formulating it as an equation and simply write that "it can be formulated thus: $AB\cdot \mid \cdot Ab$ (where the sign $\cdot \mid \cdot$ means reciprocal exclusion …)". In the following it is then stated that the principle of duality expresses a relation between two combinations (not, as was previously stated, between two propositions). No reservations are now needed. Therefore, the characteristics of the vertebrate type must either occur together with the characteristics of the mammal type or together with those that do not belong to this type. Now the formulation applies for both higher and lower concepts.

[508]

Concerning the use made of the principle of duality, the new formulation is quite suitable for arranging combinations employed in indirect inference. Also the proofs concerning direct inference (p. 37) could be clearly presented, as it is here simply a question of removing one of the two possible combinations.

Your letter did not take away my misgivings. If we distinguish at all between the principle of contradiction and the principle of duality, the latter is undoubtedly *presupposed* when we assume that we need not consider other combinations than AC and Ac. When you return the enclosed sheet, I would be grateful to hear whether you have any objections to this new formulation.

Yours sincerely,
Harald Høffding

HØFFDING TO BOHR, 1 January 1922
[Handwritten]

Carlsberg, Valby,
d. 1. Januar 1922

Kære Professor Niels Bohr!

Først og fremmest ønsker jeg Dem et godt Nytaar for Dem og Deres, saavel som for Deres betydningsfulde Arbejde. Men dernæst har jeg en Anmodning til Dem fra M. Xavier Leon, Udgiver af det meget ansete filosofiske Tidsskrift Revue de Métaphysique et de Morale, i hvilket foruden Filosofer ogsaa Naturforskere og Matematikere, saaledes Henri Poincaré, har skrevet. Anmodningen gaar ud paa et Exposé af Deres Teori om Atomernes Bygning. De kan godt skrive den paa Engelsk; man vil da i Paris faa den oversat paa Fransk.

Tidsskriftet, der væsentlig holdes op ved Xavier Leon's Offervillighed, yder ikke Honorar, men det vil være af stor Interesse for dets udsøgte Læserkreds, om De vilde imødekomme Anmodningen.

Med venligst Hilsen

Deres hengivne
Harald Høffding.

Translation

Carlsberg, Valby, January 1, 1922

Dear Professor Niels Bohr,

First of all, I wish a Happy New Year for you and yours, as well as for your important work. Then, I have a request to you from M. Xavier Leon, the editor of the highly esteemed philosophical journal, Revue de Métaphysique et

[509]

de Morale, in which not only philosophers, but also natural scientists and mathematicians, for instance Henri Poincaré, have written. The request concerns an account of your theory of the constitution of atoms. You are welcome to write it in English; it will then be translated into French in Paris.

The journal, which to a great extent is kept going by the generosity of Xavier Leon, does not give an honorarium, but it would be of great interest to its select circle of readers if you should accept the request.

With kind regards

Yours sincerely,
Harald Høffding.

HØFFDING TO BOHR, 20 September 1922
[Handwritten]

Carlsberg d. 20. Septe 1922.

Kære Professor Bohr!

Som jeg engang i Sommer nævnte for Dem, vilde jeg gerne spørge Dem om en Ting i Anledning af Deres Afhandling om "Atomernes Bygning og Stoffernes fysiske og kemiske Egenskaber"[41].

Jeg har lagt Mærke til, at De gennemgaaende bruger Udtryk, der tyder paa et *Analogi*forhold (ikke et Identitetsforhold) mellem Atomernes Bygning og de faktisk foreliggende fysiske og kemiske Data. Saadanne Udtryk er "Belysning" (p. 1), – "Forklaring eller rettere Forstaaelse" (p. 33), – "fortolke" (p. 36), – "som Spektret lærer os, og som Atommodellen gør forstaaeligt" (p. 45).

Mit Spørgsmaal er, om Udtrykket Analogi ikke vilde være det sammenfattende Udtryk for de Vendinger, De bruger paa de afgørende Punkter. Al "Forstaaelse" beror – afset fra den rene Logik – paa Analogi, og Videnskab er en strengt rationel Gennemførelse af Analogier mellem forskellige Erkendelsesomraader. Saaledes er der jo ifølge Hjelmslev[42] (og ifølge Zeuthen[43]

---

[41] Jul. Gjellerups Forlag, Copenhagen 1922. Translated into English as *Essay III: The Structure of the Atom and the Physical and Chemical Properties of the Elements* in *The Theory of Spectra and Atomic Constitution. Three Essays by Niels Bohr*, Second Edition, Cambridge University Press, Cambridge 1924, pp. 61–138. Reproduced in Vol. 4, pp. [183]–[256] (Danish original) and [257]–[340] (English translation).

[42] Johannes Hjelmslev (1873–1950), professor of mathematics at the University of Copenhagen 1917–1943.

[43] Hieronymus Georg Zeuthen (1839–1920), professor of mathematics at the University of Copenhagen 1883–1910.

i hans sidste Afhandling[44]) et Analogiforhold mellem Aritmetik og Geometri, og jeg tror, denne Opfattelse kan gennemføres overalt, saaledes ogsaa ved Forholdet mellem Aandsvidenskab og Naturvidenskab. Jeg gør saa en principiel Forskel mellem Analogier, der kan rationelt gennemføres, og andre Analogier, der væsentlig staar som Udtryk for poetiske eller religiøse Følelser og ikke kan konsekvent fastholdes.

Jeg vil naturligvis ikke drage Dem ind i en filosofisk Diskussion, men jeg vilde kende Deres Synsmaade, før jeg benytter Deres Afhandling i et Arbejde om Begrebet Analogi, som jeg er beskæftiget med.   Med venligst Hilsen

Deres hengivne   Harald Høffding

Translation

Carlsberg, September 20, 1922.

Dear Professor Bohr,

As I mentioned to you sometime this summer, I should like to ask you about one thing in connection with your treatise about "The Structure of the Atom and the Physical and Chemical Properties of the Elements"[41].

I have noticed that in most cases you use expressions indicating a relation of *analogy* (not of identity) between the constitution of the atoms and the actually available physical and chemical data. Examples of such expressions are "elucidation" (p. 1) – "explanation or rather understanding" (p. 33) – "interpret" (p. 36) – "as the spectrum tells us and the atomic model renders understandable" (p. 45).

My question is whether the expression analogy would not be the expression epitomizing the terms you apply at the crucial points. All "understanding" – save pure logic – depends on analogy, and science is a strictly rational implementation of analogies between different fields of knowledge. Thus, according to Hjelmslev[42] (and according to Zeuthen[43] in his last paper[44]) there is a relation of analogy between arithmetic and geometry, and I believe that this view may be implemented everywhere, hence also with regard to the relation between the humanities and the natural sciences. Thus, I make a fundamental difference between analogies that can be implemented rationally and other

---

[44] Most likely, H.G. Zeuthen, *Sur l'origine de l'algèbre*, Kgl. Dan. Vid. Selsk., Mat.–fys. Medd. **II**, 4 (1919) 70pp.

analogies that are primarily expressions of poetic or religious emotions and cannot be upheld consistently.

Naturally, I do not wish to involve you in a philosophical discussion, but I should like to know your point of view before making use of your treatise in a paper on the concept of analogy that I am working on.   With kind regards

Yours sincerely, Harald Høffding

BOHR TO HØFFDING, 22 September 1922
[Carbon copy]

[København,] 22. September [19]22.

Kære Professor Høffding,

Mange Tak for Deres venlige Brev, der interesserede mig meget. Det af Dem betonede Forhold angaaende Analogiens Rolle i videnskabelige Undersøgelser er utvivlsomt et væsentligt Træk ved alle naturvidenskabelige Studier, selv om det ikke altid træder klart frem. Ofte er det vel muligt at benytte et Billede af geometrisk eller aritmetisk Art, der paa saa tydelig Maade dækker det behandlede Problem i det Omfang, der er under Diskussion, at Betragtningerne nærmest faar rent logisk Karakter. I Almindelighed, og navnlig paa nye Arbejdsomraader, maa man imidlertid stadig holde sig Billedets øjensynlige eller eventuelle Utilstrækkelighed for Øje og være tilfreds, naar blot Analogien træder saa stærkt frem, at Billedets Anvendelighed eller rettere Frugtbarhed i det benyttede Omfang er udenfor Tvivl. Et saadant Forhold gælder ikke mindst paa Atomteoriens nuværende Standpunkt. Her er vi i den ejendommelige Situation, at vi har tilkæmpet os visse Oplysninger om Atombygningen, der tør betragtes at være lige saa sikre som nogen af Naturvidenskabens Kendsgerninger. Paa den anden Side møder vi Vanskeligheder af en saa dybtliggende Natur, at vi ikke aner Vejen til deres Løsning; efter min personlige Opfattelse er disse Vanskeligheder af en saadan Art, at de næppe lader os haabe inden for Atomernes Verden at gennemføre en Beskrivelse i Rum og Tid af den Art, der svarer til vore sædvanlige Sansebilleder. Under disse Omstændigheder maa man naturligvis stadig holde sig for Øje, at man opererer med Analogier, og den Takt, hvormed disse Analogiers Anvendelsesomraader i det enkelte Tilfælde afstikkes, er af afgørende Betydning for Fremskridtet.

Hvis det maatte forekomme Dem ønskeligt at høre lidt nærmere om noget bestemt af de Eksempler, De hentyder til i Deres Brev eller andre, er jeg naturligvis beredt til efter Evne at bistaa med enhver Oplysning.

Jeg sender indlagt et lille Hefte[45], som indeholder en tysk Udgave af det af Dem omtalte Foredrag, samt to tidligere Foredrag, ikke i den Hensigt yderligere at trætte Dem med fysiske Enkeltheder, men blot fordi De i Indledningen vil finde et Par Bemærkninger, der selv om de ikke i filosofisk Henseende har nogen afklaret Form, dog vil vise Dem, hvormeget det omhandlede Forhold ligger mig paa Sinde.

Med mange venlige Hilsener fra
min Hustru og mig selv,
Deres ærbødigst hengivne
[Niels Bohr]

Translation

Copenhagen, September 22, 1922

Dear Professor Høffding,

Thank you very much for your kind letter which I found very interesting. The relation you emphasize with regard to the role of analogy in scientific investigations is without doubt an essential feature of all natural science studies, although it does not always manifest itself clearly. It may often be possible to use a picture of a geometrical or an arithmetical kind that covers the problem dealt with, within the range under discussion, in so clear a manner that the considerations almost attain a purely logical character. However, in general, and especially in new fields of work, one must constantly bear in mind the apparent or the possible inadequacy of the picture and be satisfied as long as the analogy is so manifest that the utility, or rather the fruitfulness, of the picture, within the range where it is used, is beyond doubt. Such a relation holds not least at the present stage of the atomic theory. Here we are in the peculiar situation that we have fought our way to some information about the constitution of the atom, which must be considered just as certain as any of the facts of natural science. On the other hand, we meet difficulties of such a profound nature that we have no idea of the way to their solution; my personal opinion is that these difficulties are of such a kind that they hardly allow us to hope, within the world of atoms, to implement a description in space and time of the

---

[45] N. Bohr, *Drei Aufsatze über Spektren und Atombau*, Friedr. Vieweg, Braunschweig 1922.

kind corresponding to our usual sensory images. Under these circumstances one must, of course, continually bear in mind that one is employing analogies, and the discretion with which the areas of application of these analogies are defined in every single case is of decisive importance for making progress.

If you wish to hear in somewhat greater detail about any particular example you refer to in your letter, or about other examples, I am naturally prepared to help you with any information to the best of my ability.

I enclose a small booklet[45] containing a German version of the lecture you mention as well as two earlier lectures, not with the purpose of burdening you further with details of physics, but just because you will find in the introduction some remarks which, although philosophically speaking they do not have a clarified form, nevertheless will show you how important the matter under discussion is to me.

> Many kind regards from
> my wife and myself,
> Yours respectfully and sincerely,
> [Niels Bohr]

PASCUAL JORDAN

BOHR TO JORDAN, 25 January 1930
[Carbon copy]

[København,] 25. Januar [19]30.

Lieber Jordan,

Wie ich eben Born geschrieben habe, haben Sie beide durch die Übersendung Ihres schönen Buches[46] und freundlichen Briefe mir eine grosse Freude bereitet. Zur selben Zeit wie ich sehr gerührt bin über die Weise, in welcher meine alten Bestrebungen auf diesem Gebiete, wo alle die kühnsten früheren Träume jetzt so weit übertroffen sind, gedenkt werden, bin ich von Bewunderung durchdrungen über die Klarheit und Kraft, womit in Ihrem Werke die jetzige wunderbare Entwicklung dargestellt wird.

Es war mir eine ganz besondere Freude, bei Ihrem letzten Besuch in Kopenhagen mit Ihnen über die allgemeinen philosophischen Fragen zu sprechen, die uns die gegenwärtige Lage der Physik aufdrängt. Aus den Naturwissenschaften haben Sie vielleicht schon gesehen, dass ich in einem Vortrag diesen Som-

---

[46] M. Born and P. Jordan, *Elementare Quantenmechanik*, Julius Springer, Berlin 1930.

mer bei der skandinavischen Naturforscherversammlung in Kopenhagen einige Bemerkungen über die biologischen Grundfragen gewagt habe[47]. Ich bin von diesen Fragen sehr erfüllt, habe doch noch nicht die Physik ganz vergessen. Zur Zeit bin ich beschäftigt mit dem Versuch einer Darstellung der offenen Fragen der relativistischen Quantentheorie. Jetzt wo Sie Ihre Arbeit so nahe an Dänemark haben, können wir vielleicht gelegentlich auf einen Besuch in Kopenhagen hoffen, wo unsere Diskussion fortgesetzt werden könnte.

Meine Frau und ich senden Ihnen unsere herzlichsten Glückwünsche zu Ihrer Verlobung und hoffen Sie beiden einmal hier in Kopenhagen begrüssen zu können.

Mit herzlichsten Grüssen von uns allen,

Ihr

[Niels Bohr]

Translation

[Copenhagen,] January 25, [19]30

Dear Jordan,

As I have just written to Born, both of you have given me great pleasure by sending me your beautiful book[46] and kind letters. Just as I am deeply moved by the way in which my old endeavours in this field, where all the boldest earlier dreams have now been far surpassed, are remembered, I am also full of admiration for the clarity and forcefulness with which the current wonderful development is presented in your work.

During your last visit in Copenhagen, it was a special pleasure for me to talk with you about the general philosophical questions that the present situation in physics forces upon us. In the Naturwissenschaften you may perhaps already have seen that in a lecture in Copenhagen last summer at the Scandinavian Meeting of Natural Scientists, I ventured some remarks about the fundamental questions in biology[47]. I am deeply engrossed in these questions, but have not yet forgotten physics entirely. For the time being, I am busy with an attempt

---

[47] N. Bohr, *Die Atomtheorie und die Prinzipien der Naturbeschreibung*, Naturwiss. **18** (1930) 73–78. Published in English as N. Bohr, *The Atomic Theory and the Fundamental Principles Underlying the Description of Nature* in *Atomic Theory and the Description of Nature*, Cambridge University Press, Cambridge 1934 (reprinted 1961), pp. 102–119. The latter volume is photographically reproduced as *Atomic Theory and the Description of Nature, The Philosophical Writings of Niels Bohr, Vol. I*, Ox Bow Press, Woodbridge, Connecticut 1987. The English version of Bohr's article is reproduced in Vol. 6, pp. [236]–[253].

at an account of the unresolved questions in relativistic quantum theory. Since you now work so close to Denmark, perhaps we can hope that you on occasion pay a visit to Copenhagen where our discussions could be continued.

My wife and I send you our most cordial congratulations on your engagement and hope to welcome you both in Copenhagen sometime.

With cordial greetings from us all.

Yours

[Niels Bohr]

JORDAN TO BOHR, 20 May 1931
[Typewritten]

PROF. DR.
PASCUAL JORDAN   ROSTOCK, DEN 20.5.1931.
UNIVERSITÄT ROSTOCK   LOIGNYSTR. 10

Lieber Herr Bohr!

Als ich neulich die Freude hatte, mich mit Ihnen in Kopenhagen über die Quantentheorie und über ihre Bedeutung für die Biologie usw. unterhalten zu können, erzählte ich Ihnen, daß ich an einem Aufsatz über diesbezügliche Fragen schriebe. Ich habe nun diesen Aufsatz – von dessen Manuskript ich Ihnen beiliegend eine Abschrift sende[48] – tatsächlich fertig gemacht, und zwar so, wie es von vornherein geplant war. Aber ich muß gestehen, daß mir gerade unsere Unterhaltungen in Kopenhagen wieder zum Bewußtsein gebracht haben, wie sehr meine immer sehr auf "Deutlichkeit" und auf eine gewisse "radikale" Formulierungsschärfe abzielende Darstellungsweise im Nachteil ist gegenüber Ihrer vorsichtigeren und doch so treffenden Ausdrucksform. – Ich bin sehr gespannt auf Ihren Aufsatz über Biologie usw., von dem Sie mir erzählten!

Viele herzliche Grüße und bitte eine freundliche Empfehlung an Frau Prof. Bohr!

Stets Ihr ergebener
P. Jordan

---

[48] The manuscript, *Statistik, Kausalität und Willensfreiheit*, is deposited at the NBA.

Translation

Rostock, May 20, 1931

Dear Mr Bohr!

When I recently had the pleasure to converse with you in Copenhagen about the quantum theory and about its importance for biology etc., I told you that I was writing a paper on these questions. I have indeed now completed this paper – I enclose a copy of the manuscript[48] – just as I had planned beforehand. But I must admit that especially our conversations in Copenhagen made me aware once again of how much my way of presentation, always aiming at "clarity" and a certain "radical" sharpness of expression, compares unfavorably to your more cautious and yet so apposite form of expression. I am very eager to see your paper about biology etc. that you told me about!

Many cordial greetings and, please, my best respects to Mrs Professor Bohr,

Always your devoted
P. Jordan

BOHR TO JORDAN, 5 June 1931
[Carbon copy]

[København,] 5. Juni [19]31.

Lieber Jordan,

Vielen Dank für Ihren freundlichen Brief mit Ihrem schönen Aufsatz über Kausalität und Willensfreiheit, den ich vor einigen Tagen hier fand bei meiner Rückkehr von einer kleinen Reise in Pfingsten. Die Übereinstimmung in unserer allgemeinen Einstellung war mir eine grosse Freude, sowie die Sympathie die Sie meinen Bestrebungen aller Unvollkommenheiten der Ausdrucksweise zu trotze, entgegenbringen. Eben deshalb möchte ich mich aber gestatten, bei völliger Schätzung der Klarheit Ihres kräftigeren Stils, Sie auf einigen Punkten aufmerksam zu machen, die ich etwas anders formulieren würde.

Ein solcher, aber sehr kleiner, Punkt ist Ihr Vorbehalt, was die zukünftige Entwicklung der Atomphysik in ihrem Verhalten zu der klassischen Beschreibungsweise betrifft, und der mir nicht ganz so selbstverständlich vorkommt, als Sie es bezeichnen. Die Situation in der Atomphysik ist doch eine wesentlich andere als diejenige, die in der klassischen Physik in Betracht kommt. Es ist ja nicht nur so, dass wir gelernt haben, dass jede Beobachtung mit einem Eingriff in die Phänomene verknüpft ist; vielmehr haben wir erkannt, dass der ganze Beobachtungsbegriff eine Trennung zwischen Objekt und

[517]

Beobachtungsmittel, die mit der Vorstellung abgeschlossener Systeme unvereinbar ist, verlangt. Es ist eben mit Hinblick hierauf, dass ich in unseren Diskussionen so grosses Gewicht darauf gelegt habe, dass die Benutzung des Raum–Zeitbegriffs immer auf der Heranziehung von nicht zu Systemen gehörenden Massstäben und Uhren basiert sein muss. Wir haben hier mit einer weitgehenden aber doch nicht durchgreifenden Analogie zur Relativitätstheorie zu tun, denn jedenfalls in der speziellen Relativitätstheorie ist die fruchtbare und lehrreiche Betonung der Willkür der klassischen Begriffstellung nicht prinzipiell mit ihrer Unhaltbarkeit verbunden, wie es in der Atomphysik der Fall ist. Wie Sie selber betonen, wäre es ja denkbar, dass weitere Erfahrungen uns über das Vorhandensein eines ausgezeichneten Koordinatensystems belehren könnten, aber eine analoge Möglichkeit findet sich meines Erachtens in der Atomphysik nicht. Durch die weitergehende Analyse des Beobachtungsbegriffs sind wir hier auf noch tiefer liegende Erkenntnisprobleme gestossen, die bei der Beschreibung des Erfahrungsinhalts, wie sich dieser auch gestalten wird, nicht ausser Acht gelassen werden darf. Es handelt sich ja garnicht darum, dass Gesetzmässigkeiten[49] [in der Atomphysik nicht bestehen, sondern dass der Art der Getzmässigkeit] zu Folge die Situation eine wesentlich andere ist als diejenige, auf welche die Idealisation Anwendung finden kann, die man gewöhnlich als Kausalität bezeichnet.

Es ist diese Einstellung, die für meine Äusserungen im Planckheft der Naturwissenschaften über die psychischen Probleme massgebend war[50]; von der psychologischen Analyse sind wir ja eben mit Situationen der erwähnten Art vertraut. Ich weiss wohl, dass es sich auch hier nur um kleine Nuancen der Ausdrucksweise handelt; ich möchte Sie aber in dieser Verbindung darauf aufmerksam machen, dass Ihre Besprechung des Parallelismus des physischen und psychischen Geschehens vielleicht zu einem Missverständniss meiner Auffassung Anlass geben könnte. Meine Betonung der formalen Ähnlichkeit zwischen dem Wellenpartikelproblem und psychologischen Grundproblemen zielte ja nicht auf eine enge Analogie zwischen dem psychophysischen Parallelismus und der Wellen–Partikel-Dualität hin, sondern vor allem auf die

---

[49] The subsequent passages in square brackets were left out by mistake in the letter sent to Jordan. It was entered in Bohr's handwriting in the carbon copy, now in the BSC, after he discovered the mistake when discussing the matter with Oskar Klein, who was visiting. See the letter from Bohr to Jordan, 14 June 1931, BSC.

[50] N. Bohr, *Wirkungsquantum und Naturbeschreibung*, Naturwiss. **17** (1929) 483–486. Translated into English as *The Quantum of Action and the Description of Nature* in Bohr, *Atomic Theory and the Description of Nature*, ref. 47, pp. 92–101. The latter version is reproduced in Vol. 6, pp. [208]–[217].

Möglichkeit, gegenseitige Belehrung aus physikalischen und psychologischen Untersuchungen zu erlangen. Bei dem Parallelismus selber handelt es sich natürlich um eine besondere Komplementarität, die nicht durch Gesetzmässigkeiten von einseitigem physikalischem oder psychologischem Charakter begriffen werden kann. Dies war eben der Grund für die tastende, aber, wie ich fürchte, daher sehr dunkle, Weise, in der ich mich über diese Frage im Planckheft wie auch in dem späteren in den Naturwissenschaften erschienenen Vortrag zur skandinavischen Naturforscherversammlung[51] ausgedrückt habe. Wie Sie aus unseren Kopenhagener Diskussionen wissen, glaube ich indessen eine Abrundung der Beschreibung des Sachverhalts erreicht zu haben durch die weitere Verfolgung des in den erwähnten Artikeln betonten Versagens des Beobachtungsbegriffs bei den lebenden Organismen, dem durch das Töten bei Heranziehung der Beobachtungsmittel eine prinzipielle Grenze gestellt ist. Was mir besonders am Herzen liegt, ist zu betonen, dass für die Gesetzmässigkeiten, denen die Lebenserscheinungen unterliegen, im erkenntnistheoretischen Sinne ungezwungen Platz geschaffen werden kann in unserem Weltbild. Ich bin ja damit einverstanden, dass es vor allem die prinzipielle Begrenzung der Anwendbarkeit des Kausalbegriffs auf den anorganischen Makrokosmus ist, die uns die nötige Freiheit liefert, und dass man in diesem Sinne die Akausalität als Merkmal des Lebens bezeichnen kann. Die eigentümliche Reaktion der Organismen ist aber aufs engste mit den biologischen nicht mechanisch greifbaren Gesetzmässigkeiten verbunden, und unterscheidet sich dabei prinzipiell von den bei den Untersuchungen von Schwankungserscheinungen benutzten technischen Verstärkeranordnungen. Wie die Stabilität der atomaren Phänomene untrennbar mit der in dem Unsicherheitsprinzip formulierten Begrenzung der Beobachtungsmöglichkeiten zusammenhängt, ist die Eigenart der Lebenserscheinungen meiner Auffassung nach mit der prinzipiellen Unmöglichkeit einer Feststellbarkeit der physikalischen Bedingungen, unter denen das Leben sich abspielt, verknüpft. Kurz könnte man vielleicht sagen, dass die atomare Statistik das Verhalten der Atome unter wohl definierten äusseren Bedingungen betrifft, während wir bei den Organismen nicht im atomaren Massstabe ihren Zustand definieren können. Wenn man es so formuliert, ist man freigestellt, ob man die korrespondenzmässige Quantenstatistik zum anorganischen oder organischen Gebiet rechnet. Sie nimmt in unserem Weltbild eine Zwischenstellung ein zwischen dem Gebiet der Anwendbarkeit der Idealisation des Kausalbegriffs und der teleologischen Berücksichtigung der Lebenserscheinungen.

---

[51] Bohr, *Die Atomtheorie*, ref. 47.

Wie schon gesagt, sind alle meine Bemerkungen in diesem Brief nur als Ausdruck aufzufassen für die Freude an unseren gemeinsamen Interessen und dem Wunsch, dass die Verschiedenheit der Ausdrucksweisen zur allseitigen Klärung der Problemstellung beitragen wird. Wegen Beschäftigung damit, meiner Einstellung zu den bei der letzten Konferenz[52] diskutierten offenen Atomproblemen eine befriedigende Form zu geben, bin ich leider noch nicht dazu gekommen, eine Darstellung meiner Ansichten über die biologischen Probleme fertig zu bringen. Ich hoffe aber, dass ich im Laufe dieses Monats das Manuskript eines Vortrags, den ich vorigen Sommer in Edinburgh gehalten habe, und wo ich etwas näher über meine allgemeine Einstellung zu diesen Problemen eingegangen bin, zu vollenden[53]. Sobald er fertig ist, werde ich Ihnen eine Kopie davon zugehen lassen. Inzwischen schicke ich Ihnen gleichzeitig mit diesem Briefe ein Exemplar einer vor ein Paar Jahren erschienenen dänischen Aufgabe[54] dreier Naturwissenschaften-Artikel, in deren Einleitung ich vielleicht in einigen Punkten besseren Ausdruck gefunden habe für die Triebe, die, wie ich fühle, uns gemeinsam sind, als es mir in diesem Brief möglich war.

Mit vielen herzlichen Grüssen und Wünschen für Sie und Ihre Familie, auch von meiner Frau,

Ihr
[Niels Bohr]

Translation

[Copenhagen,] June 5, [19]31

Dear Jordan,

Many thanks for your kind letter with your beautiful article about causality and freedom of the will which I found here a few days ago on my return from a short trip during Whitsun. The agreement in our general approach has been a great pleasure to me, as has the understanding you show for my endeavours in spite of all imperfections in my way of expression. Just for this reason I allow

---

[52] The conference at Easter 1930 was the second of the informal conferences held at Bohr's institute in Copenhagen beginning in 1929.

[53] Manuscript, *Philosophical Aspects of Atomic Theory* (lecture given on 26 May 1930, on the award of the James Scott Prize). Bohr MSS, microfilm no. 12. Summary published in Nature **125** (1930) 958. The summary is reproduced in Vol. 6, p. [352].

[54] N. Bohr, *Atomteori og Naturbeskrivelse. Festskrift udgivet af Københavns Universitet i Anledning af Universitetets Aarsfest November 1929*, Copenhagen 1929.

myself, however, while fully appreciating the clarity of your more forceful style, to draw your attention to some points that I would express somewhat differently.

One such point, albeit a very small one, is your reservation concerning the future development of atomic physics in relation to the classical mode of description, which does not seem quite as evident to me as you describe it. Clearly, the situation in atomic physics is essentially different from that which prevails in classical physics. Not only, of course, have we learnt that every observation involves a disturbance of the phenomena; we have furthermore realized that the whole concept of observation requires a separation between the object and the means of observation that is incompatible with the idea of closed systems. It is precisely in this context that I have placed so much emphasis in our discussions on the fact that the use of the space–time concept must always be based on the use of scales and clocks that do not belong to the system. We are dealing here with a far-reaching, yet not complete, analogy to relativity theory, for at least in special relativity theory the fruitful and instructive emphasis on the arbitrariness of classical concepts is not connected in principle with their untenability, as is the case in atomic physics. As you emphasize yourself, we might imagine that additional experience could enlighten us about the existence of an absolute coordinate system, but in my opinion an analogous possibility does not exist in atomic physics. Through the far-reaching analysis of the concept of observation we have here encountered still deeper epistemological problems, which cannot be neglected in the description of our experience, however this description is formed. For the question is not at all that laws[49] [do not exist in atomic physics, but rather that,] in view of [the character of the laws,] the situation is entirely different from that to which the idealization usually called causality can be applied.

It is this viewpoint which determined my remarks about psychological problems in the Planck issue of the Naturwissenschaften[50]; by way of psychological analysis, we are familiar precisely with situations of the kind mentioned. I know well that we are dealing also here with small nuances in the way of expressing the matter; however, in this connection I want to draw your attention to the fact that your discussion of the parallelism of physical and psychical events could perhaps give rise to a misunderstanding of my point of view. My emphasis on the formal similarity between the wave–particle problem and fundamental problems in psychology does of course not aim at a narrow analogy between psychophysical parallelism and the wave–particle duality, but above all at the possibility of gaining mutual elucidation from physical and psychological investigations. Parallelism itself involves of course a special complementarity that cannot be understood by means of laws of a one-sided physical or psychologi-

[521]

cal kind. This was, precisely, the reason for the tentative, but hence I am afraid very vague, way in which I expressed myself about this question in the Planck issue as well as in the lecture at the Scandinavian Meeting of Natural Scientists published later in the Naturwissenschaften[51]. As you know from our discussions in Copenhagen, I believe to have reached a fully developed description of the matter by pursuing further the renunciation, emphasized in the mentioned articles, of the concept of observation as regards living organisms, for which concept killing by the application of the means of observation sets a limit in principle. I am particularly concerned to emphasize that, epistemologically speaking, a natural place can be found in our world view for the laws obeyed by the phenomena of life. I agree of course with the point that it is above all the fundamental limitation of the applicability of the causality concept in the inorganic macrocosmos which gives us the necessary freedom, and that in this sense acausality can be regarded as a characteristic of life. The peculiar reaction of the organism is, however, connected in the closest possible way with the laws of biology, which cannot be comprehended mechanically, and is therefore different in principle from the technical amplification devices used in the study of fluctuation phenomena. Just as the stability of the atomic phenomena is inseparably connected with the limitation of observation possibilities expressed by the uncertainty principle, so in my view the pecularities of life phenomena are connected with the impossibility in principle of ascertaining the physical conditions under which life exists. Briefly, one could perhaps say that atomic statistics deals with the behaviour of atoms under well-defined external conditions, whereas we cannot define the state of the organism on an atomic scale. If one formulates it thus, it does not matter whether one regards correspondence-like quantum statistics as belonging to the inorganic or the organic domain. In our world view, it takes an intermediate position between the domain of the applicability of the idealization of the causality concept and the teleological account of life phenomena.

As I have said already, all my remarks in this letter are only to be understood as an expression of my pleasure in our mutual interests and my wish that the difference in mode of expression will contribute to the general clarification of the problem. Because of my effort to give my view on the unresolved atomic questions discussed at the last conference[52] a satisfactory form, I have unfortunately not yet reached the point of finishing an account of my views about biological problems. I hope, however, that in the course of this month I will complete the manuscript of a lecture, which I gave last summer in Edinburgh, where I went in somewhat greater detail into my general view of these problems[53]. As soon as it is finished I shall send you a copy. Meanwhile,

I send you at the same time as this letter a copy of a Danish edition[54], which appeared a couple of years ago, of three articles in the Naturwissenschaften, in the introduction to which I have perhaps in some instances found a better formulation, than was possible for me in this letter, for the enthusiasm I feel we have in common.

With many kind regards and best wishes to you and your family, also from my wife.

Yours
[Niels Bohr]

JORDAN TO BOHR, 22 June 1931
[Typewritten]

Physikalisches Institut
    der Universität   ROSTOCK, den 22.6.[19]31.
Abteilung für theoretische Physik

Lieber Herr Bohr!

Vielen herzlichen Dank für Ihren lieben ausführlichen Brief, in welchem Sie so freundlich auf meinen Aufsatz eingegangen sind!; und auch für die nachträgliche Berichtigung[55]. Es war mir eine große Freude, Ihre Teilnahme und auch Ihre wesentliche Zustimmung zu meinen Ausführungen zu erfahren, die ja ihrerseits garnichts anderes sind, als der Versuch, einige von Ihnen selbst in den letzten Jahren erhaltenen Anregungen etwas ausführlicher auszuarbeiten. Was nun im Einzelnen Ihre kritischen Anmerkungen betrifft, so bin ich so sehr unter dem Eindruck der Richtigkeit dieser Bemerkungen, daß ich meine Zustimmung ganz summarisch dazu ausdrücken kann. Ich hoffe, in der Korrektur meines Aufsatzes an den fraglichen Stellen etwas ändern zu können. Bezüglich der biologischen Problematik war mir bereits zum Bewußtsein gekommen, daß die Charakterisierung der Organismen als Systeme, bei welchen sich die Akausalität der atomaren reaktionen bis zu makroskopischer Akausalität steigert, doch insofern unzureichend ist, als es ja eben die von Ihnen hervorgehobenen Verstärkerröhren gibt; ich hatte mich zunächst zu beruhigen versucht mit der Erwägung, daß es wohl kein Zufall ist, daß Verstärkerröhren nur von Menschen hergestellt, und nicht etwa als Naturerzeugnisse irgendwo gefunden werden. Vielleicht ist diese Bemerkung – die aber natürlich nicht im Gegensatz stehen soll zu den tiefer gehenden Erläuterungen des Wesens

---

[55] Letter from Bohr to Jordan, 14 June 1931. See ref. 49.

des Biologischen in Ihrem Briefe – nicht ganz belanglos. Ich denke, daß diejenigen Gebiete in einem Organismus, in welchen tatsächlich sehr starke Abweichungen von den "anorganischen" Gesetzmäßigkeiten vorliegen, recht klein, und vor allem *energetisch* sehr unbedeutsam sind. Also würde ich es für nicht unwahrscheinlich halten, daß *ein Teil* der organischen Reaktionen in einer der Wirkungsweise der Verstärkerröhre ähnlichen Weise funktioniert. Ich möchte also – ganz roh gesprochen – drei verschiedene Zonen des Organismus unterscheiden (zwischen denen natürlich in Wirklichkeit keine scharfe Grenze bestehen wird): Die Zone der eigentlichen "Zentren" des Lebendigen, auf welche Ihre Charakterisierung zutrifft; dann die Zone der "Verstärkerorgane"; endlich die Zone der "Werkzeugorgane", deren Reaktionsweise sich nicht wesentlich von den in der Natur vorkommenden anorganischen Reaktionen unterscheidet. Treffender, als meine Charakterisierung des Lebendigen durch die Steigerung der Akausalität ins Makroskopische, würde es wohl sein, den Charakter dieser "Verstärkung" als eine teleologische *Integration* zu betonen; man muß ja wohl den Unterschied der Quantenreaktionen von der organischen eben darin sehen, daß im "Zwischengebiet" dieser Quantenreaktionen die Kausalität aufgehört hat, während die Teleologie noch nicht angefangen hat; die regellos-statistischen Reaktionen der einzelnen Atome im Gebiet der Quantenphysik erscheinen im Organismus irgendwie "zusammengefaßt" zu einer harmonischen Gesamtwirkung; dabei ist mir Ihre Formulierung überaus einleuchtend, daß es im lebendigen Organismus (oder, im Sinne des obigen Versuchs einer detaillierten Darstellung: in den "Zentren" des Lebendigen) nicht mehr möglich sein dürfte, eine Zustandsbeschreibung im atomaren Maßstab durchzuführen – welche eben mit einer "Tötung" gleichbedeutend wäre.

Ich fühle sehr die Unvollkommenheit meiner Ausdrucksweisen, welche weit davon entfernt sind, wenigstens meine eigene Meinung unmißverständlich wiederzugeben. Aber man wird ja erst allmählich für diese so neuartigen Vorstellungen eine angemessene Sprache finden. Daran zweifle ich nicht, daß die Biologie ganz entscheidende Anregungen von der Physik zu erhalten hat, und daß es, um diese Entwicklung in Gang zu bringen, zunächst das Beste sein wird, wenn wir Physiker versuchen, auf "spekulativem" Wege die Anwendungsmöglichkeiten der neuen physikalischen Ergebnisse und Gesichtspunkte recht ausführlich zu diskutieren – obwohl natürlich in einem späteren Stadium die Arbeit vor allem auf den Biologen ruhen und in enger Zusammenwirkung von Theorie und Experiment geschehen muß.

Ich muß aber endlich erzählen, daß sich für die Publikation des Aufsatzes Hindernisse zu ergeben scheinen. Natürlich kommen nur die "Naturwissenschaften" als Zeitschrift in Frage; und nun höre ich soeben aus Berliners

Büro (er selber ist z.Z. verreist), daß Berliner[56] findet, die Naturwissenschaften hätten schon reichlich viel über Kausalität gebracht; er hat deshalb bereits einen Aufsatz von Reichenbach[57] für lange Zeit zurückgestellt. Ich werde ihm nun jedenfalls noch einmal schreiben, beurteile die Sachlage aber nicht allzu optimistisch.

Viele herzliche Grüße – bitte auch an Klein, wenn er vielleicht jetzt noch in Kopenhagen ist!

Stets Ihr

P. Jordan

Translation

Rostock, June 22, [19]31

Dear Mr. Bohr,

Many cordial thanks for your dear and detailed letter in which you so kindly respond to my article!; and also for the subsequent correction[55]. It was a great joy to learn about your sympathy and your general approval of my remarks, which of course for their part are nothing but an attempt to develop in somewhat greater detail some of your own suggestions over the last years. As regards in particular your critical remarks, I am so impressed with the correctness of these comments that I can express my agreement quite summarily. I hope to be able to make some changes in the relevant places in the proofs of my paper. Concerning the biological problems, I had already realized that the characterization of organisms as systems in which the acausality of atomic reactions expands to macroscopic acausality is nevertheless insufficient insofar as precisely the amplifier tubes that you point to do indeed exist; in order to reassure myself, I first tried to consider that it can hardly be accidental that amplifier tubes can only be man-made and cannot be found anywhere as products of nature. Perhaps this remark – which, however, of course must not stand in contradiction to the deeper-going explanations in your letter of the essence of what is biological – is not quite without interest. I imagine that those domains of the organism in which we actually find very strong deviations from the "inorganic" laws are quite small and, in any case, *energetically* very insignificant. Therefore I do not find it improbable that *a part* of the

---

[56] Arnold Berliner (1862–1942), German physicist who became editor of Die Naturwissenschaften upon his retirement.

[57] Hans Reichenbach (1892–1953), German mathematician and philosopher who emigrated to the United States in 1938. See above, *Introduction* to Part I, p. [16].

organic reactions exhibit a mode of action similar to that of amplifier tubes. Roughly speaking, I should like to distinguish between three different zones of the organism (between which of course no sharp boundary will exist in reality): The zone of the proper "centres" of living beings, for which your characterization holds; then the zone of the "amplifier organs"; and finally the zone of the "tool organs", whose mode of reaction does not differ essentially from the inorganic reactions occurring in nature. It might well be more to the point than my characterization of living beings by the expansion of acausality into the macroscopic, to emphasize the character of this "amplification" as a teleological *integration*; possibly, one must understand the difference between quantum and organic reactions by the fact that causality has ceased in the "intermediate sphere" of these quantum reactions, whereas teleology has not yet begun; the irregular-statistical reactions of the single atoms in the domain of quantum physics appear in the organism as somehow "integrated" to give a harmonious total effect; here, your formulation is entirely convincing to me, namely that in the living organism (or, in the sense of the above attempt at a detailed account: in the "centres" of living beings) it should no longer be possible to carry out a description of the situation on an atomic scale – which would mean the same as "killing".

I strongly feel the imperfection in my way of expression, which is far from rendering even my own opinion unambiguously. But we will of course only gradually find an appropriate language for these so novel ideas. I have no doubt that biology will receive quite decisive stimulation from physics and that in order to set this development in motion it would be best if we physicists next try in considerable detail to discuss, in "speculative" ways, the possibilities for applying the new results and viewpoints from physics – although, obviously, at a later stage the work must above all be in the hands of the biologists and take place in close interplay between theory and experiment.

Finally, however, I must report that obstacles seem to have arisen with regard to the publication of my article. Naturally, the only possible journal is the "Naturwissenschaften"; and now I just heard from Berliner's office (he is presently away on a trip himself) that Berliner[56] thinks that the Naturwissenschaften has already published plenty on causality; he has therefore already put aside an article by Reichenbach[57] for a long time. In any case, I shall write to him once more, but am not too optimistic about the situation.

Many cordial greetings – please also to Klein, if he is still in Copenhagen!

Always yours
P. Jordan

BOHR TO JORDAN, 23 June 1931
[Carbon copy]

[København,] 23. Juni [19]31.

Lieber Jordan,

Mit allem, was Sie in Ihrem freundlichen Brief über Ihre Ansichten erzählen, stimme ich ganz bei. Dass es bei biologischen Funktionen, soweit es den Anreiz betrifft, sich weitgehend um Erscheinungen handelt, die dem Funktionieren von gewöhnlichen Verstärkeranordnungen ähnlich sind, und dass zugleich der Unterschied zwischen Organismen und solchen Verstärkeranordnungen ist, dass bei dem ersten es sich um eine Selbststabilisierung handelt, während für das andauernde Funktionieren der letzteren menschlichen Eingriffe immer nötig sein werden, sind Gedanken, mit denen ich seit lange vertraut bin und [die ich] mit dänischen Biologen in den letzten Jahren oft diskutiert habe. Es ist eben die Bestrebung derartiger Sachverhältnisse zu berücksichtigen, die den Hintergrund bilden für meine allgemeine Ausdrucksform, die vor allem auf die erkenntnistheoretische Analyse unserer allgemeinen Situation hinzielt. Wie Sie wissen, war es aber seit langem meine Absicht, ausführlicher um die Einzelheiten meiner Ansichten über die biologischen Probleme zusammenzustellen, ohne [dass] ich doch dazu käme wegen der ständigen Diskussionen der offenen Atomfragen. Im besonderen hätte ich gehofft, mit einer solchen Darstellung ein kleines Büchlein abzuschliessen, das eine deutsche Herausgabe[58] der dänischen Universitätsschrift[59], die ich Ihnen vor ein Paar Wochen schickte, bringen sollte. Eine solche Herausgabe war bei Springer schon ein Jahr im Satz, ohne dass ich trotz dringender Aufforderung mich entschliessen konnte, es auszuschicken, weil ich fürchtete, dass es zu wenig Neues enthielt. Dem freundlichen Rat zufolge, besonders von Ehrenfest, den wir eben hier zu Besuch gehabt haben, habe ich mich doch jetzt entschlossen, das Büchlein erscheinen zu lassen, wobei ich nur in einem Addendum[60] am Ende der Übersetzung der dänischen Einleitung ein Paar Bemerkungen über die biologischen Fragen zugefügt habe, die sich den Schlussbemerkungen in der Einleitung und dem im Büchlein auch

---

[58] N. Bohr, *Atomtheorie und Naturbeschreibung*, Julius Springer Verlag, Berlin 1931. Subsequently published in English as Bohr, *Atomic Theory and the Description of Nature*, ref. 47.

[59] Bohr, *Atomteori og Naturbeskrivelse*, ref. 54.

[60] N. Bohr, *Addendum* in Bohr, *Atomtheorie und Naturbeschreibung*, ref. 58, pp. 14–15. The English version of the "Addendum" was printed in Bohr, *Atomic Theory and the Description of Nature*, ref. 47, pp. 21–24, and is reproduced in Vol. 6, pp. [299]–[302].

abgedruckten Vortrag vor [der] skandinavischer Naturforscherversammlung[61] anschliesst. In wenigen Tagen hoffe ich Ihnen ein Exemplar des Ganzen schicken zu können.

Mit vielen herzlichen Grüssen von uns allen,

Ihr

[Niels Bohr]

Translation

[Copenhagen,] June 23, [19]31

Dear Jordan,

I quite agree with everything you say about your viewpoints in your kind letter. That with regard to the stimulation of biological functions we have to do largely with phenomena similar to the functioning of ordinary amplifier arrangements, and that at the same time the difference between organisms and such amplifier arrangements is that in the former we have to do with self-stabilization, whereas for the continuous functioning of the latter, human intervention will always be required, are thoughts with which I have been familiar for a long time, and which I have often discussed with Danish biologists over the last years. It is just the endeavour to take such facts into account which forms the background for my general mode of expression, which aims above all at the epistemological analysis of our general situation. As you know, it has, however, been my intention for a long time to bring together more thoroughly the details of my views about biological problems, but I have not come around to it because of the continuing discussion of the unresolved atomic questions. In particular I had hoped to end a small booklet, constituting a German edition[58] of the Danish university volume[59] that I sent you a few weeks ago, with a presentation of this kind. Such an edition has been in press at Springer already for a year, but in spite of urgent requests I could not make up my mind to let it appear because I was afraid that it contained too little that is new. Following kind advice, especially from Ehrenfest who has just visited us here, I have nevertheless now decided to let the booklet appear; only in an addendum[60] at the end of the translation of the Danish introduction have I added a couple of remarks about the biological questions, which relate to the concluding remarks in the introduction and the lecture, also printed in the

---

[61] Bohr, *Die Atomtheorie*, ref. 47.

[528]

booklet, at the Scandinavian Meeting of Natural Scientists[61]. In a few days I hope to be able to send you a copy of the entire work.

Many cordial greetings from us all,

Yours
[Niels Bohr]

JORDAN TO BOHR, 26 November 1932
[Typewritten]

Lieber Herr Bohr!

In den "Naturwissenschaften" ist jetzt kürzlich mein quantenmechanisch-philosophischer Aufsatz erschienen[62], den Sie ja im Manuskript[63] schon vor längerer Zeit gesehen hatten. Gegenüber dem damaligen Manuskript hat er einige Aenderungen erfahren, für welche ich mir Ihre damalige freundliche Kritik zunutze gemacht habe: einige Punkte, zu denen Sie Bedenken äußerten, sind gestrichen worden, und andere sind geändert. Ferner ist ein Kapitel über die biologische Bedeutung der Qu.mech. neu hinzugekommen. Endlich habe ich es für richtig gehalten, noch etwas deutlicher, als in der ursprünglichen Fassung, zu betonen, daß alle meine Bemerkungen nur ein Versuch sind, Dinge zu wiederholen und ausführlicher auszusprechen, die Sie bereits in Ihren "vier Aufsätzen"[64] gesagt hatten. Uebringens habe ich über diese Ihre Aufsätze gerade kürzlich eine Buchbesprechung in der Phys. Zeitschr. erscheinen lassen[65].

Ich schicke Ihnen ein Exemplar meines Aufsatzes und hoffe, daß er Ihnen als ein einigermaßen *zutreffender* Kommentar Ihrer Gedanken erscheinen wird.

Mit vielen herzlichen Grüssen

stets Ihr aufrichtig ergebener

26.11.[19]32

P. Jordan

---

[62] P. Jordan, *Die Quantenmechanik und die Grundprobleme der Biologie und Psychologie*, Naturwiss. **20** (1932) 814–821.
[63] See ref. 48.
[64] Bohr, *Atomtheorie und Naturbeschreibung*, ref. 58.
[65] Phys. Z. **33** (1932) 671–672.

Translation

Dear Mr Bohr!

My quantum-mechanical–philosophical article[62], the manuscript[63] for which you have seen a long time ago, has just now appeared in the "Naturwissenschaften". As compared to the manuscript from that time, it has undergone some changes, for which I have taken advantage of your kind criticism from that time: Some points which you considered doubtful have been stricken and others have been changed. Furthermore, a new chapter about the importance of quantum mechanics for biology has been added. Finally, I have considered it correct to emphasize even somewhat more clearly than in the original version that all my remarks are only an attempt to repeat and to express more clearly matters that you had already stated in your "four essays"[64]. Incidentally, I have just recently published a book review of your essays in Phys. Zeitschr.[65]

I am sending you a copy of my article and hope that it will seem to you a reasonably *accurate* comment on your thoughts.

With many cordial greetings

always your truly devoted

November 26, [19]32   P. Jordan

BOHR TO JORDAN, 27 December 1932
[Carbon copy]

[København], 27. Dezember [19]32.

Lieber Jordan,

Ich schäme mich sehr, dass ich auf Ihren freundlichen Brief noch nicht geantwortet habe, aber ich war im letzten Monat mit verschiedener Arbeit ganz in Anspruch genommen. Es freute mich sehr, dass Ihr schöner Artikel über die quantenmechanischen und biologischen Fragen schliesslich erschien, und es war mir auch eine grosse Freude, mit Ihrem Referat meines Büchleins Bekanntschaft zu machen und die sympatische Einstellung zu fühlen, mit der Sie aller Unvollkommenheiten der Darstellung zum Trotze meinen allgemeinen Bestrebungen entgegenkommen. Wie Sie schreiben, handelt es sich in der Tat um einen gewissen Stil, der dem Leser sicherlich nicht geringe Schwierigkeiten darbietet, der aber mein einziges Mittel ist, ihm einen Eindruck zu geben, von was mir auf Herzen liegt. Ausser allgemeinen Artikeln, von denen ich Sonderdrucke beilege, habe ich diesen Sommer einen englischen Vortrag über die biologischen Probleme an einem internationalen medizinischen Lichtkongress

[530]

in Kopenhagen gehalten, der in wenigen Wochen in dem Kongressbericht unter dem Titel "Light and Life" herauskommt, und von denen Weisskopf[66], der jetzt hier arbeitet, in Begriff ist eine deutsche Übersetzung für "Die Naturwissenschaften" herzustellen[67]. Ich hoffe, Ihnen in wenigen Wochen einen Sonderdruck davon schicken zu können und werde interessiert sein zu erfahren, was Sie von der Darstellung denken. Es mag Sie auch interessieren zu hören, dass Rosenfeld und ich eine grössere Arbeit über die Messbarkeit von elektromagnetischen Feldgrössen, die bald in der Zeitschrift für Physik erscheinen wird, abgeschlossen haben[68]. Es ist uns gelungen, eine völlige Harmonie zwischen dem quantenelektrodynamischen Formalismus und der Messbarkeit nachzuweisen, aber erst nachdem wir eine Anzahl von Schwierigkeiten und scheinbaren Paradoxien überwinden mussten. Nicht nur erwies sich die Kritik von Landau und Peierls[69] unbegründet, sondern es war vor allem wesentlich, anstatt von den Werten von den Feldgrössen in Raum–Zeitpunkten immer Mittelwerten dieser Grössen über endliche Raum–Zeitbereiche heranzuziehen. Wir gehen dabei zweckmässig von den von Ihnen und Pauli ursprünglich aufgestellten Vertauschungsrelationen aus, und es zeigt sich unter anderen, dass der von Heisenberg in seinem Buch über die physikalischen Prinzipien der Quantentheorie[70] besprochene Nichtvertauschbarkeit der instantanen Werte ortogonaler elektrischer und magnetischer Feldkomponenten für die Prüfung des Formalismus nicht zweckmässig sind. In der Tat sind je zwei der Messung zugänglichen Feldmittelwerte über einen und denselben Raum–Zeitbereich immer vertauschbar. Es wäre schön, ob wir einmal über diese Sachen sowie über die Paradoxien der relativistischen Quantenmechanik, über die ich viel gedacht habe, und die wir auch in der Abhandlung berühren, einmal näher sprechen könnten. Hoffentlich werden Sie im Stande sein, diesmal an unserem

---

[66] Victor Weisskopf (b. 1908), Austrian-born physicist and one of Bohr's closest collaborators. Bohr helped Weisskopf emigrate to the United States in 1938.

[67] See ref. 22.

[68] The paper referred to, which was not published in Zeitschrift für Physik, is N. Bohr and L. Rosenfeld, *Zur Frage der Messbarkeit der elektromagnetischen Feldgrössen*, Kgl. Dan. Vidensk. Selsk., Mat.–fys. Medd. **12**, no. 8 (1933). Reproduced in Vol. 7, pp. [57]–[121] (German original) and [123]–[166] (English translation).

[69] L. Landau and R. Peierls, *Erweiterung des Unbestimmtheitsprinzips für die relativistische Quantentheorie*, Z. Phys. **69** (1931) 56–69. The relevant part of the English translation of this article is reproduced in Vol. 7, pp. [231]–[238]. The general background is described by Jørgen Kalckar in his *Introduction* to Part I of Vol. 7, pp. [3]–[30].

[70] W. Heisenberg, *Die physikalischen Prinzipien der Quantentheorie*, Hirzel Verlag, Leipzig 1930. Printed in English as W. Heisenberg, *The Physical Principles of Quantum Theory*, University of Chicago Press, Chicago 1930 (reprinted by Dover, New York 1949 and 1967).

üblichen kleinen Ostern-Konferenz[71] teilzunehmen. Sobald ich näheres über den Zeitpunkt weiss, werde ich Ihnen darüber schreiben.

Bei dieser Gelegenheit schicke ich Ihnen nur noch die herzlichsten Wünsche für das neue Jahr für Sie und Ihre Familie von uns allen,

Ihr
[Niels Bohr]

Translation

[Copenhagen,] December 27, [19]32

Dear Jordan,

I feel quite ashamed for not yet having answered your kind letter, but in the last month I have been entirely occupied with various tasks. I am very glad that your beautiful article about quantum mechanical and biological questions finally appeared, and it was also a great pleasure to become acquainted with your review of my booklet and to feel the sympathetic attitude with which you meet my general endeavours in spite of all the imperfections in the presentation. As you write, we have to do with a certain style which certainly presents the reader with no small difficulties, but this is my only means of giving him an impression of what is on my mind. Besides some general articles of which I enclose reprints, I gave a lecture in English this summer at an international medical congress in Copenhagen about the biological problems, which will appear in a few weeks in the congress proceedings under the title "Light and Life", and of which Weisskopf[66], who is presently working here, is about to prepare a German translation for the "Naturwissenschaften"[67]. I hope to be able to send you a reprint of it in a few weeks and will be interested to learn what you think of the presentation. You may also be interested to hear that Rosenfeld and I have completed a large paper on the measurability of electromagnetic field quantities, which will soon appear in the Zeitschrift für Physik[68]. We have succeeded in proving complete harmony between the quantum-electrodynamical formalism and measurability, but only after we had been forced to overcome a number of difficulties and seeming paradoxes. Not only did the criticism by Landau and Peierls[69] turn out to be unjustified, but it was above all essential always to use, instead of the values of the field quantities in space–time points, mean values of these quantities over finite space–time domains. To this end we start usefully from the commutation relations originally suggested by you

---

[71] See ref. 52.

and Pauli, and it turns out among other things that the non-commutability of the instantaneous values of orthogonal electrical and magnetic field components discussed by Heisenberg in his book about the physical principles of quantum theory[70] is not useful for testing the formalism. In fact, two of the mean field values available for measurement over the same space–time domain always commute. It would be nice if we could speak sometime in greater detail about these matters, as well as about the paradoxes of relativistic quantum mechanics, to which I have given much thought and which we also touched upon in our paper. Hopefully, you will be able to participate this time in our traditional little Easter conference[71]. As soon as I know more about the date, I will write to you about it.

I use this opportunity to send you and your family my best wishes for the New Year from us all,

Yours
[Niels Bohr]

OSKAR KLEIN

BOHR TO KLEIN, 19 January 1933
[Carbon copy]

[København,] 19. Januar [19]33.

Kære Klein,

Tak for Dit rare Brev. Det skulde more mig meget engang at se det Stykke[72], som Du fortalte, Du havde skrevet om min lille Artikel[73]. Paa Opfordring af Redaktionen for "Naturens Verden" har jeg i forrige Uge foretaget en dansk Oversættelse af Artiklen, hvoraf jeg indlagt sender den første Korrektur, som jeg lige har faaet. Paa Grund af den store Forskel mellem dansk og engelsk var Oversættelsen nemlig ikke helt let, og paa nogle Steder har jeg derfor ændret nogle Sætninger paa temmelig fri Maade, og jeg vilde meget gerne høre, om Du synes, at det er Forbedringer eller ikke, og om der er andre Punkter, hvor Du synes, at Meningen kunde være bedre udtrykt. Et Spørgsmaal er, om Du synes, at det er fornuftigt at spille saa meget paa Ordene rationel og irrationel, som jeg

---

[72] Probably O. Klein, *Biologi og atomfysik* (Biology and atomic physics), Svenska Dagbladet, 11 January 1933.
[73] See ref. 22.

har gjort i Oversættelsen. Anledningen for overhovedet at indføre disse Ord var Henvisningen til Benyttelsen af teleologiske Argumenter i Fysiologien, som jeg ikke syntes var helt tydelig eller i det mindste svær at oversætte. Rud Nielsen[74], som jeg netop har vist Korrekturen til, synes, at Forbedringerne paa forskellige Steder ikke var helt uvæsentlige, og hvis Du synes det samme, vil jeg ved hans Hjælp tage Hensyn til dem i den engelske Udgave i Nature i den Forstand, at den danske Tekst betragtes som Originalen. Ogsaa for Oversættelsen til tysk vil dette vistnok betyde en Lettelse. Jeg skammer mig meget over at blive ved at plage Dig med Artiklen, men det hænger sammen med de mange taknemmelige Tanker, jeg stadig sender Dig for Din storartede Hjælp med den.

Det var rart, at Du besluttede Dig til straks at sende Din Artikel til Zs. f. Phys.[75], og vi glæder os alle til snart at kunne studere den. Jeg selv er ikke kommet meget videre med de andre Sager, siden Du rejste, især fordi Rosenfeld paa Grund af Sygdom i sin Familie ikke, som jeg havde haabet, kunde komme herop i Januar, for at vi sammen kunde gøre Afhandlingen færdig. I Dag er imidlertid Bloch kommet, og jeg glæder mig til at prøve paa i de første Dage at komme lidt dybere ind i hele Supraledningsproblemet. Jeg skal snart skrive igen og fortælle om, hvordan det gaar dermed.

Med mange venlige Hilsener til Jer alle,

Din
[Niels Bohr]

Translation

[Copenhagen,] January 19, [19]33

Dear Klein,

Thank you for your nice letter. It would amuse me very much to see the piece[72] you told me you had written about my little article[73]. At the request of the editors of "Naturens Verden" I made a Danish translation of the article last week; I enclose the first proofs, which I have just received. Because of the great difference between Danish and English, the translation was not so easy, and in some places I have therefore altered some sentences quite freely, and I would very much like to know whether you think these are improvements or not, and whether there are other places where you think the meaning could be better

---

[74] Jens Rud Nielsen (1894–1979), Danish–American physicist. See Vol. 5, p. [318].
[75] O. Klein, *Zur Frage der quasimechanischen Lösung der quantenmechanischen Wellengleichung*, Z. Phys. **80** (1933) 792–803.

expressed. One question is whether you think it is sensible to make such a play on the words rational and irrational as I have done in the translation. The reason for introducing these words at all was the reference to the use of teleological arguments in physiology, which I did not think was quite clear, or at least was difficult to translate. Rud Nielsen[74], to whom I have just shown the proofs, thinks that the improvements in some places were not quite unimportant, and if you think the same, I shall – with his assistance – take them into consideration in the English version in Nature, in the sense that the Danish text be regarded as the original. This will probably also make the translation into German easier. I am very ashamed about continuing to bother you with the article, but this is due to the many grateful thoughts I still send you for your splendid help with it.

It was good that you decided to send your article immediately to Zs. f. Phys.[75], and we are all looking forward to be able to study it soon. As for myself, I have not made much progress with the other things since you left, especially because Rosenfeld, due to illness in his family, has not been able to come here in January as I had hoped, so that we could finish the article together. However, Bloch has arrived today, and I am looking forward to trying in the first few days to delve a little deeper into the whole problem of super-conductivity. I shall write again soon and tell you how it turns out.

With many warm greetings to you all,

Yours
[Niels Bohr]

BOHR TO KLEIN, 6 March 1940
[Carbon copy]

[København,] 6. Marts [19]40.

Kære Oskar,

Mange Tak for Dit rare Brev. Hvordan har Du dog kunnet faa fat paa de mange Avisudklip[76], det havde jeg aldrig tænkt, men jeg var vældig glad for dem. Jeg har allerede sendt flere af dem til interesserede Venner; saaledes har

---

[76] O. Klein, *Bibeln och våra dagars naturforskning* (The bible and contemporary scientific research), Göteborgs Handels- och Sjöfartstidning, 18 December 1939.

jeg i Dag sendt et til en teologisk Professor Sigmund Mowinckel i Oslo[77], som jeg traf og talte meget med paa Høstbjerg-Mødet i Januar[78].

Med største Glæde har jeg ogsaa gennemlæst Din omarbejdede Artikel til "Ord och Bild"[79] og har som altid beundret Din Evne til samtidig tankevækkende og stemningsfuld Fremstilling. Jeg vil dog gerne gøre Dig opmærksom paa, at der er enkelte Sætninger, som jeg er bange for vil kunne misforstaas. Det drejer sig om Diskussionen af Forholdet imellem Sjæl og Legeme. Mine Bemærkninger om Komplementaritetsforholdet mellem de Situationer, hvor Ordene Instinkt og Fornuft kan bruges paa formaalstjenlig Maade, tilsigtede jo, ganske som Du saa klart fremstiller det, at mane til Forsigtighed ved Sammenligninger mellem Dyre- og Menneskelivet. Alligevel blev jeg bange for, at Din Fremstilling et Øjeblik kunde bringe Læseren til at tænke, at det ved Instinktlivets Udfoldelse drejede sig om et rent biologisk Problem, medens det ved Fornuftens Brug drejede sig om et direkte spirituelt Indgreb overfor de materielle biologiske Processer. Jeg tænker jo naturligvis slet ikke paa, at der her skulde være nogen Forskel i Opfattelsen mellem os, men blot paa den Misforstaaelse som Udtrykket "en uhemmet biologisk Funktion forudsætter Fraværelsen af Bevidsthed" (Side 10 Linie 7 fra neden) maaske kunde give Anledning til. Naar jeg i den gamle Artikel Lys og Liv[80] og ved senere Lejligheder har talt om en Uddybning af den psycho-fysiske Parallelteori paa Baggrund af Komplementaritetssynspunktet, mente jeg naturligvis ikke, at der er Tale om nogensomhelst Begrænsing af Parallelismen selv, men snarere om en Erkendelse af hidtil upaaagtede Paralleller mellem Forudsætningerne for de biologiske og psychologiske Foreteelsers rationelle Analyse. Jeg tænker slet ikke paa at foreslaa nogen større Forandring i Din Fremstilling, men blot paa den eventuelle lidt forsigtigere Formulering af visse Sætninger paa lignende Maade som de Smaaændringer, vi allerede talte om under Dit sidste Ophold hernede.

Som jeg skrev sidst, tænker vi stadig med største Glæde paa Dit Besøg, der var en stor Opmuntring for mig. Sørgeligvis er dog endnu ikke vort fælles Haab

---

[77] Sigmund Mowinckel (1884–1965), professor of the theology of the Old Testament at the University of Oslo, 1922–1954. A carbon copy of Bohr's letter to Mowinckel is retained at the NBA.

[78] The meeting took place in Høsbjør, north of the Norwegian town Hamar, on 6 January 1940. It was one of a series of regional meetings for Scandinavia, held under the auspices of the League of Nations and devoted to the problems facing international scientific cooperation in general, and Finnish scientists in particular, because of the war situation.

[79] *Bibeln och våra dagars naturforskning*, Ord och Bild **50** (1941) 471–478.

[80] See ref. 22.

om en snarlig Bedring af Forholdene i Verden gaaet i Opfyldelse, og jeg frygter for, at vi alle skal igennem mørke Tider, før dette sker. Der er dog intet andet at gøre end at holde Modet oppe og hjælpe hverandre med at beskæftige Tankerne med Spørgsmaal, der har saa lidt som muligt med de øjeblikkelige Forhold at gøre.

Med mange venlige Hilsener til Jer alle fra hele Familien,

Din

[Niels Bohr]

Translation

[Copenhagen,] March 6, [19]40

Dear Oskar,

Many thanks for your nice letter. How you could get hold of the many newspaper cuttings[76] I cannot imagine, but I was very pleased to receive them. I have already forwarded several of them to interested friends; thus I have sent one today to Professor Sigmund Mowinckel, a professor of theology in Oslo[77], whom I met and spoke with at length at the Høstbjerg meeting in January[78].

I have also read with greatest pleasure your revised article for "Ord och Bild"[79], and as always I admired your ability to make a presentation that is at the same time thought-provoking and evocative. I would like to point out to you, however, that there are a few sentences which I fear might be misunderstood. They concern the discussion about the relation between soul and body. My remarks about the complementarity relation between those situations where the words instinct and reason can be used in a meaningful way, were intended, precisely as you so clearly describe it, to urge care in comparisons between animals and humans. I nevertheless became afraid that your presentation could for a moment make the reader think that the manifestation of instinct involved a purely biological problem, whilst the use of reason involved direct spiritual intervention as regards the material biological processes. Naturally, I do not at all think that there is any difference of opinion between us, but I think only of the misunderstanding that the expression, "an uninhibited biological function presupposes the absence of consciousness" (page 10, line 7 from the bottom) might perhaps give rise to. When in the old article, Light and Life[80], and on later occasions, I have spoken of an elaboration of the theory of psycho–physical parallelism on the basis of the complementarity viewpoint, I naturally did not mean that there is a question of any limitation of the parallelism itself, but rather of a recognition of previously unnoticed parallels between the pre-conditions for the rational analysis of biological and psychological phenomena. I do not at all mean to suggest any great change in your presentation, but only a perhaps slightly more careful formulation of certain sentences, similar to the small changes we already discussed during your last stay here.

As I wrote last time, we still remember with the greatest pleasure your visit, which was a great encouragement to me. Regrettably, however, our common hope for a speedy improvement in world affairs has not yet been fulfilled, and I fear that we all must go through dark times before this happens. Nevertheless, there is nothing we can do but keep up our courage and help each other with occupying our minds with questions having as little as possible to do with the immediate situation.

With many warm greetings to you all from the whole family,

Yours,

[Niels Bohr]

OTTO MEYERHOF

BOHR TO MEYERHOF, 5 September 1936
[Carbon copy]

[København,] 5. September [19]36.

Lieber Professor Meyerhof,

Ich bedaure sehr, dass ich wegen unerwarteter und unaufschiebbarer Pflichten noch keine Zeit finden konnte, eine Zusammenfassung meiner Bemerkungen auf dem Philosophenkongress[81] auszuarbeiten. Ich muss mich daher beschränken, in diesem Brief die Hauptpunkte zu erwähnen, für die Sie sich interessieren mögen.

Erstens zielen meine Bemerkungen im alten Aufsatz "Licht und Leben"[82], von dem ich ein Exemplar beilege, nur darauf hin, die alte Streitfrage des Vitalismus gegen Mechanismus von dem Standpunkt aus zu beleuchten, den

---

[81] The second international Unity of Science congress held in Copenhagen in June 1936, where Bohr gave a lecture, subsequently printed as N. Bohr, *Causality and Complementarity*, Phil. Sci. **4** (1937) 289–298. Bohr's lecture is reproduced on pp. [39]–[48]. The general background is described above, *Introduction* to Part I, pp. [20] ff.

[82] See ref. 22.

die neuere Entwicklung der Physik uns in Bezug auf das Verlangen einer "Erklärung" aufzwingt. Es handelt sich dabei ja keineswegs um eine wohldefinierte Begrenzung der Beobachtungsmöglichkeiten biologischer Erscheinungen und überhaupt um irgend eine Begrenzung der Beschreibung solcher Erscheinungen mit Hilfe physikalischer Begriffe, sondern nur um die Betonung, dass wir in der Beobachtungsfrage, wie in der ganzen biologischen Beschreibungsweise, mit einer Situation zu tun haben, die sich von derjenigen wesentlich unterscheidet, wo eigentliche physikalische Fragestellungen direkt auf eine sogenannte Erklärung hinzielen. Wie ich nur kurz im Aufsatz angedeutet habe, mag die allgemeine Belehrung der Atomphysik hier durch folgende Reihenfolge dargestellt werden:

a) Fragestellung der klassischen Physik, wo es sich immer um ein wohldefiniertes abgeschlossenes System handelt, dessen Zustände ohne Berücksichtigung der Beobachtungsmöglichkeiten beschrieben werden.

b) Quantenphysik, wo wir noch mit einem wohldefinierten System zu tun haben, wo aber die unvermeidliche Wechselwirkung zwischen Objekt und Messinstrument schon in der Beschreibung des Zustands einen wesentlichen Zug der Komplementarität bedingt.

c) Eigentliche biologische Probleme, wo schon das materielle System, in welchem sich die Erscheinung abspielt, nicht im atomistischen Sinne streng abgegrenzt werden kann, indem es ein wesentlicher Zug dieser Erscheinung und ihrer Beobachtung ist, dass es unmöglich ist, zu unterscheiden, welche Atome zum lebenden Organismus gehören, und welche im Begriff sind, in ihm aufgenommen oder von ihm ausgeschieden zu werden.

Der Sinn dieser Reihenfolge sollte sein, dass ebenso wie in der Atomphysik wegen der prinzipiellen auf der Wechselwirkung zwischen Objekt und Messmittel beruhenden Begrenzung der gewohnten Forderung einer Kausalbeschreibung für neuartige Gesetzmässigkeiten, wie sie in der Stabilität der Atome zutage treten, rationell Platz geschaffen wird, so dürfte der noch weiter gehende Verzicht auf Beobachtung und Definition, den das Aufrechterhalten des Lebens der Objekte uns aufzwingt, jeden scheinbaren Gegensatz zwischen Physik und Biologie zu vermeiden gestatten. Ich hoffe, dass Sie sehen werden, dass diese Einstellung eine viel weniger gefährliche ist als Sie im ersten Augenblick gefürchtet haben, und dass sie vor allem keinerlei Kompromiss mit dem antirationalistischen Vitalismus darstellt. Andererseits dürfte sie dazu geeignet sein, gewisse Vorurteile in der sogenannten mechanistischen Auffassung zu vermeiden, indem sie jedes Verlangen auf eine Analogie zwischen der Existenz des Lebens selbst und klargestellten Eigenschaften physikalisch–chemischer Systeme als irrational abweist. Eben dabei kann es nicht genügend stark betont werden, dass dieser Sachverhalt keinerlei Einschränkung von physikalisch–

[539]

chemischen Beschreibungs- und Untersuchungsmethoden enthält, sondern nur in der Anwendung dieser, die einzige nie zu erschöpfende, Quelle unserer Kenntnis und Verständniss der biologischen Erscheinungen sieht. Insbesondere möchte ich erwähnen, dass jede Rede von einem Versagen thermodynamischer Gesetzmässigkeiten im organischen meiner Ansicht nach auf groben Missverständnissen der Sachlage beruht.

Was meine Bemerkungen über die psychischen Probleme betrifft, handelt es sich ja zuerst nur um die rein formale Analogie zwischen den, wie ich glaube, von allen Psychologen anerkannten und von W. James so eindrucksvoll dargestellten Schwierigkeiten der Beobachtung und Analyse psychischer Erlebnisse und der Sachlage in der Atomphysik. Es ist aber keineswegs meine Ansicht, dass es sich hier um irgend eine direkte Verbindung zwischen diesen beiden Fragenkomplexen handelt, und im Aufsatz war es vor allem meine Absicht, gegen den Missbrauch der Quantenphysik im spiritualistischen Sinne zu warnen, den sogar hervorragende Physiker wie Eddington meiner Ansicht nach begehen. Das sogenannte Problem der Willensfreiheit löst sich, wie ich glaube, ebenso wie die oben berührten Paradoxien der Kausalität und des Vitalismus, durch eine genügend scharfe Kritik der Fragestellung. Eine vollständige deterministische Beschreibung des Verhaltens eines Menschen würde eine prinzipiell ausgeschlossene Kenntnis sogenannter innerer und äusserer Voraussetzungen erfordern, und eben in dieser Sachlage dürfte das unmittelbare Willensgefühl seinen natürlichen Spielraum finden. Der neueste Versuch von Jordan, bei Weiterführung solcher Überlegungen sogar Platz zu finden für die sogenannten parapsychischen Erscheinungen, dürfte aber meiner Ansicht nach ganz verfehlt sein, weil er nicht nur im schroffen Gegensatz zu der streng physikalischen Beschreibung des Verhaltens von Organismen, sondern auch im Widerspruch zu einer rationellen Verfolgung des psycho–physischen Parallelismus stehen dürfte. Meiner Überzeugung nach sind die parapsychischen Erscheinungen ein Trugbild, das eben durch die Fehlerquellen der Beobachtung und Deutung psychischer Erlebnisse vorgetäuscht wird.

Hoffentlich werden Sie in diesen Andeutungen eine Stütze finden für die in meinem vorigen Brief ausgesprochene Überzeugung, dass unsere Einstellungen aller Verschiedenheit der Ausdrucksweise zum Trotz wesentlich dieselben sein dürften.

Mit freundlichen Grüssen
Ihr ergebener
[Niels Bohr]

Translation

[Copenhagen], September 5, [19]36

Dear Professor Meyerhof,

I very much regret that due to unexpected obligations which cannot be delayed, I have not yet been able to find time to work out a summary of my remarks at the philosophers' congress[81]. I must therefore confine myself in this letter to mentioning the main points that might interest you.

First, my remarks in the old essay "Light and Life"[82] (of which I enclose a copy) only aim at elucidating the old dispute of vitalism versus mechanism from the viewpoint forced upon us by the recent development in physics with respect to the demand for an "explanation". Indeed, it is not here a question at all of a well-defined limitation of the possibilities for observing biological phenomena, or generally of any limitation of the description of such phenomena by means of physical concepts, but only of emphasizing that as regards the question of observation, just as in the entire biological mode of description, we have to do with a situation fundamentally different from that in which the questions of physics proper aim directly at a so-called explanation. As I have indicated only briefly in the essay, in this connection the general lesson of atomic physics can be described by the following sequence:

a) Questions in classical physics, where we always deal with a well-defined, closed system whose states can be described without taking the possibilities of observation into account.

b) Quantum physics, where we still have to do with a well-defined system, but where the unavoidable interaction between object and measuring instrument causes a fundamental trait of complementarity even in the description of the state.

c) Truly biological problems, where even the material system in which the phenomenon takes place cannot be limited strictly in an atomistic sense, because it is a fundamental trait of this phenomenon and its observation that it is impossible to distinguish between those atoms that belong to the living organism and those that are in the process of being taken up in it or excreted from it.

The meaning of this sequence should be that, just as in atomic physics a rational place is found for new kinds of laws – as they appear in the stability of atoms, because of the limitation (imposed in principle by the interaction

[541]

between object and measuring instrument)[83] of our usual requirement for a causal description – so an even greater renunciation of observation and definition, required by the need to keep the objects alive, allows us to avoid any apparent contradiction between physics and biology. I hope that you will see that this approach is much less dangerous than you feared at the first moment and that above all it represents no compromise whatsoever with anti-rationalistic vitalism. On the other hand, it may well be suited to avoid certain prejudices in the so-called mechanistic conception, in that it dismisses as irrational any demand for an analogy between the existence of life itself and well-defined properties of physico-chemical systems. Just for this reason, it cannot be emphasized strongly enough that this circumstance in no way involves a limitation of physical–chemical methods of description or investigation, but merely represents, in the application of these methods, the only inexhaustible source of our knowledge and understanding of biological phenomena. I wish to mention in particular that, in my opinion, any talk of a failure of the laws of thermodynamics in the organic world is due to grave misunderstandings of the situation.

As regards my remarks on psychical problems, we are indeed first of all dealing with the purely formal analogy between the difficulties of observation and analysis of psychical experience – which I believe are recognized by all psychologists and are so impressively described by W. James – and the situation in atomic physics. However, it is not at all my view that we are here dealing with any direct relationship between these two sets of problems, and in the essay it was above all my intention to warn against the abuse of quantum physics in a spiritualistic sense, as committed in my view even by eminent physicists such as Eddington. The so-called problem of the freedom of the will is solved, as I believe, just as the above-mentioned paradoxes of causality and vitalism, through a sufficiently sharp critique of how the question is posed. A completely deterministic description of human conduct would require a knowledge of the so-called internal and external preconditions, which are in principle mutually exclusive and precisely in this situation should allow natural room for the immediate feeling of will. The most recent attempt by Jordan to find room even for so-called parapsychological phenomena by extending these considerations, would, however, in my opinion be totally mistaken, because it would not only be in stark conflict with a strictly physical description of the behaviour of organisms, but also contradictory to a rational treatment of psychophysical parallelism. It is my conviction that parapsychological phenomena are a figment

---

[83] Parentheses added by editor for easier reading.

of the imagination conjured up simply by the sources of error in the observation and interpretation of psychical experience.

Hopefully, you will find support in these suggestions for the conviction expressed in my previous letter that regardless of differences in the way of expression, our approaches might be essentially the same.

With kind regards
Your devoted
[Niels Bohr]

WOLFGANG PAULI

BOHR TO PAULI, 31 December 1953
[Carbon copy]

[København,] 31. december 1953.

Kære Pauli,

Før årsskiftet vil jeg gerne sende dig en lille hilsen og takke for alt venskab og opmuntring i det gamle år. Fra Kleins, som for nogle uger siden var et par dage her i København på hjemrejsen fra Zürich, fik vi jo både venlige hilsner og gode efterretninger fra dig og Franca. Især var Margrethe og jeg så glade for at høre, at Franca nu igen er helt rask, og at I begge, som Kleins fortalte, var i det mest strålende humør. Jeg var jo også meget interesseret i at forstå, at du foruden med kvanteproblemerne igen beskæftiger dig nærmere med de almindelige biologiske og psykologiske spørgsmål, som i det forløbne år jeg også selv har tænkt meget over.

Jeg hørte fra Klein om din interesse for en ældre bog[84] af en omtrentlig navnefælle af dig om Darwinisme og Lamarckisme og har siden selv kigget nærmere på den. Det drejer sig jo om synspunkter, som utvivlsomt rummer dybe sandheder og som har været mig velbekendt fra ungdommen af, da min fader, som vi vist ofte har talt om, var stærkt optaget af sådanne spørgsmål og i sine skrifter ved forskellige lejligheder har udtrykt sig på meget træffende måde om situationen for den biologiske forskning. Det, som imidlertid først og fremmest ligger mig på sinde, er en stillingtagen til de store fremskridt inden for biologi og psykologi i den sidste menneskealder og især i denne forbindelse

---

[84] A. Pauly, *Darwinismus und Lamarckismus. Entwurf einer psychophysischen Teleologie*, E. Reinhardt, Munich 1905.

[543]

at udnytte den almindelig belæring, som atomfysikkens udvikling har givet os med hensyn til beskrivelsen af fænomener, der på grund af deres helhedspræg ikke kan behandles ved hjælp af de idealisationer, der ligger til grund for den mekaniske naturopfattelse.

Netop fordi livet, i modsætning til fænomenerne i den såkaldte livløse natur, karakteriseres ved organismernes selvstændighed og evne til at tilpasse sig til de ydre omstændigheder på en måde, der tjener til livets opretholdelse, forplantning og udvikling, er det jo en rent logisk fordring, at mekanistiske og vitalistiske eller, som mange biologer foretrækker at sige, finalistiske betragtningsmåder fremtræder som komplementære i den objektive beskrivelse af det organiske liv. I en vis forstand må man i elementerne i den mekaniske beskrivelse søge et udgangspunkt for korrespondensbetragninger på lignende måde som de mekaniske grundbegreber benyttes ved formuleringen af de kvantemekaniske lovmæssigheder vedrørende fremkomsten af individuelle processer, der ikke selv har noget modstykke i den mekaniske naturopfattelse. Det er dog selvfølgelig ikke her nogen simpel analogi, det drejer sig om, men om kravet til en yderligere vidtgående generalisation af naturbeskrivelsen med plads for livets karakteristiske træk.

Et hovedpunkt er naturligvis omstændighederne i hvert enkelt tilfælde for fænomenernes fremkomst og iagttagelse, hvorimellem der jo rent logisk ikke kan skelnes. I modsætning til de strengt afsluttede iagtagelsesresultater i de simple experimenter med atomare partikler må vi således ved livsfænomenerne tage det bestandige stofskifte i betragtning, der kræves for organismernes opbygning og fornyelse og for deres forsyning med den for deres funktioner nødvendige fri energi. Netop disse forhold synes også direkte at pege på muligheden for en for tilpasningen til de ydre kår tjenlig organisk udvikling. Utvivlsomt er jo arternes ejendommelige bestandighed nøje forbundet med den atomare stabilitet af cellernes kromosomer; ligeså væsentlig er det imidlertid at betænke, at vi i en levende organisme aldrig har at gøre med kromosomerne alene, men at de mindste enheder, der kommer i betragtning ved forplantningen er de celler, hvori kromosomernes forunderlige spaltninger og sammenkoblinger foregår ved hjælp af de nødvendige næringsstoffer og enzymer. Ud fra den omstændighed, at kønscellerne først modnes på så sent et tidspunkt af individets liv, at krav til tilpasning til ændrede ydre forhold allerede kan have meldt sig, ligger det derfor nær at formode, at sådanne ændringer ikke alene afspejler sig i fænotypen, men at genotypens bestandighed kun er en første tilnærmelse, og at den kan undergå gradvise sækulære ændringer af en for tilpasningen til milieuet formålstjenlig karakter.

Sådanne bemærkninger kunne måske forekomme meget tågede, men det punkt, jeg stadig søger at støtte mig til, er de for enhver definition krævede permanente træk hos fænomenerne, som har deres analogi i de irreversible forstærkningseffekter, hvorpå de atomare processers iagttagelse beror. Sådanne effekter møder vi jo overalt i det organiske liv, og som du ved, lægger jeg ved beskrivelsen af de til livet knyttede psykiske fænomener hovedvægten på, at det ved såkaldte bevidste oplevelser drejer sig om noget, der kan huskes og derfor må svare til særlige situationer i organismen, der netop har den som grundlag for iagttagelse krævede permanente karakter. Rent behavioristisk betragtet er jo også denne situation et særligt klart udtryk for organismernes muligheder for den mest praktiske eller formålstjenlige måde at udnytte alle veje til livets udfoldelse.

I sådan forbindelse har det moret mig igen at tænke lidt over brugen af ord som instinkt og fornuft, således som jeg kom ind på i den gamle afhandling[85] om forholdet mellem menneskekulturerne, især hvad angår mange væsentlige forskelle i dyrs og menneskers situation. Medens de nedarvede instinkter jo åbenbart har deres aftryk i den enkelte celle eller i det mindste i det befrugtede æg, drejer det sig ved såkaldt fornuft jo om vekselvirkning mellem store cellekomplexer i nervesystemet, der efter sagens natur ikke kan fuldt aftrykkes i den enkelte celle og derfor kun huskes af individet selv, hvad der også af praktiske grunde er ganske tjenlig for den organiske udvikling og menneskelivets bekvemmeligheder. Tænk hvad det ville betyde af forvirring og besvær, om man gennem arv var belastet med forfædrenes mere eller mindre tilfældige oplevelser og filosofiske vildfarelser. For kulturens stadige højnelse er det jo fuldt tilstrækkeligt, at den enkeltes ydre og indre oplevelser kan meddeles til andre, og om man ønsker det, nedtegnes for efterslægten, ved hjælp af et sprog som individet vel har nedarvede muligheder for at tilegne sig, men som ethvert barn må lære på ny, hvad der jo også giver stadige muligheder for sprogenes egen udvikling.

Det er jo ikke meningen her at komme ind på alle de gamle sager, som vi så ofte har talt om, men blot at fortælle at jeg atter beskæftiger mig nærmere dermed og synes at det virkelig er muligt stadig at nå til større klarhed, i det mindste i den forstand, som vi så mange gange har spøget med med henblik på Schillers Sprüche des Confucius[86]. Heller ikke behøver jeg vist at

---

[85] N. Bohr, *Natural Philosophy and Human Cultures*, ref. 38.

[86] "The Sayings of Confucius", poem by Friedrich Schiller in two parts, written in 1795 and 1799, dealing with the concepts of time and space, respectively. It ends with the words: "Nur die Fülle führt zur Klarheit, // Und im Abgrund wohnt die Wahrheit." (The editor's attempt at a translation: "only fullness leads to clarity, // and in the abyss truth resides.") See *Schillers Werke, Nationalausgabe*, Vol. 2, Part 1, Weimar 1983, pp. 412–413. This was one of the poems that Bohr liked to read aloud to his colleagues at informal gatherings.

sige, at det trods den benyttede sprogbrug ikke er hensigten at forgrove og forenkle den filosofiske indstilling ved tilslutning til en behavioristisk skole eller anden ideologi, der igennem selve sit navn henviser til tilhængernes begrænsede udsyn, men blot at benytte ord, der peger på de selv for den mest frigjorte indstilling nødvendige komplementer. Navnlig tror jeg, at det er muligt netop med sådan betoning at opnå et omfattende synspunkt, ikke for den aprioriske begrundelse, men for den objektive beskrivelse såvel af den organiske udvikling som af vor situation som bevidste væsener. Min optagethed med disse spørgsmål hænger sammen med, at jeg netop for tiden arbejder på en afhandling[87], som jeg håber snart at kunne sende dig og som skulle danne en forhåbentlig noget afklarende slutning i en sådan samlet udgave af mine artikler fra de senere år, som du så ofte har rådet mig til. Jeg forbereder også at forsyne udgaven med kommentarer, der skulle lette læseren en tilegnelse af indstillingen og navnlig advare ham imod nogle af de misforståelser, som man så sørgeligvis ofte træffer.

Imidlertid sender jeg som et lille livstegn et særtryk af en mindre artikel om fysikken og religionerne, som netop er udkommet i et festskrift til Johs. Pedersens 70 års fødselsdag[88]. Jeg er bange for, at den vil være en skuffelse for dig, fordi jeg har måttet indskrænke mig til nogle få antydninger og slet ikke rigtig kunne komme ind på de forskellige punkter, hvortil der er lagt op i artiklens indledning. Hovedhensigten var jo, som du vil forstå, så tydeligt som muligt at fremstille den store forskel i den filosofiske baggrund på Descartes' og Spinozas tid og den udvikling vi lever i, og jeg havde især håbet at få lejlighed til at gå langt mere direkte ind på de erkendelsesteoretiske eller rettere logiske vanskeligheder, som enhver forestilling om et personificeret forsyn rummer. Som sagt blev det kun til et lille brudstykke, som jeg måtte jage med at gøre færdig i sidste øjeblik før jeg igen skulle rejse til et af de mange CERN-møder i Genève og derfra videre til Palæstina og Grækenland. Begge steder havde Margrethe og jeg mange interessante og tankevækkende oplevelser og fik stærke indtryk både af de store kulturminder og de vanskelige nutidsproblemer,

---

[87] Most likely, Bohr refers here to his Steno Lecture, originally given in 1949, which he at this time was starting to develop into a published version. See *Steno-Forelæsning i Medicinsk Selskab (II)*, Bohr MSS, microfilm no. 22. The lecture, *Physical Science and the Problem of Life*, was eventually published as the concluding essay in Bohr, *Atomic Physics and Human Knowledge*, ref. 11, pp. 94–101. Reproduced on pp. [116]–[123].

[88] N. Bohr, *Physical Science and the Study of Religions*, Studia Orientalia Ioanni Pedersen Septuagenario A.D. VII id. Nov. Anno MCMLIII, Ejnar Munksgaard, Copenhagen 1953, pp. 385–390. Reproduced on pp. [275]–[280].

og vi håber at det ikke skal vare længe før vi får lejlighed til at tale nærmere med dig og Franca om det alt sammen.

Efter hjemkomsten havde vi et kort men overmåde rart besøg af Oppenheimers, der bl.a. fortalte om, at de venter dig og Franca til Princeton i foråret. Vi har selv måttet udskyde vort planlagte besøg i Princeton til efteråret, men det ville jo være umådelig morsomt, om I kunne komme til København, når I kommer tilbage og du igen kunne give et foredrag til CERN-gruppen her, der til den tid vedblivende vil være samlet. Iøvrigt havde vi for nogle uger siden også et besøg til gruppen af Heisenberg, der fortalte os om sine nye og dristige tanker vedrørende en sammenfattende teori for de elementære partikler. Jeg tror nok at jeg forstår pointen, men er næppe i stand til at gennemskue de indviklede matematiske beregninger, og jeg er spændt på at høre hvad du mener om det. Hovedhensigten med dette vist alt for lange brev er jo også ikke så meget at fortælle om mig selv og mine egne uforgribelige meninger, men at lokke dig at fortælle lidt om dit arbejde og jeres planer, og især om dine tanker om de almindelige filosofiske spørgsmål.

Med de venligste hilsner og beste ønsker for det nye år til dig og Franca fra Margrethe og

din gamle
[Niels Bohr]

Translation

[Copenhagen,] December 31, 1953

Dear Pauli,

Before New Year I would like to send you a small greeting and to thank you for all your friendship and encouragement in the past year. From the Kleins who were here in Copenhagen for a couple of days some weeks ago on their way home from Zurich, we received both kind greetings and good news from you and Franca. Margrethe and I were especially pleased to hear that Franca is now completely well again, and that both of you, as the Kleins told us, were in the very best spirits. I was also very interested to learn that you, besides quantum problems, are now working in more detail with general biological and psychological questions, which I too have thought about a good deal in the past year.

I heard from Klein about your interest in an old book[84] by a nearly namesake of yours about Darwinism and Lamarckism, and have since looked at it more carefully myself. What is involved are views which undoubtedly hold deep

[547]

truths and with which I have been quite familiar since my youth, because my father, as we certainly have discussed often, was very occupied with such questions and in his writings on various occasions expressed himself quite strikingly about the state of biological research. However, what primarily concerns me is how to make up one's mind about the great progress in biology and psychology in the last generation and, especially in this connection, to utilize the general lesson that the development of atomic physics has taught us with regard to the description of phenomena which, because of their feature of wholeness, cannot be dealt with by means of the idealizations that make up the basis for the mechanical conception of nature.

Precisely because life, in contrast to phenomena in so-called inanimate nature, is characterized by the independence of the organisms and their ability to adapt to external circumstances in a way that serves the sustenance, reproduction and development of life, it is a purely logical requirement that mechanistic and vitalistic or, as many biologists prefer to say, finalistic viewpoints show themselves to be complementary in the objective description of organic life. In a certain sense one must seek in the elements of the mechanical description a starting point for correspondence viewpoints in the same way as the fundamental principles of mechanics are used in the formulation of the laws of quantum mechanics regarding the occurrence of individual processes which themselves do not have any parallel in the mechanical conception of nature. However, it is of course not here a question of a simple analogy, but of the requirement for an even more far-reaching generalization of the description of nature with room for the characteristic features of life.

A major point is naturally the circumstances in each individual case for the occurrence and observation of phenomena, between which, of course, for purely logical reasons, no distinction can be made. In contrast to the rigorously closed observational results in the simple experiments with atomic particles, we must thus, in the case of life phenomena, take into account the continuous metabolism necessary for the constitution and renewal of organisms and for the supply of free energy required for their functioning. Precisely these circumstances also seem to point directly to the possibility of an organic development suitable for adaptation to external conditions. The constancy peculiar to species is undoubtedly closely connected to the atomic stability of the chromosomes of the cells; however, it is just as important to bear in mind that in a living organism we are never dealing with chromosomes alone, but that the smallest units that can be considered in reproduction are those cells in which the wonderful divisions and couplings of chromosomes take place by means of the necessary nutrients and enzymes. Because of the circumstance that gametes mature only at such a late stage in the life of the individual that requirements for adaptation to changed external conditions may already have occurred, it therefore seems natural to presume that such changes are not only reflected in the phenotype but that the constancy of the genotype is only a first approximation, and that it can undergo gradual secular changes of a character suitable for adaptation to the environment.

Such remarks might perhaps appear to be very vague, but the point that I still seek to rely on is the permanent traits of the phenomena necessary for any definition, which have their analogy in the irreversible amplification effects on which the observation of atomic processes depend. We meet such effects everywhere in organic life, and as you know, in the description of psychical phenomena connected to life, I put the main emphasis on the fact that so-called conscious experiences involve something that can be remembered and hence must correspond to special situations in the organism which have precisely the permanent character required as a basis for observation. Seen from a purely behaviouristic standpoint, this situation too is a particularly clear expression of the possibilities of organisms to utilize all paths to the unfolding of life in the most practical or suitable fashion.

In this connection it has amused me to think a little once more about the use of words such as instinct and reason, which I touched upon in the old paper[85] on the relation between human cultures, especially as regards many important differences in the situation of animals and humans. Whilst the inherited instincts obviously have their imprint in the individual cell or at least in the fertilized egg, in the case of so-called reason it is a question of interaction between large groups of cells in the nervous system, which in the nature of the matter cannot be fully imprinted in the individual cell and may thus be remembered only by the individual itself, a circumstance which also for practical reasons is quite beneficial for organic development and the pleasures of human life. Think of the confusion and difficulty there would be if one was encumbered through heredity with the more or less fortuitous experiences and philosophical delusions of one's ancestors. For the continual raising of the level of culture it is quite sufficient that the external and internal experiences of the individual can be communicated to others and, if so desired, written down for posterity by means of a language which it must be assumed that a person has inherited possibilities for acquiring, but which every child must learn anew, thus providing continual opportunities for the development of the languages themselves.

It is not my intention here to enter into all the old matters that we have discussed so often, but just to say that I am again working with them in more detail and think that it is really possible to achieve still greater clarity, at least in the sense that we have joked about so many times with regard to

Schiller's "Sprüche des Confucius"[86]. Nor do I need to say that despite the terminology used, the intention is not to coarsen or simplify the philosophical attitude by joining a behaviourist school or any other ideology which by its very name indicates the limited vision of its supporters, but just to use words that point to the complements necessary even for the most liberal attitude. In particular, I think that it is possible with precisely such an emphasis to achieve a comprehensive viewpoint, not for the a priori justification but for the objective description of both organic development and our situation as conscious beings. My preoccupation with these questions is connected with my work just now on a paper[87] which I hope to be able to send to you soon, and which should comprise a hopefully somewhat clarifying conclusion in a collected edition of my articles from the last years of the kind that you have so often recommended. I am also preparing to furnish the edition with comments, which should make it easier for the reader to understand the approach and in particular to warn him against some of the misunderstandings which unhappily one so often meets.

In the meanwhile, as a small sign of life I am sending you a reprint of a short article about physics and religions which has just appeared in a *Festschrift* for Johannes Pedersen's 70th birthday[88]. I am afraid it will be a disappointment to you, as I have had to confine myself to a few intimations and have not at all really been able to discuss the various points anticipated in the introduction to the article. As you will understand, the main purpose was to present as clearly as possible the great difference between the philosophical background at the time of Descartes and Spinoza and the development we now experience, and I had especially hoped to have the opportunity of going much more directly into the epistemological, or rather logical, difficulties implied by any idea of a personified providence. As I said, the result was only a little fragment, which I had to rush to finish at the last moment before I was going to travel once more to one of the many CERN meetings in Geneva and from there on to Palestine and Greece. In both places Margrethe and I had many interesting and thought-provoking experiences and received strong impressions of both the great cultural monuments and the difficult contemporary problems, and we hope that it will not be too long until we have the chance to speak about it all in greater detail with you and Franca.

After our return we had a short but extremely pleasant visit from the Oppenheimers, who among other things told us that they are expecting you and Franca in Princeton in the spring. As for ourselves, we have had to postpone our planned visit to Princeton until the autumn, but it would of course be immensely enjoyable if both of you could come to Copenhagen when you return, and if you could again give a talk to the CERN group here, which will still be gathered at that time. Incidentally, a couple of weeks ago we also had a visit to the group by Heisenberg, who told us about his new and daring ideas concerning a comprehensive theory for elementary particles. I think that I understand the point, but I am hardly able to grasp the complicated mathematical calculations, and I am eager to hear what you think about it. The main purpose of this probably much too long letter is not so much to tell you about myself and my own humble opinions, but to tempt you to tell us a little about your work and the plans of both of you, and especially about your thoughts about general philosophical questions.

With warmest greetings and best wishes for the New Year to you and Franca from Margrethe and

your old

[Niels Bohr]

PAULI TO DELBRÜCK, 16 February 1954[89]
[Carbon copy]

THE INSTITUTE FOR ADVANCED STUDY

PRINCETON, NEW JERSEY

February 16, 1954

Dear Max:

It was really very kind of you to have sent me the copy of your lecture which I return herewith according to your wish. Of course it interested me very much. I had heard about the Watson–Crick paper[90]. The proposed structure of DNA appeals indeed very much to the sense of mathematics with its pair of two complementary one-dimensional tapes and with its foundation on the "holy Four" of the Pythagoreans. I also heard that one has started to worry about how the tapes unwind and separate prior to duplication. I wonder whether the presence of the living cell is essential for this process or whether it can also take place as an ordinary chemical process without contact with any living material. The latter seems at least to be thinkable. The result of Hershey and Chase that this same

---

[89] The carbon copy of the present letter, now in the BSC, was originally enclosed with Pauli's letter to Bohr, 19 February 1954 (next letter).

[90] J.D. Watson and F.H.C. Crick, *A Structure for Deoxyribose Nucleic Acid*, Nature **171** (1953) 737–738.

substance DNA is also the effective one in the bacteriophages was, however, quite new to me just as the other content of your lecture[91].

The comparison at the end of your lecture of present biology with physics of the 1890's is also very stimulating (particularly if one is, like myself, somewhat skeptical on the development of physics in the next years). It gives us hope to see in biology, in not too distant a future, also the analogue to the years 1900–1905 in physics, when the break between quantum phenomena and classical physics became definitely clear. Similarly, in biology the definite break with our present physics, though in the sense of a "rational generalization", must and will come into the day light.

I am very much looking forward to seeing you and Franca and I reserve for you the evening of March 10 in New York. Place and time we shall fix later. I will come back to it.

I like to add a few words more here which may facilitate it for you to "handle me". What interests me mostly are the characteristics of the processes in living organisms which cannot be explained by the present physics (and chemistry) based on our known quantum mechanics, and which distinguishes them from the ordinary general physico–chemical process. I have the impression that these characteristics must be in the *correlation* of the different processes, in other words in an organization of all of them. Therefore, I asked myself whether, where chance appears in the usual quantum mechanical laws, in the living organisms an additional interdependence occurs. It is true that the laws of genetics are *statistical* laws, just as the laws of quantum-physics, and I know well how you have established a logical connection between these two facts by your model for mutations in your older papers with Timofeef-Resovsky. But also in the case of the mutations just the question arises whether here a characteristic additional interdependence of different genes comes into play, too, which is alien to the simple probability laws for the rate of the occurrence of the ordinary transitions between stationary states in the usual quantum physics.

With such vague ideas in my mind I am of course getting angry if biologists try to use the general concept "chance" in order to explain phenomena which are so typical for living organisms as, for instance, those appearing in the biological evolution. I am of course aware that everybody who speaks of chemical reactions of the genes is obliged to bring this into agreement with our last knowledge on the chemical constitution of the genes. But although there

---

[91] A.D. Hershey and M. Chase, *Independent Functions of Viral Protein and Nucleic Acid in Growth of Bacteriophage*, Journal of General Physiology **36** (1952) 39–56.

exists already a tradition of the biologists of California to disagree with Bohr, (a tradition which was founded by the late *Morgan*) it cannot be denied that Bohr is right, when he emphasizes that the smallest living unit is the cell and not the isolated chromosome. This is all the more so, as there are different persons who discuss the working hypothesis that the heredity on the way over the cytoplasma rather than over the cell-nucleus contains those processes which may be directed so that they could reflect an hereditary influence of the environment on the organism.

However this may be, some biologists seem to admit that there is indeed something essential still missing for the explanation of the empirical data on the biological evolution. I believe therefore that you can do something very useful by explaining to us physicists life and to the biologists chance – but, for heaven's sake, don't do it the other way round!

In this sense "Auf wiedersehen".

Yours old
W. Pauli

PAULI TO BOHR, 19 February 1954
[Typewritten with handwritten postscript]

THE INSTITUTE FOR ADVANCED STUDY
PRINCETON, NEW JERSEY

SCHOOL OF MATHEMATICS   February 19, 1954

Dear Bohr:

It was a real great pleasure to get your long New Year's letter and I have a rather bad conscience that it is only now that I am able to answer it, at least in some provisional way. Particularly your remarks on biology interested me so much that I started with reading literature about the subject, all the more as I have here much more time for it than in Zürich. Then I always postponed it to write to you until I had read "one more book (or paper)". One of the most interesting articles I have studied is one of *C.H. Waddington*: "The evolution of adaptations" in the periodical ENDEAVOUR, *12*, 134, (1953). The author, Professor in Edinburgh, belongs to those biologists who admit that there is something essential still missing in our understanding of evolution.

[553]

I also read several articles and one book of the geneticist *R. Goldschmidt*[92], who, deviating from the current view, assumes the occurrence of bigger sudden jumps, called by him "systemic mutations", "macromutations", or "saltations", which should be responsible for the biological evolution. I was, however, not convinced by this particular story, because these bigger jumps, if not deadly, seem to me much too unlikely to be explained by "chance" (a favorite word of all Darwinists which they use in a very loose and superficial way), except if these big changes would be more or less continuously preceded and prepared by a series of other smaller changes, which are not externally visible. With the latter idea I approach your own hypothesis "that also the genotype can undergo gradual secular changes of a kind, which serves the purpose of the adaptation to the environment" (p. 2 of your letter)[93]. Nevertheless not only the word "gradual", but also the word "secular", here used by you, seem to me open to a critical discussion. Regarding the latter, one has to bear in mind that the natural unit of time in biology is the number of generations, which for a given time is so different for different species, rather than the physical time. If the effect of small changes of the genotype considered by you actually exists, one should therefore expect that it would be much quicker and therefore also easier visible in organisms producing many generations in a relatively short time as, for instance, Drosophila or bacteria.

On the other hand Robert Oppenheimer, as well as a few biologists, seems to think that the heredity on the way over the cytoplasma, which is much rarer than the usual one over the cell nucleus, is just the one which in the long run can give rise to directed processes reflecting hereditable influences of the environment. But, at present, the empirical material to support such a far-reaching assumption, is extremely scarce, if at all existing. Nevertheless this guess seems to be logically related to your remark, that "in a living organism we are never dealing with the chromosomes alone, but that the smallest units, to be considered in connection with the propagation, are the cells" – a remark which seems to me to cover an essential part of the situation.

As I know that your opinions have a certain weight to Max Delbrück and also because he, as an ex-physicist may have a particular talent for explaining to us the present situation in biology, I have sent to him your letter with a

---

[92] Richard Benedict Goldschmidt (1878–1958), German biologist who emigrated to the United States in 1935. Goldschmidt wrote several articles and books on genetics.

[93] In this and subsequent quotations from Bohr's letter of 31 December 1953 (p. [543]), Pauli uses his own translation into English.

message through Weisskopf, who just went to Pasadena for a temporary visit. Delbrück's answer contained the brief and sad remark about your letter, "I think it's awful". But he also wrote that he will make a journey to Europe in March and I have a date with him in New York on March 10 where I hope to hear more about his views and ideas. Fortunately he will also pass Copenhagen, in order to see there some biologists and on this occasion he also hopes to see you "to tell you the facts of life, the known ones and the unknown ones".

He also has sent to me a paper on bacteriophages, of which I enclose a summary. He seems to be very excited and enthusiastic about the Watson–Crick paper, which you probably know and which really appeals to me. Moreover, I am enclosing for you the copy[94] of a letter of mine to Delbrück, which contains more about my own feelings on the relation between physics and biology and on the problems by which my interest seems to be attracted with increasing strength. So it seems that this discussion will still continue for quite a while.

* * *

I also enjoyed very much reading your article on "Physical science and the study of Religions"[95], but as you have already foreseen, I had hoped to hear from you more about "the logical difficulties, which every perception of a personified providence meets". These difficulties have also been discussed inside Christianity, particularly in the 17th Century, when Calvin's doctrine of predestination raised much opposition. This doctrine appears to me as the theological analogy to the philosophical determinism, which generalized the Newtonian mechanics beyond the limits of its applicability. To the opposition against Calvin's doctrine belonged also the Platonist–Christian Henry More, the fatherly friend of Newton. Fierz[96] has studied his writings in the course of his historical work about Newton and he told me last autumn that More stated explicitly: "There are certain events which cannot be known before even by God, because they are not determined in advance. If God would know them, he would know something false, which is impossible".

* * *

[94] The preceding letter.
[95] See ref. 88.
[96] The Swiss physicist Markus Fierz (b. 1912), Pauli's assistant from 1936 to 1940, was at this time professor at the University of Basle.

Heisenberg's paper[97] eventually arrived here about two weeks ago. Then I had many discussions about it, particularly with Dyson[98], Källén[99] and Thirring[100] and it seems that there are indeed strong objections against some of his mathematical assumptions. I am going to write to Heisenberg in these days about these objections and we shall hear then what he will have to reply.

The Oppenheimers, who enjoyed very much their trip to Europe and particularly their visit in Copenhagen, are sending their heartiest regards. Robert was also very interested in the content of your letter to me.

I just heard from Lund about the plan of a conference there in July 1–5, in which also Rydberg's work will be commemorated in a particular talk by you[101]. For me there is the difficulty with the date that the summer term in Zürich will not yet be finished in these days. But I hope to be able to make myself free to hear at least your talk on Rydberg, if I shall not be able to be present at the whole conference. Edlén[102] also asked me in a letter whether I too will tell something on Rydberg in coordination with your talk. I do not know whether this would be practical and I would like to hear your opinion about it. I know a little bit particularly on Rydberg's works on the periodic system of the elements including his errors.

Franca and I are sending our warmest regards to yourself, Margrethe and also to Aage and to his wife. (I hope to hear from Weisskopf more about his ideas on Aage's work, when he will be back East.) We plan to be back in Zürich around the middle of April.

Yours old,

W. Pauli

P.S. I would be glad if *Alder*[103] could report to me on his plans for a Doctor Thesis and on his choice of a subject for it.

---

[97] Probably W. Heisenberg, *Zur Quantisierung nichtlinearer Gleichungen*, Nach. Akad. Wiss. Göttingen, Math–Phys. Kl. **8** (1953) 111–127.

[98] Freeman Dyson (b. 1923), British–American physicist.

[99] Gunnar Källén (1926–1968), Swedish physicist.

[100] Walter Eduard Thirring (b. 1927), Austrian physicist.

[101] N. Bohr, *Rydberg's Discovery of the Spectral Laws*, Proceedings of the Rydberg Centennial Conference on Atomic Spectroscopy, Lunds Universitets Årsskrift. N.F. Avd. 2. Bd. 50. Nr 21 (1955) 15–21. Reproduced on pp. [373]–[379].

[102] Bengt Edlén (1906–1993), Swedish physicist.

[103] Kurt Alder (b. 1927), Swiss physicist.

PAULI TO BOHR, 26 March 1954
[Handwritten]

THE INSTITUTE FOR ADVANCED STUDY
PRINCETON, NEW JERSEY

SCHOOL OF MATHEMATICS   March 26, 1954

Dear Bohr,

I am sending to you just a few informal lines in answer to your short letter of March 3[104]. – Meanwhile I saw Delbrück and I also received from him a short and happy report about your meetings in Copenhagen. He said, that you had changed your mind in these biological questions soon after you had sent the letter to me. This seems to me quite satisfactory, as I am now rather convinced that there is no causal connection between the happenings in a gene and the environment (milieu) of the organism. After I talked with Delbrück and also with other biologists I see much better where the real problems are. Still, I am disturbed by the fact, that due to our ignorance of the connection between chemical events within genes or chromosomes on the one hand, and the visible characters of an organism on the other hand, it does not seem to be possible to *estimate* theoretically the probability (chance) of the events which were important in the biological evolution*. Therefore the word "chance" as it is used by modern "selectionists" is still *empty*. In order to bring "selection" into play, the "selectionists" have to assume before the occurrences of events (changes in the genes and chromosomes) which are entirely outside the range of any theoretical understanding. And – "wo die Begriffe fehlen, stellt das wort 'Chance' zur rechten Zeit sich ein"[105]. – Of course, I would be very interested to hear what *you* think now on these questions.

*　*　*

---

[104] BSC.

[105] The quotation is drawn from Johann Wolfgang von Goethe's Faust. See, for instance, *Goethes Faust, Gesamtausgabe*, Insel Verlag, Frankfurt am Main 1983, p. 195: "Den eben, wo Begriffe fehlen, // Da stellt ein Wort zur rechten Zeit sich ein." English translation: "For if your meaning's threatened with stagnation, // Then words come in to save the situation". See Johann Wolfgang von Goethe, *Faust, Part One* (trans. Philip Wayne), Penguin Books, London 1949, p. 97. In his German rendition in the present letter, Pauli replaced "words" with "the word chance".

Meanwhile another of our common friends passed through New York on his trip from California to Europe, and this is Otto Stern. You will see him soon** in Copenhagen (he seems to be eager to live in a *Hotel* there) and to "introduce him", I wish only to emphasize that he is *not* identical with Einstein – in this sense, that he is <u>*not*</u> interested in Einstein's old "story" that quantum-mechanics is not a complete description of "nature". It will be good therefore if you could in discussions with *Stern* avoid to repeat always the sentence (as you did the last time): "but *Einstein* said ...". Please remember: *Stern ≠ Einstein*.

The time of my departure is now approaching and we plan to be back in Zürich on April 11th. Perhaps you can write to me to Zürich, perhaps I shall also see you at one of these Geneva-meetings. I would like to hear from you also on the Lund-Conference in July and your lecture on Rydberg. (By the way, M. Riesz[106] is just now in Princeton).

At the M.I.T. I heard most illuminating things on Aage's work from Weiss-kopf and Villars[107].

Regards to him, to Margrethe and to yourself (also from our common friends in Princeton) from both of us.

Yours old,

W. Pauli

* For instance: that a reptile gets feathers (archeopteryx) or that the horse got longer legs and a cloven foot.

** Presumably on his way *back* from Stockholm.

BOHR TO PAULI, 6 April 1954
[Carbon copy]

[København,] Den 6. april 1954.

Kære Pauli,

Jeg var vældig glad at få dit brev af 26. marts få dage før jeg igen må rejse til Genève til et møde i det europæiske råd for atomkerneforskning[108]. Siden jeg skrev sidst, har vi haft besøg af Delbrück der, som du véd, bliver i Göttingen denne sommer. Besøget gav anledning til mange lærerige og interessante diskussioner også med danske biokemikere og biologer. Det synes virkelig, at vi begynder at få et grundlag for en udredning af nogle af stadierne

---

[106] Marcel Riesz (1886–1969), Swedish mathematician.
[107] Felix Villars (b. 1921), American physicist, born and educated in Switzerland.
[108] That is, CERN.

i den for celledelingen ejendommelige rytme, men samtidig forekommer det mig også, at helhedstræk træder stedse klarere frem. Jeg føler mig faktisk stadig mere overbevist om, at de almindelige betragtninger om den organiske udvikling, som jeg skrev til dig om i januar, træffer et ganske væsentlig punkt. Hvor afgørende det end på den ene side er, at der fra celle til celle videreføres et direktiv gennem fordoblingen af simple kæder af veldefinerede kemiske molekyler, synes det klart, at der under den embryologiske udvikling stadig er en tilbagekobling, der sikrer individets helhed. Førend de særlige reduktionsprocesser, der bestemmer kønscellernes dannelse, optræder, må der således være modtaget budskab om organismens udvoksning og vel også om de krav, livet stiller til dens reaktioner over for de ydre forhold. Du ser at jeg ingenlunde føler mig slået af marken ved den mekanistiske skoles kritik, men at jeg blot prøver at lære så meget som muligt om de forskellige sider af situationen. Faktisk må jeg tilstå at jeg, hvor diletantisk det end vil blive betragtet af mange fagbiologer, stadig søger at drømme mig dybere ind i livets rytme og det hele vekselspil i organismerne, og at jeg føler, at det, hvor beskedent resultatet end vil blive, nærmer sig til at man kan opnå en rimelig epistemologisk indstilling til væsentlige hovedtræk ved livets problem. Jeg håber at kunne give en mere klar fremstilling af sådanne spørgsmål i et foredrag "On the Unity of Knowledge" ved "The Bicentennial Celebrations" ved Columbia universitetet til efteråret[109]. Da manuskriptet til foredraget helst skal afleveres allerede i midten af juni, er jeg begyndt på forberedelserne og skal prøve at gøre manuskriptet færdig så snart jeg kommer hjem fra rejsen i tilknytning til hvilken Margrethe og jeg håber at få et par ugers meget tiltrængt ferie i Italien.

Jeg håber meget at vi snart skal få en lejlighed til rigtig at tale sammen om alle disse spørgsmål, og jeg forstod på Gustafson[110], at der er mulighed for at du allerede i maj vil komme på besøg i Lund; vi regner med at du så også vil besøge CERN-gruppen i København. I håb om snarligt gensyn og med de venligste hilsner til Franca og dig selv fra Margrethe og

din gamle
[Niels Bohr]

PS: Desværre vil jeg være rejst inden Stern kommer hertil på vej til Stockholm,

---

[109] Manuscript, *Unity of Knowledge*. Bohr MSS, microfilm no. 21. The article was subsequently published as N. Bohr, *Science and the Unity of Knowledge* in *The Unity of Knowledge* (ed. L. Leary), Doubleday & Co., New York 1955, pp. 47–62. Reproduced on pp. [83]–[98].

[110] Torsten Valdemar Gustafson (1904–1987), Swedish physicist. Professor at the University of Lund from 1939.

men både Margrethe og jeg håber meget, at han kommer hertil snart igen efter vi er kommet tilbage.

Translation

[Copenhagen,] April 6, 1954

Dear Pauli,

I am very pleased to have received your letter of 26 March a few days before I have to go to Geneva again to a meeting in the European council for nuclear research[108]. Since I last wrote, we have had a visit from Delbrück who, as you know, is staying in Göttingen this summer. The visit gave rise to many instructive and interesting discussions also with Danish biochemists and biologists. It really seems as though we are beginning to get a basis for an explanation of some of the stages in the rhythm peculiar to cell division, but at the same time it also appears to me that features of wholeness are becoming ever clearer. In fact, I feel ever more convinced that the general reflections about organic development, about which I wrote to you in January, hit quite an important point. However important it is, on the one hand, that a directive is transmitted from cell to cell via duplication of simple chains of well-defined chemical molecules, it seems clear that during the embryological development there is continually a feedback ensuring the wholeness of the individual. Prior to the appearance of the special reduction processes determining the formation of the gametes, a message must thus have been received about the maturing of the organism and indeed also about the requirements life sets on its reactions with regard to external conditions. You see that I do not feel in any way defeated by the criticism of the mechanistic school, but that I am just trying to learn as much as possible about the various aspects of the situation. Actually, I must confess that, however amateurish it might be regarded by many professional biologists, I am still trying to dream myself deeper into the rhythm of life and the entire interplay within organisms, and I feel, however modest the result will be, that the possibility is approaching of achieving a reasonable epistemological approach to essential distinctive features of the problem of life. I hope to be able to give a clearer account of such questions in a lecture, "On the Unity of Knowledge", at the Columbia University Bicentennial Celebrations in the autumn[109]. As the manuscript for the lecture should preferably be submitted already in the middle of June, I have begun the preparations and will try to finish the manuscript as soon as I return from the journey, which Margrethe and I hope to combine with a couple of weeks' much-needed holiday in Italy.

I very much hope that we will soon have an opportunity really to discuss all these questions, and I understood from Gustafson[110] that there is a possibility that you will visit Lund already in May; if so, we trust that you will also visit the CERN group in Copenhagen. In the hope of seeing you soon, and with kindest wishes to Franca and yourself from Margrethe and

your old
[Niels Bohr]

PS. Unfortunately I will have left before Stern comes here on his way to Stockholm, but both Margrethe and I greatly hope that he will soon come here again after our return.

BOHR TO PAULI, 7 February 1955
[Carbon copy]

[København,] 7. februar 1955.

Kære Pauli,

Jeg er ked af ikke at have skrevet så længe, men min tid har været meget optaget siden Margrethe og jeg kom hjem fra Amerika lige før jul. Det var hyggeligt at være ved Princeton Instituttet igen, og det var en særlig glæde at se hvor roligt og værdigt Robert har taget atomenergikommissionens afgørelse. Både han og Kitty var faktisk gladere end de længe har været, og det har betydet en stor opmuntring for dem at føle hvor gode venner de har ikke alene blandt fysikerne, men også i vide kredse af det amerikanske samfund, for hvilke den urimelige sag har været et stort chock. For Robert selv håber vi nu alle at han vil få mere ro til videnskabeligt arbejde, og han er jo som hele kredsen i Princeton dybt interesseret i de mange problemer som experimenterne med elementarpartiklerne stadig gør mere aktuelle. Mens jeg var derovre havde vi også mange diskussioner om de almindelige biologiske og psykologiske spørgsmål, i hvilke blandt andre Delbrück deltog. I den sidste tid er jeg mere og mere optaget af spørgsmålet om den organiske udvikling og har i den forbindelse tænkt nærmere over irreversibiliteten med hensyn til hvilken der består en større principiel forskel mellem situationen i fysikken og biologien end man ofte gør sig klart. Jeg arbejder på en artikel[111] om disse

---

[111] This most likely also refers to the Steno Lecture. See ref. 87.

spørgsmål, som jeg håber snart[112] at kunne sende dig. Idag vedlægger jeg imidlertid kun et eksemplar af en tale[113], som jeg holdt ved den i forbindelse med Columbia Bicentennial sidste oktober afholdte konference om Unity of Knowledge og hvor repræsentanter for naturvidenskaberne såvel som de humanistiske videnskaber deltog. Trods vanskelighederne for en fysiker ved stadig kort at måtte gentage kendte ting forsøgte jeg efter bedste evne at give udtryk for en almindelig logisk indstilling og jeg skal være taknemmelig for at høre om du er helt utilfreds med resultatet af mine bestræbelser. Iøvrigt længes vi alle meget efter at du snart vil besøge os i København og tale til CERN gruppen og at Franca, som jeg håber er helt rask igen, vil komme med.

Med de hjerteligste hilsner og bedste ønsker for det nye år til jer begge,

Din gamle
[Niels Bohr]

Translation

[Copenhagen,] February 7, 1955

Dear Pauli,

I am sorry that I have not written for such a long time, but my time has been very occupied since Margrethe and I came home from America just before Christmas. It was nice to be at the Princeton Institute again, and it was a special pleasure to see how calmly and nobly Robert has accepted the decision of the atomic energy commission. Both he and Kitty were in fact happier than they have been for a long time, and it has been a great encouragement for them to feel that they have such good friends, not only among physicists, but also in wider circles of American society, for whom the preposterous case has been a big shock. As for Robert himself, we all now hope that he will have more peace for scientific work, and like the whole circle at Princeton he is of course deeply interested in the many problems that the experiments with elementary particles make ever more immediate. While I was there we also had numerous discussions about general biological and psychological questions, in which Delbrück, among others, took part. Recently I have become increasingly absorbed by the question of organic evolution, and in this connection I have thought more closely about irreversibility, with regard to which there is a greater fundamental difference between the situation in physics and in biology

---

[112] Pauli refers to this word ("snart") in his subsequent letter.
[113] See ref. 109.

than one is often aware of. I am working on an article[111] about these questions which I hope to be able to send you soon[112]. Today, however, I only enclose a copy of a talk[113] I gave at the conference about Unity of Knowledge held in connection with the Columbia Bicentennial last October, where representatives of the sciences as well as of the humanities participated. Despite the difficulties for a physicist continually having briefly to repeat known facts, I tried to the best of my ability to express a general logical approach, and I should be grateful to hear whether you are totally dissatisfied with the result of my efforts. Otherwise, we are all greatly longing for you to visit us soon in Copenhagen and talk to the CERN group, and for Franca, whom I hope is quite well again, to come too.

With kindest regards and best wishes for the New Year to both of you,

Your old,

[Niels Bohr]

PAULI TO BOHR, 15 February 1955
[Typewritten with handwritten corrections]

Physikalisches Institut
der Eidg. Technischen Hochschule
Zürich

ZÜRICH 7/6, February 15th, 1955
Gloriastraße 35

Dear Bohr,

It is with great pleasure that I received your nice letter and above all, the text of your lecture on "Unity of Knowledge". The general outlook of it is of course the same as mine. Under your great influence it was indeed getting more and more difficult for me to find something on which I have a different opinion than you. To a certain extent I am therefore glad, that eventually I found something: the definition and the use of the expression "detached observer", which appears on page 10 above of your lecture and which reappears on page 13 in connection with biology. According to my own point of view the degree of this "detachement" is gradually lessened in our theoretical explanation of nature and I am expecting further steps in this direction.

1) As you will see in the reprint[114] on my lecture on "probability and physics", which I have sent to you, it seems to me quite appropriate to call the conceptual description of nature in classical physics, which Einstein so emphatically wishes to retain, "the ideal of the detached observer". To put it drastically the observer has according to this ideal to disappear entirely in a discrete manner as hidden spectator, never as actor, nature being left alone in a predetermined course of events, independent of the way in which the phenomena are observed. "Like the moon has a definite position" Einstein said to me last winter, "whether or not we look at the moon, the same must also hold for the atomic objects, as there is no sharp distinction possible between these and macroscopic objects. Observation cannot *create* an element of reality like a position, there must be something contained in the complete description of physical reality which corresponds to the *possibility* of observing a position, already before the observation has been actually made." I hope, that I quoted Einstein correctly; it is always difficult to quote somebody out of memory with whom one does not agree. It is precisely this kind of postulate which I call the ideal of the detached observer.

In quantum mechanics, on the contrary, an observation hic et nunc changes in general the "state" of the observed system in a way not contained in the mathematically formulated *laws*, which only apply to the automatical time dependence of the state of a *closed* system. I think here on the passage to a new phenomenon by observation which is technically taken into account by the socalled "reduction of the wave packets". As it is allowed to consider the instruments of observation as a kind of prolongation of the sense organs of the observer, I consider the impredictable change of the state by a single observation – in spite of the objective character of the result of every observation and notwithstanding the statistical laws for the frequencies of repeated observation under equal conditions – to be *an abandonment of the idea of the isolation (detachment) of the observer from the course of physical events outside himself.*

To put it in nontechnical common language one can compare the role of the observer in quantum theory with that of a person, who by its freely chosen experimental arrangements and recordings brings forth a considerable "trouble" in nature, without being able to influence its unpredictable outcome and results which afterwards can be objectively checked by everyone.

---

[114] W. Pauli, *Wahrscheinlichkeit und Physik*, Dialectica **8** (1954) 112–124.

Probably you mean by "our position as detached observers"[115] something entirely different than I do, as for me this new relation of the observer to the course of physical events is entirely *identical* with the fact, that "our situation as regards objective description in this field of experience" gave rise to the demand "of a renewed revision of the foundation for the unambiguous use of our elementary concepts", logically expressed by the notion of complementarity.

2) Passing now from physics to other sciences like psychology and particularly biology I am most interested in your approach, which certainly seems to me to go in the right direction. Without entering a discussion of the dependence of such concepts as "snart"[116], not only on the state of motion but also on the psychological attitude of the observer, I am very much looking forward to your article on the organic evolution which you announced in your letter.

In discussions with biologists I met large difficulties when they apply the concept of "natural selection" in a rather wide field, without being able to estimate the probability of the occurrence *in a empirically given time* of just those events, which have been important for the biological evolution. Treating the empirical time scale of the evolution theoretically as infinity they have then an easy game, apparently to avoid the concept of purposiveness. While they pretend to stay in this way completely "scientific" and "rational", they become actually very irrational, particularly because they use the word "chance", not any longer combined with estimations of a mathematically defined probability, in its application to very rare single events more or less synonymous with the old word "miracle". I found for instance *H.J. Muller* very characteristic for this school of biologists (see also his recent article "Life" in *"Science"*, issue of January 7, 1955, which certainly contains very interesting material), but also our friend Max *Delbrück*. With him this is combined with vehement emotional affects and a permanent threat to run away which I interpret as obvious signs of overcompensated doubts.

You can imagine how much better than "natural selection" sounds for me "natural evolution"[117] which I never heard before from you and which you use

---

[115] The three quotations in this paragraph are taken from Bohr's manuscript *Unity of Knowledge*, ref. 109. They were subsequently reproduced verbatim in Bohr's published article of the same name, ref. 109, p. 54, this volume, p. [90].

[116] "Snart" is the Danish word for "soon". Pauli is being ironic as regards Bohr's letter of 7 February 1955, where Bohr writes, "I am working on an article about these questions which I hope to be able to send you soon" (see ref. 112).

[117] This and the next quotation are taken from Bohr's manuscript and were repeated verbatim in Bohr's article, this volume, pp. [97] and [93]. See ref. 115.

now on page 19 of your lecture. I hope that your announced article will tell us more about your use of the latter concept.

Concluding this letter, I add some remarks about your sentence on page 14 concerning the "medical use of psychoanalytical treatment in curing neurosis". I am quite glad about this sentence, as logic is always the weakest spot of all medical therapeuts, who never learned the rigorous logical demands of mathematics.

Historically the word "the unconscious" was used by German philosophers of the last century, particularly by *E. von Hartmann*, (also E.G. Careus), developing further older allusions of Leibniz and Kant. The Psycholamarckist *A. Pauly*[118], on whom we spoke already, quoted von Hartmann in 1905 (Freud was not known to him), when he called processes of biological adaptation, already in plants, an *"unconscious* judgement of the psyche of the organisms". In this way however, only a new name was introduced, which did not explain anything. Freud was the first who made practical applications of the unconscious replacing hereby this word by "subconsciousness", which you also apply. With this change of the word Freud wanted to emphasize that all "contents of the subconsciousness" were earlier in the consciousness and had been suppressed ("verdrängt") afterwards. In this way Freud's subconsciousness was like a bag containing a finite number of objects. The purpose of the psychoanalytical treatment was therefore to make this bag again empty by upheaval of the suppression.

To this restricted concept of subconsciousness C.G. Jung is among others in opposition since about 1913. He reestablished the older word the unconscious of the philosophers emphasizing that every change of consciousness for instance in a medical treatment, also changes backwards the unconscious, which therefore can never be made "empty of contents", only a small part of which has ever been in consciousness. The aim of the medical treatment according to Jung and his school is therefore the establishment of a correct and sound "equilibrium between consciousness and the unconscious", like an equilibrium between two powers. This process in which this equilibrium is reached or reestablished, they also call "the assimilation or integration of the unconscious to the consciousness".

I only refer here historically a situation without identifying myself with this kind of terminologies, which seem to me rather far from logical clarity. The Jung-school is more broad minded than Freud has been, but correspondingly

---

[118] August Pauly (1850–1914), German zoologist. See ref. 84.

also less clear. Most unsatisfactory seems to me the emotional and vague use of the concept of "Psyche" by Jung, which is not even logically selfconsistent.

I am very glad about the prospect of a visit in Copenhagen in the autumn of this year, when also your 70th birthday will be celebrated. Francas treatment is not yet finished entirely, but she is much better and there is much hope, that she also will be able to go to Copenhagen this next time.

With all good wishes from both of us to yourself, Margrethe and the whole family,

yours old

W. Pauli

Where your lecture on "Unity of Knowledge" will be printed in case I like to quote it?

BOHR TO PAULI, 2 March 1955.
[Carbon copy]

[København,] March 2, 1955

Dear Pauli,

On my return from the CERN meeting in Geneva, I am writing to thank you for your letter of February 15th. It was very good of you to write me so carefully about your reaction to my article and, as always, you touch upon a very central point. A phrase like "detached observer" has of course like all words different linguistic and emotional aspects, but using it in connection with the phrase "objective description", taken as theme of the discussion on Unity of Knowledge, it had to me a very definite meaning. In all unambiguous accounts it is indeed a primary demand that the separation between the observing subject and the objective content of communication is clearly defined and agreed upon. The aim of the article is just to stress that this condition is indispensable in all scientific knowledge, including biology and psychology, while in art as well as in religious belief one allows oneself to neglect or rather tacitly to shift such separation. In this connection, the historical information in your letter about the use of terminology by psychologists was very valuable to me, and I was glad that you on the whole sympathize with my approach. Indeed, contrary to what some of our common friends seem to believe of me, I have always sought scientific inspiration in epistemology rather than in mysticism, and how

[567]

horrifying it may sound, I am at present endeavouring by exactitude as regards logic to leave room for emotions.

It is on this background that it seems to me very important that we fully understand each other in questions of terminology. Of course, one may say that the trend of modern physics is the attention to the observational problem and that just in this respect a way is bridged between physics and other fields of human knowledge and interest. But it appears that what we have really learned in physics is how to eliminate subjective elements in the account of experience, and it is rather this recognition which in turn offers guidance as regards objective description in other fields of science. To my mind, this situation is well described by the phrase "detached observer", and it seems to me that your reference to our controversy with Einstein is hardly relevant in this connection. Just as Einstein himself has shown how in relativity theory "the ideal of the detached observer" may be retained by emphasizing that coincidences of events are common to all observers, we have in quantum physics attained the same goal by recognizing that we are always speaking of well defined observations obtained under specified experimental conditions. These conditions can be communicated to everyone who also can convince himself of the factual character of the observations by looking on the permanent marks on the photographic plates. In this respect, it makes no difference that in quantum physics the relationships between the experimental conditions and the observations are of a more general type than in classical physics. I take it for granted that, as regards the fundamental physical problems which fall within the scope of the present quantum mechanical formalism, we have the same view, but I am afraid that we sometimes use a different terminology. Thus, when speaking of the physical interpretation of the formalism, I consider such details of procedure like "reduction of the wave packets" as integral parts of a consistent scheme conforming with the indivisibility of the phenomena and the essential irreversibility involved in the very concept of observation. As stressed in the article, it is also in my view very essential that the formalism allows of well defined applications only to closed phenomena, and that in particular the statistical description just in this sense appears as a rational generalization of the strictly deterministic description of classical physics.

I am eager to learn your reaction to these points as I feel that it is essential, not least for the approach to the wider problems on which we are working, to be as precise as possible in terminology, and above all to avoid any vagueness as to the demands of objective description. It was a great joy to learn that we can expect a visit of you and Franca in the autumn and perhaps there is an opportunity of meeting you even earlier, since I am invited to give a talk in Basel at the end of March on the general epistemological problems.

With kindest regards and best wishes to you both from us all,

Yours, in every way, old

[Niels Bohr]

PAULI TO BOHR, 11 March 1955
[Typewritten with handwritten corrections]

Physikalisches Institut
der Eidg. Technischen Hochschule
Zürich

ZÜRICH 7/6, March 11th, 1955.
Gloriastraße 35

Dear Bohr,

I find your letter of March 2nd very youthfull, which is just the reason that it is not easy for me to answer. Although we have the same view "as regards the fundamental physical problems which fall within the scope of the present quantum mechanical formalism"[119] and although I agree with some parts of your letter, the situation is now complicated by your use in a publication of a phrase like "detached observer" (without comment!) which I used already in some publications in a very different way. I believe that this should be better avoided to prevent a confusion to the readers* and I don't cling at all to particular words myself. I also felt, already before your letter arrived, that my brief characterisation of the observer in quantum theory as "non-detached" is in one important respect misleading. As is well known to both of us, it is essential in quantum mechanics that the apparatus can be described by classical concepts. Therefore the observer is always entirely detached to the *results* of his observations (marks on photographic plates etc.), just as he is in classical physics. I called him, however, in quantum physics *"non-detached"*, *when he chooses his experimental arrangements*. I still believe today, that this more restricted use of my terminology is very good and that it has been unhappily obscured in your article in a non-logical way!

I shall try to make my point logically clear, by defining my concepts, replacing hereby the disputed phrase by other words. As I was mostly interested in

---

[119] Quotation from previous letter.

the question, *how much informative reference to the observer an objective description contains*, I am emphasizing that a communication contains in general *informations on the observing subject*.

Without particularly discussing the separation between a subject and the informations about subjects (given by themself or by other persons), which can occur as elements of an "objective description", I introduced a concept "degree of detachment of the observer" in a scientific theory to be judged on the kind and measure of informative reference to the observer, which this description contains. For the objective character of this description it is of course sufficient, that every individual observer can be replaced by every other one which fulfills the same conditions and obeys the same rules. In this sense I call a reference to experimental conditions an "information on the observer" (though an impersonal one), and the establishment of an experimental arrangement fulfilling specified conditions an "action of the observer" – of course not of an individually distinguished observer but of "the observer" in general.

In physics I speak of a detached observer in a general conceptual description or explanation only then, *if it does not contain any explicit reference to the actions or the knowledge of the observer*. The ideal, that this should be so, I call now "the ideal (E)" in honor of Einstein. Historically it has its origin in celestial mechanics.

There is an important *agreement* between us that we find Einstein not consequent in this formulation of the "ideal E". Indeed, there is no a priori reason whatsoever to introduce here a difference between the *motion* of the observer on the one hand, and the realization of specified experimental conditions by the observer on the other hand. If Einstein were consequent he had to "forbidd" also the word coordinate system in physics as being "not objective". That the situation in quantum mechanics has a deep similarity with the situation in relativity is already shown by the application of mathematical groups of transformation in the physical laws in both cases.

In this way I reached the conclusion to distinguish sharply between the "ideal of an objective description" (in a wider sense) on the one hand (which I warmly supported just as you do) and the "ideal of the detached observer" on the other hand (which I rejected as much too narrow).

What really matters for me is not the word "detached" but the more *active* role of the observer in quantum physics, which is *already implied* in your constatation of the "indivisibility of the phenomena and the essential irreversibility involved in the very concept of observation"[120]. According to quantum physics

---

[120] Ibid.

the observer has indeed a new relation to the physical events around him in comparison with the classical observer, who is merely a spectator: The experimental arrangement freely chosen by the observer lets appear *single* events *not* determined by laws, the ensembles of which are governed by *statistical* laws. In this way one obtains just the logical foundations of an "*objective* description" of the "incidents" (Eingriffe) which the quantum-mechanical observer makes in its surroundings with his experimental arrangements. [Attention: there is *no* logical contradiction between a word like "trouble" and the possibility of its "objective" observation and description!][121]. It is not relevant to me, if you say the same thing using *different* terminologies (but please use then really *different* words than I). They will only confirm my statements again as all these statements on the observer are part of an "objective description".

I confess, that very different from you, I do find sometimes scientific inspiration in mysticism** (if you believe that I am in danger, please let me know), but this is counterbalanced by an *immediate* sense for mathematics. The result of both seems to be my kind of physics, whilst I consider epistemology merely as a logical comment to the application of mathematics in physics***. Thus when I read a sentence as "how to eliminate subjective elements in the account of experience"[122] my immediate association is "group theory" which then determines my whole reaction to your letter. Although the first step to "objectivity" is sometimes a kind of "separation", this task excites in myself the vivid picture of a superior *common order* to which all subjects are subjugated, mathematically represented by the "laws of transformations" as the key of the "map", of which all subjects are "elements".

I hope that it will be possible to find a terminology which will turn out to be satisfactory for both of us, but it is no hurry with it. I propose to resume this discussion only when your *new* article will be ready, which I am eagerly awaiting. It will show me your terminologies in more general cases of objective descriptions, of which I am most interested in the application to biology, in connection with your new expression "natural evolution".

From March 16th till about 27th I am away in Germany and Holland and when I come back I hope either to see you or to hear from you (I wrote to Basel to get informations on your lecture there)****.

Hoping that you will in future (just as I do myself) enjoy the enrichment coming from the different kind of access to science by different scientists, expressed in different, but not contradicting terminologies, I am sending, also

---

[121] This and the subsequent pair of square brackets (in a footnote) are Pauli's.

[122] Quotation from previous letter.

in the name of Franca, all good wishes to yourself, to Margrethe and to the whole family.

as yours complementary old

W. Pauli

* An explaining remark about it in your *new* article would be most welcome!

** By the way: the "Unity" of everything has always been one of the most prominent ideas of all mystics.

*** We are here *both* in our letters in a realm of informations on the writing subject, which do *not* belong to the *"objective* content of the communications".

**** Meanwhile I heard from P. Huber[123], in Basel, [*Fierz*[124] is in the United States], that your lecture there is on March 30. On this date I am very glad, because I shall be back from my trip by then.

Paa Gensyn![125]

BOHR TO PAULI, 25 March 1955
[Carbon copy]

[København,] March 25, 1955

Dear Pauli,

Thanks for your letter of which I was glad to see that even if I am old you do not feel that I am yet so petrified that we cannot have such animated and fruitful discussions as in our younger days. You are certainly right that, as regards many personal utterances, in my letters like in yours, we are not detached on-lookers, although of course we have so much in common that it is a pure discussional accident which words, like mysticism or logical systematism, the one or other of us uses for mutual educational purposes. I also read with great pleasure your beautiful Columbia radio-lecture[126] which I had not seen before. Of course, I appreciate the background for your use of the phrase "detached observer" on that occasion, but in my article I was using the phrase in a more generalized

---

[123] Paul Huber (1910–1971), Swiss physicist.

[124] See ref. 96.

[125] Roughly translated into English as, "I hope to see you soon!"

[126] W. Pauli, *Matter* in *Man's Right to Knowledge*, International Symposium Presented in Honor of the Two-Hundredth Anniversary of Columbia University, 1754–1954, H. Muschel, New York 1954, pp. 10–18. Pauli informed Bohr of this publication in a separate undated note (wrongly dated in the BSC as "between 6 and 26 April 1954"), which may have been an enclosure to his letter of 11 March 1955.

(or, if you prefer, more limited) sense suited to point to the characteristics of our position in science and art.

To characterize scientific pursuit I did not know any better word than detachment, especially in connection with psychological studies. As regards quantum mechanics and biology, I wanted to stress the difficulties which even in these fields have had to be overcome to reach the detachment required for objective description or rather for the recognition that in such fields we meet with no special observational problem beyond the situations of practical life to cope with which the word "observer" has been originally introduced. To my mind, the lesson was merely that continued exploration of the regularities of nature only gradually should teach us of the necessary caution in looking for unambiguously communicable experience. As you, I am of course prepared to change terminology when it is clear that this will promote common understanding, but before any of us decides on such steps, I wish to call your attention to the pure scientific aims I have perhaps not sufficiently clearly presented in my article[127].

To make myself more clear, I may for a moment remind of the days of so-called "classical" physics and "critical" philosophy, when in the description of the course of events the role of the tools of observation was disregarded and space–time coordination and causality were considered a priori categories. You are certainly right that Einstein is not consequent when speaking of the ideal of detached observer and neglecting his own wisdom of relativity, which Eddington poetically described by the picture of how long man traced a footprint in the sand until he recognized that it was his own. Seriously, I mean that you are yourself as inconsequent in stressing the difference in such respect between classical and quantum mechanics. It is true that, before the epistemological aspects of the observational problem were so widely cleared up, a certain confusion was prevalent, but after the thorough lesson which we have received, the whole situation including that of classical mechanics appears in a new light. Though in a vast field of experience one could neglect the interaction between what was regarded as separate objects of investigation and tools of observation, one often overlooked our reach of interfering with the course of events through our freedom of choosing the experimental arrangement. Indeed, in those days, relying upon the deterministic and reversible character of the mechanical description, one might rather have thought that such influencing within a large scope was possible in unlimited detail.

On the basis of the recognition of the limited divisibility of elementary phenomena we have, however, obtained a more generalized description embracing

---

[127] See ref. 109.

new fundamental regularities of nature, the orderly comprehension of which in principle implies statistical account even as regards reversibility, and which for the exhaustion of knowledge demands mutually exclusive experimental arrangements. The point which I especially wanted to stress in the article is that, just by avoiding any such reference to a subjective interference which would call for misleading comparison with classical approach, we have within a large scope fulfilled all requirements of an objective description of experience obtainable under specified experimental conditions. The freedom of the choice of the experimental arrangement is indeed common to classical and quantum physics and, considering all aspects of the situation, we may say that in both cases a sharp separation between the "observer" and the "phenomena" is retained. The difference is only that in quantum phenomena we have for their definition to include the description of the whole experimental arrangement and that we have less possibility of influencing the course of events.

Still, if the study of natural phenomena were exhausted by simple experience, one might not take questions of terminology too serious, at any rate within the scope in which order is already obtained. I want, however, to challenge you whether you really mean that, in the description of proper biological phenomena, we have to do with an even greater interference with events on the part of the observer than that you want to stress in quantum mechanics. To my mind, the situation is entirely opposite, since the characteristics of our position in biological studies is just the impossibility without excluding the display of life to arrange the experimental conditions required for well defined mechanistic description. It is of course true that physiological research just consists in studying the reactions of the organism to experimental conditions open to our choice, but it appears to me to be practical as well as rational to include such reactions under varied specified conditions in an exhausive account of organic life. A further point which in this connection is on my mind is to stress that, in the description of the characteristic properties of the organism, reversal of events is logically excluded, and just this circumstance is of course of fundamental importance for speaking of "natural evolution".

Perhaps you may think that all this talk has very little to do with our dispute and that I am only going around the issue. It is, however, far from my intention and I have even, without any success, made a great effort to find another word than "detached", better suited to express our position as observers. Indeed, I really think that our divergency is more related to the use of the word observation itself, with which I simply understand a recording which is unambiguously communicable in common language without requiring any further creative treatment. How I might relate such utterance with my favoured comparison of mathematics to a game with the only rule of not explicitly referring to oneself, I leave to you to ponder about. I shall here not go further in repeating things which are not new to any of us and postpone the battle about the word "detached" to our meeting in Basel in preparation of which I only wanted to remind of our resources for defence as well as attack, irrespective of the word "old"[128].

With kindest greetings from home to home,

and paa gensyn[130]

Yours ever

[Niels Bohr]

EDGAR RUBIN

BOHR TO RUBIN, 20 May 1912

[Handwritten]

Hulme Hall, Manchester. 20-5-12.

Kære Edgar!

Du maa undskylde at jeg igen forstyrrer Dig med mit Vrøvl; jeg har heller slet ikke Tid selv; men jeg kan ikke faa Ro i mit Sind, før jeg har spurgt Dig, om der er nogensomhelst Mening i at sige noget saadant: Genkendelse (der efter sin Karakter er rent kvalitativ; og hvor der er et nyt ydre Indtryk til at bryde Hemningerne) afhænger primært kun af Tilstanden paa de 2 Tidspunkter, og kun sekundært af Tilstanden imellem (igennem en betinget eller naturlig Udviskning); medens Reproduction (der maaske er mere kvantitativ (selektiv) ???, og hvor Hemningerne skal brydes "indefra") primært afhænger baade af Tilstanden paa Indtrykkets Tidspunkt og af Hemninger foranlediget af Tilstanden i Reproductions-Øjeblikket og Tiden imellem. (Reproduction kan jo fremhjælpes med stimulerende Midler som Alkohol eller Hypnose (i saadanne Tilfælde, hvor Genkendelse vilde finde sted ???)). Kære Edgar, Du maa ikke tro, at jeg ikke forstaar, at saadant noget kun kan siges, med Chance for at være rigtigt eller sammenhængende, efter et virkeligt Studium og Kritik, eller at jeg er bleven en vild og rasende Dialektiker; Sagen er snarere, at det er

---

[128] The BSC contains two different versions of this last paragraph of the letter. The editors are grateful to CERN Archivist Roswitha Rahmy for confirming that the version here reproduced was the one that Pauli actually received.

[130] See ref. 125.

saa længe siden jeg har haft Lejlighed til at tale om saadan noget, og at mit ringe Kendskab i det engelske Sprog ikke tillader mig Dyrkelsen af nogen slags Dialektik her; og at det derfor var som at lægge en Lunte til en Tønde Krudt. Jeg har som sagt kun skrevet det, fordi jeg ikke har Tid til at lade være med at glemme det, indtil jeg engang paa en Tur faar at vide hvormeget jeg har misforstaaet af det.

De venligste Hilsener fra Niels.

Jeg begynder igen, men blot for at sige, at det jo kun er *en* Maade af mange at prøve at sige det paa og at jeg allerede tror at jeg ved, paa hvilket Punkt det er galt.
(Undskyld Skriften, og bryd Dig ikke om dersom Du ikke kan læse det.)

Translation

Manchester, May 20, 1912

Dear Edgar,

Forgive me for disturbing you once again with my nonsense; I do not at all have time for it myself either; but my mind will not be at rest until I have asked you whether it makes any sense at all to say something like this: recognition (which according to its nature is purely qualitative; and where there is a new external impression to break down inhibitions) depends primarily only on the state at the 2 points in time, and only secondarily on the state in between (via conditioned or natural effacement); whilst reproduction (which perhaps is more quantitative (selective) ??? and where inhibitions must be broken down "from within") primarily depends both on the state at the point in time of the impression and on inhibitions brought about by the state at the moment of reproduction and the time in between. (Reproduction can, of course, be facilitated by stimulants such as alcohol or hypnosis (in such cases where recognition would take place ???)). Dear Edgar, you must not think that I do not understand that such a thing only can be said with a chance of it being correct or coherent after a veritable study and critique or that I have become a wild and raving dialectician; the truth is rather that it is such a long time since I have had the opportunity to speak about such things; and that my poor command of the English language does not allow me to practise any kind of dialectics here; and that it therefore was like setting a fuse to a barrel of gunpowder. I have, as I said, only written it down because I haven't got the time to stop forgetting it, until one day on a walk I will hear how much of it I have misunderstood.

Kindest regards from Niels

I am starting again, but just to say that it is indeed only *one* way out of many of trying to say it, and that I already think I know on which point it is wrong. (Please excuse my writing, and don't worry if you can't read it.)

INVENTORY OF RELEVANT
MANUSCRIPTS
IN THE NIELS BOHR ARCHIVE

INTRODUCTION

The following documents are mainly from the years 1928–1962, the period covered by the present volume.

Unless otherwise noted, the folders listed form part of the Bohr MSS, the originals of which are retained in the NBA. In each case, the microfilm number is provided (e.g. "mf. 11").

The titles of the folders have been assigned by the cataloguers of the collection, as have all dates in square brackets. Unbracketed dates are taken from the manuscripts themselves.

Whenever an item has been reproduced in the present volume, the relevant page numbers are provided in the margin. When only an excerpt has been printed, these numbers are followed by the letter E. In the one case when a tape-recorded lecture by Bohr has been retranscribed by the editors, this is indicated with an R in the margin.

Each item is provided with an explanatory note in small print. A page number at the end of such a note (e.g. "Cf. p. [221]") indicates that the item provided the basis for the publication reproduced in this volume beginning on that particular page.

1   *Tale ved Studenterjubilæet*   21 September 1928

Typewritten, carbon copy, 25 pp., Danish and English, mf. 11.

Speech at the reunion of graduating students from 1903. Manuscript, photostat and translation. Cf. p. [223].

xxxv E   2   *Kausalität und Objektivität*   [1929]

Typewritten and handwritten [N. Bohr and O. Klein], 47 pp., German and Danish, mf. 12.

Drafts and notes for unpublished papers. Main titles: "Kausalität und Objektivität" (Causality and Objectivity) and "Quantentheorie und Anschaulichkeit" (Quantum Theory and Visualizability). One page is dated 11 September 1929.

3   *Philosophical Aspects of Atomic Theory*   26 May 1930

Typewritten, carbon copy, handwritten [Margrethe Bohr], 50 pp., English, mf. 12.

(a) Transcript of address delivered before the Royal Society of Edinburgh on 26 May 1930 at the award of James Scott prize. (b) Notes for the lecture.

4   *International High School*   3 August 1931

Typewritten, carbon copy, with handwritten corrections [unidentified], 20 pp., English, mf. 12.

Envelope labelled: "Stenografisk Referat af Foredrag ved Internationale Højskole i Helsingør 3–8–1931. Philosophical aspects of atomic theory." (Shorthand report of lecture delivered at the International High School at Elsinore, 3 August 1931).

5   *H. Høffding's Views on Physics and Psychology*   [August 1932]

Typewritten, carbon copy, with handwritten corrections [N. Bohr], 7 pp., English, mf. 13.

Notes of lecture given before the 10th International Psychology Congress at Carlsberg Honorary Mansion, Copenhagen, August 1932.

6   *Activities of Life and Thermodynamics*   [late summer 1934]

Handwritten [O. Klein and Betty Schultz], 5 pp., English and Danish, mf. 13.

Notes and parts of drafts of unpublished paper prepared in connection with a discussion in Nature.

7   *Lecture before Warburg Society*   [Spring 1936]
Handwritten [N. Bohr and F. Kalckar], 2 pp., English, mf. 14.

Outline of lecture on some humanistic aspects of atomic theory. Note (N. Bohr): "Disposition til foredrag i Warburg Society i London, Foråret 1937 [1936] … i diskussionen deltog Rutherford." (Outline of lecture before the Warburg Society in London, Spring 1937 [1936] … Rutherford took part in the discussion).

8   *Causality and Complementarity*   [21 – 26 June 1936]
Typewritten, carbon copy, 23 pp., English, mf. 14.

Draft and manuscript of report of lecture delivered before the Second International Congress for the Unity of Science, Copenhagen, 21 – 26 June 1936. Cf. p. [37].

9   *Journey to U.S.A. and Japan*   [17 January – June 1937, 20 November]
Carbon copy and handwritten [Hans Bohr, N. Bohr and F. Kalckar], 50 pp., English and Danish, mf. 14.

Brief notes concerning the journey to U.S.A., Japan and China, 1937. (a) Lecture notes, almost all dated. (b) Envelope labelled: "Foredrag holdt ved en Film fra Rejsen 20 Nov 1937" (Talk given at the showing of a film from the journey, 20 November 1937).

10   *Tribute to Rutherford*   20 October 1937
Handwritten [N. Bohr and L. Rosenfeld], 3 pp., English, mf. 14.

Photoprint of manuscript of tribute to Rutherford, Bologna, 20 October 1937 on the announcement of Rutherford's death.

11   *Relations to Rutherford*   [November 1937]
Carbon copy, handwritten [N. Bohr], 6 pp., English and Danish, mf. 14.

(a) Notes and part of draft for the obituary notice in Nature (Suppl.) **140** (1937) 1048. (b) Title: "Notes about my relations to Rutherford".

12   *Foredrag om Rutherford*   22 November 1937
Carbon copy, 3 pp., Danish, mf. 14.

Notes of lecture about Rutherford, delivered before the Physical Society, Copenhagen, 22 November 1937.

13   *The Causality Problem in Atomic Physics*   [April–October 1938]

Typewritten, carbon copy and handwritten [L. Rosenfeld and un-identified], 174 pp., English and Danish, mf. 15.

Notes, drafts and manuscript for a report of the lecture given at the Physical Congress, organized by the International Institute of Intellectual Co-operation in Warsaw 30 May–3 June 1938. The manuscript exists in two copies, one of them containing an additional footnote on p. 1. Part of the material is dated, the dates being in the period 28 April to 27 October 1938.

14   *Videnskapen og Livet*   [Spring 1939]

Typewritten, 21 pp., Norwegian, mf. 16.

Shorthand report of lecture on "Science and Life" given at Nansenskolen, Oslo.

15   *Analysis and Synthesis*   [1939–1942]

Typewritten, carbon copy and handwritten [Ernest Bohr, N. Bohr, Aage Bohr, S. Rozental, L. Rosenfeld and unidentified], English and Danish, 164 pp., mf. 16.

Notes and drafts for unpublished paper on analysis and synthesis. Almost all pages are dated.

16   *Kommentarer til filosofisk Afhandling*   [1939–1942]

Handwritten [S. Rozental], 4 pp., Danish, mf. 16.

(a) Notes commenting on a philosophical paper. (b) Sheet with N. Bohr's comments on some lines in the novel "Point Counter Point" by Aldous Huxley, 1928.

17   *Analyse og Syntese i Naturvidenskaben*   3 December 1940

Handwritten [N. Bohr], 2 pp., Danish, mf. 16.

Notes. Outline of lecture with title: "Filosofisk Forening Lund 3 December 1940 Analyse og Syntese i Naturvidenskaben". (Philosophical Society, Lund, 3 December 1940. Analysis and Synthesis in Science).

18   *Musikaften*   18 June 1941

Typewritten, carbon copy, 1 p., Danish, mf. 16.

Title (Margrethe Bohr): "Tale holdt af Niels ved en Musikaften den 18de Juni 1941" (Speech given by Niels at a musical evening, 18 June 1941).

19   *Erkendelsesproblemet i Atomfysikken*   20 February 1942
Typewritten, 12 pp., Danish, mf. 16.

Title: "Erkendelsesproblemet i Atomfysikken. Foredrag i 'Parentesen'" (The epistemological problem in atomic physics. Lecture given in Parentesen (the association for physics students at the University of Copenhagen)).

20   *Dansk Kultur*   7 June [1942]
Handwritten [Aage Bohr], 2 pp., Danish, mf. 16.

Notes for article "Dansk Kultur" (Danish Culture). Cf. p. [251].

21   *Foredrag for Kreds af Skolefolk*   [August 1942]
Handwritten [Aage Bohr], 4 pp., Danish, mf. 16.

Outline of lecture delivered before a group of educationalists.

22   *Høffdings efterladte Manuskript*   19 March 1943
Typewritten, carbon copy, 2 pp., Danish, mf. 16.

Address given at the meeting of the Royal Danish Academy of Sciences and Letters in connection with the reading of a manuscript left upon the death of Harald Høffding. Cf. p. [323].

XXVIII E

23   *Fysik og Biologi*   26 March 1946
Typewritten, 13 pp., Danish, mf. 17.

Outline and shorthand report of lecture on physics and biology; delivered before the Biological Society [Copenhagen].

24   *Atomet og Mennesket*   [1 May 1946]
Handwritten [Aage Bohr and S. Rozental], 4 pp., Danish, mf. 17.

Title: "Atomet og mennesket" (Atoms and man). Notes for a lecture given at the University of Copenhagen. Three pages dated 24 April or 1 May 1946.

25   *Epistemological Problems in Science*   11 June 1946
Handwritten [Aage Bohr and S. Rozental], 2 pp., English, mf. 17.

Notes, dated 10 June 1946, for lecture given at the University of Lund, Sweden, 11 June 1946 [1947].

26   *10. Skandinaviske Matematikerkongres*   29 August 1946
Typewritten, 9 pp., Danish, mf. 17.

Title: "Foredrag ved Matematikerkongressen 1946 holdt paa Carlsberg 29/8 1946."
(Lecture given at the Congress of Mathematicians held at Carlsberg 29 August
1946). Shorthand report with some gaps.

27   *Observational Problem in Atomic Physics*   [24 September
1946]
Typewritten, carbon copy and handwritten [unidentified] 29 pp.,
English and Danish, mf. 17.

(a) Title: "Observational Problem in Atomic Physics". Outline of lecture given 24
September 1946 at the Princeton Bicentennial Conference 23–25 September 1946.
(b) Transcript of dictograph record of the lecture and of the preceding words of
introduction.

28   *Foredrag for de studerende*   [November 1947]
Typewritten and handwritten [S. Rozental], 24 pp., Danish, mf. 17.

(a) Notes for lecture given to the students at the University of Copenhagen, dated
7 and 8 November 1947. (b) Shorthand report of the lecture.

29   *Steno-Forelæsning i Medicinsk Selskab (I)*   [February 1949]
Typewritten with handwritten corrections [N. Bohr and unidenti-
fied], 15 pp., Danish, mf. 18.

Report of Steno Lecture, February 1949, delivered before the Medical Society,
Copenhagen.

30   *Atomerne og vor erkendelse*   1 April 1949
Typewritten, carbon copy and handwritten [Margrethe Bohr, N.
Bohr, S. Rozental, Betty Schultz and unidentified] 157 pp., Danish,
mf. 18.

Notes concerning lecture "Atomerne og vor erkendelse" (Atoms and human
knowledge) given in the programme for schools on Danish radio 1 April 1949
and on Norwegian radio 7 April 1949. Published as feature article in the Danish
national daily Berlingske Tidende 2 April 1949. Published in abridged form in Vor
Viden **33** (1950) 123.

31   *Gifford Lectures*   21 October–11 November 1949

Typewritten and handwritten [S. Rozental and L. Rosenfeld], 37 pp., English, mf. 19.

Notes for the Gifford Lectures, a series of ten lectures given at the University of Edinburgh 21 October – 11 November 1949. Three versions of a typewritten synopsis of the lectures with the title: "Synopsis of Lectures. CAUSALITY AND COMPLEMENTARITY. Epistemological Lessons of Studies in Atomic Physics." Handwritten notes with dates in the period 25 October 1948–13 September 1949.

[174]–[181]

32   *Summary of Gifford Lectures*   [1949–1950]

Typewritten and handwritten [Margrethe Bohr, L. Rosenfeld and S. Rozental], 26 pp., English and Danish, mf. 19.

Notes for talk with title "Causality and Complementarity", intended to be broadcast by the British Broadcasting Corporation. Dated between 27 December 1949–25 January 1950.

33   *Mindefest for H.C. Ørsted*   9 March 1951

Typewritten, with handwritten corrections [unidentified], 23 pp., Danish, mf. 19.

Title: "Tale ved mindefest for H.C. Ørsted, Københavns universitet, den 9. marts 1951" (Speech at the commemoration for H.C. Ørsted, the University of Copenhagen, 9 March 1951). On the occasion of the 100th anniversary of Ørsted's death. A few minor changes introduced for the printing of the address are included. Cf. p. [341].

34   *Atomic Physics and the Problem of Life*   29 March 1951

Handwritten [Aage Bohr], 3 pp., English, mf. 19.

Outline of a lecture with the title: "Atomic Physics and the Problem of Life". Microbiological Symposium, 29 March 1951, held at the Institute for Theoretical Physics, University of Copenhagen.

35   *Ny Forskning og gammel Visdom*   [3 August 1951]

Typewritten and handwritten [Margrethe Bohr, N. Bohr and un-identified], 72 pp., Danish, mf. 19.

Typewritten part marked: "Ny Forskning og gammel Visdom." (New research and old wisdom.) (a) Outline and (b) Report of lecture given at the University of Reykjavik, 3 August 1951. (c) Notes made for publication in an Icelandic pamphlet. Apparently not published.

36   *Svar til Oluf Rothe om viljens frihed*   20–25 August 1951
Typewritten, carbon copy and handwritten [Aage Bohr], 28 pp.,
Danish, mf. 19.

(a) Press cutting: "Niels Bohr og viljens frihed" (Niels Bohr and freedom of the
will) by Oluf Rothe in the Danish national daily newspaper "Information", 18
August 1951. (b) Draft of newspaper article, intended as a reply. Dated between
20–25 August 1951.

37   *2. International Poliomyelitis Conference* [3 September 1951]
Typewritten, 6 pp., English, mf. 19.

Address at the opening session of the Second Poliomyelitis Conference (3–7
September 1951). Cf. p. [65].

38   *Objektivitet, kausalitet og komplementaritet*   [1952]
Typewritten, carbon copy and handwritten [N. Bohr and A. Pe-
tersen], 29 pp., Danish and English, mf. 20.

Notes for unpublished article. Title of typed part: "Objectivitet, kausalitet og
komplementaritet" (Objectivity, causality and complementarity). Part of the material
is dated 8–13 March 1952.

39   *Atomernes verden og den menneskelige erkendelse*   1952
Carbon copy and handwritten [A. Petersen and S. Hellmann],
33 pp., Danish, mf. 20.

Notes and manuscript with title: "Atomernes verden og den menneskelige
erkendelse" (The world of atoms and human knowledge). Typed pages dated
29–30 October.

40   *Congress of Radiology*   19 July 1953
Typewritten, 8 pp., English, mf. 20.

Two slightly different manuscripts for the opening address at the 7th International
Congress of Radiology. 19 July 1953. Cf. p. [73].

41   *Physical Science and the Study of Religions*   7 November
1953
Typewritten, carbon copy and handwritten [A. Petersen and un-
identified] 81 pp., English and Danish, mf. 20.

Notes for "Physical Science and the Study of Religions". The material is dated
between 15 June and 22 October 1953. Cf. p. [273].

42   *Rydberg's discovery of the spectral laws*   [1 July 1954]

Typewritten, carbon copy and handwritten corrections [A. Petersen and unidentified], 24 pp., English, mf. 20.

Apart from minor differences, the manuscript of: "Rydberg's discovery of the spectral laws". The carbon copy does not contain all the corrections. Cf. p. [371].

XXXVII, XLVII E   43   *Unity of Knowledge*   28 October 1954

Typewritten, carbon copy and handwritten [N. Bohr, A. Petersen and L. Rosenfeld], 262 pp., English and Danish, mf. 21.

Notes, drafts and manuscript of the address with the title: "Unity of Knowledge", delivered at the Conference on the Unity of Knowledge, arranged in connection with the Bicentennial of Columbia University at Arden House, New York, 28 October 1954. Most of the material is dated between 13 March 1954 and 6 January 1955. Cf. p. [79].

44   *Physical Science and Man's Position*   9 August 1955

Typewritten, carbon copy and handwritten [Margrethe Bohr, Aage Bohr, J. Lindhard, A. Petersen and S. Rozental], 248 pp., English and French, mf. 21.

(a) Notes for the printed version of Bohr's address at the International Conference on Peaceful Uses of Atomic Energy, arranged by the United Nations 9–19 August 1955. (b) French translation of the manuscript. (c) Text of a paragraph inserted in the address but not included in the published version. Part of the material dated between 1 July–4 August 1955. Cf. p. [99].

45   *Atomerne og den menneskelige erkendelse*   14 October 1955

Typewritten, carbon copy and handwritten [Aage Bohr, A. Petersen and L. Rosenfeld], 113 pp., Danish, mf. 21.

Notes etc. for lecture at the meeting of the Royal Danish Academy of Sciences and Letters, 14 October 1955. Part of the material dated between 29 September 1955 and 25 September 1956. Two copies of the draft dated 14 October 1955.

46   *Fjerde Nordiske Psykologmøde*   [25 June 1956]

Handwritten [A. Petersen], 19 pp., Danish, mf. 22.

Notes for the lecture given at the Fourth Scandinavian Meeting of Psychologists (25–30 June 1956). The material is dated between 6–25 June 1956.

47 *Atoms and Human Knowledge, Nicola Tesla* [10 July–7 November 1956]

Typewritten, carbon copy and handwritten [A. Petersen], 37 pp., English and Danish, mf. 22.

(a) Tribute at opening ceremony of The Nicola Tesla Conference, Belgrade, 10 July 1956. (b) Manuscripts. Title: "Atoms and Human Knowledge". Footnote: "Except for certain points, which have been elaborated, this article presents the content of the author's address delivered without manuscript at the Nicola Tesla congress, July 1956". One of the manuscripts is dated 7 November 1956. (c) Notes for the report of the lecture. Dated 27 October 1956.

48 *Human Genetics Congress* [1 August 1956]

Typewritten, carbon copy and handwritten [A. Petersen], 2 pp., English, mf. 22.

Part of draft of address given at the opening session of the First International Congress of Human Genetics (1–6 August 1956). Dated 26 July 1956.

XLVIII E

49 *Steno-Forelæsning i Medicinsk Selskab (II)* [August 1957]

Carbon copy and handwritten [A. Petersen], 43 pp., Danish, mf. 22.

Notes for the article "Fysikken og livets problem" (Physical Science and the Problem of Life) prepared for publication in 1957 (Schultz). The article is based on the Steno Lecture delivered before the Medical Society, Copenhagen in February 1949. Part of the material dated between 25 August 1953 and 2 August 1957. Cf. p. [113].

50 *Lecture at Macalester College* 11 December 1957

Carbon copy with handwritten corrections [A. Petersen] and handwritten [Margrethe Bohr], 23 pp., English, mf. 22.

(a) Notes of lecture given at Macalester College, Minnesota, 11 December 1957, at the award of the honorary degree of doctor of science. Words of introduction and conclusion by President C.J. Turck and Dean Huntley du Pre, respectively. (b) Notes for lecture. Dated 10 December 1957.

51 *Lecture at the University of Oklahoma* [13 December 1957]

Typewritten, carbon copy, 17 pp., English, mf. 22.

Title: "Atoms and Human Knowledge". Notes of lecture given at the University of Oklahoma. Cf. p. [191].

52 *Atomic Physics and Human Knowledge* 1957

Handwritten [Aage Bohr, Ernest Bohr, A. Petersen], 109 pp., Danish and English, mf. 22.

Various notes related to: "Analysis and Synthesis", "Light and Life", "Position and Terminology in Atomic Physics". Drafts for Introduction and Preface to "Atomic Physics and Human Knowledge" (Wiley 1958), "Postscriptum to Discussion with Einstein" dated 8 August 1957. Cf. p. [107].

[182]–[190] E,R 53 *Karl Compton Lectures M.I.T.* [23 May–28 August 1958]

Handwritten [N. Bohr, Margrethe Bohr and A. Petersen], 165 pp., English and Danish, mf. 22.

Lecture notes on "Quantum Physics and the Notion of Complementarity" or "The Philosophical Lesson of Atomic Physics". Six lectures delivered at the Massachusetts Institute of Technology in November 1957. Later prepared for publication though never published. (a) First lecture "Classical physics and atomic theory", 5 November 1957. (b) Second lecture "Elements of quantum theory", 8 November 1957. (c) Third lecture "Development of quantum mechanics", 14 November 1957. (d) Fourth lecture "Observational problem and epistemological clarification, part I, 18 November 1957. (e) Fifth lecture "Observational problem and epistemological clarification, part II, 21 November 1957. (f) Sixth lecture "Complementarity (applied also to Biology and Psychology)", 26 November 1957. (g) Notes prepared for possible publication, "The Philosophical Lesson of Atomic Physics", as well as what appear to be the original notes for lecture one.

54 *Philosophical lesson* [3 January–13 February 1958]

Handwritten [A. Petersen], 22 pp., English and Danish, mf. 23.

Rough notes on the relation of atomic physics to philosophical issues. Includes programme for an "Institute for the Study of Man."

55 *Diverse manuskripter* 1958

Typewritten and handwritten [N. Bohr, Margrethe Bohr, S. Hellmann], 28 pp., English, mf. 23.

Various manuscripts. (a) The Unity of Knowledge. Talk at Iowa State University, January 1958. (b) Atoms and Human Knowledge. Talk at Roosevelt University, 4 February 1958. (c) Notes for talk at Houston, Texas.

56 *Rutherford Lecture* [November 1958]

Handwritten [L. Rosenfeld and A. Petersen], typewritten and carbon copy, 38 pp., English and Danish, mf. 23.

(a) Handwritten notes in the period preceding the lecture, 8–19 November 1958.

(b) Typescript of (part of) the lecture in which Bohr reminisces about Rutherford and the period in Manchester in 1911. (c) *The Physical Society Bulletin*, November 1958, announcing the lecture, to be given at Imperial College, London on 28 November 1958.

57 *Quantum Physics and Biology* [29 July to 23 November 1959]

Handwritten [N. Bohr, A. Petersen, T. Weis-Fogh and S. Hellmann] and typewritten, 166 pp., English and Danish, mf. 23.

Notes and drafts of talk at Symposium on Models in Biology, Bristol, 7 September 1959, later prepared for publication.

58 *Models in Biology* [22 November 1959 to 18 February 1960]

Typewritten and handwritten [L. Rosenfeld, Ernest Bohr and A. Petersen], 26 pp., English and Danish, mf. 23.

Notes and drafts from Symposium on Models in Biology, Bristol, 7 September 1959. Cf. p. [125].

59 *Physical Models and Living Organisms* [1 June 1960]

Typewritten with handwritten corrections [A. Petersen], 10 pp., English, mf. 24.

Remarks [at Johns Hopkins anniversary] based on 1959 Symposium on Models in Biology, Bristol. Cf. p. [133].

60 *Quantum Physics, Biology, Psychology* [10–27 May 1960]

Typewritten and handwritten [N. Bohr, Margrethe Bohr, L. Rosenfeld and A. Petersen], 22 pp., English, mf. 24.

(a) Three copies of Bohr's remarks "Quantum Physics and Biology" at Bristol Symposium with minor alterations. (b) Preliminary draft of a revision of this publication dated 10–27 May 1960. Cf. p. [125].

61 *Farmaceutkongres* [29 August 1960]

Typewritten with handwritten corrections [A. Petersen, J. Kalckar and E. Rüdinger], 34 pp., Danish and English, mf. 24.

Drafts of address delivered at the International Congress of Pharmaceutical Sciences, Copenhagen, 29 August 1960. Cf. p. [145].

62   *Germanistkongres, 2. Internationale*   [22 August 1960]

Typewritten with handwritten corrections, 24 pp., English, mf. 24.

Drafts of an address delivered at the opening of the Second International Germanist Congress, Copenhagen, 22 August 1960. Cf. p. [139].

63   *Kulturkongres (Fond. Européenne)*   [31 August to 21 October 1960]

Handwritten [N. Bohr and Margrethe Bohr] and typewritten, 200 pp., English and Danish, mf. 24.

Notes and drafts of address "The Unity of Human Knowledge" delivered at the congress arranged by La Fondation Européenne de la Culture, Copenhagen, 21 October 1960. Cf. p. [155].

[416]–[420] E   64   *Rutherford Memorial Lecture (1–4)*   [29 July–August 1961]

Handwritten [N. Bohr, L. Rosenfeld, A. Petersen, E. Rüdinger, S. Hellmann] and typewritten, 132 pp., English and Danish, mf. 25.

Notes and drafts of an elaborated version for publication of lecture delivered to the Physical Society of London, 28 November 1958. Delivered at the Rutherford Jubilee Conference, Manchester, 5 September 1961. Cf. p. [381].

65   *Diverse manuskripter*   [March and August 1961]

Typewritten, 16 pp., English and German, mf. 25.

Various typescripts for publication. (a) Recollections of Toshio Takamine, prepared for a memorial volume. Dated March, 1961. (b) German translation of "The Unity of Human Knowledge": "Die Einheit menschlicher Erkenntnis" for Zeitschrift Europa, August 1961.

66   *Heisenberg Festskrift*   [24 March and November 1961]

Handwritten [N. Bohr, J. Kalckar, L. Rosenfeld, S. Hellmann] and typewritten, 28 pp., English, Danish and German, mf. 25.

Notes and drafts of "Die Entstehung der Quantenmechanik" (The Genesis of Quantum Mechanics) for the Heisenberg *Festschrift*, November 1961. Proofs. Cf. p. [421].

[207]–[216] E   67   *Taler og forelæsninger*   [2 September–11 October 1961]

Handwritten [L. Rosenfeld and J. Kalckar] and typewritten, 46 pp., Danish and English, mf. 25.

Drafts and manuscripts of (a) Talk on the occasion of Carlsberg brewer J.C. Jacobsen's 150th birthday anniversary, 2 September 1961; (b) Talk at Rutherford Jubilee International Conference, September 1961; (c) Welcome to Hevesy at the anniversary of the Danish Society of Engineers, 5 October 1961; (d) Address to the University of Brussels on the award of a doctor's degree, 11 October 1961.

68   *Sonningprisen*   [19 April 1961]
Handwritten [N. Bohr] and typewritten, 13 pp., Danish, mf. 25.

Notes and drafts of the talk on the receipt of the Sonning Prize, Copenhagen University, 19 April 1961. Cf. p. [281].

69   *Köln, Indvielse af Genetisk Institut*   [13 and 22 June 1962]
Typewritten and handwritten [E. Rüdinger, N. Bohr, Aage Bohr and M. Delbrück], 67 pp., English, Danish and German, mf. 26.

Drafts and notes for talk at the inauguration of Institute for Genetics, Cologne, 22 June 1962, "Light and Life Revisited". Cf. p. [161].

INDEX

Subjects which appear throughout the volume are not listed. These include atomic physics/theory, atomic nucleus, biology, classical mechanics/physics/theory, experiment/measurement/observation, living organism, physics, quantum mechanics/physics/theory and quantum of action. However, since the present volume is devoted to Bohr's philosophy and in particular to the complementarity concept, the term "complementarity" *is* indexed, with subterms referring to Bohr's various applications of this term. Similarly, some of Bohr's characteristic terms – such as "lesson" and "conditions" as well as concepts (e.g., "love" and "justice") which Bohr introduced as complementary – have been indexed. Some such terms – e.g., "phenomenon" – have been indexed only when given a particular meaning by Bohr.

All persons mentioned in the running text (other than Niels Bohr) are indexed, whereas institutions and places are only indexed selectively.

When a term on a page is found only in a footnote or in a picture caption, the page number in the index is followed by the letter n or p, respectively.

It is hoped that the cross references will help the reader identify subjects listed under headings which differ from the terms actually used in the text.

Printed and bound by CPI Group (UK) Ltd, Croydon, CR0 4YY
03/10/2024
01040333-0017